ST/ESA/SER.R/133

Department of Economic and Social Affairs
Population Division

Population Distribution and Migration

Proceedings of the United Nations
Expert Group Meeting
on Population Distribution and Migration
Santa Cruz, Bolivia,
18-22 January 1993

Convened in preparation for the
International Conference on Population
and Development,
Cairo, 5-13 September 1994

United Nations New York, 1998

NOTE

The designations employed and the presentation of the material in the present publication do not imply the expression of any opinion whatsoever on the part of the Secretariat of the United Nations concerning the legal status of any country, territory, city or area or of its authorities, or concerning the delimitation of its frontiers or boundaries.

The term "country" as used in the text and tables also refers, as appropriate, to territories or areas.

In some tables, the designations "developed" and "developing" economies are intended for statistical convenience and do not necessarily express a judgement about the stage reached by a particular country or area in the development process.

The views expressed in signed papers are those of the individual authors and do not imply the expression of any opinion on the part of the United Nations Secretariat.

Papers have been edited and consolidated in accordance with United Nations practice and requirements.

ST/ESA/SER.R/133

UNITED NATIONS PUBLICATION
Sales No. E.98.XIII.12
ISBN 92-1-151324-3

PREFACE

The Economic and Social Council, in its resolution, 1991/93 of 26 July 1991, decided to convene the International Conference on Population and Development in 1994, with population, sustained economic growth and sustainable development as the overall theme. By its resolution 1992/37 of 30 July 1992, the Council accepted the offer of the Government of Egypt to host the Conference and decided to hold it at Cairo from 5 to 13 September 1994.

At the request of the Council, the Secretary-General appointed the Executive Director of the United Nations Population Fund (UNFPA) to serve as the Secretary-General of the Conference and the Director of the Population Division of the Department of International Economic and Social Affairs* of the United Nations Secretariat as Deputy Secretary-General.

Also in its resolution 1991/93, the Council authorized the Secretary-General of the Conference to convene, as part of the preparations for the Conference, six expert group meetings corresponding to the six groups of issues that it had identified as those requiring the greatest attention during the forthcoming decade. One of those six meetings was on population distribution and migration; it was convened at Santa Cruz, Bolivia, from 18 to 22 January 1993. The Meeting was organized by the Population Division in consultation with UNFPA.

Contained in this volume are the report and recommendations of the Meeting and the papers submitted to the Meeting. These materials will not only make a valuable contribution to the 1994 Conference itself but also serve as useful tools for future research on the interrelations between population distribution and migration and further work of the United Nations in that area.

It is acknowledged with appreciation that the Government of Bolivia, which hosted the Meeting, contributed significantly to both the substantive and the organizational aspects of the Meeting. Also, full recognition is due to the experts and other participants for their significant and constructive contribution to the work of the Meeting.

---

*Now the Department for Economic and Social Affairs.

CONTENTS

PART THREE. SOCIAL, ENVIRONMENTAL AND POLICY ASPECTS OF POPULATION DISTRIBUTION AND INTERNAL MIGRATION

PART FOUR. INTERNATIONAL MIGRATION TRENDS AND PROSPECTS

PART FIVE. SOCIAL, ECONOMIC AND POLITICAL ASPECTS OF INTERNATIONAL MIGRATION

PART SIX. DISCUSSION NOTES

TABLES

Figures

Boxes

No.

Explanatory notes

Symbols of United Nations documents are composed of capital letters combined with figures.

The following symbols have been used in the tables throughout this report:

Two dots (..) indicate that data are not available or are not separately reported;
An em dash (—) indicates that the amount is nil or negligible;
A hyphen (-) indicates that the item is not applicable;
A minus sign (−) before a number indicates a decrease;
A point (.) is used to indicate decimals;
A slash (/) indicates a crop year or financial year, for example, 1988/89;
Use of a hyphen (-) between dates representing years (for example, 1985-1989), signifies the full period involved, including the beginning and end years.

Details and percentages in table do not necessarily add to totals because of rounding.

Reference to "dollars" ($) indicates United States dollars, unless otherwise stated.

The term "billion" signifies a thousand million.

The following abbreviations have been used in this report:

AIDS	acquired immunodeficiency syndrome
APEC	Asia-Pacific Economic Cooperation
ASEAN	Association of South-East Asian Nations
BMR	Bangkok Metropolitan Region
CELADE	Centro Latinoamericano de Demografía
CONAPO	Consejo Nacional de Población (Mexico)
CRE	Commission for Racial Equality (United Kingdom)
CSR	Convenio Simón Rodríguez
DHS	Demographic and Health Survey
EAM	Estatuto Andino de Migraciones
ECA	Economic Commission for Africa
ECE	Economic Commission for Europe
ECLAC	Economic Commission for Latin America and the Caribbean (CEPAL)
ECOWAS	Economic Community of West African States
EFTA	European Free Trade Association
EPZ	export processing zone
ESCAP	Economic and Social Commission for Asia and the Pacific
ESCWA	Economic and Social Commission for Western Asia
EU	European Union
EVD	Extended voluntary departure
FAO	Food and Agriculture Organization of the United Nations
FELDA	Federal Land Development Agency (Malaysia)
FUR	functional urban region
GATS	General Agreement of Trade in Services
GATT	General Agreement on Tariffs and Trade
GCC	Gulf Cooperation Council
GDP	gross domestic product

GNP	gross national product
GSP	Generalized System of Preference
Habitat	United Nations Centre for Human Settlements
HIV	human immunodeficiency virus
IIASA	International Institute for Applied Systems Analysis
ILO	International Labour Organization
IOM	International Organization for Migration
IRCA	Immigration Control and Reform Act (United States of America)
IUSSP	International Union for the Scientific Study of Population
JITCO	Japan International Training Cooperation Organization
MERCOSUR	Mercado Común del Cono Sur
NAFTA	North American Free Trade Association
NIEs	newly industrialized economies
OAS	Organization of American States
OAU	Organization of African Unity
ODA	official development assistance
OECD	Organisation for Economic Cooperation and Development
RURE	Regional and Urban Restructuring in Europe
SAARC	South Asian Association for Regional Cooperation
SOPEMI	Continuous Reporting System on Migration
TPS	temporary protected status
UDEAC	Union Douanière et Economique de l'Afrique Centrale
UNCHS	United Nations Centre for Human Settlements (Habitat)
UNCTAD	United Nations Conference on Trade and Development
UNDP	United Nations Development Programme
UNEP	United Nations Environment Programme
UNESCO	United Nations Educational, Scientific and Cultural Organization
UNFPA	United Nations Population Fund
UNHCR	United Nations High Commissioner for Refugees
UNICEF	United Nations Children's Fund
UNRWA	United Nations Relief and Works Agency for Palestine Refugees in the Near East
USCR	U.S. Committee for Refugees
WFS	World Fertility Survey
WHO	World Health Organization

POPULATION DISTRIBUTION AND MIGRATION: THE EMERGING ISSUES

*United Nations Secretariat**

A. THE DIMENSIONS OF MIGRATION

The processes of migration and population distribution are intricately related to important economic, social, cultural, political and environmental factors that have brought them to the forefront of international consciousness. Indeed, during the past decade, concern about migration, its causes and consequences, has grown substantially in both developed and developing countries. However, such concern has not yet elicited a concerted effort to fill the many gaps existing in the knowledge and understanding of the migration process. In terms of measurement alone, recent and reliable estimates at the world level do not exist. Yet migration, internal or international, involves millions of people. Census data referring largely to the 1970s and early 1980s have been used to estimate the migrant stock at the world level for 1985 (Skoog, 1994). The estimate, which includes the number of refugees reported that year by the Office of the United Nations High Commissioner for Refugees (UNHCR), amounts to over 105 million persons, most of whom are identified as migrants because they lived outside of their country of birth. Estimates at the regional level indicate that Europe and North America were hosting the largest numbers of international migrants (23 million and 20 million, respectively). In the developing world, Southern Asia occupied first place with 19 million international migrants, followed by Northern Africa and Western Asia which, in combination, had over 13 million migrants, and by sub-Saharan Africa with 11 million. Eastern and South-eastern Asia, combined, had about 8 million international migrants, and Latin America and the Caribbean were hosting another 6.5 million. Oceania closed the list with nearly 4 million.

In comparison, the number of internal migrants is at least an order of magnitude higher. In India alone, the 1981 census enumerated about 200 million people as migrants (defined as persons born outside their district of residence) and, according to estimates of the evolution of the world's urban population over time, the population of urban areas grew by about 500 million persons during 1975-1985 with about half of that gain being accounted for by net rural-urban migration or reclassification during that period. Since the limited evidence available indicates that rural-urban migration accounts for a relatively low proportion of all migration flows in many countries (United Nations Secretariat, 1993), all forms of migration are likely to involve significantly higher numbers of persons, probably in the range of 750 million to 1 billion people moving during 1975-1985. Indeed, it is worth underscoring that rural-rural migration tends to predominate in countries that are still mostly rural (such as India, Thailand and most of those of Africa) and that urban-urban migration is the dominant form in countries that are highly urbanized (such as those of the Americas and Europe).

Despite the numerical importance of rural-rural and urban-urban migration, attention continues to be focused mostly on rural-urban flows and even migration per se has received less attention than population redistribution, particularly the continued growth of urban areas in the developing world. At the level of the world as a whole, the population is expected to grow by some 919 million between 1985 and 1995, of which 729 million, or nearly 80 per cent, will be absorbed by urban areas (United Nations, 1991a and 1991b). Developing regions will account for 93 per cent of total population growth during 1985-1995 and their urban areas will absorb 90 per cent of the growth expected in all urban areas, that is, some 660 million persons. Although by 1995 the urban population of the world as a whole is still expected to account for less than half of the world's total population (48 per cent, to be exact), by the year 2000 the 50 per cent mark will have been surpassed. There will remain, however, a large differential between developed and developing regions in terms of the proportion of the population living in urban areas: 74 versus 41 per cent, respectively, in 1995. Even by 2025, only 61 per cent of

*Population Division of the Department of Economic and Social Affairs.

the population of developing countries is expected to live in urban areas as compared with 83 per cent in developed countries.

The structure of urban systems at the national level has not changed markedly in most countries during the past decade and continued urban growth is expected to maintain, at least until the end of the century, the pattern of primacy that characterizes many developing countries. The tendency towards population concentration has contributed to the rising number and size of megacities. It is estimated that by the year 2000, there will be 28 cities with populations of at least 8 million, of which 22 will be located in the developing world. The continued growth of already large urban agglomerations in developing countries certainly poses enormous challenges to their planners and administrators.

In the developed world, total population growth has been low and is expected to be lower still during what remains of the twentieth century. Growth will be concentrated in urban areas, with rural areas as a whole losing population. The tendency towards population deconcentration that was observed during the late 1960s, 1970s and early 1980s in several developed countries seems to have abated and major urban agglomerations are again growing. Although by the year 2000 the developed world will still have six megacities, managing large urban agglomerations is not a problem per se. Rather, one of the main challenges facing developed countries is the maintenance of a sufficiently flexible labour force, in terms of both geographical mobility and malleable skills, that can adapt to the economic restructuring that is already under way. The role that foreign labour has played or is likely to play as restructuring proceeds in the context of an ageing population is a major issue in most developed countries. Indeed, it is the growing concern of the developed world about the future prospects for international migration that is driving much of the debate about migration taking place in the international community today.

B. The international context

The past decade has witnessed changes in the international context that have important implications for migration, both internal and international. During the 1980s, Japan consolidated its position as an economic power, outperforming the United States of America in terms of trade, productivity, level of foreign reserves and share of foreign assets of financial institutions from industrialized countries. The United States, though still the main economic power in the world, has seen its position eroded by a decade of trade and budget deficits. In Europe, the need to compete more effectively in the economic arena with Japan and the United States has fuelled the drive towards the achievement of a Single European Market where both full economic unity and a large measure of political unity are the eventual if elusive goal. Lastly, the collapse of the Union of Soviet Socialist Republics (USSR), brought about in large part by economic difficulties, has not only changed the balance of power in the world but also brought forth a number of old and newly independent countries seeking to participate in the capitalist experience and to become part of the global economy.

In the developing world, during the 1980s the newly industrialized economies (NIEs)—Hong Kong, the Republic of Korea, Singapore and Taiwan Province of China—experienced rates of economic growth unparalleled in any other region. The second generation NIEs—Indonesia, Malaysia and Thailand—experienced comparable rates of growth during the late 1980s (United Nations, 1992). Such dramatic growth owes much to the high levels of foreign capital flowing to those countries since the 1970s. Thus, during 1970-1982, the seven NIEs plus the Philippines received nearly US$14 billion in foreign direct investment from Japan alone (Johnson, 1991). Although most of the foreign direct investment of developed countries has been and still is directed to other developed countries, starting in the late 1960s several developing countries began to attract foreign direct investment from developed countries seeking the advantage of cheap labour in countries eager to expand exports (Bailey and Parisotto, 1993). One such policy, the creation of export processing zones, has been instrumental both in attracting foreign investment and in fostering internal migration.

In addition to the seven NIEs, China has made important strides in terms of economic growth during the 1980s and India recorded higher average annual economic growth during the 1980s than during the 1970s (United Nations, 1992). The oil-rich countries of Western Asia, though generally experiencing relatively high rates of economic growth, were affected by the low oil prices prevailing during most of the 1980s and even more so by the costs of the Persian Gulf conflict. In the rest of Asia, economic growth was sluggish and the economies in transition, such as Mongolia, Viet Nam and the other countries of Indochina, suffered from the disruption of trade resulting from the dismantling of the Council for Mutual Economic Assistance and the collapse of the Soviet Union.

For Latin America and Africa, the 1980s are known as the "lost decade". During the 1970s, the strength of most Latin American economies compared favourably with that of other developing regions. Latin American gross national product (GNP) grew at a nominal annual rate of 16 per cent, slightly higher than that for East Asia (Johnson, 1991). During 1980-1988, however, the nominal rate of GNP growth in Latin America dropped to less than 3 per cent annually and would have been negative if adjusted for inflation. The high levels of debt incurred during the 1970s contributed to such dismal economic performance. Problems with debt-servicing forced many countries to adopt stringent austerity programmes that included reductions in imports, reduced government spending, privatization of government-run enterprises and constriction of the domestic money supply to fight inflation. Although by the early 1990s such adjustment and stabilization efforts were proving successful in reducing inflationary pressures and improving fiscal balances in most countries of the region, such an outcome had been obtained at the expense of widening the gap between the rich and the poor. The economic difficulties of Latin American countries were further exacerbated by the failure of the United States to act as the economic engine for the region. Indeed, the latter's persistent budget deficits, by reducing investment capital in the rest of the world, have been an economic drain, particularly for developing countries.

In Africa, economic conditions are dire. During the 1980s, even nominal GNP rates of growth were negative. Despite Africa's richness in natural resources, capital flows have largely evaded the continent since the 1970s. Lack of sufficient energy resources has contributed to the economic stagnation of the region. When 70 per cent of a continent's energy is derived from wood, the potential for industrialization is low and that for environmental degradation is high. Sub-Saharan Africa, in particular, is heavily dependent on official development assistance (ODA). In the 1990s, both Africa's traditional exports and sources of financial assistance have become less secure. Because African member States of the Lomé Conventions have duty-free access to the European Union, the latter has absorbed the majority of sub-Saharan Africa's exports (60 per cent in 1987) but the consolidation of the Single European Market may bring about a revision of such preferences (Johnson, 1991). Development assistance to Africa is being curtailed both because of the recession affecting developed countries and owing to the perceived need to provide more aid to former Eastern-bloc countries. In the meantime, Africa's population continues to grow at very high rates and, with the cold war over, internal conflict has flared up in a growing number of African countries.

C. Migration in the new world context

The changes outlined above have led to a new world order characterized by the growing interdependence of the major economic powers and by the adoption of "global" market strategies to promote economic growth. The latter are based on a new economic ideology that promotes the reduction of government and the limitation of its role as investor or producer. Government's main task is propounded as being the maintenance of a free market, the sine qua non for the efficient use of private capital. Globalization arises because, to achieve a truly competitive world market, Governments must open their economies so that exports become the engines of growth. The growing tendency of both developed and developing countries to adopt such strategies has led to major transformations in the modes of production that have important implications for migration and population distribution.

Among the institutional changes associated with economic restructuring, the increasing mobility of capital is a crucial factor. Such mobility, both territorial and sectoral, has been facilitated by two interrelated processes: the modernization and expansion of financial systems, and the concomitant modernization and territorial diffusion of information technology. Such developments have made possible new modes of enterprise organization and management, allowing the relocation of different production processes to wherever competitive advantages assure the greatest profitability of capital. Through increasing segmentation and vertical disintegration, many enterprises have attained greater territorial diffusion. Large enterprises have increasingly assumed a multisectoral and multiregional character. Locational decisions have been determined by factors likely to raise profitability, including access to natural resources, cheap labour, adequate markets and better communication to external markets. Thus, in countries that are mainly exporters of natural resources, new investment has tended to locate itself near the sources of such resources or in the places where they are exported. In countries where manufactured goods constitute a significant proportion of exports, the location of new plants has been determined by the availability of adequate managerial capacity, a favourable labour market, access to adequate means of communication and effective linkages to import and export markets. To the extent that those locations have been different from those favoured by previous development strategies that promoted industrialization for import substitution, new patterns of population distribution have emerged. That is the case, for instance, of Mexico where the expansion of the *maquiladora* (in-bond assembly) industry along the border with the United States has exerted a powerful attraction on migrants from the interior, thereby causing cities along the northern border to grow very rapidly. Thus, in some cases, the new economic strategies have stimulated the growth of certain medium-sized cities, though it is worth underscoring that such an outcome is the result of a growing transformation of the global economy rather than of specific population distribution policies (De Mattos, 1994).

To the extent that the new economic strategies have led to economic recovery in certain countries, they have increased internal demand for a number of products and thus fostered their production, usually in locations near the main internal markets, namely, the large agglomerations and the medium-sized cities located on their immediate periphery; that is to say, the economic advantages associated with the concentration of population in major agglomerations are not being eroded by the increasing territorial diffusion of production. Indeed, the new mode of development implies that unequal regional growth and population distribution will persist as long as areas lacking the necessary advantages fail to attract investment capital. Even within major agglomerations, restructuring has meant an increase in social inequalities stemming from the decline of formal sector employment, falling real wages within that sector and an increase in the informal sector, particularly that involved in industrial activities (United Nations, 1992).

Not only have the changes described led to increasing regional disparities within countries, but those between countries have also been growing, particularly within the developing world, as some countries are more successful than others in establishing effective links with the global market. One strategy used by a number of countries to attract foreign direct investment and foster exports is the creation of export processing zones (EPZs), that is, industrial sites operating under special legislation that allows the duty-free import of raw materials for the assembly and manufacture of goods that are destined primarily for export. Usually, a set of tax advantages and guarantees regarding the repatriation of profits are granted to the enterprises establishing themselves in EPZs. Although not all investment in EPZs is made by multinationals, it is estimated that in 1986 multinationals accounted for nearly two thirds of the 1.9 million jobs in EPZs (ILO, 1988). In 1986, 176 EPZs were in operation in 47 countries: 8 in Africa, 20 in Latin America and 19 in Asia (Kreye, Heinrichs and Frobel, 1987). Brazil, Mexico, Puerto Rico, the Republic of Korea and Singapore were the countries having the largest number of workers employed in EPZs. An outstanding feature of employment in EPZs is the high proportion of women involved: in most countries, over 60 per cent of EPZ workers are women and, according to the limited evidence available, a sizeable proportion of them are migrants, often from rural areas (Zlotnik, 1993). Indeed, growing opportunities for the employment of women in manufacturing are related to the increase in economically motivated migration among women in certain developing countries, particularly the second generation NIEs (United Nations, 1993).

It has been argued that the movement of capital to labour is a strategy to prevent the movement of labour to capital, that is, the international migration of labour to capital-rich countries. Japan has pursued such a strategy by promoting the establishment of Japanese subsidiaries abroad, particularly in the NIEs. Although the strategy appears to have been successful both in promoting the economic development of the receiving countries and in reducing labour needs in Japan, it has not prevented international migration altogether. Since Asian NIEs such as Hong Kong and Singapore had small populations to start with, they have long been importers of foreign labour—the former from neighbouring China, the latter from Malaysia—and both have experienced considerable emigration, particularly Hong Kong in view of its 1997 devolution to China. Among the other NIEs, Indonesia, the Republic of Korea and Thailand have been important labour exporters, especially to the oil-producing countries of Western Asia, and the Republic of Korea has been a major source of emigrants to the United States.

The experience of those countries appears to confirm the findings of the Commission for the Study of International Migration and Cooperative Economic Development established by the Congress of the United States in 1986. Among other things, the Commission concluded that "the development process is itself destabilizing and probably enhances rather than reduces emigration pressures in sending countries over the short to medium term. Rapid population growth, the transformation of traditional agriculture, profound normative and cultural changes, urban growth, increases in educational attainment, improvements in communications and transportation, and the consolidation of social networks in sending and receiving locations accompany development and encourage population displacements" (United States Commission for the Study of International Migration and Cooperative Economic Development, 1990, p. E-2).

The export of capital has not prevented Japan from attracting migrants. Between 1980 and 1990, the foreign population legally registered in that country rose from 783,000 to 1,075,000 and most of the increase was accounted for by the rise in the number of Asians, particularly nationals of China and the Philippines. There were also sizeable increases in the number of ethnic Japanese from Brazil and Peru who qualify for long-term residence status according to the 1990 amendments to the Immigration Control and Refugee Recognition Act (Zlotnik, 1994). In addition, there is evidence suggesting that the number of foreigners working illegally in Japan is increasing. Some estimates put the number of irregular migrants in the country at about 300,000, most of them employed in low-wage, low-status or dangerous jobs shunned by the Japanese. As other developed countries have discovered, some jobs, particularly those in the construction or service sectors, cannot be exported.

Not only is the path towards economic development likely to increase emigration pressures, but it may also lead to immigration as intercountry disparities grow. In Asia, for instance, some of the larger NIEs—Malaysia and Taiwan Province of China—have been attracting irregular migrants from other countries in the region. In Europe, countries like Italy or Spain became important receivers of undocumented migrants as their economic conditions improved and labour shortages developed in selected sectors. Such developments occurred largely at the margin of government intervention, mostly because Governments of countries that had long been labour exporters were not prepared for their transformation into labour importers. In Europe, the revision of international migration policies has been slow and has generally relied on ad hoc regularization drives to gain control over undocumented migration.

This discussion indicates that development and the policies or strategies followed to foster it have important impacts on migration, both internal and international. The modernizing influences associated with development contribute to the movement of people. As traditional agrarian economies are transformed, market transactions increasingly take the place of social obligations and the use of modern equipment reduces reliance on labour-intensive farming, thereby leading people to leave rural areas and go to the city in search of better economic opportunities. Ideally, successful development would provide suitable opportunities for most migrants in the form of jobs in the formal sector of the economy. At present, however, rural-urban migration in many developing countries is propelled less by the opportunities in urban areas than by the difficult conditions in rural areas, particularly the widespread scarcity of land exacerbated by rapid population growth. Urban markets are unable to generate the jobs required by the growing number of largely unskilled workers converging towards them. Dual labour markets develop, with relatively few people employed in modern sector jobs and most workers remaining outside the economic mainstream. Lack of opportunities at home make the possibility of working abroad for higher wages very attractive and increasing numbers of persons are taking that route despite the restrictions imposed by many receiving countries. The growing migration flows between countries create new linkages between them that often condition policy responses to both migration and development.

D. The effects of migration on development

Emigration has helped ease some of the problems confronting countries engaged in the development process. Thus, the emigration of economically active persons, even if temporary, helps relieve labour market pressure and migrant remittances are an important source of foreign exchange for many developing countries. It is estimated that in 1989 migrant remittances amounted to US$ 65 billion, a figure that compares favourably with the US$ 47 billion provided as official development assistance (ODA) by Organisation for Economic Cooperation and Development (OECD) member States to developing countries that year (IOM, 1994). Although there is no consensus about whether or not remittances are used in a manner that accelerates development, there exists no doubt that they contribute to improving the living standards of family members left behind and generally are used rationally by migrants and their families provided viable investment opportunities exist. For some countries, the economic impact of remittances has been estimated to be considerably larger than that of foreign direct investment. In the Philippines, for instance, the US$ 1.7 billion received as remittances from some 650,000 Filipinos abroad in 1989 would generate US$ 3.9 billion of economic growth assuming a multiplier effect of 2.3. Since every dollar of remittance-generated economic activity is estimated to be equivalent to that produced by $5.40 of capital investment, the remittance level for 1989 would be equivalent to about US$ 21 billion of investment. The actual level of foreign direct investment received by the Philippines in 1989 was only of US$ 850 million (IOM, 1994). Although such estimates fail to take account of the multiplier effect of foreign direct investment and the opportunity costs associated with emigration, they nevertheless indicate that in the short term the export of labour provides an attractive option for developing countries.

Among the possible negative effects of migration on development, one that raises considerable concern is the loss of skilled personnel, especially of those possessing the managerial, professional and technical skills needed for the acquisition and use of technology. Africa in particular is estimated to have lost a very considerable proportion of its highly skilled workers through migration and this phenomenon certainly hinders its development prospects. Although developed countries are generally assumed to be the destination of skilled personnel from developing countries, there is evidence suggesting that an increasing number of skilled workers are moving from one developing country to another (Kritz and Caces, 1992). The oil-rich countries of Western Asia have been a major destination for such migrants, but so have certain Asian NIEs, such as Singapore. The transfer of technology has also involved the migration of professional and technical personnel from developed to developing countries, particularly in relation to the operation of multinational corporations; that is to say, a virtuous circle is likely to arise in cases where development strategies are successful: economic growth attracts both the capital and the personnel that help maintain such growth. Conversely, however, lack of economic growth prevents the accumulation of both the capital and human resources needed to fuel development.

Similar issues may be raised with respect to internal migration, though in that context the basic question is whether migration aids or hinders the attainment of balanced regional development. This question has been considered mostly in regard to rural-urban migration and the concerns raised parallel those related to international migration. Thus, rural-urban migration has been claimed to reduce the development potential of rural areas by depriving them of their most dynamic human elements, namely, the young and relatively better educated. Remittances are viewed as a generally positive contribution of migrants to the families they leave behind, but caveats are being raised about their potential to increase income disparities or to be used mainly for consumption rather than investment purposes. Although there are no measures of the general macroeconomic impact of remittances by internal migrants, studies in certain communities have indicated that remittances constitute relatively high percentages of rural household incomes and are therefore instrumental in assuring the survivorship of those households. As in the case of international migration, it is difficult to separate the impacts of migration on development from those of the latter on the former. Both "exogenous" economic and demographic factors are likely to condition the effects of migration on economic growth. In Africa, for instance, where the rural population continues to increase, reliance on subsistence agriculture is heavy and educational levels are low, rural out-migration is unlikely to have major effects in depleting human capital in rural areas. At the other end of the spectrum, in Japan the out-migration of the younger generations, particularly young women, who are no longer attracted by agricultural work has had a considerable effect on rural areas. That such developments have prompted some Japanese farmers to import foreign wives is yet another example of the interrelations between internal and international migration.

E. Nation rebuilding

The late 1980s witnessed a series of events whose impact on the structure of the nation State system has major implications for migration. In particular, the reconfiguration of nation States in Europe and the former Soviet Union has implied a redefinition of the term "migration" and a reformulation of the distinction between internal and international movements. The developments in Europe are especially interesting because they juxtapose the movement towards greater integration of nation States that led to the creation of the European Union (EU) with the disintegration of nation States occurring among the former Eastern-bloc countries.

Within the European Union, the Single European Act, the first major amendment to the Treaty establishing the European Economic Community (Treaty of Rome) (United Nations, 1958) since its adoption in 1957, came into force on 1 July 1987 and established in its article 13 that the Treaty establishing the European Economic Community (EEC Treaty) should be supplemented by the following provision (article 8 A): "The internal market shall comprise an area without internal frontiers in which the free movement of goods, persons, services and capital is ensured in accordance with the provisions of this Treaty" (Commission of the European Communities, 1986). The Act further stated that the goal was to establish progressively the internal market over a period expiring on 31 December 1992. As part of this process, freedom of movement and residence was extended beyond workers who were nationals of member States to the majority of European Community citizens. Thus, the self-employed, students and retirees who were citizens of a member State and members of their families were granted freedom of residence in any member State other than their own provided they would not become a burden for the receiving State (Hovy and Zlotnik, 1994).

These changes took place at a time when international migration to the European Union was increasing. Indeed, partly because of the recession that had affected European countries during the early 1980s and partly because of the policies favouring the return migration of foreign workers and their families implemented by several countries during that period, net migration to the main receiving countries in the European Union was very low during 1980-1984 (Zlotnik, 1991). However, as of 1984 or 1985 an increasing trend set in, fuelled by continued family reunification, the admission of migrants granted a special status because of their ancestry (for example, ethnic Germans who qualified automatically for German citizenship in the Federal Republic of Germany) and the growing number of asylum-seekers, a significant proportion of whom were allowed to stay. In addition, during the 1980s, Italy, Spain and, to a lesser extent, Greece—all former labour-exporting countries within the European Union—became the receivers of sizeable flows of undocumented migrants originating mostly in Africa (OECD, 1990).

Consequently, as the European Union project for a frontier-free Europe was taking shape, concern about the ability of member States to control the Union's external borders grew. Such concern gave rise to several initiatives aimed at strengthening that ability, mainly through the coordination of visa policies and of regulations on the granting of asylum and refugee status. Both initiatives, however, were largely pursued at the margin of Community institutions. First, on 14 June 1985, Belgium, France, the Federal Republic of Germany, Luxembourg and the Netherlands signed an agreement at Schengen (Schengen Agreement) on the gradual removal of controls at their common borders. The Schengen group was later enlarged to include Italy, Spain and Portugal and an Application Convention was signed in mid-1990. The second initiative concerned coordination regarding asylum applications and culminated on 15 June 1990 with the adoption by the European Council of the Dublin Convention relating to the determination of the State responsible for examining applications for asylum lodged in one of the member States of the European Communities.

Those developments plus the establishment of a European citizenship by the Maastricht Treaty on European Union (7 February 1992) did not, however, go far enough in terms of eliciting the response of coordination of international migration policies among member States. In fact, the position of most member States of the European Union is that the abolition of border controls shall not affect their right to take such measures as they consider necessary to control immigration from third countries. Indeed, the movement of third-country nationals, asylum-seekers and immigration issues are areas tied intrinsically to the concepts of sovereignty and national identity concepts that nation States are loath to relinquish. Nevertheless, member States recognize, albeit reluctantly, that a common European Union strategy regarding migration that respects domestic interests and national sovereignty is necessary to control the effects of domestic migration policy on the common labour market, to meet the challenges posed by the increasing number of asylum-seekers and undocumented migrants, and to face the problems related to the adjustment and integration of third-country nationals that are increasingly subject to marginalization (Callovi, 1992).

As the first institution in the post-national era, the European Union believes that international migration is an integral issue of foreign policy, particularly in relation to policies regarding Mediterranean countries, the African, Caribbean and Pacific countries (Lomé Convention), and Central and Eastern Europe. Thus it was that the ministers of immigration for the twelve European Union member States agreed in 1991 to pursue a work programme aimed at achieving the harmonization of migration and asylum policies of member States. Although progress in that direction has been slow, harmonization is likely to result in relatively stringent policies towards migration, since member States generally favour the maintenance of low migration levels from third countries, the reduction of undocumented migration and a speedy adjudication of asylum applications with effective restrictions on the stay of those who do not qualify for refugee status.

The EU's concerns about increasing migration pressures are further justified by the developments in its immediate vicinity, particularly as former Eastern-bloc countries undergo the necessary adjustments to pass from a command to a market economy and new nation States emerge from the collapse of the old. As Brubaker (1992) noted: "The breakup of the Soviet Union has transformed yesterday's internal migrants, secure in their Soviet citizenship, into today's international migrants of contested legitimacy and uncertain membership". Although in recent years emigration from the successor States of the Soviet Union has been relatively limited, involving mostly persons belonging to Soviet nationalities that have immigration and citizenship rights in external homelands (for example, ethnic Germans, Jews, Greeks and so on), if economic and political dislocation increases, outflows may rise, particularly among those related by intermarriage to persons who can claim an external homeland. In addition, the internationalization of previously internal migration has transformed certain Soviet citizens into foreign minorities in the successor States. Such transformation has touched Russians in particular, about 25 million of whom lived in non-Russian former Soviet republics (Brubaker, 1992). The possibility that a substantial number of them might choose or be forced to return to the Russian Federation if deprived of citizenship or other rights in their place of current residence cannot be discarded.

In contrast with the breakup of the former Soviet Union and, more recently, that of Czechoslovakia, which have been mostly peaceful, that of the former Yugoslavia has resulted in prolonged conflict between rival ethnic groups. As in other such cases, conflict itself has led to considerable population displacement, further exacerbated by explicit strategies of "ethnic cleansing" aimed at dislodging unwanted ethnic groups from a specific territory. So far, however, most displaced persons from the former Yugoslavia have remained within the country's former borders. Thus, of the estimated 2 million displaced persons in mid-1992, about 1.7 million remained in Bosnia and Herzegovina, or in Croatia and Slovenia. By July 1993, the number of persons in need of assistance in the former Yugoslavia had grown to 3.6 million, 2.3 million of whom were in Bosnia and Herzegovina. Other European countries have been reluctant to grant asylum to former Yugoslavs for fear of being faced with still larger inflows of persons who are likely to remain. Thus, Austria, the Netherlands and Switzerland tried to close their borders to asylum-seekers from Bosnia and Herzegovina by requiring a visa. In other cases, citizens from the former Yugoslavia have not been allowed to apply for asylum but have been granted temporary protection. As of 1992, Germany was hosting 220,000 citizens from the former Yugoslavia, Sweden another 74,000, Austria 73,000, Switzerland 70,000 and Hungary over 40,000 (United Nations, 1994). Rather than admit more persons in need of asylum, Western countries have preferred to finance relief assistance for displaced persons remaining in Croatia and Slovenia as well as for those besieged in Sarajevo (Suhrke, 1994).

Not only are the prospects for resolving the conflict in the former Yugoslavia still dim, but there is also cause for concern regarding the stability of other successor States in the region (the conflict between Armenia and Azerbaijan, for instance, continues). Even the peaceful reunification of Germany has not been devoid of strains, as extremist groups in the still impoverished eastern part of the country vent their anger and frustration against foreigners and asylum-seekers. Such developments are used to justify further the need to control migration, that is, to limit the inflow of "unwanted" migrants to Western-bloc countries.

F. Forced migration

As of early 1993, there were an estimated 19 million refugees in the world, most of them hosted by developing countries, and that number did not include the 2.6 million Palestinian refugees under the United Nations Relief and Works Agency for Palestine Refugees in the Near East (UNRWA). Indeed, one of the major developments in international population flows during the late 1980s was the steady increase of the refugee population, which doubled between 1985 and 1992, passing from 8.5 million to 17 million.

During the cold war era, persons leaving Communist countries were generally admitted as refugees by Western-bloc countries, particularly the United States, and many refugee outflows resulted from the indirect confrontation of the superpowers in developing countries. Consequently, the end of the cold war brought the hope of reducing refugee flows by solving existing conflicts and establishing the conditions for a lasting peace that would encourage the repatriation of refugees. However, although encouraging developments have taken place in countries like Afghanistan, Cambodia, El Salvador, Ethiopia and Mozambique, peace remains fragile and the return of stability to the major refugee-generating regions is still a far-off prospect. In addition, as the world moves away from the bipolar cold war era, the deterrents to induce caution among developing countries are weakening and expansionary pursuits through aggression are more likely to be acted out. In contrast, the growing economic interdependence among the advanced industrialized nations, their tradition of political democracy and the existence of nuclear weapons are lessening their security dilemma and, although conflict between them will not disappear, it is not likely to be resolved militarily (Goldgeier and McFaul, 1992). Military intervention by the main powers in developing countries will only occur if their vital interests are threatened, as they had been by the Iraqi invasion of Kuwait. Otherwise, those powers are likely to remain on the sidelines, as has already been the case in respect of the civil wars in Liberia, the Sudan and even Ethiopia. Such prospects bode ill for the large numbers of refugees and displaced persons that will continue to be generated by open conflict in developing countries.

During the 1980s, the rise of refugee flows worldwide led to a new development that, compounded by the end of the cold war, is having profound effects on the refugee system, namely, the increasing number of persons seeking asylum in Western-bloc countries. Although low by world standards, the number of asylum-seekers in European countries rose from 50,000 in 1983 to over 400,000 in 1991. Countries that had admitted relatively small numbers of mostly prescreened refugees for resettlement were suddenly confronted with a continuous spontaneous flow of asylum claimants. The fact that numerous asylum-seekers originated in countries that had supplied labour to Western Europe suggested that asylum was being used as a back door to immigration. However, most asylum-seekers originated in Eastern-bloc countries (the traditional sources of refugees under cold war tenets) or in developing countries affected by civil strife, war or violent inter-ethnic conflict. Fearing the consequences of uncontrolled migration, Western-bloc countries in Europe engaged in a process geared towards the restrictive streamlining of the asylum system in the region. The Schengen Agreement and the Dublin Convention can be seen as part of that process. Together they restrict access as regards seeking asylum, narrow the eligibility criteria applied to those who are eventually granted asylum, shorten the determination process leaving few avenues for appeal, and facilitate the deportation of rejected claimants (Suhrke, 1994). When the cold war ended, Eastern-bloc countries in Europe were brought into the emerging asylum regime so that better control of East-West flows could be gained. In addition, institutionalized informal consultations were used to get Australia, Canada and the United States to participate in deliberations on asylum and immigration policy (Suhrke, 1994).

The overseas developed countries have also been receiving increasing numbers of asylum-seekers and, although less averse to receiving immigrants than the European countries, they have been seeking ways to control that unwanted inflow in order not to jeopardize their normal immigration programmes. The United States, being strongly influenced by foreign policy considerations, has dealt in a variety of ways with potential asylum-seekers reaching its territory. Thus, it has been fairly generous in granting asylum to Cubans and Nicaraguans (while the latter were under a leftist regime). In contrast, Salvadorans have generally been denied asylum, though the United States legislature granted them temporary protected status through recently approved legislation. Lastly, Haitians have largely been prevented from reaching United States territory and lodging asylum requests (United States Committee for Refugees, 1992). Such lack of consistency diminishes the ability of the United States to influence the ongoing transformation of the asylum system.

In Asia, the continued outflow of Vietnamese asylum-seekers has been met with increasingly negative reactions from the main receiving countries. In fact, the move towards restricting the asylum system may be said to have started in that region. The adoption of individual adjudication procedures has resulted in the rejection of the vast majority of refugee claims lodged in South-East Asian countries and forced repatriation has been considered, though relatively few asylum-seekers have actually been repatriated (United Nations, 1994). It is thus clear that, at a time when processes resulting in forced migration are increasing, the better-off countries are sealing their borders as best they can. The rationale to do so is often based on the argument that most asylum-seekers are in fact "economic refugees" and do not deserve protection. Given that most situations leading to massive population outflows involve both economic deprivation and political oppression or civil violence, that argument seems disingenuous.

More recently, a new type of forced migrant has begun to attract attention: the so-called environmental refugee. The term refers to persons impelled to move for a variety of reasons, ranging from sudden natural disasters to prolonged drought or land degradation through over-exploitation (The Population Institute, 1992). The spatial dislocation of such persons can be temporary or long term. Given the overarching meaning of the term, the numbers involved are difficult to estimate, especially since most of those persons remain in their own countries. In that sense, they are akin to the members of another category of forced migrants, the internally displaced, who move in order to avoid persecution or violence but do not manage to leave their own countries or are unwilling to do so. A common trait of these two types of forced migrant is that they are in need of assistance and often of protection as well. However, either because of internal conflict or as a result of economic stringencies, their own Governments are unwilling or unable to provide such assistance or protection. Moreover, calls for concerted and consistent action by the international community have largely failed to elicit a response.

G. The realm of policy

1. *International migration*

Since policies regarding immigration embody the belief that sovereign nation States have the prerogative to decide who can reside in their territories and under what conditions, they have usually been formulated on a unilateral basis with little regard for their international repercussions. In the context of a world characterized by greater interrelatedness, however, nation States are finding increasingly that it is not in their interest to operate independently from one another, especially with respect to migration issues. Consequently, there has been a trend, especially among developed countries, towards devising new mechanisms for the coordination or even the harmonization of migration policies and control measures. The process is most advanced in Europe, where a series of conferences and continuous consultations have been institutionalized under the so-called Vienna and Berlin processes, named thus after the ministerial meetings that took place in those cities in 1991. The aim of those processes is to share information and devise mechanisms to control East-West migration, undocumented flows and the movement of potential asylum-seekers by eliciting the cooperation of several Eastern and Central European countries.

In recent years there has also been a growing awareness among Western-bloc countries that policies other than those explicitly related to migration may have major impacts on migration flows. Thus, the United States Commission for the Study of International Migration and Cooperative Economic Development (1990) concluded that it was important for the United States to ensure that the migration implications of economic and foreign policy were taken into account explicitly in the formulation of the latter. The Commission also emphasized that expanded trade between countries of emigration and the United States was the single most important strategy to reduce migration pressures in the long run. Similar suggestions have been made in other forums (IOM, 1994), but developed market economies have found it difficult to reduce trade barriers for those goods that developing countries can produce competitively, especially farm products and textiles. There are, however, some hopeful developments. The successful conclusion of the Uruguay Round of multilateral trade negotiations under the General Agreement on Tariffs and Trade (GATT) is one such development, particularly in view of the fact that since the Uruguay Round started in 1986 a dozen developing countries have joined GATT and others have started negotiations to do so (United Nations, 1992). Another is the establishment of a North American free trade zone under the North American Free Trade Agreement (NAFTA) that has been concluded between Canada, Mexico and the United States. However, although NAFTA has the potential to increase trade between the main emigration country in the region, Mexico, and its northern neighbours, NAFTA's long-term impact on migration is uncertain. In addition, migration was not considered explicitly in negotiating the Agreement.

The extent to which ODA can be used to reduce migration pressures in developing countries has also been an issue considered in international forums. Since migration is a heterogeneous process having different characteristics in different regions, if ODA is to reduce migration pressures, strategies for its use must be adapted to the various contexts. In the Caribbean, for instance, the promotion of ecotourism could both create jobs and help conserve the environment; in Mexico, ODA could be targeted to the promotion of regional development and of small businesses; in Northern Africa and Turkey, reduction of population growth could be the main goal; and in Southern Asia, ODA should preferably be devoted to agricultural projects aimed at reducing rural out-migration (Kuptsch and Herman, 1992). Given that these activities fall well within the range of those already being financed by ODA, that most are likely to exert a negative impact on migration only in the long run and that ODA levels remain modest (amounting to about 0.35 per cent of gross domestic product (GDP) of donor countries), the short-term effects of ODA on migration pressures are likely to be marginal. In the long run, an effective reduction of population growth would probably contribute the most to reducing emigration.

At the level of the international community, one of the major developments in terms of setting standards regarding the rights of migrants has been the adoption by the United Nations General Assembly of the International Convention on the Protection of the Rights of All Migrant Workers and Members of Their Families in 1990 (Assembly resolution 45/158 of 18 December 1990, annex). The Convention establishes for the first time an international definition of the term "migrant worker", of the terms referring to different categories of migrant workers (those in a regular situation and those in an irregular situation), and of the term "members of their family" (of such workers). It extends basic human rights to all migrant workers, whether in a regular or in an irregular situation, and provides that documented migrant workers and members of their families shall enjoy equality of treatment with nationals of the States of employment in a number of legal, economic, social and cultural areas. The Convention thus seeks to prevent or eliminate the exploitation of all migrant workers. It also establishes some means to review the compliance of nations in upholding the rights of migrant workers. Twenty member States must ratify the Convention for it to enter into force. Although ratification is not expected to be swift, the very existence of the Convention represents a major step forward in formalizing the responsibility of receiving States in upholding the rights of migrants and assuring their protection. The Convention has even more relevance at present when anti-migrant sentiment is growing in a number of Western countries.

2. *Internal migration and population distribution*

For over 30 years, Governments facing the problems associated with rapid urban population growth have been taking measures to slow it, particularly by reducing or reversing the flows of migrants to urban areas in general or to certain urban agglomerations in particular. Policies aimed at influencing migration have taken mainly three forms: (*a*) those intending to transform the rural economy so as to retain people in rural areas; (*b*) those intending to control in-migration to large cities; and (*c*) those aimed at redirecting migration from large to medium-sized or small cities. Despite the difficulties in assessing the effectiveness of the different policies, it is generally agreed that most of them have had only limited success in reducing urban growth or in changing the pattern of population distribution.

There are several reasons for such a disappointing performance. A major one is the lack of consistency between the migration-oriented policies and other national policies, including those relative to trade, industrialization or investment in infrastructure. Decentralization measures, for instance, are unlikely to be effective when the pursuit of export-led growth dictates that industry should be located in the major agglomerations, the latter being the sites with the best communications. Inconsistent policies often result from a lack of political commitment to a better distribution of economic activities within the country. In addition, inadequate information about the determinants of migration or about its heterogeneity often leads to policies based on erroneous or simplistic assumptions (ILO, 1994).

Despite the constraints facing the Governments of developing countries, especially those undergoing structural adjustment, it is important to stress that the most effective strategy to improve population distribution is to adopt a balanced approach that promotes simultaneously the economic development of rural areas, the improvement of employment and living conditions in large cities, and the growth of small and intermediate urban centres. In so doing, it is essential that urban efficiency be improved by, among other things, reducing food subsidies, pricing public services realistically and eliminating subsidies to private investors. Although the elimination of such "urban biases" is unlikely to reduce the attractiveness of large agglomerations in the short run, their indirect effects on migration will be important.

The promotion of rural development should be a goal in itself, irrespective of its possible impact on migration. Rural development strategies should be combined with policies promoting the growth of small towns and intermediate urban centres, so that the latter may provide markets for agricultural products and be the centres of agro-processing and other small-scale industries. The provision of public services in rural areas can be facilitated by the development of such towns. Efforts to improve the access of rural residents to health and educational services are crucial in improving the quality of their lives and reductions in fertility related to the improved provision of family planning services are likely to reduce migration pressures in the medium term.

Indeed, given the high rates of natural increase still prevalent in the urban areas of many developing countries, reductions of urban fertility would also contribute significantly to controlling the growth of cities. The urban poor, in particular, require access to education, health and family planning services. The challenge is to find the means of improving the incomes and productivity of the urban poor while facing the fiscal constraints now common in developing countries. The evidence suggests, however, that most countries are not meeting such a challenge.

REFERENCES

Bailey, Paul J., and Aurelio Parisotto (1993). Multinational enterprises: what role can they play in employment generation in developing countries? In *The Changing Course of International Migration*. Paris: Organisation for Economic Cooperation and Development.

Brubaker, W. Rogers (1992). Citizenship struggles in Soviet successor states. *International Migration Review* (Staten Island, New York), vol. 24, No. 2 (summer), pp. 269-291.

Callovi, Giuseppe (1992). Regulation of immigration in 1993: pieces of the European Community jig-saw puzzle. *International Migration Review* (Staten Island, New York), vol. 24, No. 2 (summer), pp. 353-372.

Commission of the European Communities (1986). Single European Act. *Bulletin of the European Communities* (Brussels), Supplement 2/86.

De Mattos, Carlos Antonio (1994). The moderate efficiency of population distribution policies in developing countries. Chapter X of the present volume.

Goldgeier, J. M., and M. McFaul (1992). A tale of two worlds: core and periphery in the post-cold war era. *International Organization* (Geneva), vol. 46, No. 2 (spring), pp. 467-491.

Hovy, Bela, and Hania Zlotnik (1994). Europe without internal frontiers and international migration. *Population Bulletin of the United Nations* (New York), No. 36. Sales No. E.94.XIII.12.

International Labour Organization (1988). *Economic and Social Effects of Multinational Enterprises in Export Processing Zones*. Geneva: ILO.

__________ (1994). Migration and population distribution in developing countries: problems and policies. Chapter XX of the present volume.

International Organization for Migration (IOM) (1994). Migration and development. Chapter XXV of the present volume.

Johnson, H. J. (1991). *Dispelling the Myth of Globalization: The Case for Regionalization*. New York: Praeger.

Kreye, O., J. Heinrichs and F. Frobel (1987). Export processing zones in developing countries: results from a new survey. Multinational Entreprises Programme Working Paper, No. 43. Geneva: International Labour Office.

Kritz, Mary M., and Fe Caces (1992). Technology transfers and migration. In *International Migration Systems: A Global Approach*, Mary M. Kritz, Lin L. Lim and Hania Zlotnik, eds. Oxford: Clarendon Press.

Kuptsch, C., and B. Herman (1992). Two essays on using ODA to reduce emigration. World Employment Programme Working Paper. Geneva: International Labour Office.

Organisation for Economic Cooperation and Development (1990). *SOPEMI: Continuous Reporting System on Migration 1989*. Paris: OECD.

Skoog, Christian (1994). The quality and use of census data on international migration. Paper presented at the XIII World Congress of Sociology, Bielefeld, Germany, 18-23 July.

Suhrke, Astri (1994). Safeguarding the institution of asylum. Chapter XVI of the present volume.

The Population Institute (1992). Desperate departures: the flight of environmental refugees. *Towards the 21st Century* (Washington, D.C.), No. 4.

United Nations (1958). *Treaty Series*, vol. 298, No. 4300.

__________ (1991a). *World Population Prospects, 1990*. Sales No. E.91.XIII.4.

__________ (1991b) *World Urbanization Prospects, 1990*. Sales No. E.91.XIII.1.

__________ (1992). *World Economic Survey, 1992*. Sales No. E.92.II.C.1 and Corr. 1and 2.

__________ (1993). *Internal Migration of Women in Developing Countries*. Sales No. E.94.XIII.3.

__________ (1994). *World Population Monitoring, 1993: With a Special Report on Refugees*. Sales No. E.95.XIII.8.

United Nations Secretariat (1993). Types of female migration. In *Internal Migration of Women in Developing Countries*. Sales No. E.94.XIII.3. New York: United Nations.

United States Commission for the Study of International Migration and Cooperative Economic Development (1990). *Unauthorized*

Migration: An Economic Development Response. Washington, D.C.: United States Government Printing Office.

United States Committee for Refugees (1992). *World Refugee Survey 1992.* Washington, D.C.: United States Committee for Refugees.

Zlotnik, Hania (1991). South-to-North migration since 1960: the view from the North. *Population Bulletin of the United Nations* (New York), Nos. 31/32. Sales No. E.91.XIII.18.

__________ (1993). Women as migrants and workers in developing countries. In *Sociology and Social Development in a Time of Economic Adversity*, James Midgley and Joachim Singelmann, eds. *International Journal of Contemporary Sociology* (Joensuu, Finland), vol. 30, No. 1, special issue, pp. 39-62.

__________ (1994). Comparing immigration in Japan, Europe and the United States. *Population and Environment: A Journal of Interdisciplinary Studies* (New York), vol. 15, No. 3 (January), pp. 173-187.

Part One

REPORT AND RECOMMENDATIONS OF THE MEETING

I. REPORT OF THE MEETING

A. BACKGROUND

The Economic and Social Council, in its resolution 1991/93 of 26 July 1991, recalling its resolution 1989/91of 26 July 1989 in which it had decided to convene an international meeting on population in 1994, under the auspices of the United Nations, decided that the meeting should henceforth be called the International Conference on Population and Development, emphasized that population, sustained economic growth and sustainable development would be the overall theme of the conference, and decided that the objectives of the Conference should be, *inter alia,* to contribute to the review and appraisal of the progress made in reaching the objectives, goals and recommendations of the World Population Plan of Action[1] and to identify instruments and mechanisms in order to insure the operational implementation of the recommendations. In the same resolution, the Council authorized the Secretary-General of the Conference to convene six expert group meetings as part of the preparatory work.

Pursuant to that resolution, the Secretary-General of the Conference convened the Expert Meeting on Population Distribution and Migration in Santa Cruz, Bolivia, from 18 to 22 January 1993. The Meeting was organized by the Population Division of the former Department of Economic and Social Development of the United Nations Secretariat in consultation with the United Nations Population Fund (UNFPA). The Bureau of the Meeting consisted of Mr. Aristide Zolberg as Chairman, Ms. Lin Lean Lim, Mr. Robert Obudho and Ms. Nasra Shah as Vice-Chairmen and Mr. Alfredo Lattes as Rapporteur. The participants, representing different geographical regions, scientific disciplines and institutions, included 16 experts invited by the Secretary-General of the Conference in their personal capacities; and representatives of the five regional commissions, the United Nations Centre for Human Settlements (Habitat), the Office of the United Nations High Commissioner for Refugees (UNHCR), the International Labour Organization (ILO), the Food and Agriculture Organization of the United Nations (FAO) and the World Health Organization (WHO). Also represented were the following intergovernmental and non-governmental organizations: the International Organization for Migration (IOM), the Organisation for Economic Cooperation and Development (OECD), the International Union for the Scientific Study of Population (IUSSP), Population Action International, the Population Institute, the International Institute for Applied Systems Analysis (IIASA), the Center for Migration Studies and the World Council of Churches (WCC).

As a basis for discussion, the 16 experts had prepared papers on the main agenda items in order to provide a framework for discussion. The views expressed by the experts were their own and did not necessarily represent the views of their Governments or organizations. The Population Division of the Department of Economic and Social Development had prepared a background document for the meeting, entitled "Population distribution and migration: the emerging issues". Discussion notes were provided by the United Nations Office at Vienna, the Economic Commission for Africa (ECA), the Economic Commission for Europe (ECE), the Economic and Social Commission for Asia and the Pacific (ESCAP), the Economic Commission for Latin America and the Caribbean (ECLAC), the Economic and Social Commission for Western Asia (ESCWA), the United Nations Centre for Human Settlements (Habitat), UNHCR, ILO, FAO, WHO, IOM and The Population Institute.

B. OPENING STATEMENTS

Opening statements were made by the Minister of Planning, the Secretary-General of the International Conference on Population and Development, the Deputy Secretary-General of the Conference, the Director of the Centro Iberoamericano de Formación para el Desarrollo and the Vice-President of Bolivia.

All speakers emphasized the importance of migration in the modern world and particularly its contribution to urbanization. The Secretary-General of the Conference noted that urbanization was an integral part of the development process but that there were considerable differences between the urbanization process being experienced by developing countries during the second half of the twentieth century and that experienced by the developed world a century before. In particular, the cities of developing countries had to absorb greater numbers of migrants and generally had smaller productive bases than their counterparts in the developed world of the nineteenth century. Furthermore, most developing countries had high proportions of their urban populations concentrated in a single city. Evidence suggesting that in recent years the small and medium-sized cities of some developing countries were growing more rapidly than large urban agglomerations was considered a positive sign indicating a more balanced development process within countries.

The Secretary-General of the Conference further noted that rural-urban migration was only one of the forms of internal migration that were related to development. Rural-rural movements were sizeable in many countries and their interrelations with the environment were of growing importance, particularly where resettlement in frontier areas rich in biodiversity was involved.

The Secretary-General of the Conference remarked that international migration had historically been generally perceived as a positive process enabling the development of many of what were today's industrialized countries and opening new possibilities for millions of people. During the past decade, however, the changes taking place in Eastern Europe and the growing economic and demographic disparities between developed and developing countries had considerably increased the perceived potential for international migration, thereby causing concern in the main receiving countries, particularly those in Europe. International migration was characterized as a sensitive political issue whose discussion was often hampered by the lack of information on both the size and nature of migration flows. Mention was made in particular of refugees and undocumented migrants, representing two types of flows whose magnitude was frequently the subject of speculation.

The Secretary-General of the Conference stressed that, despite the concerns raised by South-to-North migration, most international migrants found themselves in the developing world. Migration between developing countries involved all types of flows, but those of refugees and migrant workers were singled out. The increasing participation of women among the latter and the need to ensure that they were protected against abuse were underscored. It was noted that migrants' remittances were a major source of foreign exchange earnings for several countries whose policies often favoured the export of manpower.

Noting that developed countries experiencing low fertility were facing the prospects of a smaller labour force, whereas developing countries were having difficulty coping with the continued growth of theirs, the Secretary-General of the Conference suggested that there might be a convergence of interests if only immigration were not such a contentious issue. Given the increasing pool of potential migrants and the forces leading to globalization, migration pressures were judged to be on the increase. Trade liberalization, especially in agricultural products and those with a high labour content, was cited as a more effective means of reducing those pressures than current levels of international assistance for developing countries.

The Secretary-General of the Conference concluded by underscoring the need to collect, analyse and exchange data on all types of international migration. She recognized migration as a major issue for the 1990s and stressed the need to establish how migrants and their families could best contribute to development, how their integration into host societies could be enhanced, and how ignorance and prejudice jeopardizing their welfare could be overcome.

C. Summary of the papers and discussion

1. *Overview of the main issues*

The presentation of the paper entitled "Population distribution and migration: the emerging issues", prepared by the Population Division, highlighted the sheer scale of internal and international migration. For the world as a whole, net internal migration of all types—urban-rural, rural-urban, rural-rural and urban-urban—was estimated to have involved roughly some 750 million to 1 billion persons during 1975-1985. Although rural-urban migration continued to be the focus of most research and policy concern, rural-rural flows were significantly higher in countries that were still mostly rural (such as India, Thailand and most of Africa) and urban-urban migration was the dominant form in highly urbanized countries such as most of those in the Americas and Europe. International migration, although involving a smaller number of persons, was also significant. Census data, referring mostly to the 1970s and early 1980s, indicated that the number of persons enumerated outside their country of birth (or citizenship) amounted to some 77 million, a figure that represented a lower bound for the stock of international migrants worldwide.

To set migration issues in context, a review was made of the economic trends experienced by the main world regions during the 1980s. Among the industrialized countries, the consolidation of Japan's position as an economic power was noted as well as the move towards further economic unification by the European Community. The positive performance of the newly industrializing countries of Asia was contrasted with that of the rest of the developing world. In particular, the low economic growth experienced by most of Latin America and Africa was highlighted. The growing tendency of Governments to adopt "global" market strategies to promote economic growth was considered to have important implications for migration and population distribution. As enterprises became increasingly multiregional, their locational decisions affected both internal and international migration and, since those decisions were determined by the competitive advantages of different locations, they usually led to unbalanced regional growth and population distribution. The differential success of countries and regions within countries in becoming linked to the global market was increasing economic disparities and thereby fuelling migration. As a consequence, urban primacy still characterized many countries and the number of large urban agglomerations continued to grow, particularly in developing countries.

It was noted that the strategy of moving capital to labour was not entirely successful in stemming migration pressures. Thus, several of the Asian newly industrializing economies (NIEs) experienced significant migrant outflows as they pursued export-led economic development. Their experience confirmed that the nature of the development process was destabilizing and that it was likely to increase migration pressures in the short-to-medium term.

Changes in the nature and configuration of the nation State were also identified as having important implications for migration. The drive towards greater unification among the member States of the European Community was contrasted with the disintegration of nation States occurring among former Eastern-bloc countries. Such changes were blurring the distinction between internal and international migration, as citizens of member States gained greater freedom of movement and establishment within the Community, while Russians in the non-Russian successor states of the Soviet Union became international migrants. Disintegration of nation States could lead to sizeable migration flows, particularly when conflict was involved, as in the case of the former Yugoslavia. Indeed, involuntary migration, particularly that of refugees, asylum-seekers and internally displaced persons had increased during the 1980s and was likely to remain of major importance during the 1990s. Given the negative reactions that the growing number of asylum-seekers were eliciting in several world regions, it was important to ensure that the right to asylum be safeguarded.

With respect to migrants' rights, mention was made of the adoption by the United Nations General Assembly of the International Convention on the Protection of the Rights of All Migrant Workers and Members of Their Families in 1990 (Assembly resolution 45/158 of 18 December 1990, annex). The Convention extended basic human rights to all migrant workers, irrespective of their legal status, and provided that documented migrant workers and members of their families should enjoy equality of treatment with nationals of the States of employment in a number of legal, economic, social and cultural areas.

The discussion stressed the need to consider different types of migration separately. Particular mention was made of return migration, and the differentiation between temporary and permanent movements, and between voluntary and forced movements within countries. In a growing number of countries, internal conflict was causing the displacement of sizeable populations whose plight demanded more attention. With respect to the category of "environmental refugees", caution was urged in the use of the term since in most instances environmental factors were only one of a host of causes leading to migration. It was further stated that many of the migrants thus labelled were not even outside of their country of nationality and therefore could not be considered refugees. Since the definition of "refugee" contained in the 1951 Convention relating to the Status of Refugees[2] did not include environmental factors, UNHCR had no mandate for their protection. It was important, nevertheless, to take account of the environmental consequences of forced migration, whether it occurred within or between countries.

Participants noted that, in contrast to previous assessments, migration was increasingly recognized as a logical strategy of individuals to enhance their opportunities or assure their survival. Economic growth entailed both urbanization and the movement of labour. Although in many countries the Government was reducing its planning role, it nevertheless had to meet the needs and priorities of people and enterprises in the locations where they chose to establish themselves.

Lastly, although economic disparities were acknowledged as one of the major determinants of migration, the importance of growing demographic imbalances was also underscored. Because of high population growth, many developing countries, particularly those in Africa, were facing labour force increases that they would have difficulty absorbing. Such prospects suggested that migration pressures would increase.

2. *Patterns of population distribution and development*

The presentation of the paper entitled "Global urbanization: towards the twenty-first century" stressed that the process of urbanization was an intrinsic part of economic development. In comparing the experience of developed and developing countries, the difference in the magnitude of the process was highlighted. Thus, whereas the urban areas of developed countries had absorbed about 100 million persons during 1815-1915, urban areas in developing countries were expected to accommodate about 1.5 billion additional persons during 1990-2010. In addition, while developed countries had generally had the means to build adequate infrastructure while urbanizing, most developing countries, particularly those in Africa and Asia, experienced serious unmet infrastructure needs. Two aspects of globalization were having important impacts on urbanization: the transactional revolution involving more efficient flows of information and capital, and the new division of labour associated with the free movement of capital to maximize accumulation. Those processes had led to a greater centralization of the urban system and the emergence of transaction nodes facilitating the movement of people, information, capital and commodities. Successful nodes were often mega-urban regions that had carved a niche in the global market by providing certain goods and services.

Despite these important changes, policy was still coloured by a nineteenth century anti-urban perspective and was often geared to controlling the growth of large cities. Evidence to the contrary notwithstanding, it was widely assumed that the spatial separation of urban and rural activities would persist, and paradigms of the urban transition derived from the Western experience of the past century were still considered valid. There was a need to modify them and to take account of the diverging urbanization patterns that the uneven incorporation of developing countries into the world economic system had brought about.

The discussion underscored that, in many countries, the economic and social differences between urban and rural areas were becoming less marked. A revision of the definitions of rural and urban areas seemed necessary if the needs of planners were to be met. There was a need to identify functional regions, especially in light of the increasing significance of rural-urban interactions. The livelihood of many households depended on both urban and rural activities. In certain rural villages of Malaysia, for instance, households derived less than half of their income from agriculture. The role of different types of migration in fostering rural-urban interactions was emphasized, but much remained to be known about the economic, social and environmental implications of such interactions.

Some participants suggested that the sheer size of cities involved diseconomies, while others argued that the existence of such diseconomies had not been proved. The issue was stated further in terms of the question "Diseconomies for whom"? While private capital benefited from economies of scale in large urban centres, the poor often had to bear the diseconomies associated with crowded housing, health hazards and lack of infrastructure. In Africa, in particular, where urbanization was often not driven by economic growth, cities faced severe problems. Furthermore, in most developing countries the informal sector was absorbing increasing numbers of urban workers, particularly women. There was a tendency to casualize labour by shifting production from the factory to the informal sector through subcontracting. However, it was noted that the informal sector was very heterogeneous and that in certain of its subsectors earnings were higher than those in the formal sector.

The consideration of population distribution patterns in specific world regions followed. In the presentation of the paper "Population distribution patterns in developed countries", it was noted that the increasing concentration of the urban population in large urban centres had been considered a normal concomitant of urbanization until data for the 1960s showed that in several developed countries larger metropolitan areas had lost population, in relative terms, to smaller urban centres. That phenomenon, known as counter-urbanization, was formally defined by Fielding in terms of the existence of a negative relationship between size of place and rate of population growth. The evidence showed that counter-urbanization had been fairly widespread in the developed world during the 1960s and 1970s: only the Eastern European countries appeared not to have experienced it. However, during the 1980s, the trend changed in key countries. Thus, in the United States of America during the 1980s large metropolitan areas again grew more rapidly than non-metropolitan areas.

Five sets of factors were said to explain the rise and fall of counter-urbanization since the 1960s. The first was related to the increasing mobility of capital which led to the relocation of manufacturing from major urban centres to smaller cities and towns. The second, labelled "rural-resource development", involved the specialization of certain rural areas in particular branches of rurally based economic activity. The third involved changing residential preferences favouring smaller towns or rural areas over major urban centres. The fourth was related to the fact that residential preferences were specific to certain population subgroups and thus led to overall changes in preferences as the composition of the total population evolved. The last factor involved counter-urbanization's being related in various countries to changes in the size, nature and distribution of housing stock. The evidence also suggested that government actions played an important role in promoting population deconcentration and in phasing it out. However, it was difficult to predict how population distribution would evolve in developed countries in the future. Cyclic changes were judged possible and it was suggested that the impact of international migration on population distribution might be especially significant if overall population growth continued to be low. Participants argued that the effect of international migration had in all probability already been significant in stopping counter-urbanization, since in some European cities international migrants constituted 10-20 per cent of the population.

The discussion centred on a few factors judged to be essential in assessing the possible evolution of population distribution in developed countries. Shifts in the age structure stemming from sustained low population growth could have important implications for population distribution if different birth cohorts had different residential preferences. The impact of Governments' non-spatial policies and sectoral priorities was considered to have greater influence on population distribution than their explicit spatial (or territorial) policies. For instance, the location of defence industries away from major population centres contributed to counter-urbanization. Changes in the distribution of power and resources among national, state and local Governments also had important spatial implications.

The need to adopt more precise definitions for the study of population distribution in developed countries was stressed. The use of functional urban regions seemed necessary, but it demanded suitable databases and concepts that facilitated international comparisons. The use of geographical information systems integrating population data seemed desirable, especially because the urban agglomerations of interest often consisted of several interacting cities.

The Expert Group proceeded to consider population distribution issues relative to the main developing regions. The presentation of the paper entitled "Population distribution patterns and development in Africa" noted the diversity characterizing the continent in terms of population distribution, changes over time and the factors influencing internal migration. Although Africa was the least urbanized region of the developing world, most African countries were undergoing rapid urbanization. The high fertility characterizing African countries made major contributions to the growth of urban areas, and rural-urban migration continued to be significant. African Governments were faced with the need to develop comprehensive population distribution policies. The quality of urban management had to be improved and an effective partnership had to be developed between central and local Governments. It was urgent to devise methods to provide low-cost urban services and ways of recovering investment costs so that services could be expanded, although it was recognized that subsidies might be needed to ensure that the poorest groups had access to needed services. Given the importance of the rural sector in African countries, the development of economic and institutional linkages between rural and urban areas was needed to foster synergistic interactions between rural and urban development in the region.

The presentation of the paper on "Population distribution policies and development in Asia" noted that Asia's level of urbanization was still low, but that the region's average concealed large variations between countries. Asia comprised the most populous countries in the world as well as rather small countries characterized by great economic dynamism. The countries of the Association of South-East Asian Nations (ASEAN) in particular were likely to maintain a rapid rate of urbanization in conjunction with sustained economic growth. In the most populous countries, high rural population densities and changes in agricultural production had the potential of fuelling rapid urbanization and posed enormous challenges for the achievement of sustainable development.

In Asian countries, as in other countries of the developing world, population and economic activity tended to be concentrated in one or a few large cities. The concentration of the urban population in a single city was less marked in the most populous countries of the region (China, India and Indonesia) and in countries that had pursued active rural development policies, such as Malaysia and Sri Lanka. In addition, in a number of countries the proportion of the urban population in the primate city had been declining. The growth rates of several Asian mega-cities had also declined somewhat, partly because of the redistribution of production and population outside the metropolitan area's boundaries to areas that were still functionally linked to the city. In a sense, therefore, the slow-down recorded was artificial. The diseconomies of scale in some mega-cities had encouraged investment to seek new locations. In addition, structural adjustment and government policies that removed incentives for the location of industries in mega-cities, increased prices for agricultural products and promoted exports had contributed to reducing mega-city growth. Decentralization policies were also a factor encouraging the growth of medium-sized cities at the expense of larger ones.

The presentation of the paper on "Population distribution and development in Latin America" noted that Latin America was the most urbanized region of the developing world, having experienced rapid urbanization during the twentieth century that had led to the concentration of both people and economic activities in a few large cities. During 1925-1975, the total population of the region had tripled, while the urban population increased eightfold. In the 1990s, the number of people living in poverty in urban areas surpassed that in rural areas. As in other developing regions, there was considerable diversity in the population distribution of the various Latin American countries, arising from historical and structural differences. With regard to primacy and the growth of the largest cities within each country, over half of the countries of the region were expected to show some reversal of population concentration in the largest cities. Recent evidence indicated that the primacy of Buenos Aires was declining and that, during the 1980s, Mexico's three largest cities had experienced lower growth rates than expected. Such trends were likely related to the economic changes experienced by the region as structural adjustment led to lower industrialization, the casualization of employment and a reduction of the attractiveness of large cities for migrants. Such structural changes had also given rise to new forms of territorial mobility, particularly those of a temporary nature which did not necessarily lead to population redistribution in the traditional sense. Yet, temporary migrants had an impact on the economic and social functioning of the areas they linked. Policy makers, planners and researchers had to consider all types of movements in devising social and economic policies aimed at achieving sustainable development.

The presentation of the report on "Migration to the City of Santa Cruz, Bolivia" highlighted many of the social, economic and political issues raised in the regional papers. The population of Santa Cruz had grown very rapidly, especially during 1976-1992. Despite an annual growth rate of 6.7 per cent, certain social indicators had improved: there was a reduction in the proportion illiterate and a decline in infant mortality. However, city and regional authorities had experienced considerable difficulties in ensuring that the provision of basic infrastructure and services kept pace with population growth and the outskirts of the city were still characterized by poor housing and high infant mortality.

The discussion stressed that urbanization was an inevitable part of development and, consequently, that it was futile for Governments to insist on stopping rural-urban migration. Nevertheless, it was recognized that Governments could reduce pressures for rural out-migration by promoting the use of labour-intensive technologies and by raising the prices of agricultural products. Instances were cited, however, of successful rural development projects that had improved agricultural productivity but did not slow migration to urban areas and that may have even accelerated it by improving the economic status of rural dwellers and increasing their aspirations.

The need to integrate population distribution policies (including urbanization policies) into national development strategies was underscored. In doing this, it was important to keep in mind that rural and urban development were two sides of the same coin. Strategies that emphasized one at the expense of the other were doomed to failure. In many developing countries, the rural population was expected to keep on growing and the need to improve rural economic opportunities was urgent. Measures to strengthen urban-rural economic interactions and to improve rural infrastructure so as to increase productivity were considered desirable. Whereas the high urbanization rates experienced by certain regions had brought to the forefront of the urban agenda a series of problems related to poverty and the environment, there seemed to be an unwarranted disregard for similar problems in rural areas, where they were equally widespread.

Participants stressed the need for Governments of developing countries to strengthen the capacity, competence and accountability of city and municipal authorities. It was judged that a key task of Governments was to address the social and economic needs of their populations, whether in rural or urban areas. It was particularly important to work towards the alleviation of poverty, which was growing in both rural areas and cities. The practice of withholding investment on city infrastructure so as not to attract the rural poor was criticized. Decentralization policies that involved the relocation of manufacturing industries from large to medium-sized cities were said to contribute to growing unemployment in the former since the service sector was not capable of absorbing as many workers. The provision of adequate services in cities demanded the removal of general subsidies, the adoption of progressive taxing schemes and the true pricing of urban services coupled with subsidies for the poor. Given the potential for rapid urbanization, especially in regions where the rural population was still growing rapidly, there was an urgent need to improve urban infrastructure and to provide access to services to all population subgroups, particularly the poor. To gain control of urban growth, particularly in regions where general population growth was high, attention had to be paid to policies aimed at reducing natural increase.

The discussion addressed the question of political will and how such will could be developed and maintained to implement policies consistently. In particular, although the role of community-based organizations and non-governmental organizations was widely regarded as important and in need of support from Governments and international organizations, little was known as to their role in influencing political will.

In considering issues related to large urban centres, participants noted the emergence of complex forms of urban agglomerations that fell outside the conventional definitions used to study them. An example was the development of urban or quasi-urban settlements along the main roads or highways linking major cities. Another was the polygonal configuration of cities surrounding a major metropolitan area and linked to it by a variety of economic and social interactions, as was the case of Mexico City and its surrounding medium-sized cities.

In ascertaining the impact of development on population distribution, note was taken of how its effects might change over time. In Latin America, in particular, when economic growth was based on import substitution, industrialization was the engine of growth of most large cities. With the adoption of export-oriented policies in the 1970s and 1980s, the centres of economic activity changed and population growth accelerated in urban centres located near ports, borders or export-producing sites (mining centres, for instance). Thus, the recent growth of certain medium-sized cities in countries such as Chile or Mexico was closely related to the new mode of economic development espoused by the Government.

Mention was made of population distribution issues within urban agglomerations. In some agglomerations, the population at the core was ageing while the mean age of the population in suburban areas was declining. In others, important socio-economic differences existed between the different subregions within an agglomeration. Measures such as the eviction of poor groups from inner-city tenements or from squatter settlements were not judged to be effective in controlling city growth.

In a number of countries, internal strife had led to increased migration. Although that migration often took the form of rural-rural movements, forced migration between urban areas and from rural to urban areas was also common. The population that was thus internally displaced was often in need of protection and assistance. Yet there was no internationally established mechanism to provide such protection. A related problem was the capacity of war-torn countries to ensure the safety and provide adequate infrastructure and services to repatriating refugees. Repatriation was likely to be successful only if assistance could be provided to reintegrate refugees in their areas of origin.

3. *Policies affecting internal migration and population distribution*

The presentation of the paper entitled "The social and environmental problems associated with rapid urbanization" highlighted four issues: the main environmental problems and their causes; who was most vulnerable to them; the extent to which migrants as compared with non-migrants were vulnerable, and the key policy issues involved in addressing environmental problems. The large scale and wide range of environmental problems evident in the cities of developing countries were reviewed in conjunction with their serious health impacts. Biological pathogens (mainly excreta-related, insect-borne, airborne or food-borne) and the lack of access to safe and sufficient water were generally the most serious threats to health, although exposure to chemical pollutants and physical hazards because of housing deficiencies also contributed significantly to psychosocial disorders in many cities.

Migrants were not necessarily the group most vulnerable to such hazards, since vulnerability was influenced, among other things, by age, health status, socially determined roles, and lack of means to avoid a certain hazard or to cope with the impact of illness or injury. In most cities, the single most important factor determining vulnerability was income, although certain age and occupational groups among the low-income population were especially vulnerable. Because of gender differentiation in work undertaken, access to services, income sources and access to shelter, women were particularly at risk in respect of certain environmental hazards.

Participants noted that the underlying cause of most of the problems mentioned was not so much rapid urbanization as the incapacity of Governments to cope with the rapid economic, social and demographic changes that accompanied it. Thus, some of the most rapidly growing cities did not have serious environmental problems, whereas urban centres that were growing more slowly often had more serious problems. In addition, poorer groups in cities (including poor migrants) often bore a disproportionate share of the costs of environmental problems. It was stressed that developing the capacity and competence of city and municipal authorities to manage urban development, control pollution and ensure sufficient investment in infrastructure and services was the main strategy to reduce the impact of environmental problems on the health of urban dwellers.

Under certain circumstances, migrant groups could be particularly vulnerable: for example, migrants were especially susceptible to diseases that were endemic in the area of destination because they had not been exposed to such diseases in the area of origin. To the extent that certain types of migrants were more likely to join the ranks of the urban poor than others, they were also particularly vulnerable to the health hazards posed by environmental problems. Yet the main policy issue was considered to be the reduction of poverty and the vulnerability of *all* urban dwellers to environmental hazards, whether they were migrants or not.

Several measures were suggested to improve the plight of the urban poor, including facilitating their access to credit, providing them with technical assistance to improve their housing, and promoting the creation of partnerships between neighbourhood groups and non-governmental organizations so that strategies to gain access to water, sanitation or garbage collection could be devised collectively. Such measures had a greater probability of being successful if local or municipal government was strengthened by allowing it to collect taxes and giving it greater responsibility for the welfare of local communities while at the same time ensuring its accountability to the governed. It was recognized that the urban poor often had difficulties in securing adequate land sites on which to establish themselves. In many cities, the poor, though numerous, occupied only small proportions of the city's territory. Community organization and effective leadership were judged necessary to improve the plight of squatters, particularly if they had to fight for land rights.

The paper presented on "Population distribution policies and their impact on development" contrasted the spatial implications of two of the most common development strategies. The first was that adopted by many developing countries between 1930 and the early 1970s. Based on Keynesian theories of economic development, that strategy involved substantial state intervention to promote industrialization for import substitution and to direct population redistribution. Such a strategy, termed "peripheral Fordism", was widely applied throughout Latin America where it led to unbalanced capital accumulation that in turn gave rise to unbalanced population distribution by, among other things, reinforcing the primacy of the largest cities. To combat the inequitable effects of capital accumulation, Governments adopted territorial policies that sought to promote growth in peripheral regions but that rarely had much impact, partly because they were seldom accorded priority.

The second strategy, which had already been widely implemented in developing countries by 1990, greatly reduced the role of the State and ascribed a larger role to market forces both for production and for population redistribution. Economically, it involved an outward orientation with priority given to exports. While in the previous strategy the combination of state support and national capital was seen as the main source of productive investment, in the new one international investment was sought and encouraged. The State, with a reduced role, concentrated its efforts on maintaining fiscal austerity and balanced budgets. The key goal was the competitive incorporation of the country into the new international division of labour. The new strategy had important spatial implications since, within each national economy, rapid urbanization tended to occur in those locations that had served as growth poles with respect to production, the transport of goods for export and associated services. Concomitantly, the centres where import substitution industries had been concentrated sometimes declined in importance. The integration of an economy with the world market usually served as a cleansing mechanism that inexorably excluded those activities that could not be modernized or compete in external markets or with imported products. Such changes had important spatial implications. Within countries, certain social groups and particular locations benefited while others lost ground. The structure of employment was becoming increasingly polarized, with few earning high salaries and the majority earning low salaries. The number of women among the lowly paid tended to increase. In countries undergoing such changes, the rate of population growth of the main urban agglomeration often declined.

Participants noted the contradiction inherent in the adoption of the new strategy by Governments that also claimed to have ambitious goals regarding spatial distribution policies. The more economies became export-oriented, the greater the influence of global forces on the spatial distribution of production and, consequently, of population. Increasingly, mega-urban regions around the world were being shaped by global and national forces. In addition, a reduction of regional inequalities in per capita income did not necessarily imply greater social equality, since poorer groups in peripheral regions often received few benefits from new investments. Governments were often too ready to import, uncritically, strategies conceived elsewhere to foster economic development and direct territorial planning.

4. *Internal migration and its implications for development*

In the presentation of the paper entitled "Migration as a survival strategy: the family dimensions of migration", it was noted that, despite the deficiencies of available information on migration, both internal and international, the evidence pointed to the fact that migration had increased considerably both in scale and in complexity during the 1980s. Movement away from the place of origin, either of a permanent or a temporary nature, had become an option to improve the life chances of a wider spectrum of the population of developing countries. The changing economic, social and political context of developing countries had contributed to increasing the scale of population mobility. However, to understand the evolution of mobility, it was important to supplement explanations at the macrolevel with an understanding of how microlevel processes were determining who migrated and how. There was growing evidence that migration resulted from decisions made by families rather than individuals and that migration was often a family's strategy to ensure survival and minimize risks. Through migration, a family could diversify its sources of income, in terms of both location and type of work. In some societies, migration decisions within the family were taken mostly by men (the patriarch, for instance) and they often involved the temporary migration of unmarried offspring whose earnings were to supplement family income. In some contexts, single women were increasingly selected for migration since their income-earning opportunities, particularly in export-oriented industries, had been growing.

It was noted that the traditional roles of families and households changed as urbanization proceeded. Studying the role of families and kin groups in the migration process was important in understanding how migration networks developed. Networks facilitating migration involved not only family members and kin, but also a variety of agents, recruiters, lawyers or other middlemen who actively encouraged and sustained migration. Networks were thus becoming increasingly institutionalized and commercialized, and were able to sustain population flows even when the economic conditions giving rise to them had changed. Their operation was also fairly resilient to government action. The importance of improving the understanding of how networks got established and evolved was stressed.

The discussion underscored the need to gather better data and develop methods that would permit a more accurate assessment of the scale of population mobility and its growing complexity. More attention had to be devoted to the social and economic dimensions of migration, including those at the family level and those shaping migration networks. With regard to international migration, networks both depended on and contributed to the emergence of transnational groups that could straddle several countries. The existence of such groups fostered the further exchange of information, capital and people. Participants stressed the positive aspects of network operation and warned against the tendency to criminalize their informal elements.

Because of the lack of appropriate information, the extent of temporary migration and its implications for development could not be established with certainty. Temporary movements were said to be more common than those leading to permanent relocation, but it was recognized that what started as temporary might become permanent. In international migration in particular, temporariness could be a function of whether or not migrants were allowed to stay. When migration was part of a family's survival strategy, temporary movements were more likely to be preferred since they were more conducive to maintaining the links between migrant and family members left behind. That was particularly the case when women migrated on their own. Indeed, the increasing participation of women in autonomous migration, whether internal or international, had the potential of changing gender relations within the family.

5. *Economic aspects of international migration*

In the presentation of the paper entitled "Growing economic interdependence and its implications for international migration", it was noted that in the past four decades increased volumes of international trade and investment had generally not proved to be substitutes for the movement of labour. In a world characterized by growing interdependence, there were competing tendencies towards globalization and regionalization, both of which impinged on migration pressures. Growing economic interdependence among nation States fostered and was fostered by international migration, but the latter was both an opportunity and a source of vulnerability for the interdependent States. In cases where barriers had been raised against economic integration or where poor countries had been involuntarily delinked from the more advanced countries, unauthorized migration had been an increasingly significant form of adjustment.

Two main explanations were offered for the fundamental paradox that the economic integration of countries would, initially, increase rather than reduce migration pressures. First, modalities of socio-economic development associated with rising interdependence among countries were essentially disruptive and dislocating, often leading to considerable internal movements and precipitating in some instances increases in international migration. Second, international trade and foreign investment created bridges between trading and investing partners that activated the flows of labour, both legal and illegal. Once activitated, networks based on family, community or employment relations would sustain migration flows.

The implications of economic interdependence on international migration had to be examined within a dynamic systems framework, which took into account the fact that migration movements had been initiated and sustained by various dynamic processes that included trade, foreign direct investment, foreign aid and flows of technology whose direction and extent were determined within a global framework of historical, cultural, economic and political ties.

Geopolitical realignments and new economic arrangements were configuring the world into regional blocs that combined an outward with an inward orientation, sometimes translated as free movement of persons within combined with barriers to movement from without. Economically, the world was moving along a three-speed path, with the OECD countries and the newly industrializing economies of Asia moving forward; low-income countries, especially those in Africa, sliding backwards; and a third group of countries, including most of Latin America, China and India, having somewhat better prospects as long as they could maintain strong linkages with the world economy. Thus, the economic differentials both among developing countries and between most of the developing and the developed world were expected to increase. Demographic imbalances were also large and expected to remain so. Such disparities were expected to lead to increasing migration pressures. To the extent that barriers to legal, authorized migration continued to grow, such pressures would give rise to unauthorized migration, particularly between countries already linked by social networks or to those where the casualization of labour made possible the creation of labour-market niches for undocumented migrants.

In the presentation of the paper on "The economic implications of migration" the focus was on the economic responses to the creation of a common market and the role played by migration. Given that the common market arrangements envisaged for Southern American States did not involve mechanisms to effect transfers and subsidies between Member States, such arrangements would benefit more the relatively richer nations at the expense of the poorer ones. In particular, the former would be better able to attract from poorer nations the skilled personnel needed to foster investment and development. Poorer countries, such as Bolivia, had therefore an interest in going beyond mere common market agreements establishing free trade zones. A more complete economic union was necessary to enhance development prospects.

In discussing the economic aspects of migration, it was important to distinguish between short- and long-term effects. Thus, it was possible for increased trade and foreign direct investment not to affect migration pressures in the short term while at the same time having a strong influence in reducing those pressures in the long run. Increasingly, trade liberalization was being seen as a means of fostering development and thus contributing, over the long run, to reducing migration. The cases of Mexico, Turkey and Northern Africa, for instance, deserved closer scrutiny to ascertain the impact of trade and foreign direct investment on both internal and international migration flows.

It was noted that trade liberalization, per se, did not guarantee that all developing countries would participate equitably in the world economy. Delinking would continue to occur, with some countries remaining relatively marginalized. Migration pressures were consequently unlikely to disappear even if complete trade liberalization was achieved. A complex set of strategies was needed if migration flows were to be reduced significantly.

Participants underscored the importance of taking account of political issues when discussing the economic aspects of migration, since there were often tensions between the economic and political interests of nation States. Attempts to convert human beings into labour so that it could be exchanged with minimal social consequences had largely failed. Given their low population growth, industrialized countries were facing the prospects of a declining labour force but insisted that they would not again resort to foreign labour. Yet, they could not entirely close their borders, if only because tourism and business travel kept on increasing. Unauthorized migration was likely to continue, especially if unmet labour needs continued to exist at the low end of the scale in the better-off countries.

6. *International migration in a changing world*

In the presentation of the paper on "The integration and disintegration of nation States and their implications for migration", it was noted that, historically, national boundaries had been temporary and that border changes had often resulted from armed conflict and entailed population movements. Situations in which forced population movements were likely to arise generally involved such groups as (*a*) colonizing populations stranded as minorities in new States; (*b*) trading or administrative intermediaries in new States; and (*c*) the classical "national minorities" in new States. Russians in the Baltic States and the newly independent States of Central Asia would belong to the first category, but their forced resettlement, although potential, had largely not taken place as yet. A review of other cases where population "unmixing" had occurred indicated that it had traditionally been approved by the international community. In some instances, the forceful relocation of population groups might be inevitable; the issue was whether it could happen without violence.

At the other end of the spectrum, the movement towards integration, particularly economic integration, by several groups of countries raised the issue of the relativization of the nation State. Thus, the economic integration being pursued by the European Community implied that the nation State would lose control over certain key economic and social aspects. Interestingly, the question of controlling population flows had emerged as an important obstacle to the further political and economic integration of the Community. By the time of the meeting, the application of the provisions of the Single European Act had been formally postponed and ratification of the Maastricht Treaty was by no means certain. Concerns about ensuring an adequate control of the external Community borders were contributing to those developments. For integration to proceed, effective border control and stringent restrictions on external migration seemed to be necessary.

The discussion noted that the disintegration of nation States could lead to the emergence of "new" minorities as the basis for demographic accounting changed. The case of Hungarians in Slovenia was mentioned. When the latter was part of Yugoslavia, Hungarians had constituted a small proportion of the total population of the country. However, in independent Slovenia, Hungarians constituted almost one tenth of the total population. Such change could give rise to inter-ethnic tensions.

In the presentation of the paper on "The process of integration of migrant communities", it was noted that "integration" was the general term used to refer to the process by which migrants became incorporated into the host society. Four types of migrant incorporation were distinguished: assimilation, integration, exclusion and multiculturalism. Assimilation was the one-sided process of adaptation of migrants to the local community by which they became indistinguishable from the majority of the population. It was consistent with policies of "benign neglect" whereby the State left matters largely to market forces. Integration involved a process of mutual accommodation between migrants and the host society. It generally involved the removal of barriers to individual participation through equal opportunities- and affirmative action-type legislation. Exclusion involved the incorporation of migrants into only selected areas of society. Migrants were denied access to other areas mainly through legal mechanisms. Multiculturalism involved the transformation of immigrant populations into ethnic communities that remained distinguishable from the host society but whose members were granted equal rights in most spheres of life. Both exclusion and multiculturalism led to the creation and maintenance of ethnic communities. Multiculturalism was judged to be the best model for incorporation, although it needed to be adapted to each set of circumstances.

Integration was the incorporation model most commonly followed by immigrant-receiving societies, sometimes including some multicultural elements. In the 1980s, more clearly multicultural models were adopted by Australia, Canada and Sweden. Countries such as Germany or Switzerland still favoured exclusionary models. Government policies and historical factors influenced the relative success of the models followed, but in most receiving countries ethnic group formation was taking place. An important element influencing incorporation and ethnic group formation was the transformation of temporary migrants into long-term or permanent settlers. Governments that had insisted on making the migration of labour strictly temporary were less likely to facilitate the full incorporation of the migrants who had remained. The position of the latter was particularly vulnerable, being subject to hostility, discrimination and residential segregation.

Cross-national comparisons indicated that in all countries there were major groups that had not yet become integrated. The process of ethnic group formation was largely determined by the actions of the State during the early stages of migration. The best chances for successful integration lay in policies that facilitated permanent settlement, family reunion and access to citizenship. Ethnic group formation benefited from support for migrant associations, social networks, the use of the migrants' language and the maintenance of cultural values. Successful integration depended on active state policies, especially in the areas of housing, employment, education and language training, access to health and social services. Special measures had to be taken in regard to female migrants, who were more likely to be isolated and marginalized. It was essential to adopt and implement legislation to combat all forms of racism and violence against migrants.

The discussion underscored the importance of time in assessing the success or failure of any incorporation model. In many instances, successful incorporation became evident only as of the second generation. In terms of timing, three stages of incorporation could be distinguished: incorporation in the economic or labour-market areas occurred first, followed by integration through access to social services, education and housing, with full social and cultural incorporation occurring only in the long term. Access to citizenship was felt to be essential in regard to achieving the third stage.

It was recognized that the issues regarding the integration of migrant populations varied from region to region. In Western Asia, the enclave development strategy followed by the oil-rich countries had resulted in the segregation of most migrants, an effort had been made to enforce the rotation of foreign labour and there was no intention to incorporate long-term foreign residents into society. In Israel, the integration of large numbers of Soviet Jews conflicted with the interests of the Palestinian population in the West Bank and Gaza. In Europe, Islam was increasingly seen as a marker for differentiation rather than integration and the trend towards European integration was having largely negative effects on the incorporation prospects of most migrants from outside the region. European Governments argued that the prospects for migrant integration in the region depended on effective border control and the limitation of future migration inflows. In Africa, long-standing refugee populations included a significant number of second- and third-generation refugees, but few countries of asylum had taken steps to grant them citizenship.

In the presentation of the paper on "The future of South-to-North migration", it was noted that migration from developing to developed countries had been increasing and was expected to surpass migration between developed countries in every region, even though migration from former Eastern-bloc countries was still an important component of migration flows, particularly in Europe. East-to-West migration was expected to abate, however, in the future. Migration from developing countries was already a major component of the flows directed to the main countries of permanent resettlement and even Japan, the archetypal closed society, was experiencing significant inflows of migrants from developing countries.

It was argued, citing the report of the United States Commission for the Study of International Migration and Cooperative Economic Development, that the development process led to greater migration pressures, at least in the short-to-medium term. To the extent that those pressures resulted in South-to-North migration and that developed countries continued to raise barriers against those flows, unauthorized migration was likely to increase.

In Europe, asylum-seekers constituted an important type of unwanted migration. Their numbers, which had fluctuated considerably during the 1980s, included large proportions of persons from developing countries. The increasing number both of asylum-seekers and of undocumented migrants in certain European countries indicated that policies aimed at maintaining a zero net migration balance had failed. However, given the political and economic realities of the world as a whole, such policies were unrealistic. Since migrant inflows could not be totally avoided, making some allowance for them seemed imperative. The adoption of explicit quotas for the admission of migrants was cited as a possible strategy to enhance control over such flows.

The discussion noted the difficulties in making assumptions about future South-to-North migration. Although its volume was expected to increase, actual outcomes depended on future labour-market needs in developed countries, the potential for chain-migration through family reunion, the effectiveness of restrictive admission policies in the receiving countries, the occurrence of destabilizing events in developing countries, prospects for economic development in the main countries of origin, and the possibility of increasing linkages between the developed world and developing countries that remained marginalized. Differentials in population growth, although clearly affecting the potential for migration, were judged to have relatively weak linkages with actual migration flows.

No firm conclusion could be reached regarding the future labour needs of developed countries, particularly those of Western European countries. It was argued that, through greater productivity, increases in the retirement age and the incorporation of women into the labour force, those countries could supply their labour needs even if population size declined. On the other hand, it was recognized that there were certain jobs that natives were increasingly unwilling to perform. Yet, even if there was a need for foreign labour, European demand was limited while the potential supply from developing countries was very large. European countries were consequently reluctant to open their doors formally to workers from developing countries for fear of attracting too many.

The need to devise strategies that would reduce migration pressures in the developing world was underscored. Cooperation between specific sending and receiving countries was cited as a possibility and the agreement reached between Albania and Italy to reduce outflows from the former was given as an example. However, it was recognized that such agreements could jeopardize the situation of bona fide refugees who might be deprived of the possibility of seeking asylum.

7. *International migration between developing countries*

The paper presented on "Migration between Asian countries and its likely future" indicated that migration within the Asian region was predominantly directed to the oil-producing countries of Western Asia and towards Japan and the newly industrializing economies of East and South-East Asia. Data on such movements, however, were rare and often inconsistent. According to different sources, for instance, the number of foreign workers in Saudi Arabia was 1.3 million in 1984 or 3.5 million in 1985, a discrepancy that could hardly be attributed to changes over time. In view of such problems, conclusions could only be tentative.

Migration to Western Asia, which had started long before the oil price hike of the early 1970s, had increased considerably after that and undergone a shift in composition from Arab to Asian sources. In addition, Asian sources had been further diversified during the late 1970s to include increasing numbers of workers from East and South-East Asia as opposed to those originating in Southern Asia (India and Pakistan). In 1988, it was estimated that one quarter of the migrant workers to Western Asia originated in the Philippines.

During the 1980s, growing labour force needs associated with rapid economic growth transformed certain countries in East and South-East Asia into destinations for migrant labour. Thus Japan, Malaysia, the Republic of Korea and Taiwan Province of China were reported to be hosting sizeable numbers of largely undocumented migrants. Hong Kong and Singapore had a longer history of importing foreign labour, the former mostly from China, the latter from Malaysia.

In Asia, the migration policies of receiving countries generally had three goals: to limit migration, to eliminate illegal migration and to reinforce migration regulations meant to ensure the quality of migrant workers. Countries of origin, on the other hand, aimed at increasing the number of migrant workers abroad, protecting the rights of migrant workers, and regulating migration so as to prevent the exploitation of migrants by recruiting agents. An important concern of countries of origin was to protect the increasing number of women who engaged in temporary worker migration, mostly as domestic servants. The lower cost involved in female as compared with male migration was seen as a cause of the increase in female labour migration. Although some countries had attempted to restrict female migration, the restrictions imposed had had more of a symbolic than a real impact.

In 1989, Asia hosted more than 7 million refugees, a majority of whom originated in Afghanistan. The second largest refugee population in Asia consisted of the more than 2 million Palestinian refugees under the United Nations Relief and Works Agency for Palestine Refugees in the Near East (UNRWA). South-East Asia, in contrast, had produced during 1975-1988 an estimated 1.5 million Indo-Chinese who had been resettled outside the region, particularly in the United States of America.

Regarding future prospects, continued and increasing migration within Asia was expected, mostly for the following reasons: the high population growth rates experienced by many countries had produced considerable labour surpluses, there was already an efficient recruitment industry that fostered migration, and both individuals and Governments of countries of origin benefited from migration. Furthermore, continued economic growth in low-fertility countries, particularly Japan and the newly industrializing economies of Asia, was expected to produce labour shortages during the next decade. However, if those countries continued to limit legal migration, the undocumented type would increase.

In the presentation of the paper on "Migration between developing countries in the African and Latin American regions and its likely future", the limited availability and poor quality of migration data in sub-Saharan Africa were highlighted. On the basis of data on the foreign-born gathered by censuses, it was estimated that, during the late 1970s, about 8 per cent of sub-Saharan Africa's population (or 35 million people) consisted of persons born outside the country in which they were enumerated. The Office of the United Nations High Commissioner for Refugees (UNHCR) statistics further showed that, in 1991, there were over 5 million refugees in the region.

Using similar data for Latin America, it was estimated that around 1980 less than 2 per cent of the region's population was foreign-born, although only 41 per cent of the foreign-born originated in the region. In contrast with migration in the African region, a country outside the region, namely the United States, was the major destination of Latin American migrants, most of whom originated in Mexico.

There were important differences within each region. In sub-Saharan Africa, Western Africa showed the highest concentration of migrants, most of whom moved for economic reasons. Eastern Africa, in contrast, hosted large refugee populations. In Southern Africa, migration flows involved the highly organized migration of labour to the Republic of South Africa plus growing refugee populations in the subregion. In Middle Africa, migration could be characterized as being largely male and temporary, but there was considerable diversity between countries. Although the data available did not permit an adequate assessment of trends, it was known that some countries, such as Ghana and Nigeria, had become migrant-sending countries after having been, during different periods, important countries of destination. In Latin America, only a few countries, Argentina, Brazil and Venezuela, were hosting large numbers of migrants. Central America was the subregion hosting the largest number of refugees, although their numbers had been decreasing. However, the internally displaced population in that subregion was said to have grown, mostly because of internal strife. Despite the weaknesses of the data available, the evidence suggested that there had been an increase in intraregional mobility in both sub-Saharan Africa and Latin America.

Both growing economic differentials and political instability had contributed to the increase of intraregional flows. The latter was more likely to continue fuelling migration in sub-Saharan Africa than in Latin America; but to the extent that development accelerated more in certain countries than others, especially in Latin America, development differentials were likely to induce further migration. Most intraregional migration in Africa and Latin America had occurred at the margin of government policies. That situation was not expected to change since African and Latin American Governments were thought to be less likely than those of other world regions to intervene in controlling or impeding migration. Indeed, few had explicit policies regarding international migration, and Latin American countries in particular considered their immigration levels too low and wished to increase them. As in the case of intra-Asian migration, experts expected that migration within Africa or Latin America would increase as the opportunities to migrate to developed countries became increasingly restricted.

The migration of the highly skilled was considered in the African and Latin American contexts. Its effects were not necessarily negative when there was an adequate supply of such personnel, but development could be hindered if the scale of emigration of the highly skilled was large and critical sectors of the economy were affected. The same was true, however, of the emigration of less skilled workers when such workers were in short supply. In Latin America especially, the migration of the highly skilled was increasingly being viewed in the context of regional development and the need to formulate policies for the training, development and use of human resources was recognized.

It was difficult to assess the effects of regional trade accords, such as the Economic Community of West African States (ECOWAS) and the Andean Pact, on migration. In both cases, economic factors would have been likely to induce migration, even in the absence of agreements allowing the free movement of people. In addition, the type of free movement that was actually allowed involved in most instances severe restrictions on the right of establishment or the right to exercise an economic activity. Thus, existing or planned trading blocs still had a long way to go before the free movement of labour became a reality within their borders.

The discussion stressed the significance of migration flows between developing countries and the variety characterizing such movements. The paucity of data regarding migration to developing countries was considered a serious drawback, particularly because important changes in trends were likely to go undetected for long periods, if not permanently. Mention was made, for instance, of countries in Latin America, such as Argentina or Venezuela, that seemed to have ceased to be attraction poles for migrants during the 1980s. It was also suggested that the scale of temporary international migration was growing in the region. There was, however, little solid evidence to validate those claims. Similarly, regarding the movement of skilled personnel, Australian statistics indicated that Australian professionals were increasingly going to work in developing countries, such as Malaysia and Singapore. Lack of data from the receiving countries precluded a better assessment of those trends.

The evolution of international migration in Asia was considered to provide important examples of the effects of development. Thus, labour migration from the Republic of Korea had abated as development proceeded. This was probably occurring in Malaysia and the question was whether it would occur in more populated countries, such as India or Pakistan. In addition, important economic interactions likely to fuel both migration and development were taking hold of various sets of countries. The triad constituted by Indonesia, Malaysia and Singapore provided one example, as did Hong Kong and China, and Thailand and the Indo-chinese countries.

In Western Asia, the oil-producing countries continued to demand foreign labour, although in some of them demand had shifted from blue-collar to service-sector workers. The impact of migration on both their development prospects and those of the countries of origin was likely to vary from case to case, particularly because of competition among the latter. According to some participants, the prospects for effective cooperation among sending countries to protect the rights of their expatriate workers did not seem promising.

There was considerable variation in the policy responses to migration by developing countries. Among receiving countries, those in Western Asia and Singapore, for instance, controlled international migration strictly, whereas most countries in Africa or Latin America had fairly lax migration controls. Among countries of origin, a variety of policies and strategies had been used to foster worker migration, prevent the migration of those with needed skills or facilitate the return and reintegration of migrant workers. The latter aspect of migration and its implications for development were judged to deserve greater attention, in particular when the return of workers was prompted by deteriorating circumstances in the receiving countries, for instance, those brought about by the invasion of Kuwait or by the economic difficulties experienced by Venezuela.

With regard to forced migration, the end of the cold war was judged to hold positive prospects for resolving long-standing conflicts in Latin America and reducing the intervention of the superpowers in the region. Such developments would facilitate the return of refugees to their countries of origin and reduce the possibilities of future refugee outflows. The prospects for Africa, however, were less rosy. States that had been artificially sustained by cold war enmities were likely to collapse and produce significant refugee outflows. Such developments would generally stem from both the economic and political disintegration of existing States, often exacerbated by environmental problems. The latter, however, would rarely be the sole cause of forced migration and, once more, would not warrant the use of terms such as "environmental refugee".

Concluding the discussion on migration between developing countries, a paper on "International migration policies in Bolivia" was presented. It indicated that Bolivia was undertaking a regularization drive for undocumented migrants and that draft legislation regarding migration was being considered by Congress. Bolivia favoured the admission of international migrants who would help people the country as well as contribute to development and to the rational use of natural resources.

8. *Refugees and asylum-seekers*

In the presentation of the paper entitled "Safeguarding the right to asylum" a distinction was made between the right to seek and the right to enjoy asylum. The 1951 United Nations Convention relating to the Status of Refugees did not establish an individual's right to seek asylum; rather it established the right of States to grant asylum. In addition, the Convention failed to provide adequate protection for victims of war or generalized violence. At the regional level, however, the Organization of African Unity (OAU) 1969 Convention Governing the Specific Aspects of Refugee Problems in Africa dealing with asylum and refugees in Africa and the Cartagena Declaration on Refugees had widened the definition of "refugee" to include those categories.

The refugee regime that had been established in 1951 experienced a period of expansion in terms of who was covered by it until the late 1970s. Demand expanded, but so did the institution of asylum. Most of the expansion took place in Africa. In the developed world, refugee flows were largely equated with East-to-West flows or outflows from Communist countries that were acceptable as part of the cold war. During the 1980s, however, as the number of refugees and asylum-seekers continued to grow, the expanding demand for asylum was met with an increasingly restrictive response, particularly from developed countries. Faced with potentially large numbers of persons who could make a reasonable claim for asylum under prevailing international instruments, receiving countries were adopting a series of measures to restrict access to asylum. Furthermore, in the cases of Iraq and Yugoslavia, refugee outflows had been prevented by internalizing asylum and keeping would-be refugees in "safe" zones within their own countries.

Several policy implications of the current crisis of the asylum regime were identified, including the need to maintain the right to asylum, to consider ways of integrating refugee and immigration policy, to deal with large refugee outflows through the institution of temporary safe haven, to establish regional regimes that would strengthen burden-sharing with respect to the protection and assistance of refugees, to monitor the size of refugee stocks and flows, and to devise new ways of dealing with the underlying causes of refugee flows.

The paper entitled "Changing solutions to refugee migrations" focused mostly on Africa to discuss the prospects and challenges faced by those seeking solutions to the problem of the growing number of refugees in the continent. It was noted that the situation of refugees in Africa was changing. In earlier decades most countries of asylum had been fairly generous in admitting and assisting refugees; however, their generosity was running out. In Khartoum, for instance, refugees were being blamed for falling wages and rising rents. In rural areas, as the number and length of stay of refugees increased, their competition with the local population grew, giving rise to antagonism. Given those developments, it was urgent to find innovative ways of fostering the settlement of long-term refugees in countries of asylum, particularly by regularizing their status as long-term residents or by facilitating naturalization. Assistance should aim at making refugees self-sufficient and at minimizing their competition with the local population. It was also necessary to channel assistance both to refugees in rural areas and to the increasing number living in urban areas. Although in some countries, refugees were forcefully kept in rural areas or returned there if found in urban areas, increasing numbers were settling spontaneously in cities and were in need of assistance.

Repatriation was described as the favourite solution for the plight of refugees. It was recalled that the 1990s was named the "Decade of repatriation" by the United Nations High Commissioner for Refugees and participants alluded to the many successful repatriation drives that had already taken place. Since resettlement opportunities in third countries of asylum were diminishing and access to asylum in developed countries was becoming increasingly difficult, repatriation was seen as the most viable solution in the future. Nevertheless, repatriation had its drawbacks. Cases in which repatriation had been less than voluntary were cited. In addition, in some countries repatriation had taken place even when the conflicts leading to refugee outflows were still far from over. Other crucial issues regarding repatriation remained to be resolved. UNHCR was usually in charge of aiding in the repatriation, but it had no mandate to provide assistance for the long-term reintegration of returning refugees. There was a need to establish the point at which the needs of repatriates became developmental rather than humanitarian and to provide them with the necessary support. In doing so, it was important to balance the needs of repatriates with those of the local population. The issue of whether returning refugees would choose or be able to return to their areas of origin rather than to different destinations in their country was raised as was the difficult choices facing second-generation refugees who might have very weak ties with the country of origin.

The discussion reviewed the evolution of the refugee regime in relation to the general immigration policies favoured by developed countries, especially during the cold war period, and noted that one of the best ways of controlling migration was at the point of origin by instituting visas, controlling access to means of transportation and so forth. Such methods were increasingly used to restrict access to asylum. In addition, in Europe there was a strong movement favouring the harmonization of asylum policies. The question was whether such harmonization would institute a minimum or a maximum common denominator. The move towards harmonization had some positive aspects, including the fact that it was likely to introduce greater predictability in the adjudication procedure. To maintain the integrity of the asylum system, it was argued that countries had to adopt consistent treatments for those asylum-seekers who were not granted refugee status. When 80 per cent of those rejected nevertheless stayed in the receiving country, control was lost and there was little incentive to maintain a costly adjudication system.

Concern about the increasing number of internally displaced persons in need of protection was raised. International law provided a very weak basis for their protection, since it again involved a conflict between human rights and state rights, specifically, a State's sovereignty. Nevertheless, it was agreed that the plight of the internally displaced deserved more attention from the international community, particularly after internal safe havens had been used to protect would-be refugees, as in the case of the Iraqi Kurds.

Mention was made of the debate underlying the decision to internalize asylum. When refugees flows arose as part of ethnic cleansing, the international community was reluctant to validate such a strategy by providing external safe havens for the population being expelled. However, by maintaining would-be refugees within their countries or immediate areas of origin, the international community had less incentive to combat the root causes of the conflict. Furthermore, it seemed unethical to add the burden of stopping ethnic cleansing to the plight of expelled persons or those fleeing persecution.

The possibility of instituting adequate burden-sharing mechanisms at the regional level was considered. The model established by the Comprehensive Plan of Action[3] regarding Vietnamese refugees was judged to be a successful example of such burden-sharing. In Central America, however, it had proved more difficult to share responsibility and decision-making. Some participants were skeptical about the possibilities of burden-sharing among European countries. It was nevertheless stressed that if viable strategies were to be found to combat the root causes of refugee movements, more countries would have to integrate their refugee policy with other foreign-policy issues and strive to elicit greater international cooperation.

Lastly, it was noted that statistical information on refugees was very weak. The estimates provided by Governments were often unrealistic and lacked a scientifically acceptable basis. Although it was recognized that accurate statistical accounting was difficult when refugee flows occurred rapidly and involved large numbers of people, better methods had to be devised to gather refugee statistics, particularly regarding long-standing refugee populations. Better statistics on repatriation were also sorely needed. It was important to sensitize those involved in refugee assistance about the importance of statistics and to enlist their cooperation.

NOTES

[1]*Report of the United Nations World Population Conference, Bucharest, 19-30 August 1974* (United Nations publication, Sales No. E.75.XIII.3), chap. I.

[2]United Nations, *Treaty Series*, vol. 189, No. 2545.

[3]Report of the Secretary-General on the Office of the United Nations High Commissioner for Refugees: International Conference on Indo-Chinese Refugees (A/44/523), annex, sect. II.

II. RECOMMENDATIONS

A. POPULATION DISTRIBUTION AND INTERNAL MIGRATION

The World Population Plan of Action[1] and the recommendations on its further implementation underscore the need to integrate population distribution policies into overall development planning with the aim of promoting a more equitable regional development. Among the strategies proposed are the use of incentives to reduce undesired migration, the reduction of rural-urban inequalities, the avoidance of "urban biases" stemming from economic or social policies that favour urban areas, the adoption of rural development programmes aimed at increasing rural production, efficiency and incomes, and the provision of assistance to migrant women. Many of the Plan's recommendations and those made for its further implementation remain valid, but new developments and a better understanding of the linkages between migration and development indicate other avenues for action.

Population mobility is an option to improve the life chances of a wide section of the world population. Improvements in transportation and communications, the increasing mobility of capital, the speed of transactions, the widening social networks are all contributing to increased permanent and non-permanent movements. It is now recognized that such flows are rational responses by individuals and families to interregional differences in opportunities and to the need to ensure and widen the sources of family support. Moreover, an increasing number of persons are compelled to migrate in order to ensure their own survival.

Recognizing that the free movement of people and the process of urbanization are essential elements of a productive economy, the priority is not to radically transform population distribution and population mobility patterns but rather to facilitate trends that result in improved life chances for a wide spectrum of the population and to meet the needs of people and enterprises in the locations where they establish themselves. At the same time, however, it is recognized that in many parts of the world rapid urbanization and the development of very large cities present enormous challenges to Governments in terms of their achieving urban management and provision of services without neglecting the important needs of rural populations. The recommendations that follow outline strategies to address these issues.

Recommendation 1. Population distribution policies should be an integral part of development policies. In trying to achieve a better spatial distribution of production, employment and population, Governments should adopt multi-pronged strategies such as encouraging the growth of small and medium-sized urban centres, and promoting the sustainable development of rural areas while at the same time improving employment and living conditions in large urban centres. In doing so, principles of good governance with respect to accountability and responsiveness should be adhered to.

Recommendation 2. All government policies and expenditures have some influence on the spatial distribution of population and on migration flows, with many of the strongest influences deriving not from specific urban and regional policies or spatial planning but from macroeconomic and pricing policies, sectoral priorities, infrastructure investment and the distribution of power and resources among central, provincial and local Governments. Governments should evaluate the extent to which both their spatial policies and the spatial impacts of non-spatial policies contribute to their social and economic goals.

Recommendation 3. With regard to urban areas, the main priority must be to increase the capacity and competence of city and municipal authorities to manage urban development, to respond to the needs of their citizens, especially the poor, for basic infrastructure and services, and to provide poor groups with alternatives to living in areas at risk of natural and technological disasters. To finance such infrastructure and services, Governments should consider equitable cost-recovery schemes and increasing revenues by broadening the tax base.

Recommendation 4. In order to increase administrative efficiency and improve service provision, Governments should decentralize expenditure responsibility and the right to raise revenue to regional, district and municipal authorities. Partnerships among residents, community-based organizations, local authorities, non-governmental organizations and the private sector should be fostered for rural and urban development.

Recommendation 5. Governments should make efforts to develop economic and institutional links between urban centres and their surrounding rural areas by, among other things, improving infrastructure (roads, electricity, water supply, telecommunications), expanding education and health services, and providing technical assistance for the marketing and commercialization of rural products.

Recommendation 6. In order to help create alternatives to out-migration from rural areas, Governments should not only enhance rural productivity and improve rural infrastructure and social services, but also facilitate the establishment of credit and production cooperatives and other grass-roots organizations that give people control over resources and improve their welfare. Governments should recognize and safeguard traditional rights over common lands and water resources. In addition, Governments and the private sector should collaborate in promoting vocational training and off-farm employment opportunities in rural areas, ensuring equal access for men and women.

Recommendation 7. Given that a substantial number of migrants engage in economic activities within the informal sector of the economy, efforts should be made to improve the income-earning capacities of those workers by facilitating their access to credit, vocational training, a place to ply their trade, transportation and health services, including family planning. In doing so, it is important to ensure that women and men have equal access to services.

Recommendation 8. Governments and non-governmental organizations should encourage and support group mobilization and organization by and for persons affected by migration, such as women left behind, domestic servants, workers in the informal sector and urban squatters. Such organizations can foster community participation in development and self-help programmes, mobilize savings and credit, organize for production, provide counselling and other social protection and legal services, identify problems and make them known to decision makers.

Recommendation 9. Given that in many countries high population growth in rural areas is due to natural increase and that the latter also makes major contributions to the growth of the urban population, population policies and programmes that ensure adequate access to health services and family planning should be implemented by Governments wishing to reduce urban growth.

Recommendation 10. Given the increase in migration triggered by environmental degradation, natural disasters and armed conflict, there is a need to address the underlying causes as well as to develop mechanisms to protect and aid the victims, regardless of whether they are within or outside their own country. International and regional organizations, non-governmental organizations and Governments are urged to cooperate in addressing such causes and in developing such mechanisms.

B. International migration

The World Population Plan of Action and the recommendations for its further implementation establish that international migration policies must respect the basic human rights and fundamental freedoms of individuals, as set out in the Universal Declaration of Human Rights,[2] the International Covenant on Economic, Social and Cultural Rights,[3] and the International Covenant on Civil and Political Rights[3]. It further calls for Governments of receiving countries to grant migrant workers in a regular situation (documented migrant workers) and accompanying members of their families equal treatment to that accorded to their own nationals in terms of working conditions, social security, participation in trade unions, and access to health, education and other social services.

With respect to "undocumented migrants" (defined as persons who have not fulfilled the legal requirements of the State in which they find themselves for admission, stay or exercise of economic activity), the Plan emphasizes that all measures designed to curb their numbers must respect their basic human rights. It is further suggested that the International Labour Organization (ILO) Convention concerning Migrations in Abusive Conditions and the Promotion of Equality of Opportunity and Treatment of Migrant Workers, 1975 (No. 143),[4] part I, be used to provide guidelines for the formulation of policies aimed at controlling undocumented migration.

In recent years, growing demographic and labour-market imbalances as well as increasing disparities in economic growth and development among countries and regions and major changes in world political and economic systems have contributed to intensifying migration pressures. These growing migration pressures in developing countries and in former Eastern-bloc countries have heightened concern among the main receiving countries. Such concern has been prompted, at least in part, by the negative attitudes of nationals towards migrants in receiving countries. Like internal migration, voluntary international migration is a rational response to interregional economic differences.

The recommendations presented below take account of developments since 1984 including the adoption of the 1990 International Convention on the Protection of the Rights of All Migrant Workers and Members of Their Families,[5] which established a new set of standards regarding the rights of all migrant workers.

Recommendation 11. In formulating general economic, trade and development cooperation policies, Governments of both sending and receiving countries should take account of the possible effects of such policies on international migration flows. Where trade barriers contribute to growing migration pressures in developing countries, Governments of developed countries are urged to remove them, as well as to promote investment in countries of origin and to channel development assistance to job-creating projects.

Recommendation 12. Governments of countries of origin are urged to recognize and act upon their common interests by cooperating with one another in their negotiations with labour-importing countries to adopt standardized work contracts, establish adequate working conditions and social protection measures for their migrant workers, and control illegal recruitment agents. Governments of countries of origin should appoint labour attachés in receiving countries to ensure that work contracts are honoured and to look after the welfare of their migrant workers. Advocacy organizations should also have a recognized role in protecting migrant workers' rights.

Recommendation 13. Governments of countries of origin wishing to foster the inflow of remittances and their productive use for development should adopt sound exchange-rate, monetary and economic policies, facilitate the provision of banking facilities that enable the safe and timely transfer of migrants' funds, and promote the conditions necessary to increase domestic savings and channel them to productive investment.

Recommendation 14. Governments of receiving countries should protect the rights of all migrant workers and members of their families by conforming to the guidelines established by the International Convention on the Protection of the Rights of All Migrant Workers and Members of Their Families and other relevant international instruments. It is particularly important that Governments of receiving countries ensure that all migrant workers, irrespective of whether their status is regular or irregular, be protected from exploitation by unscrupulous intermediaries, agents or employers.

Recommendation 15. Taking account of the Declaration on the Elimination of Discrimination against Women[6] and of the Convention on the Elimination of All Forms of Discrimination against Women[7], Governments of sending and receiving countries are encouraged to review and, where necessary, amend their international migration legislation and regulations so as to avoid discriminatory practices against female migrants. In addition, Governments are urged to take appropriate steps to protect the rights and safety of migrant women facing specific problems, such as those in domestic service, those engaging in out-work, those who are victims of trafficking and involuntary prostitution, and any others in potentially exploitable circumstances.

Recommendation 16. As previously recommended in the World Population Plan of Action, Governments of receiving countries that have not already done so are urged to consider adopting measures to promote the normalization of the family life of documented migrants in the receiving country through family reunion. Demographic and other considerations should not prevent Governments in the receiving country from taking such measures.

Recommendation 17. Governments of receiving countries are urged to promote good community relations between migrants and the rest of society and to take measures to combat all forms of racism and xenophobia by, for instance, adopting legislation against racism, establishing and supporting special agencies to combat racism and xenophobia, taking appropriate educational measures and using the mass media.

Recommendation 18. Governments should guarantee equal economic and social rights to long-term foreign residents and facilitate their naturalization.

Recommendation 19. Governments of countries within regions wherein exist rights of free movement of their respective citizens should extend these rights to their long-term foreign residents from third countries.

Recommendation 20. Governments should provide information to potential migrants on the legal conditions for entry, stay and employment in receiving countries.

Recommendation 21. Governments of receiving countries should increase their efforts to enhance the integration of the children of migrants ("second-generation migrants") by providing them with educational and training opportunities equal to those of nationals, allowing them to exercise an economic activity and facilitating the naturalization of those who have been raised in the receiving country.

Recommendation 22. Governments of countries both of origin and of destination are urged to promote and support migrant associations that provide information and social services and enable migrants and returning migrants, especially female migrants in vulnerable situations, to help themselves.

Recommendation 23. Governments of receiving countries should consider adopting effective sanctions against those who organize illegal migration as well as against those who knowingly employ undocumented migrants. Where the activities of agents or other intermediaries in the migration process are legal, Governments should introduce regulations to prevent abuses.

C. Refugees

The World Population Plan of Action and the recommendations for its further implementation emphasize the need to find durable solutions to the problems related to refugees and refugee movements, especially in terms of voluntary repatriation or resettlement in third countries. When neither of these two approaches is possible, the Plan advocates the provision of assistance to first-asylum countries to help them meet the needs of refugees. It also calls for Governments to accede to the 1951 Convention[8] and the 1967 Protocol relating to the Status of Refugees.[9]

The world refugee population continues to increase and, while there are prospects for durable solutions through voluntary repatriation, the growing number of persons seeking asylum are straining the asylum system. The recommendations that follow emphasize the need to preserve international protection principles, particularly the right to asylum, and the adoption of appropriate strategies relating to returnee programmes and to the assistance of vulnerable groups.

Recommendation 24. Governments, the international community and non-governmental organizations are urged to address the underlying causes of refugee movements and to take appropriate measures regarding conflict resolution, promotion of peace, respect for human rights including those of minorities, poverty alleviation, democratization, good governance and the prevention of environmental degradation. Governments should refrain from policies or practices that lead to forced migration or population movements of an involuntary nature.

Recommendation 25. Governments are urged to continue facilitating and supporting international protection and assistance activities on behalf of refugees and to promote the search for durable solutions to their plight.

Recommendation 26. Governments are encouraged to strengthen regional and international mechanisms that enhance their capacity to share equitably the protection and assistance needs of refugees.

Recommendation 27. Governments are urged to protect the right to asylum by respecting the principle of "non-refoulement" (non-turning away), granting asylum-seekers access to a fair hearing and providing temporary safe haven when appropriate.

Recommendation 28. Governments, international organizations, community-based organizations and non-governmental organizations are urged to contribute to and participate in repatriation programmes that ensure that initial rehabilitation assistance is linked to long-term reconstruction and development plans.

Recommendation 29. Given that many refugee populations in countries of first asylum have been in exile for extended periods, Governments, international organizations and non-governmental organizations are urged to assist these long-standing refugee populations in achieving self-sufficiency. Governments of first-asylum countries are invited to take steps to regularize the situation of long-standing refugees with little prospect of repatriation by facilitating their naturalization.

Recommendation 30. The international community, through the Office of the United Nations High Commissioner for Refugees (UNHCR) and appropriate relief organizations, should address the specific needs of refugee women. In particular, Governments of countries of asylum should ensure the right of female refugees to physical safety and facilitate their access to counselling, health services, material assistance, education and economic activity. Governments should ensure that female refugees have resettlement opportunities equal to those of male refugees.

D. Data and research needs regarding population distribution and migration

The World Population Plan of Action and the recommendations for its further implementation recognize that migration and urban statistics is the least developed area of population statistics and recommend that Governments should improve it by using their national population censuses, sample surveys and administrative record systems to obtain information on internal migration, urbanization and international migration. However, despite some advances, many deficiencies still remain. Furthermore, understanding of migration processes has not advanced as much as understanding of fertility and mortality. The recommendations below indicate areas that should be given priority in data gathering and research.

Recommendation 31. The United Nations system and other appropriate organizations should support and promote research on population distribution, internal and international migration and urbanization aimed at providing a sounder basis for the formulation of environmental, development and population distribution policies.

Recommendation 32. National statistical offices are urged to collect, tabulate, publish and disseminate demographic data on vital events, migration and population size and characteristics by relevant geographical areas so as to facilitate a better understanding of population change processes and their policy implications at local, regional and national levels. The dissemination of detailed census data coded for micro-regions in machine-readable form should be given priority. The United Nations is urged to foster these activities.

Recommendation 33. Recognizing the major changes that have occurred in the structure and functioning of urban systems, the United Nations and national statistical offices are urged to review the existing standard definitions and classifications of rural and urban populations.

Recommendation 34. The United Nations and appropriate national and international agencies are invited to review the adequacy of existing definitions and classifications of international migration. They should also address the problems of incorporating these into efficient data-collection systems. The development of methods allowing the estimation of undocumented migration and, where relevant, statistics on remittances should also be supported. Efforts should be made to conduct in-depth migration surveys in countries hosting sizeable migrant populations. Governments are urged to produce and disseminate statistics on international migrants classified by place of birth, country of citizenship, occupation, sex and age.

Recommendation 35. Given the deficiencies of refugee statistics and their relevance for planning refugee assistance and for understanding the consequences of refugee movements, measures to improve them should be accorded priority. Governments of countries of asylum as well as intergovernmental and non-governmental organizations dealing with refugees are urged to cooperate with the United Nations in devising and implementing procedures to register and monitor refugee populations.

Recommendation 36. Recognizing the lack of systematic data on displaced persons, Governments are urged to cooperate with the United Nations and non-governmental organizations to facilitate data collection on displaced populations and their needs.

Recommendation 37. The United Nations should promote the exchange of information on both trends and policies of international migration by the creation of working groups of national experts whose task would be to prepare periodical reports on international migration developments in each of their countries following the Continuous Reporting System on Migration (SOPEMI) model of the Organisation for Economic Cooperation and Development (OECD). Regional summaries of such reports should be produced and disseminated by the United Nations.

NOTES

[1]*Report of the United Nations World Population Conference, Bucharest, 19-30 August 1974* (United Nations publication, Sales No. E.75.XIII.3), chap. I.
[2]General Assembly resolution 217 A (III).
[3]General Assembly resolution 2200 A (XXI), annex.
[4]See *International Labour Conventions and Recommendations, 1919-1981* (Geneva, International Labour Office, 1982).
[5]General Assembly resolution 45/158, annex.
[6]General Assembly resolution 2263 (XXII).
[7]General Assembly resolution 34/180, annex.
[8]United Nations, *Treaty Series*, vol. 189, No. 2545.
[9]Ibid., vol. 606, No. 8791.

ANNEXES

ANNEX I

Agenda

1. Opening statement.

2. Election of officers and adoption of the agenda.

3. Overview of the main issues relating to population distribution and migration:

   Population distribution and migration: the emerging issues.

4. Patterns of population distribution and development:

   (*a*) Global urbanization: towards the twenty-first century;
   (*b*) Population distribution in developed countries;
   (*c*) Population distribution in Africa;
   (*d*) Population distribution in Asia;
   (*e*) Population distribution in Latin America.

5. Policies affecting internal migration and population distribution:

   (*a*) The social and environmental problems associated with rapid urbanization;
   (*b*) Population distribution policies and their impact on development.

6. Internal migration trends and its implications for development:

   Migration as a survival strategy: the family dimension of migration.

7. Economic aspects of international migration:

   Growing economic interdependence and its implications for international migration.

8. International migration in a changing world:

   (*a*) The integration and disintegration of nation States and their implications for migration;
   (*b*) The process of integration of migrant communities;
   (*c*) The future of South-to-North migration.

9. International migration between developing countries:

   (*a*) Migration between Asian countries and its likely future;
   (*b*) Migration between developing countries in the African and Latin American regions and its likely future.

10. Refugees and asylum-seekers:

   (*a*) Safeguarding the right to asylum;
   (*b*) Changing solutions to refugee movements.

11. Adoption of recommendations.

12. Closure of the Meeting.

ANNEX II

List of participants

EXPERTS

Stephen Castles, Centre for Multicultural Studies, University of Wollongong, Australia

Anthony G. Champion, Department of Geography, University of Newcastle-upon-Tyne, United Kingdom of Great Britain and Northern Ireland

Graeme J. Hugo, Department of Geography, University of Adelaide, Australia

Alfredo Lattes, Centro de Estudios de Población (CENEP), Buenos Aires, Argentina

Lin Lean Lim, Regional Office for Asia and the Pacific, International Labour Organization, Bangkok, Thailand

Carlos Antonio De Mattos, Instituto de Estudios Urbanos, Pontificia Universidad Católica de Chile, Santiago, Chile

Terence McGee, Institute of Asian Research, Department of Geography, University of British Columbia, Vancouver, Canada

Philip Muus, Department of Human Geography, University of Amsterdam, Netherlands

Robert A. Obudho, Department of Geography, University of Nairobi, Kenya

Ernesto M. Pernia, Economic Research Unit, Asian Development Bank, Manila, Philippines

John R. Rogge, Disaster Research Unit, University of Manitoba, Winnipeg, Canada

David Satterthwaite, Human Settlements Programme, International Institute for the Environment and Development, London, United Kingdom of Great Britain and Northern Ireland

Nasra M. Shah, Department of Community Medicine, Faculty of Medicine, Kuwait University, Safat

Sharon Stanton Russell, Research Scholar, Center for International Studies, Massachusetts Institute of Technology, Cambridge, Massachusetts, United States of America

Astri Suhrke, Department of Social Sciences and Development, Christen Michelsen Institute, Fantoft, Norway

Aristide Zolberg, Department of Political Science, New School for Social Research, New York, New York, United States of America

SECRETARIAT OF THE INTERNATIONAL CONFERENCE ON POPULATION AND DEVELOPMENT, 1994

Nafis Sadik, Executive Director, United Nations Population Fund (UNFPA) and Secretary-General of the Conference

Shunichi Inoue, Director, Population Division, Department for Economic and Social Information and Policy Analysis, United Nations Secretariat, and Deputy Secretary-General of the Conference

Jyoti Shankar Singh, Director, Technical and Evaluation Division, United Nations Population Fund (UNFPA), and Executive Coordinator of the Conference

German Bravo-Casas, Coordinator, World Population Conference Implementation, Population Division, Department for Economic and Social Information and Policy Analysis, United Nations Secretariat, and Deputy Executive Coordinator of the Conference

UNITED NATIONS

Department of Economic and Social Development, Population Division

Birgitta Bucht, Chief, Demographic Analysis Branch

Hania Zlotnik, Chief, Mortality and Migration Section, and Technical Secretary of the Meeting

Keiko Ono, Population Affairs Officer, Mortality and Migration Section

Bela Hovy, Associate Expert, Mortality and Migration Section

United Nations Population Fund (UNFPA), Technical and Evaluation Division, New York

Michael Vlassoff, Technical Officer

Joyce Bratich-Cherif, External Relations Adviser

UNFPA, La Paz, Bolivia

Rainer Rosenbaum, Country Director

Waldo San Martín, Programme Officer

Committee for Development Planning
Solita Monsod

Office of the United Nations Higher Commissioner for Refugees
Maryluz Schloeter-Paredes, External Relations

United Nations Office at Vienna
Elisabeth Sjöberg, Division for the Advancement of Women

United Nations Centre for Human Settlements (Habitat)
Irene Vance, Chief, Technical Adviser

Economic Commission for Africa (ECA)
Toma Makannah, Population Division

Economic Commission for Europe (ECE)
Tomas Frejka, Population Activities Unit

Economic Commission for Latin America and the Caribbean (ECLAC)
Miguel Villa, Coordinator, Population and Development Area

Economic and Social Commission for Asia and the Pacific (ESCAP)
Srawooth Paitoonpong, Population Division

Economic and Social Commission for Western Africa (ESCWA)
George Kossaifi, First Population Affairs Officer, Social Development and Population Division

SPECIALIZED AGENCIES

Food and Agriculture Organization of the United Nations (FAO)
Nadia Forni, Human Resources Institution and Agrarian Reform Division

International Labour Organization (ILO)
A. S. Oberai, Employment and Development Department

World Health Organization (WHO)
Maria Helena Henriques

INTERGOVERNMENTAL ORGANIZATIONS

International Organization for Migration (IOM)
Geneva, Switzerland
Reinhard Lohrmann, Chief, Division of Research and Forum Activities

IOM, La Paz, Bolivia
Albrecht Fuchs, Chief of Mission

Organisation for Economic Cooperation and Development (OECD)
Antonella Bassani

NON-GOVERNMENTAL ORGANIZATIONS

International Institute for Applied Systems Analysis (IIASA)
Wolfgang Lutz, Population Project

Population Action International
Robert Engelman, Director, Population and Environment Program

World Council of Churches
Rosario Sanchez

OBSERVERS

Eduardo Araujo
María Eugenia Ascarrunz
Ramón Prada Bacadiez
John R. Bermingham
Jesus Bolívar
Luz Marina Díaz
Antonio Golini
Bernard Inch
Ronald Jiménez
María Eugenia Moscoso
Juan Pita
Jaime Suárez
Lydio Tomasi
Melvy A. Vargas

ANNEX III

List of documents

Document No.	*Agenda item*	*Title and author*
ESD/P/ICPD.1994/EG.VI/1	-	Provisional agenda
ESD/P/ICPD.1994/EG.VI/2	-	Provisional annotated agenda
ESD/P/ICPD.1994/EG.VI/3	3	Population distribution and migration: the emerging issues United Nations Secretariat
ESD/P/ICPD.1994/EG.VI/4	4	Global urbanization: towards the twenty-first century Terence G. McGee and C. J. Griffiths
ESD/P/ICPD.1994/EG.VI/5	4	Population distribution patterns and development in Asia Ernesto M. Pernia
ESD/P/ICPD.1994/EG.VI/6	4	Population distribution patterns in developed countries Anthony Champion
ESD/P/ICPD.1994/EG.VI/7	4	Population distribution patterns and development in Africa Robert A. Obudho
ESD/P/ICPD.1994/EG.VI/8	6	The social and environmental problems associated with rapid urbanization David Satterthwaite
ESD/P/ICPD.1994/EG.VI/9	4	Population distribution and development in Latin America Alfredo Lattes
ESD/P/ICPD.1994/EG.VI/10	5	Migration as a survival strategy: the family dimension of migration Graeme J. Hugo
ESD/P/ICPD.1994/EG.VI/11	6	Population distribution policies and their impact on development Carlos Antonio De Mattos
ESD/P/ICPD.1994/EG.VI/12	7	Growing economic interdependence and its implications for international migration Lin Lean Lim

Document No.	Agenda item	Title and author
ESD/P/ICPD.1994/EG.VI/13	7	The integration and disintegration of nation States and their implications for migration Aristide Zolberg
ESD/P/ICPD.1994/EG.VI/14	8	Migration between Asian countries and its likely future Nasra M. Shah
ESD/P/ICPD.1994/EG.VI/15	7	The process of integration of migrant communities Stephen Castles
ESD/P/ICPD.1994/EG.VI/16	8	The future of south to north migration Philip Muus
ESD/P/ICPD.1994/EG.VI/17	8	Migration between developing countries in the African and Latin American regions and its likely future Sharon Stanton Russell
ESD/P/ICPD.1994/EG.VI/18	9	Safeguarding the right to asylum Astri Suhrke
ESD/P/ICPD.1994/EG.VI/19	9	Refugee migration: changing characteristics and prospects John R. Rogge
	INFORMATION PAPERS	
ESD/P/ICPD.1994/EG.VI/INF.1	-	Provisional organization of work
ESD/P/ICPD.1994/EG.VI/INF.2	-	Provisional list of participants
ESD/P/ICPD.1994/EG.VI/INF.5	-	Provisional list of documents
	DISCUSSION NOTES	
ESD/P/ICPD.1994/EG.VI/INF.6	-	Social and economic consequences of migration to large cities in the ESCAP region Economic and Social Commission for Asia and the Pacific
ESD/P/ICPD.1994/EG.VI/INF.7	-	Africa's refugee problem: dimensions, causes and consequences Economic Commission for Africa

Document No.	*Agenda item*	*Title and author*
ESD/P/ICPD.1994/EG.VI/INF.8	-	Health and urbanization in developing countries World Health Organization
ESD/P/ICPD.1994/EG.VI/INF.9	-	The impact of refugee flows on countries of asylum and the challenge of refugee assistance Office of the United Nations High Commissioner for Refugees
ESD/P/ICPD.1994/EG.VI/INF.10	-	Rural development, the environment and migration Food and Agriculture Organization of the United Nations
ESD/P/ICPD.1994/EG.VI/INF.11	-	Migration and population distribution in developing countries: problems and policies International Labour Organization
ESD/P/ICPD.1994/EG.VI/INF.12	-	Migration and development International Organization for Migration
ESD/P/ICPD.1994/EG.VI/INF.13	-	International migration in the ECE region Economic Commission for Europe
ESD/P/ICPD.1994/EG.VI/INF.14	-	(*a*) A gender perspective on migration and urbanization: some issues; (*b*) Crime prevention, migration and urbanization United Nations Office at Vienna
ESD/P/ICPD.1994/EG.VI/INF.15	-	Desparate departures: the flight of environmental refugees The Population Institute
ESD/P/ICPD.1994/EG.VI/INF.16	-	Arab labour migration to the Gulf Economic and Social Commission for Western Asia
ESD/P/ICPD.1994/EG.VI/INF.17	-	Population dynamics in large cities of Latin America and the Caribbean Economic Commission for Latin America and the Caribbean

Part Two

PATTERNS OF POPULATION DISTRIBUTION AND DEVELOPMENT

III. GLOBAL URBANIZATION: TOWARDS THE TWENTY-FIRST CENTURY

Terence G. McGee and C. J. Griffiths**

It is currently expected that during the first years of the twenty-first century, for the first time in human history, more than one half of the world's population will live in urban areas and that by 2025 that proportion will be nearing 58 per cent. Consequently, the world seems to be on a path leading to inevitable urbanization. However, this overall trend masks substantial differences between the major world regions, particularly between more and less developed regions.[1] By 1990 the world's population stood at an estimated 5.3 billion persons, with 4.1 billion living in developing countries and 1.2 billion in developed countries. Approximately 45 per cent of the total population of the world (2.4 billion) lived in urban places, with 1.5 billion in developing countries and 0.9 billion in developed countries. Although the proportion urban in developing countries was only 37 per cent, their urban population surpassed by a considerable margin that of developed countries where the level of urbanization[2] stood at 73 per cent. By 2025, the level of urbanization in the developing world will have risen to 61 per cent, implying that 4.4 billion people will live in the urban centres of developing countries, accounting for about 80 per cent of the total urban population in the world at that time (United Nations, 1991).

Thus, underlying the discussion of current world urbanization trends is the massive volume of population that already resides and is expected to reside in the urban areas of the developing world. The numbers involved pose major challenges for the achievement in developing countries of sustainable urbanization, that is, of the ability to manage cities so that they do not impose an undue burden on existing resources and so that they provide an adequate quality of life to their residents. The challenges ahead are even more daunting given that increasing urbanization levels are occurring even as existing urban systems remain characterized by the concentration of the population in one urban centre or a few. If such a pattern of urban concentration prevails, the developing world will experience the emergence of mega-urban regions as major components of the urban system. Of course, the existence of cities is not a new phenomenon, but the transition of developing countries from being mostly rural to being mostly urban is a comparatively recent one.

In the countries that are currently identified as developed, the levels of urbanization began to increase rapidly almost 200 years ago, so that most saw their urban population reach the 50 per cent level during the last decade of the nineteenth or in the early twentieth century. In comparison, most of today's developing countries are expected to reach similar levels of urbanization only during the first decades of the twenty-first century. In addition, between 1990 and 2005, the population living in the urban areas of the developing world is expected to increase by over 1 billion people. Such an increase will pose major problems for the maintenance and expansion of urban infrastructure, the achievement of liveable urban environments for all urban dwellers, and the pursuit of a dynamic urban economy. The present paper analyses the processes underlying the growth of global urbanization. Its aims are threefold: (*a*) to explore the urbanization process as experienced by developed and developing countries in terms of the differences and similarities; (*b*) to analyse the major demographic, economic, and social features of urbanization in developed and developing countries; (*c*) to discuss the implications of past experience on future urbanization prospects in the developing world.

A. Defining urban places and urbanization

Urbanization is conceptually assumed to consist of three components: demographic, economic and social. About each of these areas considerable debate has arisen. Social and economic issues are important but it is the demographic dimension that is crucial. However, any analysis rests upon acceptable definitions of the process

*Department of Geography, Institute of Asian Research, University of British Columbia, Vancouver, Canada.

of urbanization. Most authorities agree that it consists of at least four components: (*a*) the territorial identification of urban places; (*b*) the increase or decrease of population in urban places; (*c*) an increase in the number of people engaged in non-agricultural occupations stemming from structural changes and economic growth; and (*d*) the existence of a distinctively built environment and organization in cities that encourage ways of life that are described as urban and generally differ significantly from the mode of life in rural areas, although increased access to the mass media and other means of communication are making this distinction less relevant (Armstrong and McGee, 1986).

Clearly, the definition of *urban places* adopted by a country has major implications for the measurement of urbanization and its conceptualization as a process. When the definitions used by different countries vary, cross-country comparisons may be misleading. In many countries, urban places are identified on the basis of political or administrative criteria. Thus, in Bangladesh, urban places are those having a municipality, a town committee or a cantonment board. Other countries define urban places on the basis of population size, but whereas in Malaysia the threshold separating rural from urban places is 10,000 inhabitants, in Argentina it is 2,000, and in Nigeria, 20,000. Some countries use a combination of criteria. Japan, for instance, identifies urban places on the basis of a political definition, a population threshold and the fact that over 60 per cent of the population of a place must be engaged in non-agricultural activities (United Nations, 1991). Such discrepancies in definition underlie the estimates and projections of the urban population and of the level of urbanization made by the United Nations. Although it is recognized that such an approach is eminently pragmatic, the need to move towards the development of functional definitions of urban places that increase the validity of cross-country comparisons persists. This problem is not new (Davis, 1969), but perhaps the wider availability of detailed census information in machine-readable form may pave the way for its use by researchers in ways that obviate this long-standing problem.

B. The process of urbanization in developed and developing countries

Although the traditional classification of countries in developed and developing categories according to their different levels of socio-economic development is losing much of its validity, that distinction is still relevant for the discussion of the urbanization process from a historical perspective. Historically, today's developed countries were characterized by considerably smaller populations than currently developing nations when the former embarked on the urbanization transition, that is, when their urban places began accommodating more than half of their population. As already noted, between 1990 and 2005, the urban places of the developing world are expected to gain over 1 billion inhabitants, a figure at least 10 times larger than the estimated 100 million involved in the urban transition that took place in Europe and North America between 1815 and 1915 (Williamson, 1988).

In addition, the global context in which the urban transition of today's developing countries is taking place is quite different from that prevailing during the nineteenth century. In Europe, the expansion of industrialization and urbanization occurred in an era in which global competition was very limited and there was ready access to the credit, capital and international markets characterizing the "age of imperialism". There was of course political competition for those markets but, generally, the age of expansion permitted rapid economic growth.

European countries also had a long historical experience with city management, particularly with respect to the operation of city property markets and municipal taxation, which meant that it was possible to raise much of the capital for infrastructure development from the growing tax base of the cities. Overseas countries of immigration inherited that expertise. Of course, the process of urban domestic capital accumulation occurred gradually throughout the nineteenth century and its pace varied from country to country or even from city to city. Nevertheless such a process did enable an urban infrastructure to be created or expanded in most of the cities of today's developed countries.

Most historical evidence indicates that the population growth rates of developed countries were much lower than those being experienced today by developing countries. Thus, between 1776 and 1871, the population of developed countries grew at approximately one half the rate of that of developing countries today. In addition, the migration of Europeans to overseas colonies or former colonies greatly relieved population pressures in Europe, particularly in the urban areas of the continent, since many of the overseas emigrants would have migrated to European cities if overseas migration had not been an option (Bairoch, 1988).

Although there were some structural blockages in employment and city growth related to the rates of economic growth, the urban transition in developed countries generally involved the growth of cities populated mostly by wage-earners, engaged in factory and service occupations. The engagement of large segments of the urban population in informal activities characterized by low incomes and low productivity was not a dominant feature of the urbanization experience of developed countries except for relatively short periods (as in London between 1830 and 1870) or in the more slowly evolving economies of Southern Europe. Consequently, the urban transition in developed countries led to the expansion of a wage-earning population and thus increased demand for consumer durables. The urban transition therefore gave rise to growing urban domestic markets for the products of the industrial revolution (Armstrong and McGee, 1986).

The current phase of the urban transition in developing countries has very different historical roots. The urbanization process in developing countries has been deeply associated with their incorporation into the global economy, which is variously described as colonization or imperialism. While there have been substantial variations in the way this process has affected urbanization in individual countries and regions, which are partly related to pre-existing urbanization and the cultural resilience of certain societies in, for instance, China, Japan and Thailand (McGee, 1967), the overall impact of such incorporation was broadly similar during the period extending from 1500 onwards.

One of the major consequences of the process of incorporation was the creation of large primate cities dominating the urban hierarchies of the countries in which they were located. Thus, the literature about the history of urbanization in developing countries is very much focused on cities such as Ibadan, Jakarta, Mexico City, Nairobi and Rio de Janeiro. Because much of the economic activity centred on those cities, which were the administrative centres and conduits for the flow of raw materials to the developed world, they grew disproportionately.

The process of incorporation also created dualistic societies in which much of the rural population lived in poverty, had low productivity and often belonged to a low-status ethnic group, while the urban population consisted of a comparatively small elite comprising colonialists, foreign entrepreneurs and an indigenous oligarchy. The rest of the population in cities worked largely in the informal sector, being engaged in low paying service and cottage-industry occupations.

There was, however, considerable differentiation in terms of how the process of incorporation occurred through time and in space. The process of political decolonization, which began in Latin America in the nineteenth century but was not completed until the 1970s, resulted in different patterns of urbanization at the regional level, especially with respect to the major regions including Latin America, the Asia-Pacific region, and Africa.[3]

Table 1 shows the sharp difference in the levels of urbanization between the different developing regions in 1990. Latin America has attained levels of urbanization similar to those of developed countries. Africa and Asia, in contrast, are just entering the "accelerated" phase of the urban transition. Because Latin America has already largely experienced the urban transition, the major problems it faces relate to managing the growth of the urban population rather than to absorbing large numbers of persons moving from rural to urban areas (Roberts, 1989).

Attempts to explain the reasons for the different urbanization experiences of the developing regions include a fusion of explanations drawn from neoclassical theory, world system theory and convergence theory (Armstrong and McGee, 1986). In the current phase of accelerated urbanization in Africa and Asia, the latter will experience the largest urban growth in terms of the number of people involved and it is therefore important to highlight its experience.

C. THE DYNAMICS OF URBAN POPULATION GROWTH

The population of urban places can increase for four reasons: the balance of births and deaths is positive,

TABLE 1. TOTAL, URBAN AND RURAL POPULATION AND PERCENTAGE OF THE POPULATION LIVING IN URBAN AREAS BY MAJOR REGION, 1950-1990

Region		1950	1960	1970	1980	1990
		A. Population (millions)				
World total	Total	2 516	3 020	3 698	4 448	5 292
	Urban	734	1 032	1 352	1 757	2 390
	Rural	1 783	1 988	2 345	2 691	2 902
More developed regions	Total	832	945	1 049	1 137	1 207
	Urban	448	572	699	799	875
	Rural	384	373	350	338	331
Less developed regions	Total	1 684	2 075	2 649	3 312	4 086
	Urban	286	460	654	959	1 515
	Rural	1 398	1 615	1 995	2 353	2 571
Europe	Total	393	425	460	484	498
	Urban	222	260	307	341	366
	Rural	171	166	153	143	133
Northern America	Total	166	199	226	252	276
	Urban	106	139	167	186	207
	Rural	60	60	59	66	68
Oceania	Total	13	16	19	23	26
	Urban	8	10	14	16	19
	Rural	5	5	6	7	8
USSR (former)	Total	180	214	243	266	289
	Urban	71	105	138	167	190
	Rural	109	110	105	98	99
Africa	Total	222	279	362	477	642
	Urban	32	51	83	133	217
	Rural	190	228	279	345	425
Asia	Total	1 377	1 668	2 102	2 583	3 113
	Urban	226	359	481	678	1 070
	Rural	1 151	1 309	1 621	1 905	2 042
Latin America	Total	166	218	286	363	448
	Urban	69	107	164	236	320
	Rural	97	111	122	127	128
		B. Percentage urban				
World total		29.2	34.2	36.6	39.5	45.2
More developed regions		53.8	60.5	66.6	70.3	72.6
Less developed regions		17.0	22.1	24.7	28.9	37.1
Europe		56.5	61.1	66.7	70.4	73.4
Northern America		63.9	69.9	73.8	73.9	75.2
Oceania		61.3	66.3	70.7	71.2	70.6
USSR (former)		39.3	48.8	56.7	63.0	65.8
Africa		14.5	18.3	22.9	27.8	33.9
Asia		16.4	21.5	22.9	26.3	34.4
Latin America		41.5	49.3	57.3	65.0	71.5

Source: *World Urbanization Prospects, 1990* (United Nations publication, Sales No. E.91.XIII.11).

leading to natural increase; there is migration from rural to urban places within the country; there is international migration to the urban areas of a country; and there is territorial expansion of urban places as their boundaries are redefined or as rural areas become urbanized (this process is known as reclassification).

Although the nineteenth-century cities of Europe differed from their twentieth century counterparts in the developing world in terms of culture and political structures, they were similar regarding social conditions and demographic features. The expansion of urbanization in nineteenth-century Europe was accompanied by an expansion of industry and of wage work outside of the home. The large numbers of people migrating from rural areas to the cities were absorbed by the growing manufacturing sector. The result was a real improvement in the social and economic conditions of the urban population and further migration. Migration was thus a major contributor to the increasing population of European cities. In 1801, the population in the 10 largest cities in England represented only 16 per cent of the total population of the country, but by 1851 their share had grown to 23 per cent. The population of the United Kingdom was 67 per cent urban in 1880 and by 1980 it had reached 91 per cent. In 1880, France and the United States of America were 35 and 29 per cent urban, respectively, and by 1990 they both had over 70 per cent of their population living in urban areas.

The rapid growth of the population of European cities caused problems similar to those experienced today by the cities of developing countries. Lack of adequate housing, the growth of slums, lack of sanitation or garbage collection were all common ills of European cities. As Friedrich Engels, as cited by Abu-Lughod and Hay (1977), observed, it was common for heaps of garbage and ashes to lie in all directions in the poor neighbourhoods, and for foul liquids to be emptied before the doors of houses, where they gathered in stinking pools. Those were the places of abode of the poorest of the poor, the worst-paid workers, who lived with thieves and with the victims of prostitution indiscriminately huddled together. This description, which refers to the city of London before 1850, could apply to any number of places in the developing world today. London's population more than doubled in the 50 years between 1801 to 1851. In addition, during that period the first major waves of emigration to North America began. Between 1801 and 1901, more than 12 million people left the United Kingdom. If that population had migrated to the urban centres of their country, the annual rate of urban growth in the United Kingdom would have been over 5 per cent. The development of modern health measures began during that period, but their full implementation, which resulted in a dramatic decline of the death rate in cities, was not completed until the twentieth century and consequently did not contribute to a marked acceleration of the growth of the urban population during the nineteenth century. The rate of urban growth in the United Kingdom was estimated to be 1 per cent per annum during that period (Weber, 1965).

In contrast, the major component of urban growth in Africa and Latin America in recent times appears to be natural increase and it is expected to remain so until the next century (table 2). In Asia, net rural-urban migration and reclassification still account for over half of urban growth. However, the figures for Asia are greatly influenced by the growth of the urban population of China, which increased by 6.7 per cent per annum between 1980 and 1985, primarily because of reclassification. If China is excluded, the share of urban growth accounted for by natural increase in Eastern Asia becomes two thirds for each of three periods considered. It is likely, however, that the components of urban growth presented in table 2 overestimate the share of natural increase for the following reason: to derive those components, it was assumed that the rate of natural increase in urban areas is equal to that of the total population, while in most countries natural increase in urban areas is known to be lower than that of the total population, mainly because fertility is lower in cities than in rural areas.

According to estimates presented by Hugo (1992) for selected countries of Asia, the contribution of internal migration and reclassification was lower during the 1960s than during the 1970s in four out of the six countries with data available for the two periods (table 3). This finding would support the view that, as rates of natural increase continue to decline in the urban areas of Asian countries during the next decades, the contribution of rural-urban migration and reclassification to urban growth will increase. It should also be noted that many Asian countries, but particularly those with the largest populations, still have large numbers of people living in rural areas, so that the potential contribution of rural-urban migration to urban growth can be substantial. Currently, the urbanization process in Africa and Asia is

TABLE 2. COMPONENTS OF URBAN GROWTH IN THE MAJOR WORLD REGIONS FOR SELECTED PERIODS
(*Percentage*)

Major region	*1980-1985*		*1990-1995*		*2000-2005*	
	Natural increase	*Migration and reclassification*	*Natural increase*	*Migration and reclassification*	*Natural increase*	*Migration and reclassification*
World	57	43	58	42	58	42
More developed regions	65	35	62	38	49	51
Less developed regions	45	55	50	50	54	46
Europe	45	55	34	66	26	74
Northern America	85	15	75	25	60	40
USSR (former)	54	46	79	21	56	44
Oceania	110[a]	-10[a]	97	03	83	17
Africa	60	40	61	39	65	35
Asia	40	60	44	56	46	54
Latin America	67	33	73	27	77	23

Source: Economic and Social Commission for Asia and the Pacific, "Urbanization in Asia and the Pacific", paper prepared for the pre-Conference Seminar of the Fourth Asia-Pacific Population Conference on Migration and Urbanization: Inter-relationships with Socio-Economic Development and Evolving Policy Issues (Seoul, Republic of Korea, 21-25 January 1992).

NOTE: The contribution of natural increase is calculated by assuming that the urban population has the same rate of natural increase as the total population of each region. The contribution of net migration and reclassification is then the residual.

[a]Oceania experienced net out-migration from urban areas during 1980-1985.

TABLE 3. ESTIMATED COMPONENTS OF URBAN POPULATION GROWTH FOR SELECTED COUNTRIES IN ASIA AND THE PACIFIC, 1960-1980

Region and country	*Intercensal growth rate of urban population*		*Estimated rate of urban natural increase*		*Estimated contribution of migration and reclassification*		*Percentage contribution of migration and reclassification*	
	1960s	*1970s*	*1960s*	*1970s*	*1960s*	*1970s*	*1960s*	*1970s*
Asia								
Bangladesh	6.6	10.6	2.7	3.7	3.9	6.9	58.6	65.7
Brunei Darussalam	7.9	..	4.6	..	3.3	..	41.7	..
Hong Kong	2.6	2.8	2.2	2.3	0.4	0.4	14.3	16.1
India	3.2	3.7	2.2	2.1	1.0	1.7	..	40.2
Indonesia	3.5	5.1	2.5	2.5	1.1	2.6	31.5	50.6
Iran, Islamic Republic of	..	4.8	..	2.7	..	2.1	..	43.2
Japan	2.4	1.7	1.3	1.2	1.1	0.5	47.6	27.7
Macau	4.0	..	3.8	..	0.2	..	4.0	..
Malaysia	..	4.8	..	2.3	..	2.5	..	52.1
Maldives	..	8.7	..	3.1	..	5.6	..	64.1
Nepal	3.2	7.3	2.1	2.9	1.1	4.3	33.6	59.7
Pakistan	..	4.3	..	3.6	..	0.7	..	17.1
Philippines	..	4.3	..	2.6	..	1.7	..	40.2
Republic of Korea	6.2	5.0	2.5	2.2	3.7	2.8	60.3	56.4
Thailand	..	5.1	..	2.1	..	3.0	..	57.8
Oceania								
Australia	2.4	1.4	1.9	1.3	0.5	0.1	20.3	5.0
Fiji	..	3.1	..	2.1	..	1.0	..	33.1
New Zealand	3.5	0.4	1.7	0.2	1.8	0.2	52.4	51.6

Source: Graeme J. Hugo, "Migration and rural-urban linkages in the ESCAP region", paper prepared for the pre-Conference Seminar of the Fourth Asia-Pacific Population Conference on Migration and Urbanization: Inter-relationships with Socio-Economic Development and Evolving Policy Issues (Seoul, Republic of Korea, 1992), p. 15.

NOTE: Two dots (..) indicate that data are not available.

being fuelled by the general lack of opportunities in rural areas and the perception that opportunities are better in urban areas. Continued population growth in rural areas is one of the factors prompting out-migration and urbanization. Rural growth rates in Africa and Asia continue to be high (table 4), both because fertility continues to be substantial and as a result of major reductions in mortality following the introduction of modern health measures after the Second World War. However, the resulting rise in population growth has not been accompanied by the same economic development that was experienced by developed countries during the nineteenth century. In addition, there is currently a reduced scope for international migration and no prospects for its having a major impact in reducing the urban population in developing countries.

TABLE 4. ANNUAL GROWTH RATES OF THE TOTAL, URBAN AND RURAL POPULATIONS OF MAJOR REGIONS, 1950-1990

Region		*1950-1960*	*1960-1970*	*1970-1980*	*1980-1990*
World total	Total	1.82	2.03	1.85	1.74
	Urban	3.41	2.71	2.62	3.08
	Rural	1.09	1.65	1.37	0.76
More developed regions	Total	1.27	1.04	0.80	0.60
	Urban	2.44	2.00	1.34	0.92
	Rural	-0.30	-0.63	-0.36	-0.20
Less developed regions	Total	2.09	2.44	2.23	2.10
	Urban	4.76	3.53	3.83	4.58
	Rural	1.44	2.11	1.65	0.89
Europe	Total	0.80	0.79	0.52	0.28
	Urban	1.58	1.67	1.07	0.70
	Rural	-0.31	-0.76	-0.68	-0.78
Northern America	Total	1.79	1.31	1.06	0.91
	Urban	2.69	1.85	1.08	1.08
	Rural	-0.03	-0.08	1.02	0.41
Oceania	Total	2.21	2.03	1.65	1.50
	Urban	2.99	2.68	1.71	1.42
	Rural	0.85	0.60	1.50	1.69
USSR (former)	Total	1.74	1.25	0.90	0.83
	Urban	3.91	2.75	1.94	1.28
	Rural	0.04	-0.43	-0.66	0.03
Africa	Total	2.30	2.59	2.77	2.97
	Urban	4.61	4.80	4.72	4.95
	Rural	1.84	2.01	2.11	2.09
Asia	Total	1.92	2.31	2.06	1.86
	Urban	4.62	2.92	3.44	4.56
	Rural	1.29	2.14	1.62	0.70
Latin America	Total	2.74	2.70	2.39	2.11
	Urban	4.47	4.20	3.65	3.08
	Rural	1.31	0.98	0.40	0.04

Source: World Urbanization Prospects, 1990 (United Nations publication, Sales No. E.91.XIII.11), calculated from tables A.2, A.3 and A.4.

According to table 4, during 1950-1960 the urban populations of Africa, Asia and Latin America all increased at annual rates of about 4.5 to 4.6 per cent. These rates contrast markedly with those experienced by the urban population of more developed regions during the same period, which averaged 2.4 per cent per year, but when compared with the growth rates of the European urban population during the nineteenth century, the difference is smaller, for the urban populations of the major cities in Europe increased by about 3 per cent during the nineteenth century, a rate three times higher than that of the total population. In the United Kingdom, the population of the 10 largest cities increased at 3.5 per cent per year and, if overseas migration had not been possible, that figure would have been considerably higher.

Sharp variations in population size underlie the differences in the urbanization process experienced by Africa, Asia and Latin America. Thus, the prospects for Asia, with 3.1 billion people in 1990, differ from those of Africa, with 642 million, and Latin America, with only 448 million in 1990 (table 1). Asia's demographic profile tends to be dominated by China and India whose combined population stood at nearly 2 billion in 1990. In fact, in 1990, six countries accounted for 83 per cent of the population in Asia, namely, Bangladesh, China, India, Indonesia, Japan, and Pakistan (United Nations, 1991). Among those countries, only Japan has accomplished a successful urban transition and the remaining five present the major challenge regarding urbanization in developing Asia.

As already noted, natural increase is a major component of urban population growth in certain regions. Indeed, the urban populations of developing countries tend to be heavily concentrated in the younger ages and particularly in the main reproductive ages. The concentration of recent migrants in the young adult ages is a well-known fact. Similar age distributions prevailed in Europe during the nineteenth century, when rural-urban migration was dominated by young, underemployed agricultural workers. In European cities, however, mortality was so high that growth had to be achieved largely through migration. That is not the case in today's developing countries.

Furthermore, the high levels of natural increase prevalent in both urban and rural areas not only fuel the growth of the former, but also lead to rural-urban migration by increasing the population in rural areas that are hard put to keep on absorbing large numbers of agricultural workers.

In the history of the Western urban transition, it is generally argued that the contribution of rural-urban migration was a major factor and this process was always analysed as entailing a move from agricultural to non-agricultural employment. Currently, the redistribution of population in the developed regions generally involves city-to-city migration or urban expansion. The rural population has levelled off and has even declined in real terms as rural-urban migration ceases to be a factor in population increase in the cities. Natural increase in cities of developed nations appears to be at or below replacement.

With regard to the influence of net rural-urban migration on the growth of cities, it is difficult to obtain acceptable estimates of rural-urban migration, be it for Europe during the nineteenth century or currently for developing countries. Indirect evidence suggests that net rural-urban migration has been an important component of urban growth in developing countries since 1950. In addition, because migrants tend to be concentrated in the child-bearing ages, their impact on natural increase is also significant.

The impact of internal migration on urban growth has varied from one country to another. In Seoul, for instance, migration accounted for 55 per cent of the city's population growth during 1955-1960, when natural increase was still high. With the decline in fertility that started in 1960, the contribution of migration to the growth of the population of Seoul increased to 81 per cent during 1966-1970 (ESCAP, 1990). For Buenos Aires, net gains through international and internal migration were the main component of population growth between 1900 and 1960. Mexico City has had a more varied experience. In the 1940s, migration was the dominant component of its population growth, but during the 1950s and 1960s, natural increase dominated. Then, as fertility declined, migration again became dominant after 1975. In Jakarta, the contribution made by net migration has been declining in relative importance. In Africa, lack of data preclude an assessment of the importance of migration to urban growth. Yet, the 1970 census of Accra showed that out of the 564,000 inhabitants of the city, 360,000 had been born elsewhere (Zachariah and Condé, 1981), thus giving some indication of the importance of migration in populating the city over the long run.

In Latin America, rural out-migration resulted from a combination of technological changes introduced in agriculture and high population growth in rural areas. Yet, since the region passed the period of rapid urbanization, the importance of rural-urban migration has declined. Between 1970 and 1990, 96 per cent of the population growth in the region took place in urban areas and declines in the size of the rural population are projected after 1990. The growth rate of the urban population declined from 4.5 per cent annually during 1950-1960 to 3 per cent during 1980-1990, but that of the rural population dropped from 1.3 per cent in the 1950s to barely above zero in the 1980s (table 4). Natural increase is currently the major component of urban growth in Latin America.

Another problem in assessing the dynamics of urbanization is related to changes in the territorial configuration of urban places. In some cases, the failure of data gatherers to take into account changes in the boundaries of urban places may result in the underestimation of the urban population and consequently in the overestimation of the rural population. In Europe and Northern America, for instance, cities have expanded beyond their original borders as city dwellers move to the suburbs and continue to carry out many activities in the city proper. The continued spread of residential areas and the dispersion of employment have created urban centres linked by transportation corridors. In recent decades, there have been several attempts to conceptualize and document the phenomenon of city expansion, particularly when it involves growing linkages between previously independent urban places.

Indeed, rapid urbanization in both developed and developing countries has been associated with the concentration of the urban population in a small number of urban centres within each country. Such concentration has given rise to mega-cities, that is, cities with at least 8 million inhabitants. In 1990 an estimated 10 per cent of the world's urban population lived in mega-cities (United Nations, 1991). In 1950, only two of the world's cities, London and New York, qualified as mega-cities, but by 1990 the number of mega-cities had increased to 20, with 14 in the developing world.

As large urban agglomerations, including mega-cities, grow, they show a tendency to spread outwards, often beyond the boundaries of their so-called metropolitan areas to form what we shall hereinafter call *extended metropolitan regions*. Cities such as Jakarta, São Paulo and Tokyo are all extending their influence into the surrounding rural areas, sometimes located up to 100 kilometres from the urban core. The main difference between these urban configurations and the megalopolis typical of North America is the high population density in both the urban cores and the rural areas surrounding them (Ginsberg, Koppel and McGee, 1991).

An example of an extended metropolitan region is Singapore and its surrounding area, which includes the Malaysian state of Johor and the Indonesian archipelago of Riau. In this region, market interchanges have been growing and links with areas once considered rural are gaining strength. This extended metropolitan region is characterized by high levels of economic diversity and interaction, a high percentage of non-farm employment (over 50 per cent), and a deep penetration of global market forces into the countryside (Macleod and McGee, 1993). The expansion of Singapore's sphere of influence has resulted in greater population concentration and a change in the economic activities over the area.

D. The economic processes underlying urbanization

The shift, in today's developed countries, from an economy based on agricultural production to one based on industrial production occurred mostly during the nineteenth and early twentieth centuries. The development of factory production changed the mode of capital accumulation in Europe and allowed surplus capital to be invested both in industry and in building infrastructure in urban areas.

In developing countries, the first wave of urbanization took place when many were colonized and did not result in the development of industry and infrastructure in the cities. Instead, any surplus capital produced was exported to the colonizing nation, thus reinforcing the accumulation of capital in the urban centres of Europe. Consequently, the indigenous economy was not revolutionized in the same manner as in Europe. During the post-colonial period, developing countries have been playing catch-up, but their rapid urban population growth has generally outpaced the development of urban infrastructure and the expansion of industry. The result has often been deficit financing, as developing countries have tried to provide modern services to a growing population. In developed countries, the manufacturing and industrial sectors have traditionally provided employment for a major share of the urban population. Since 1945, however, more and more jobs are being generated by the service and financial sectors. In fact, since the 1973 economic crisis, a growing number of jobs in the industrial and manufacturing sectors have been transferred from developed to developing countries as foreign direct investment in the latter has grown. In developed countries, such changes have brought about a shift of population away from older industrial areas to those where the service and financial sectors are expanding.

In the developing world, the industrial base has always been small in comparison with the service sector, which was more thoroughly developed by the colonial powers. In the 1980s, however, the industrial sector in the low- and middle-income economies grew at 5.3 per cent per annum (World Bank, 1987). However, it is estimated that during that decade, only 23 and 13 per cent, respectively, of the labour force of middle- and low-income economies worked in the industrial sector. In fact, the agricultural sector is still a major source of employment in developing countries. In many of them, over 50 per cent of the labour force is involved in agricultural production and this figure does not reflect the high number of rural/urban dwellers in extended metropolitan regions who work both in agriculture and in the industrial or service sectors.

In the major urban areas of the European Union, two types of problems have arisen (Cheshire, 1990). First, urban centres whose economy had been centred on resource-based industries such as steel production are displaying both a reduction of population and severe urban problems. Second, some urban centres are still experiencing urban growth although they are in the midst of largely impoverished and backward rural areas. There are, however, signs of urban renewal in some areas, as the service sector increases in core cities. This trend may continue as the economic integration of Europe advances.

E. The experience of countries in the ESCAP region

Since the 1970s, the economies of the countries of the Asia-Pacific region have, as a group, been among the most dynamic in the world. The most rapid economic growth, however, has been confined to a relatively small group of countries or areas, including Japan, the newly industrializing economies (NIEs), namely, Hong Kong, the Republic of Korea, Taiwan Province of China and Singapore, and the ASEAN 4 (Indonesia, Malaysia, the Philippines and Thailand). Other countries in the region have experienced only moderate economic growth and slow structural change.

Since 1950, countries in the region have been variously exposed to development policies that evolved from those promoting import substitution, so that development could be achieved by relying on each country's endowments of natural resources and labour, to those promoting production for export to achieve export-led growth. Those espousing the latter policies have been aided by the liberalization of trade achieved through successive rounds of multilateral trade negotiations of the General Agreement on Tariffs and Trade (GATT) and an increasingly open world economy.

The NIEs have moved faster than other Asian countries on the road to economic development and by now their growth process relies heavily on industries intensive in the use of human capital and technology. Other countries face different conditions. In the ASEAN 4 countries, resource-intensive industries based on the exploitation of, among other things, petroleum, rubber or tin, have been very important. Since 1970, all four countries have had vigorous industrialization programmes fuelled to a large extent by foreign direct investment, particularly from Japan.

The countries of Southern Asia have been more inward-oriented and have protected their industry to a greater degree, thus favouring, particularly in India and Pakistan, the emergence of heavy industry and other capital-intensive industries. Yet, during the 1980s, countries in the subregion began to foster the expansion of labour-intensive industries oriented towards both local and international markets. India in particular has begun to develop a high-technology sector.

The centrally planned economies of Cambodia, China, the Lao People's Democratic Republic, Myanmar and Viet Nam have all passed through phases of State ownership and control of most of the elements of the economy but are now moving towards liberalizing their economies by reinstating market mechanisms. This change is permitting them to increase productivity and exports. However, although the reforms adopted have injected considerable dynamism to their economies, they have also complicated economic management. Relaxing central controls before macroeconomic tools are fully in place has given rise to imbalances and created strong inflationary pressures.

Lastly, the developing countries of Oceania have experienced limited structural change and still rely to a great extent on international assistance and tourism as the main sources of income. Only Fiji and Papua New Guinea, with large populations and a more adequate resource base, are moving towards a genuine structural change.

It is of interest to explore the relation between parameters indicating the level and type of economic development and urbanization. Table 5 shows to what extent different economic sectors have contributed to the gross domestic product (GDP) of selected countries of the Asia-Pacific region and how the composition of their labour force has changed. It also displays their current level of urbanization. Three different groups can be identified on the basis of the level of urbanization. The first is constituted by the developed economies of Australia, Japan and New Zealand in conjunction with the Asian NIEs. All those countries have over 70 per cent of their population living in urban areas. The second includes the ASEAN 4 countries plus Fiji and Pakistan, most of which have levels of urbanization above 30 per cent. Thailand constitutes an exception but it is included in this group because there is reason to believe that available data underestimate the true level of urbanization in that country. The third group encompasses the remaining countries, namely China and all Southern Asian countries with the exception of Pakistan. In the case of China, the level of urbanization presented in table 5 may be on the low side. Tolba (1992) has suggested that a more realistic figure would be 26.2 per cent.

The grouping proposed permits a rough assessment of the relation between economic performance and urbanization. Among the highly urbanized countries, agriculture has made a relatively small contribution to GDP since the 1960s and is currently at minimal levels. Manufacturing and industry were both important in the 1960s for most of the ASEAN 4 countries and Asian NIEs and their contribution to GDP increased significantly in the cases of the Republic of Korea and Singapore. In the rest of the highly urbanized countries, the joint contribution of manufacturing and industry to GDP remained virtually unchanged between 1960 and 1990 or declined slightly. In contrast, the contribution of services tended to increase, though exceptions exist.

Among the countries with intermediate levels of urbanization, the contribution of agriculture has generally declined dramatically between 1960 and 1990, and that of manufacturing and industry has tended to increase. The main exceptions are the Philippines, where the share of agriculture in GDP declined only slightly, and Fiji where the contribution of manufacturing failed to increase. The share of services increased in Fiji, Indonesia, Pakistan and Thailand.

Among the rest of the countries in Southern Asia and China, the patterns of structural change are even more variable. Only in China, India and Sri Lanka do manufacturing and industry account for relatively high proportions of GDP, but only India has experienced a significant drop in the share of agriculture. In Bangladesh, Myanmar and Nepal agriculture still accounts for the major share of GDP, and only Bangladesh has experienced some increase in the share of manufacturing and industry. In most of these countries, the service sector continues to make a sizeable contribution to GDP and it helps accommodate the excess labour force in urban areas.

The changes observed in terms of the labour force are similar. In the highly urbanized countries there has been a major reduction of the proportion of the population working in agriculture and, in most cases, a rise of that working in the service sector. In Australia, Hong Kong and New Zealand, the proportion of the labour force working in manufacturing and industry has declined, but none of the other three countries in this group registered reductions in both those categories.

Countries with intermediate levels of urbanization display even more variability in the patterns of change of their labour force. Again, all of them have experienced reductions in the proportion of workers in agriculture. Yet there is only a weak tendency towards the growth of the proportion of the labour force in industry or manufacturing, and greater consistency in the increasing trend displayed by the service sector. When these changes are juxtaposed with those regarding the components of GDP, it becomes clear that the growing contribution that manufacturing is making to GDP in a number of the countries in this group has not been matched by a proportionate increase in the share of the labour force devoted to manufacturing and industry. Consequently, the growing labour force has had to be accommodated by agriculture or by the service sector.

TABLE 5. DISTRIBUTION OF GDP AND OF THE LABOUR FORCE ACCORDING TO SECTOR OF ECONOMIC ACTIVITY FOR SELECTED COUNTRIES IN THE ASIA-PACIFIC REGION, 1960 AND 1990

Country	*Percentage urban in 1990*	*Sector of economic activity*							
		Agriculture		*Manufacturing*		*Industry*		*Services*	
		1960	*1990*	*1960*	*1990*	*1960*	*1990*	*1960*	*1990*
				A. *As percentage of GDP*					
Developed countries									
Australia	85	11	4	26	17	13	22	38	36
Japan	77	13	3	34	29	11	12	43	57
New Zealand[a]	87	13	8	22	20	9	8	56	64
NIEs									
Hong Kong	94	3	0	22	20	12	7	63	73
Korea, Republic of .	74	25	4	19	34	8	9	47	52
Singapore	100	4	0	12	30	6	7	79	63
ASEAN 4									
Indonesia	31	51	24	9	19	6	17	33	40
Malaysia	43	33	21	8	24	10	16	49	39
Philippines	43	26	23	20	25	8	8	46	43
Thailand	23	40	17	13	24	6	11	41	48
Pacific Islands									
Fiji	24	41	21	18	11	7	10	34	58
Southern Asia									
Bangladesh	16	57	46	5	7	2	7	36	40
India	22	50	31	14	20	6	10	30	40
Myanmar	28	33	48	8	9	4	3	55	40
Nepal	10	66	59	3	6	8	9	23	26
Pakistan	32	46	27	12	16	4	8	38	49
Sri Lanka	21	32	23	15	16	5	10	48	51
China	19	23	32	..	..	48	46	29	21
				B. *As percentage of labour force*					
Developed countries									
Australia	85	13	5	28	16	13	10	46	69
Japan	77	33	8	21	24	8	10	38	58
New Zealand	87	16	10	24	18	12	7	48	65
NIEs									
Hong Kong	94	7	1	39	30	11	9	43	60
Korea, Republic of .	74	61	20	6	28	2	7	30	45
Singapore	100	8	0.5	14	29	7	7	71	63
ASEAN 4									
Indonesia	31	68	56	5	8	2	4	25	36
Malaysia[b]	43	37	31	16	16	8	7	39	47
Philippines	43	59	45	12	11	3	5	26	39
Thailand	23	82	64	3	9	1	3	12	24

TABLE 5 (*continued*)

Country	Percentage urban in 1990	Sector of economic activity: Agriculture 1960	Agriculture 1990	Manufacturing 1960	Manufacturing 1990	Industry 1960	Industry 1990	Services 1960	Services 1990
Pacific Islands									
Fiji[c]	24	53	44	7	8	8	6	31	42
Southern Asia									
Bangladesh	16	86	56	4	10	1	3	12	24
India[d]	22	71	66	9	10	2	2	18	21
Myanmar[e]	28	69	64	7	8	2	2	22	26
Nepal[d]	10	94	91	2	0	0	0	4	8
Pakistan	32	59	51	13	13	3	7	25	29
Sri Lanka	21	49	43	9	11	3	5	39	42
China	19	83	61	12[f]	17	6	0.2	10	22

Source: Economic and Social Commission for Asia and the Pacific, *Industrial Restructuring in Asia and the Pacific, in Particular with a View to Strengthening Regional Cooperation* (ST/ESCAP/960, 1991).

[a]The figures for New Zealand refer to 1970 and 1990.
[b]The figures for Malaysia refer to 1980 and 1990.
[c]The figures for Fiji refer to 1970 and 1990.
[d]The figures for India and Nepal refer to 1960 and 1980.
[e]The figures for Myanmar refer to 1970 and 1980.
[f]This figure refers to 1980.

In the third group, Southern Asian countries continue to have sizeable proportions of their labour force in agriculture and in the service sector. In socialist economies, changes in the structure of employment are more difficult to ascertain, but it is clear that the majority of the labour force remains in the agricultural sector.

The persistence of large populations engaged in agricultural production in most of the Asia-Pacific region increases the potential for large flows of rural-urban migration. However, one may cast some doubt on the accuracy of the statistics available. Information on labour force participation is often weak or misleading because—particularly in developing countries—individuals often have more than one job. In rural areas in particular, it is increasingly common for persons to be engaged in both farm and non-farm employment. Survey data indicate that between 20 and 30 per cent of all rural workers in Asia may also have some type of non-agricultural employment that constitutes a major source of income and there are reasons to believe that that proportion may be increasing (Mukhodpadhyay and Lim, 1985).

The importance of off-farm employment in the processes of structural change is well illustrated by the experience of Japan, the Republic of Korea and Taiwan Province of China, all of which had reached full employment and were experiencing rising real wages and higher rates of GDP by the early 1980s. In all three, there was a dramatic increase of off-farm income as a result of accelerated industrialization and the implementation of the green revolution in rural areas. In Japan, the share of off-farm income rose from 54 per cent in 1965 to 79 per cent in 1980. The corresponding increase for Taiwan Province of China was from 27 to 79 per cent, whereas the increase for the Republic of Korea—from 16 to 20 per cent—was considerably more modest. A similar process is taking place in the countries with intermediate levels of urbanization. In Malaysia, for instance, it is estimated that about 40 per cent of the income of rural households comes from non-farm sources.

In Southern Asian countries and particularly in the coastal regions of China, non-farm income in rural areas is also rising and it has important implications for the urbanization process, especially because a major part of non-farm employment is being generated by the heavily populated regions surrounding the largest urban centres of the Asia-Pacific region. Non-farm employment permits not only improvement in household income, but also investment in agricultural diversification and in small industry. The symbiosis between urban centres and the rural communities located near them creates growth areas that are attractive to both indigenous and foreign capital. If the Government provides public infrastructure, such as electricity and transportation, the process of consolidation and expansion of those areas is enhanced.

The case of the Bangkok Metropolitan Region (BMR), which includes Bangkok and the five adjacent provinces, is illustrative. The BMR absorbed almost 40 per cent of Thailand's increase in manufacturing employment during 1980-1988. Almost 66 per cent of the increase in manufacturing occurred in the five adjacent provinces which, by 1988, accounted for almost a third of all employment in the BMR. At that time, the major surge of Japanese investment in manufacturing, which has been concentrated in the outer parts of the BMR and on the eastern seaboard, had not yet taken place (Kruvan, 1991). Similar developments have taken place in China, where investment by Hong Kong and others has fuelled the growth of industry in Guangdong and transformed the region between Guangzhou and Hong Kong. Taiwanese and other foreign investment is responsible for the industrial growth in Jiangsu and Fujian provinces.

Per capita income shows a strong relation with the level of urbanization. In the Asia-Pacific region, the group of high-income countries (those with a per capita income of at least US$ 3,240 in 1990) included Australia, Brunei Darussalam, Hong Kong, Japan, Macao, New Zealand, the Republic of Korea and Singapore, most of which had high levels of urbanization and whose urban population accounted for 13 per cent of that of the region. Middle-income countries, defined as having a per capita income of between US$ 500 and US$ 3,240 in 1990, included Fiji, Indonesia, Malaysia, Mongolia, Papua New Guinea, the Philippines and Thailand, whose combined urban population accounted for 11 per cent of that of the region and whose levels of urbanization were moderate. The rest, with per capita incomes of under US$ 500, were classified as low-income countries and accounted for a full 76 per cent of the urban population in the region, since they included the most populous countries in it (Afghanistan, Bangladesh, Bhutan, Cambodia, China, India, the Lao People's Democratic Republic, Maldives, Myanmar, Nepal, Pakistan, Sri Lanka and Viet Nam). Large urban agglomerations, defined as those with at least 2 million inhabitants in 1990, were largely concentrated in low-income countries (30 out of 46). Of the remaining 16, 10 were in high-income countries and 6 in the middle-income ones. The poorest countries are therefore those most likely to face serious problems regarding the management and delivery of services in their numerous large cities.

One of the most important facets of the urbanization process is the changing structure of the urban labour force. Research on the urban labour force has been dominated by models that divide it in two components: an informal sector made up of largely self-employed persons or small-scale entrepreneurs that covers virtually every type of urban activity, from manufacturing to services; and a formal sector consisting of salaried workers employed by legally chartered enterprises or by the public sector. The informal sector is usually characterized as having low productivity and producing low incomes (ILO, 1972; Hart, 1973; McGee, 1967, 1971, 1976 and 1982; McGee and Yeung, 1977; Armstrong and McGee, 1986). Since 1970, there have been major policy debates concerning the problem of labour force absorption and the capacity of Governments and the private sector to generate sufficient productive employment for the rapidly growing urban populations of developing countries. One policy position has considered the informal sector a holding sector for the surplus labour force that can provide unskilled workers with some employment opportunities. Another position has been focused on increasing employment in the formal sector by fostering rapid industrialization, particularly of the export-oriented kind. This strategy has been successful in the NIEs and in certain ASEAN countries, including Malaysia and Thailand. However, estimates of labour force growth in the Asia-Pacific region indicate that, during the 1980s, the urban labour force was growing only 1 per cent faster than the total labour force of the region (3.1 versus 2.1 per cent, respectively). During the 1990s, the difference between the two rates of growth will increase to about 1.5 per cent (3 versus 1.5 per cent), and by the first decade of the twenty-first century, the urban labour force will grow almost three times as fast as the total labour force (3 versus 1.1 per cent). Translated into numbers of people, the urban labour force of the Asia-Pacific region will grow by 328 million between 1990 and 2010. Consequently, if most countries in the Asia-Pacific region have had difficulty so far in creating employment for their rising number of urban dwellers, the prospects for the future are even more dire. It is therefore likely that the urban informal sector will continue to be a major source of employment for urban dwellers in the years to come.

F. Global processes and urbanization

Although there are many differences between the urbanization experience of developed and developing countries, a number of global processes seem to be leading to a greater homogenization of urbanization around the world. Two types of processes can be distinguished: the transactional revolution; and the impact of globalization on national space.

The transactional revolution involves fundamental changes in the means of communication, exchange and interaction between and within countries. The transactions of interest involve people, commodities, capital and information. Commuting is a form of geographical transaction involving people; the shipping of automotive parts from Thailand to Montreal is a form of commodity transaction; the transfer of funds from New York to Jakarta is a capital transaction; and Cable News Network (CNN) transmissions to all parts of the world represent informational transactions (McGee, 1985). A major recent development is that the constraints of time and distance are no longer important for the transmission of information and capital through space. However, the flows of people and commodities, though they are becoming more efficient, are still subject to a number of constraints, including those related to time and distance.

Most planners realize that the transactional revolution is strengthening the processes leading to the greater centralization of the urban system, to the extent that urban development is increasingly concentrated in a few urban regions within a country. However, the current process of urban concentration involves the creation not only of mega-cities or primate cities, but also of extended metropolitan regions that can take the form of corridors, as in the case of Jakarta-Bandung or Rio De Janeiro-São Paulo, or of sprawling urban regions, as in the case of the Bangkok Metropolitan Region.

Centralization stems from the importance of transaction nodes whose existence permits the minimization of transaction costs. Therefore, the new extended metropolitan regions are characterized by the inclusion of international container ports, international airports, industrial estates and free export zones, as well as international hotels close to multinodal offices, and retailing complexes scattered throughout urban space.

The transactional revolution leads to a form of urbanization that, at the level of individual regions, can create discontinuous patterns of land use which are generally most marked in the urban periphery where agriculture, industry, leisure activities and residential developments are juxtaposed. In the peripheral areas of the extended metropolitan regions there are often major environmental and administrative problems as well as conflicts regarding land use (McGee, 1991). Despite these difficulties, it is the emerging extended metropolitan regions in the developing world that evince the most rapid economic growth. The causes for their dynamism therefore deserve to be better understood so that policies to enhance it may be developed.

The transactional revolution is also affecting the global and regional mode of operation. At the global level, the emergence of the new international division of labour has involved the relocation of industrial activity to developing countries where labour costs are low. Such relocation has been made possible by improvements in transport and communication, as well as in production technologies. The major international investors (often in joint ventures with local entrepreneurs) have identified production niches in various locations. The industries created are producing goods both for export and for local consumption, especially if the local population is generating enough surplus capital to buy those products. The decentralization of production at the global level has been made possible by new forms of organization and marketing. During the 1980s, order-cycle times in developed countries were reduced by 400 per cent and the use of just-in-time production and delivery systems expanded. Clearly, better logistic management depends on improvements in telecommunications and information systems. To become part of the global market and maintain competitiveness, developing countries have to provide access to those systems. That is the reason for the continued success of the semiconductor assembly zone in Bayan Lepas, Penang, Malaysia, which remains globally competitive even though wages have risen.

Another facet of accelerating globalization is the emergence of global cities serving as major nodal points in the movement of capital, people and commodities. Within the ASEAN region, for instance, there is considerable competition between Bangkok and Singapore to become the major air transportation node in the region. Singapore has several advantages, being a stable city State without a large rural population. Indonesia, because of its proximity to Singapore, stands to benefit from the global expansion of Singapore in transactional terms.

Developing countries are currently competing to develop a niche in the global market. Virtually no major city today lacks a world trade centre or first-class convention facilities. Joining the global market entails important changes in cities, in terms not only of the built environment but also of establishing and expanding telecommunication networks, improving airports, and expanding the road system, especially in areas linked to the global market. Indonesia, for instance, has improved its infrastructure for tourism considerably, particularly in Bali.

The move towards globalization implies that national boundaries will become less important as barriers to the movement of goods and people. From the economic perspective, it is indeed advantageous to development to increase cross-border economic activity. In Asia, the Riau-Singapore-Johor growth triangle illustrates how the interaction of complementary economies can lead to rapid economic growth. The potential for the creation of other transnational nodes of interaction exists in the Medan-Penang-Phuket triangle; Pontianak-Kuching; and Manado-Philippines.

G. Conclusion

In comparing the urbanization experience of developed and developing countries, there are many points of convergence and others of divergence. In the future, convergence is most likely to occur in terms of the functioning of extended metropolitan regions and other "global cities". Divergence will characterize the evolution of other urban places in the urban hierarchy, since global and national forces are likely to favour the expansion of urban places devoted to particular niches, be they tourism or certain industries. A case of convergence is perhaps best illustrated by Bangkok and Los Angeles, both of which encapsulate the essence of outer-city development. Both are territorially vast, amorphous, multi-centred and unbalanced urban regions. They both have large peripheries with growing populations residing up to 100 kilometres from the core. The Bangkok Metropolitan Region houses approximately 15 million people and the Los Angeles consolidated metropolitan area contains just short of 14 million people. In addition, both cities are plagued by congestion, pollution and, perhaps most significantly, social problems linked to ethnicity, the persistence of poverty, an increase in crime, and difficulties in ensuring "social order". They provide classic examples of why a new urban agenda is needed, one that is sensitive to both spatial planning and population needs. The functioning of Bangkok and Los Angeles can teach relevant lessons about the liveability and sustainability of urban giants.

In general, assessments of future urbanization assume that the process will continue to evolve largely as it has so far. Yet, there are reasons to question that assumption. Regarding Africa and Asia, for instance, the conventional view that a clear distinction between rural and urban places will persist as the urbanization process advances needs to be evaluated. Areas where both agricultural and non-agricultural activities persist are emerging. They are usually adjacent to urban centres or are located between urban centres that interact with each other. Their emergence and mode of operation owe much to the transactional revolution and to economic changes taking place at the supranational level. Latin America also exhibits new patterns of urbanization. Some cities are evolving in ways similar to cities in developed countries but others are coming to resemble those of Asia, where the distinction between the urban and the rural is less clear, and where the sheer number of people is affecting the quality of life in the cities.

The conventional conceptualization of urbanization and the urban transition is inadequate in three respects. First, it assumes that the spatial separation of rural and urban activities will persist. Second, it assumes that population concentration leads to urbanization. Yet, in many parts of the developing world, cities exist within densely populated regions dedicated mostly to intensive agriculture. Despite their population density, those areas are not urbanized although they interact with the urban core in their midst. Third, the paradigm of the urban transition based on the experience a century ago of today's developed countries may not be transferable to today's developing countries. Indeed, the uneven incorporation of developing countries into the world economic system from the fifteenth century onward has led to new patterns of urbanization.

In terms of common traits, developed countries also experienced rapid population growth as they became urbanized, rural-urban migration was a major component of urban growth and the rural population eventually started to decline as the urbanization transition was completed. However, the size of the population in today's developing regions is considerably larger than that of developed countries a century ago. Furthermore, urbanization is now taking place at the same time as improvements in transport and communications that have led to a reduction in the constraints imposed by space and time. Consequently, developing countries are facing the double challenge of attracting the technology needed to become part of the global market and managing better the cities that are the main actors in the process of globalization. If urbanization is to serve its purpose as the basis for development, that challenge must be met.

Notes

[1]According to current United Nations practice, the less developed regions include Africa, all of Asia with the exception of Japan, Latin America and the Caribbean, and Oceania excluding Australia and New Zealand. The more developed regions include Australia, Canada, Japan, New Zealand, the United States of America, the former Union of Soviet Socialist Republics (USSR) and all of Europe.

[2]The level of urbanization is the proportion of the population living in urban areas.

[3]This paper does not deal with the particular urbanization experience of Islamic countries in Northern Africa and Western Asia. In broad terms, the historical patterns of their incorporation into the global system are similar to those of the rest of Africa or some parts of Asia. Excellent accounts of the different historical experiences of Africa and Latin America can be found in Stren (1972) and Roberts (1978).

References

Abu-Lughod, Janet, and R. Hay, eds. (1977). *Third World Urbanization.* Chicago, Illinois: Maroufa Press.

Armstrong, W., and Terence G. McGee (1986). *Theatres of Accumulation, Studies in Asian and Latin American Urbanization.* New York: Methuen.

Bairoch, Paul (1988). *Cities and Economic Development from the Dawn of History to the Present.* Chicago, Illinois: Chicago University Press.

Cheshire, Paul (1990). Explaining the recent performance of the European Community's major urban regions. *Urban Studies* (Glasgow), vol. 27, No. 3, pp. 311-333.

Davis, Kingsley (1969). *World Urbanization 1950-1970.* Berkeley, California: Institute of International Studies.

Ginsberg, N., B. Koppel and T. G. McGee, eds. (1991). *The Extended Metropolis Settlement Transition in Asia.* Honolulu: University of Hawaii Press.

Hart, K. (1973). Informal income opportunities and urban employment in Ghana. *Journal of Modern African Studies* (Cambridge, United Kingdom, and New York), vol. 2, No. 1, pp. 61-89.

Hugo, Graeme J. (1992). Migration and rural-urban linkages in the ESCAP region. Paper prepared for the pre-Conference Seminar of the Fourth Asia-Pacific Population Conference on Migration and Urbanization: Inter-relationships with Socio-Economic Development and Evolving Policy Issues, Seoul, Republic of Korea, 21-25 January.

International Labour Organization (ILO) (1972). *Employment, Incomes and Equality: A Strategy for Increasing Productive Employment in Kenya.* Geneva: ILO.

Kruvan, J. (1991). Patterns and trends of employment by location and sector. Background paper, Nos. 2 and 3. NUDF Project. Bangkok, Thailand: Development Research Institute.

Macleod, Scott, and Terence G. McGee (1993). Emerging extended metropolitan regions in the Asia-Pacific urban system: a case study of the Singapore-Johor-Riau growth triangle. In *The Asia-Pacific System: Towards the 21st Century,* Yue-man Yeung and Fu-chen Lo, eds. Tokyo: United Nations University.

McGee, Terence G. (1967). *The Southeast Asian City.* New York: Praeger.

_________ (1971). *The Urbanization Process in the Third World.* London: G. Bell and Son.

_________ (1976). The persistence of the proto-proletariat. Occupational structures and planning the future of Third World cities. *Progress in Human Geography* (Cambridge), vol. 9, No. 1, pp 1-38.

_________ (1982). Proletarianization, industrialization and urbanization in Asia. A case study of Malaysia. *Asian Studies Lecture 13.* Adelaide, Australia: Flinders University.

_________ (1985). Circuits and networks of capital: the internalization of the world economy and national urbanization. *Conference on Urban Growth and Economic Development in the Pacific Region.* Taipei, Taiwan Province of China: Academia Sinica, Institute of Economics.

_________ (1991). The emergence of Desakota regions in Asia: expanding a hypothesis. In *The Extended Metropolis Settlement Transition in Asia,* N. Ginsburg, B. Koppel and T. G. McGee, eds. Honolulu: University of Hawaii Press.

_________, and Yue-man Yeung (1977). *Hawkers in Southeast Asian Cities.* Ottawa: International Development Research Center.

Mukhodpadhyay, S., and C. P. Lim (1985). Rural non-farm activities in the Asian region: an overview. In *Development and Diversification of Rural Industries in Asia,* S. Mukhodpadhyay and C. P. Lim, eds. Kuala Lumpur: Asia and Pacific Development Centre.

Roberts, B. (1978). *Cities of Peasants.* London: Edward Arnold.

_________ (1989). Urbanization, migration and development. *Sociology Forum* (Austin), vol. 4, pp. 665-691.

Stren, R. (1972). Urban policy in Africa: a political analysis. *African Studies Review* (Cambridge), vol. 15, No. 3, pp. 489-516.

Tolba, Mostafa K. (1992). *Saving Our Planet, Challenges and Hopes.* London: Chapman and Hall.

United Nations (1991). *World Urbanization Prospects, 1990.* Sales No. E.91.XIII.11.

_________, Economic and Social Commission for Asia and the Pacific (ESCAP) (1990). *Population Research Leads,* No. 35. ISSN 0252-4503.

_________ (1991). *Industrial Restructuring in Asia and the Pacific in Particular with a View to Strengthening Regional Cooperation.* ST/ESCAP/960.

_________, Division of Industry, Human Settlements and the Environment (1992). Urbanization in Asia and the Pacific. Paper prepared for the pre-Conference Seminar of the Fourth Asia-Pacific Population Conference on Migration and Urbanization: Inter-relationships with Socio-Economic Development and Evolving Policy Issues, Seoul, Republic of Korea, 21-25 January.

Weber, A. F. (1965). *The Growth of Nineteenth Century Cities.* Ithaca, New York: Cornell University Press.

Williamson, J. G. (1988). Migration selectivity, urbanization and industrial revolutions. *Population Development Review* (New York), vol. 14, pp. 287-314.

World Bank (1987). *World Development Report, 1987.* New York: Oxford University Press.

Zachariah, K. C., and Julien Condé (1981). *Migration in West Africa. Demographic Aspects.* Washington, D.C.: Oxford University Press.

IV. POPULATION DISTRIBUTION IN DEVELOPED COUNTRIES: HAS COUNTER-URBANIZATION STOPPED?

*Anthony Champion**

It might be thought that patterns of population distribution in developed countries would be relatively stable. After all, their overall rates of population increase are low by comparison with those of the majority of developing countries. Moreover, most of the developed world has experienced a long-established process of urbanization and rural depopulation, so that the scope for further increases in the proportion urban is limited and the urban network and infrastructure are well established. However, this is a picture that is far from reality, because in the 1960s substantial changes, many of which had not been foreseen or expected, started to take place. In countries where the urban population has continued to grow more or less consistently, such growth has not been evenly spread throughout the urban system and there has been a redistribution of people among urban places, with some declining in number of their inhabitants. In addition, many countries have experienced "counter-urbanization", a phenomenon associated with a reversal of net migration gains in metropolitan areas and a turnaround in rural depopulation.

The present paper documents the trends in population distribution experienced by developed countries during the past two decades and assesses their implications for future migration patterns and the changes they entail for urban and regional systems. Particular attention is devoted to counter-urbanization and related aspects of population deconcentration, including an examination of its direct causes in terms of demographic components and a discussion of its underlying causes and consequences. It is important to recognize, however, that counter-urbanization does not represent a complete reversal of the urbanization process, as a review of recent trends in the proportion of the population living in urban areas will show.

A. Trends in urban and rural populations

Since the early days of the Industrial Revolution, population distribution trends have been closely associated with the urbanization process, that is, with the progressive increase in the proportion of the population living in "urban places". This process has now reached an advanced stage in developed countries, with an average of three out of every four persons living in urban areas, a significant increase over the level of urbanization that existed 40 years ago. In contrast, the rural population in the developed world has not only shrunk in terms of its share of the total population but has also been declining in absolute terms. Moreover, these trends are expected to continue.

As table 6 indicates, nearly 73 per cent of the population in developed countries lived in urban areas in 1990, the proportion being up by one third from its 1950 level (54 per cent). The increase in the proportion urban during 1970-1990 was less than that during 1950-1970, but it was still substantial. Moreover, the proportion urban is expected to rise by just over 6 percentage points by 2010, when roughly 4 out of every 5 persons in developed countries are expected to be urban dwellers.

Table 6 also shows the levels of urbanization reached by the various subregions and countries constituting the developed world. Australia, New Zealand and Northern Europe have maintained their position as the most urbanized regions, with urbanization levels above 70 per cent even in 1950, well over 80 per cent in 1990 and expected to reach 87 per cent in 2010. Western Europe had also reached the 80 per cent mark by 1990, followed closely by Canada, Japan and the United States of America. In Japan, the rise in urbanization has been extremely swift since 1950, as has also been the case for Southern Europe and the former USSR. However, the largest increases have taken place in Eastern Europe during both 1950-1970 and 1970-1990. Further increases of 8-10 percentage points are expected for these last three regions.

*Department of Geography, University of Newcastle-upon-Tyne, United Kingdom.

TABLE 6. PERCENTAGE OF THE POPULATION RESIDING IN URBAN AREAS, 1950-2010

	1950	*1970*	*1990*	*2010*
World total	29.3	36.6	43.1	52.8
Developed countries	54.3	66.6	72.7	79.1
Developing countries	17.0	24.7	34.3	46.8
Northern America	63.9	73.8	75.4	80.6
Canada	60.8	75.7	77.1	82.1
United States	64.2	73.6	75.2	80.4
Europe	56.2	66.6	73.4	80.0
Eastern	34.1	49.8	63.1	73.2
Northern	72.7	80.4	83.2	86.9
Southern	44.6	56.0	65.8	74.9
Western	67.1	76.1	80.0	84.2
Australia	75.1	85.2	85.2	87.0
New Zealand	72.5	81.1	83.9	86.8
Japan	50.3	71.2	77.2	81.6
USSR (former)	41.5	56.5	66.1	74.5

Source: *World Urbanization Prospects: The 1992 Revision* (United Nations publication, Sales No. E.93.XIII.11), table A.1.

Table 7 shows the rates of urban and rural population growth that account for the urbanization trends observed during 1970-1990. The developed world as a whole experienced increases in the urban population and reductions of the rural population during both 1970-1980 and 1980-1990, though the differential between the rates of growth of the two types of area narrowed somewhat after 1980, mainly as a result of slower urban growth. However, aggregate growth rates mask significant geographical variations. North America and Oceania, in particular, contrast with most other developed regions in the overall gains recorded by their rural populations over the period considered. Australia, Japan and New Zealand are also distinctive in that they experienced stronger growth of the rural population during 1980-1990 than in the previous decade. Indeed, Japan and New Zealand experienced a switch from rural decline to rural growth. Northern and Western Europe, together with the former Union of Soviet Socialist Republics (USSR), also exhibit a shift towards a slower reduction of the rural population in moving from one decade to the next. With respect to the urban population, significant reductions in the urban growth rate were experienced by most developed regions, with the exception of the United States and, to a lesser extent, Canada. During 1980-1990, only in Australia did the growth rate of the rural population exceed that of the urban population, implying that the proportion living in urban areas was declining.

The picture drawn by the data in tables 6 and 7 must be interpreted with caution, however, since the data refer to urban areas as defined by each country. Because the definitions used vary considerably between countries, both in terms of the method used to delineate places (administrative unit, built-up area, travel-to-work zone) and in relation to the basis of the definition of "urban" (population size, employment structure, threshold levels of such measures), international comparability is compromised. In addition, not all countries update their information on urban places with the same thoroughness or timeliness and, consequently, time-series data and rates of change derived from them should not be interpreted too strictly. Lastly, not too much credence should be placed in the projections beyond 1990 because in many cases the 1980-1990 trends are themselves estimates or projections that still need to be verified on the basis of the latest census data.

Nevertheless, the overall pattern revealed by tables 6 and 7 presents a relatively consistent picture, namely that the dominant trend in the more developed regions of

TABLE 7. PERCENTAGE CHANGE IN THE SIZE OF THE URBAN AND RURAL POPULATIONS, 1970-1990

	Urban population		*Rural population*	
	1970-1980	*1980-1990*	*1970-1980*	*1980-1990*
World total	29.6	30.3	14.9	11.8
Developed countries	14.2	10.4	-3.4	-2.5
Developing countries	46.0	46.9	18.1	13.9
Northern America	11.4	12.1	10.7	3.5
Canada	12.9	12.8	12.8	4.0
United States	11.3	12.0	10.5	3.5
Europe	11.4	7.8	-6.7	-6.8
Eastern	24.8	14.2	-9.0	-9.3
Northern	5.3	3.8	-6.7	-2.4
Southern	17.8	12.0	-4.2	-8.8
Western	6.7	5.0	-8.0	-3.2
Australia	17.9	15.4	12.5	21.2
New Zealand	13.6	9.5	-3.2	6.0
Japan	19.8	7.1	-7.4	1.4
USSR (former)	20.9	15.4	-5.6	-1.7

Source: Calculated from *World Urbanization Prospects: The 1992 Revision* (United Nations publication, Sales No. E.93.XIII.11), tables A.2 and A.3.

the world is towards higher levels of urbanization, that is, towards rising proportions of people living in urban places.

B. URBAN CONCENTRATION AND COUNTER-URBANIZATION

The last observation would seem to be at odds with the attention that has been given to the phenomenon of counter-urbanization over the past two decades. This contradiction, however, is more apparent than real, because the two processes are not complete mirror images of each other. They are defined according to two different geographical scales, with counter-urbanization being identified more in terms of redistribution of population down the hierarchy of urban centres than in relation to the proportion of population living in urban as opposed to rural settlements.

Increase in the degree of concentration of the urban population in the larger urban places is a traditional concomitant of urbanization, a process that normally has reinforced the trend towards greater spatial polarization of population because larger cities tend to be characterized by higher densities than small ones. It is this aspect of urbanization that no longer seems to be as universal as it once was. In the first place, individual cities have become more diffuse as the population densities of their central areas have declined and as suburbanization and the growth of satellite towns have led to the sizeable growth of their geographical area. As a result, it is now more appropriate to talk in terms of functionally defined "metropolitan areas" or "functional urban regions" (FURs) than of physically defined urban areas (Hall and Hay, 1980). Secondly, as Berry (1976) noted when he coined the term "counter-urbanization", it has been found that, even where urban places are identified on the basis of less stringent definitions, larger metropolitan areas have been losing population, at least in relative terms, to smaller urban centres (Fielding, 1982; Korcelli, 1984; Frey, 1988; Champion, 1989).

There is no doubt that, after a century or more of steady urbanization, population distribution in developed countries has come to be dominated by very large urban concentrations (Dogan and Kasarda, 1988). In the United States, 77 per cent of the people lived in the nation's 283 metropolitan areas in 1980, a proportion double the 37.5 per cent living in its 71 metropolitan areas in 1910, while the proportion in areas with 1 million residents or more almost tripled, rising from 17.5 to 49.0 per cent over the same period (Frey, 1990). In the member States of the European Union, the "millionaire cities" (FURs with at least 1 million inhabitants) accounted for one third of the total population in 1981,

TABLE 8. DEGREE OF URBAN CONCENTRATION OF SELECTED DEVELOPED COUNTRIES AS MEASURED BY THE PERCENTAGE OF THE URBAN POPULATION LIVING IN CITIES WITH AT LEAST 1 MILLION INHABITANTS IN 1990

	1965	*1990*	*Change*
Northern America			
Canada	37	39	+2
United States	49	48	-1
Eastern Europe			
Bulgaria	21	19	-2
Former Czechoslovakia	15	11	-4
Hungary	43	33	-10
Poland	32	28	-4
Romania	21	18	-3
Northern Europe			
Denmark	38	31	-7
Finland	27	34	+7
Sweden	17	23	+6
United Kingdom	33	26	-7
Southern Europe			
Greece	59	55	-4
Italy	42	37	-5
Portugal	44	46	+2
Western Europe			
Austria	51	47	-4
France	30	26	-4
Netherlands	18	16	-2
Australia	60	59	-1
Japan	37	36	-1

Source: World Bank, *World Development Report, 1992* (New York, Oxford University Press, 1992), section entitled "World development indicators", table 31.

while over two thirds of the population (69.8 per cent) lived in the 229 FURs with at least 300,000 people (Cheshire and Hay, 1989). Estimates provided by the World Bank (1992) confirm the high degree of urban concentration across the more developed world, with cities of 1 million inhabitants or more accounting for at least one quarter of urban dwellers in most countries and for over one third in some (table 8). Australia, with 3 out of every 5 of its urban dwellers living in cities with at least 1 million inhabitants, heads the list of those countries for which 1990 data are available and is followed, in order of importance, by Greece, the United States, Austria and Portugal, in all of which at least 45 per cent of the population live in such cities.

Comparison of 1990 figures with those for 1965 indicates a widespread tendency for a reduction in the proportion of the urban population living in cities with at least 1 million inhabitants (table 8). Only 4 of the 19 developed countries with data available appear to have experienced an increase in the concentration of the urban population in cities of 1 million or more dwellers. A reduction of the proportion of the urban population living in such cities may be due, however, to statistical underbounding, whereby the administrative boundaries of a city remain fixed even as the functional city grows beyond them. Nevertheless, the rather universal reduction of the weight of cities with 1 million or more inhabitants is suggestive of a widespread process of population redistribution down the urban hierarchy, either through the absolute decline of the largest cities or through the faster growth of smaller urban places.

This shift in population distribution down the urban hierarchy forms the essence of the counter-urbanization phenomenon, as first articulated in a rigorous manner by

Fielding (1982). It is now generally accepted that counter-urbanization may be deemed to be occurring when there exists across the settlement system a negative correlation between the net migration rates of places and their population sizes, where "places" are defined on the basis of relatively self-contained functional entities. The term "counter-urbanization" thus refers to a process of population deconcentration across a functionally defined urban settlement system, whereas "urbanization" is used in this context to denote a traditional pattern of rising population concentration in which larger centres grow at a faster rate than smaller ones and thereby increase their share of the population. Figure I illustrates these relationships in diagrammatic form. It also shows how Fielding (1982) saw the switch from "urbanization" to "counter-urbanization" in Western Europe between the 1960s and the 1980s, as the size of the settlements experiencing the most rapid growth slid downward in the urban hierarchy.

Over the past two decades, an extensive literature has documented the widespread shift towards counter-urbanization. Signs of such a shift were first observed in the United States where, between the 1960s and 1970-1973, non-metropolitan areas experienced a "turnaround" from recording net annual migration losses of 0.3 million to gains of 0.4 million (Beale, 1975). During 1970-1980, the population of the large metropolitan areas of the United States grew at barely half the rate of the medium-sized and smaller metropolitan areas, with non-metropolitan areas outpacing the metropolitan aggregate (Frey, 1992). For France, Fielding (1986) showed that there was a strong positivc correlation between size of place and net migration rate during 1954-1962 which changed progressively to become negative in 1975-1982. He also documented the fact that a large number of developed countries experienced a shift towards counter-urbanization or at least a slowing of the trend towards urban concentration. These developments

Figure I. Models of urbanization and counter-urbanization based on the relationship between net migration and settlement size

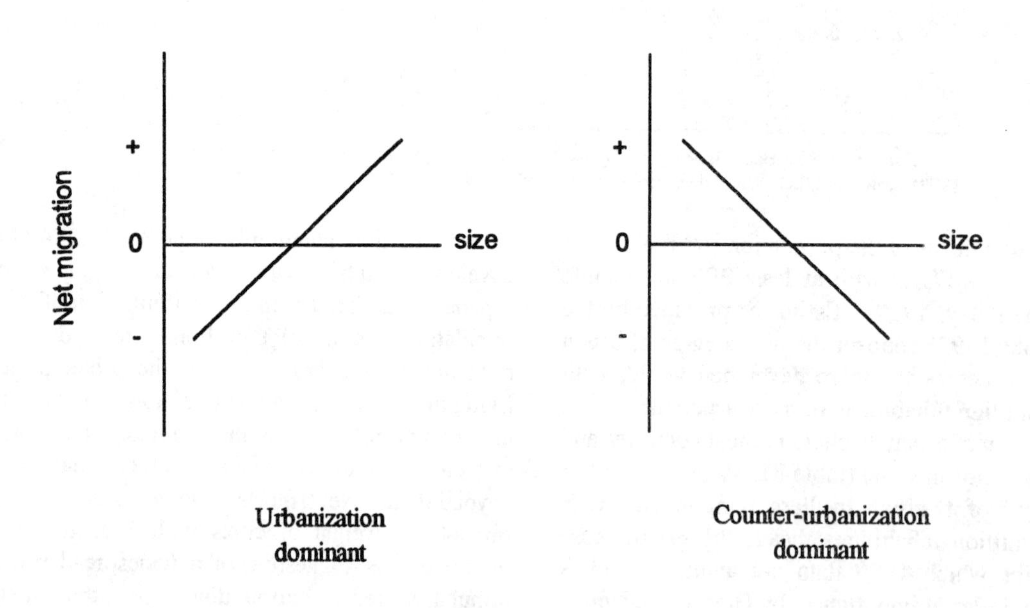

Source: Anthony J. Fielding, "Counterurbanisation in Western Europe", *Progress in Planning* (1982), pp. 8 and 10.

were confirmed by an international comparative study of migration and metropolitan decline (Frey, 1988) and by a collection of more detailed studies relating to nine developed countries, namely, Australia, Denmark, France, Germany, Italy, Japan, Norway, the United Kingdom of Great Britain and Northern Ireland and the United States (Champion, 1989). Although the scale, the timing and, to a certain extent, the nature of the phenomenon varied among countries and, in some cases, also within them, the similarities were sufficient to warrant considering counter-urbanization a universal phenomenon.

More recently, however, a different picture has been emerging. A decade ago, it was expected that the shift away from concentration in the larger metropolitan areas towards medium-sized and small settlements would accelerate through the 1980s, as illustrated in figure II. Yet, that acceleration does not seem to have occurred on a wide scale and indeed there is some evidence suggesting a revival of population growth in some of the largest cities of the developed world. The most significant reversal seems to have taken place in the United States, where trends for 1980-1990 suggest a return to the situation prevailing during the 1960s (table 9). During 1980-1990 and most notably during 1985-1990, large metropolitan areas outgrew the smaller ones (Frey and Speare, 1991; Frey, 1992). In the United Kingdom, London's rate of population loss shrank markedly towards the mid-1980s (Champion and Congdon, 1988) while, according to the latest census of France, the growth rate of the Paris agglomeration matches the national rate of population growth between 1982 and 1990 (Jones, 1991). A significant revival of metropolitan growth during the early and mid-1980s has also been noted in other countries, notably Japan and Norway, though not in Denmark or Germany (Champion, 1989).

Any attempt at a more rigorous test of recent trends in counter-urbanization requires an analysis that correlates migration and size for individual urban centres, a task that is not feasible until data from the 1990 round of censuses become available for properly defined "urban places". In the meantime, ongoing research for the European Science Foundation's project on Regional and Urban Restructuring in Europe (RURE) gives preliminary indications by correlating estimated rates of population change for the past three decades with population

Figure II. Migration and settlement size, 1950-1980: a possible sequence

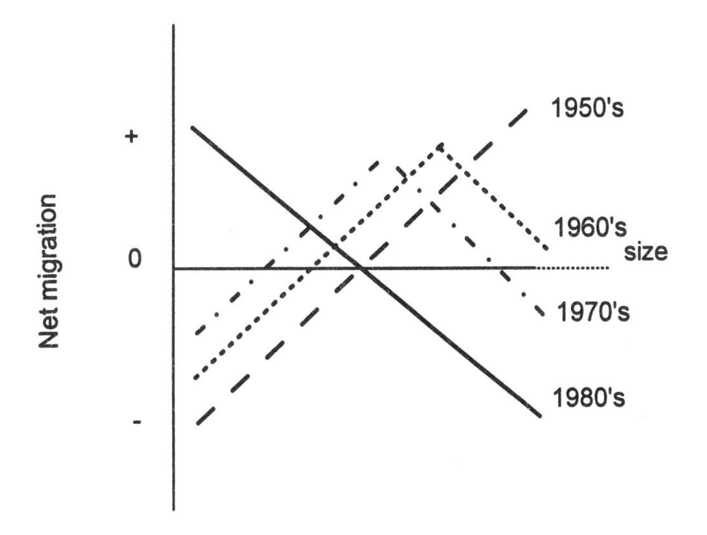

Source: Anthony J. Fielding, "Counterurbanisation in Western Europe", *Progress in Planning* (1982), p. 9.

TABLE 9. POPULATION CHANGE IN THE UNITED STATES, BY TYPE OF METROPOLITAN AREA, 1960-1990

Period	*Type of metropolitan area*		*Non-metropolitan area*
	Large	*Other*	
Percentage change			
1960-1970	18.5	14.6	2.2
1970-1980	8.1	15.5	14.3
1980-1990	12.1	10.8	3.9
1980-1985	6.0	6.1	3.6
1985-1990	5.8	4.4	0.3

Source: Adapted from William H. Frey, "The new urban revival in the United States", paper presented at the United Kingdom/United States Seminar on Migration in Post-industrial Society, Los Angeles, California, 3 and 4 August 1992.

NOTE: Metropolitan areas are defined according to constant boundaries as of 1990. Large metropolitan areas include 39 consolidated metropolitan statistical areas (CMSAs) and metropolitan statistical areas (MSAs) with 1990 populations of at least 1 million.

density at the regional level. Data assembled for 16 countries have produced the results shown in table 10. They show a predominance of positive correlations in the 1980s (10 out of the 16 countries have that feature), indicating a prevailing tendency towards greater regional concentration of population; this marks a departure from the 1970s, when only 5 out of the 16 countries had positive correlations, and a return to the 1960s pattern, when 11 of the 16 countries were in the positive group. There is also a greater uniformity of change between the 1960s and 1970s (with all but one country shifting to higher negative correlations or lower positive ones) than between the 1970s and the 1980s, when a rather evenly matched set of positive and negative shifts is observed. There is considerable regularity of differences and trends for groups of European countries. In particular, Western and Central European countries are characterized by the early onset of overall deconcentration and by its continuation as a highly significant dimension of population change into the 1980s. In the Mediterranean countries and Ireland there is a prevailing, albeit weakening, tendency towards concentration, and in Scandinavia deconcentration is slight and short-lived.

Lastly, it should be noted that these "urban system" trends have, to some extent, been echoed in studies that monitor migration balances at the broader regional scale of the core-periphery distinction within countries. Indeed, the work of Vining and Kontuly (1978) in that area first revealed the widespread nature of the swing of net migration away from the major metropolitan centres of the developed world and thereby prompted the international search for more detailed evidence on counterurbanization (Champion, 1989). The latest work of that type (Cochrane and Vining, 1988) indicates a swing back towards a more traditional picture though, as with the more disaggregated regional analyses presented in table 10, recent patterns are not characterized by the same degree of uniformity as those of earlier decades.

As shown in figure III, the majority of the country groups of the developed world can be classified as one of three types according to the relative migratory attractions exerted by their core and their periphery in the early 1950s and their trends in the 1960s; subsequently, however, the trends in the net migration rate are similar across all three types, with the upward shifts recorded by peripheral regions in the early 1970s succeeded by downward shifts by the end of that decade, a pattern followed by most countries with the exception of France and Germany. On the other hand, according to the evidence presented by Cochrane and Vining (1988), two other classification types, comprising less urbanized regions or countries, namely, Eastern Europe and the set consisting of the Republic of Korea and Taiwan Province of China have experienced increasingly faster core growth since 1950.

In sum, while the proportion of the population living in urban areas is expected to keep on rising, its distribution across the urban system and also among the broad regional divisions of countries is changing in a less stable and uniform fashion. In particular, during the late 1960s and early 1970s most countries in the developed world, with the apparent exception of Eastern Europe, experienced some degree of migratory swing away from large cities and towards smaller urban centres and rural areas, which in many cases reversed earlier patterns of metropolitan growth and rural depopulation and was associated with a relative increase in the migration-related attractiveness of peripheral regions at the expense of the national cores. Subsequent experience, still to be confirmed by detailed analysis of the 1990 round of censuses, would seem to indicate a second reversal of pre-existing patterns, but this trend does not appear to be as universal as the original turnaround.

TABLE 10. CORRELATION BETWEEN POPULATION GROWTH RATES AND POPULATION DENSITY OVER THREE DECADES FOR SELECTED EUROPEAN COUNTRIES

Country (number of regions)	*Correlation coefficient*			*Shift*	*Shift*
	1960s	*1970s*	*1980s*	*60s/70s*	*70s/80s*
Western and Central Europe					
Austria (9)	-0.48	-0.60*	-0.61*	-	-
Belgium and Luxembourg (12)	+0.07	-0.46**	-0.61**	-	-
France (96)	+0.31***	-0.03	+0.17	-	+
Germany (West) (30)	-0.48**	-0.68***	+0.00	-	+
Netherlands (11)	-0.15	-0.69**	+0.22	-	+
Switzerland (24)	+0.14	-0.36*	-0.41**	-	-
United Kingdom (65)	-0.09	-0.35***	-0.38***	-	-
Mediterranean Europe and Ireland					
Greece (51)	+0.56***	+0.29**	+0.31**	-	+
Ireland (26)	+0.61***	+0.26	+0.25	-	-
Italy (92)	+0.47***	-0.00	-0.07	-	-
Portugal (20)	+0.73***	+0.52**	+0.63***	-	+
Spain (50)	+0.78***	+0.80***	+0.21	+	-
Scandinavia					
Denmark (15)	-0.19	-0.52**	-0.45*	-	+
Finland (12)	+0.70**	+0.53*	+0.28	-	-
Norway (19)	+0.14	-0.24	+0.52**	-	+
Sweden (24)	+0.73***	-0.01	+0.45**	-	+
All countries (557)	+0.29***	-0.06	+0.06	-	+
Western and Central Europe (247)	+0.09	-0.29***	-0.08	-	+
excluding France (151)	-0.13	-0.36***	-0.31***	-	+
Mediterranean Europe and Ireland (239)	+0.52***	+0.23***	+0.07	-	-
Scandinavia including Iceland (71)	+0.26**	-0.13	+0.03	-	+

Source: C. Vandermotten for the European Science Foundation Project on Regional and Urban Restructuring in Europe.

NOTE: A plus sign (+) indicates a shift towards greater population concentration (urbanization tendency); a minus sign (-) indicates a shift towards lesser concentration (counter-urbanization). Correlations are significant at the 0.001 (***), 0.01 (**) or 0.05 (*) level. Shifts have not been tested for statistical significance.

C. IMMEDIATE CAUSES OF RECENT TRENDS

The evidence available so far on the experience of the 1980s presents a rather mixed picture, but one thing is certain: the past decade has not seen the general intensification of the tendency towards counter-urbanization that was anticipated by Fielding (1982). Recent observations pose a major research challenge, requiring not only an analysis of recent trends but also a reassessment of the findings of studies dealing with the previous decade. A first step is to examine the immediate or "proximate" causes of the observed trends, with reference to the demographic components of change. At this level of explanation, three factors are normally distinguished in the counter-urbanization literature, namely, natural increase, internal migration and international migration.

As regards the part played by natural increase, the literature attaches virtually no importance to deaths under the assumption that mortality rates are quite stable over time and exhibit relatively consistent geographical differentials which are tending to narrow. In contrast, fertility does seem to have played a part in population deconcentration over the past two decades. In particular, the decline in birth rates that occurred in most developed countries after the early 1960s played a large part in changing the growth rates of large cities from positive to negative, as natural increase became unable to offset

Figure III. Net internal migration: generalized trends for groups of countries

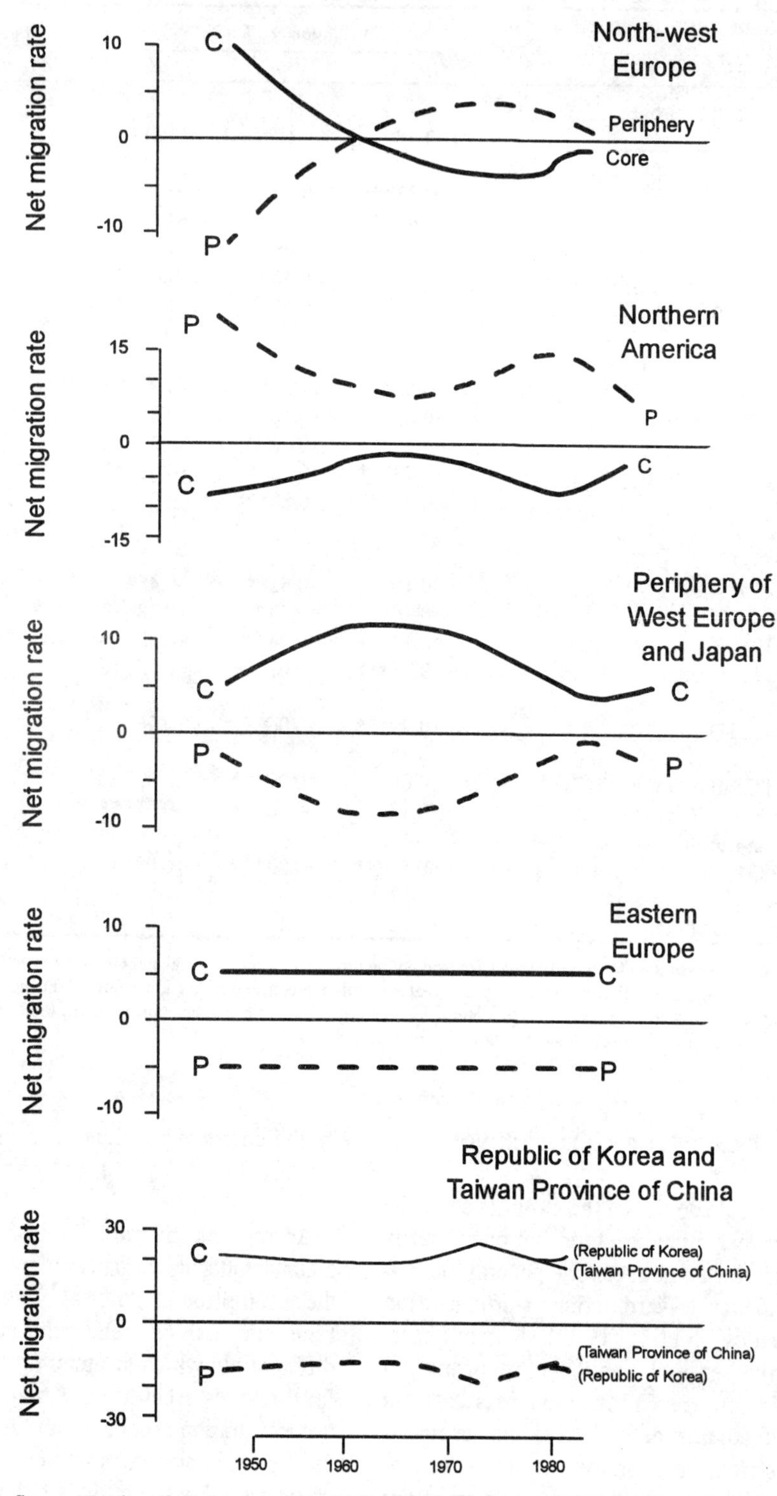

Source: Stephen G. Cochrane and Daniel R. Vining, "Recent trends in migration between core and peripheral regions in developed and advanced developing countries", Working Papers in Regional Science and Transportation, No. 108 (Philadelphia, Pennsylvania, Regional Science Department, University of Pennsylvania, 1986).

migration losses; in addition, in countries such as Italy, Japan and the United States, where fertility declined faster in large cities than in other areas, it also contributed to counter-urbanization. Generally, however, there has been a tendency for the birth rates of urban and rural areas of developed countries to converge.

Fertility trends may have also contributed to slow counter-urbanization during the 1980s. Thus, in the United States, although non-metropolitan levels of fertility continue to be higher than metropolitan levels, the 1980s saw a significant reduction of non-metropolitan fertility and a renewed convergence of the fertility rates of metropolitan and non-metropolitan areas after a decade in which metropolitan fertility rates had fallen significantly faster than those for non-metropolitan areas (Fuguitt, Beale and Reibel, 1991). In France, an analysis of 1982-1990 intercensal changes showed that natural increase in the Île-de-France region, which contains Paris, was high (Jones, 1991). Similarly, there was a significant increase in the number of births in London that helps explain its population recovery towards the mid-1980s (Champion and Congdon, 1988).

The role of international migration tended to be underestimated in the early studies of counter-urbanization, perhaps because its effects on regional and local population distribution within countries are difficult to monitor. It is known that the arrival of international migrants tends to bolster core regions within countries, especially the largest settlements in any area of destination. The reasons for this outcome are various, including the fact that core regions are often ports of entry, have more jobs available, have housing vacated by the net loss of internal migrants, and contain already established immigrant communities. The effects of international migration can be seen most clearly in the traditional countries of immigration, namely, Australia, Canada and the United States, where immigrants have tended to establish themselves in the major metropolitan centres. One possible effect of international migration on counter-urbanization occurs through the added pressure that international migrants put on cities and the economic and social costs incurred which may prompt the increased out-migration of the local population. If, in addition, international migration flows diminish, net migration from the city may become negative. In Europe, the reduction of international migration that took place in countries like France or Germany as of the mid-1970s may have contributed towards accelerating the population losses of major urban centres and thereby increased the rate of population deconcentration.

During the 1980s and particularly since 1985, international migration is likely to have contributed to the revival of metropolitan centres in developed countries. The United States in particular has experienced an important rise in immigration, most of it directed towards metropolitan areas (United Nations, 1996). In Europe, net migration gains have been registered since 1985 by a number of countries, particularly by Germany. The admission of persons having the right to citizenship (ethnic Germans in the case of Germany), of close family members of established migrants, of some migrant workers and of growing numbers of asylum-seekers all contributed to that trend (Salt, 1992; United Nations, 1996). The globalization of the world economy, linked to restructuring and the rapid growth of the financial sector, has led to the emergence of "world cities" specializing in high-level transactional activity and information-processing that are attracting all types of migrants, but particularly highly skilled personnel from all over the world (Salt, 1988). In Europe, the collapse of communist regimes and the end of the cold war have already given rise to important East-to-West migration flows, particularly of ethnic minorities having a homeland that will accept them (for example, ethnic Germans or Jews). Although the massive flows predicted have not yet materialized, international migration has indeed increased and can be expected to contribute to the growth of major cities in receiving countries.

Although natural increase and international migration have both influenced population distribution over the past two decades, it is the switches in the patterns and rates of internal migration that have been largely responsible for the observed tendency towards population deconcentration. In all those countries that experienced a significant shift towards counter-urbanization in the 1970s, the net migration from more peripheral areas to the major metropolitan centres declined steeply between the 1960s and 1970s, even if it did not actually become negative in every case. The most heavily urbanized region of each country or at least its largest city or cities experienced heavy net migration losses at some stage between the early 1960s and early 1980s: Greater London and south-east England during the late 1960s and early 1970s; the large metropolitan areas of Australia and the northern part of the United States, as well as Copenhagen and Oslo through most of the 1970s; the

Paris agglomeration and the Japanese metropolitan areas (albeit to a limited extent) mainly during the late 1970s; and the three metropolitan areas of north-west Italy during the early 1980s (Champion, 1989).

In terms of the age structure of the migration flows that led to population deconcentration, the early counter-urbanization literature identified retirement migration as an important factor, but subsequent research has shown that centrifugal tendencies have not been caused entirely, or even largely, by movements of the elderly. It is true that many of the places that grew faster in the 1970s are in areas traditionally associated with retirement migration and also that the number and general mobility of elderly people have increased over time. Nevertheless, retirement migration was already a well-established phenomenon before the 1970s. Instead, the major change occurring after the 1960s was the increasing tendency for persons in younger age groups to behave according to a "retirement pattern" in their net migration flows. Such a tendency has been documented very clearly for Germany, where the late 1970s saw a significant swing towards counter-urbanization in the net movements of people aged 30-49 and their children (Kontuly and Vogelsang, 1989). For France, Winchester and Ogden (1989) found that net in-migration to the Paris region in 1975-1982 had been restricted to adults under age 30, while Frey (1989) showed that the non-metropolitan areas of the United States switched from a net loss to a gain of people aged 25-34 between the late 1960s and the late 1970s. Frey also validated the pervasiveness of deconcentration trends among population subgroups by showing that blacks were as heavily involved in these migration switches as non-blacks.

Just as internal migration was the key element in the process of population deconcentration in the 1970s, it also would appear to have played the dominant role in the subsequent slowdown or reversal of this process. In the United Kingdom of Great Britain and Northern Ireland, for instance, the recovery of the major city regions in the 1980s corresponded to a marked narrowing of the migration-rate differentials between settlement-size categories and a tendency of convergence towards a zero net migration balance. This trend can be attributed in large part to the effect of recessionary conditions that froze up the entire migratory system. During the recovery of the mid-1980s, migration propensities again accelerated, but outflows from more peripheral regions and rural areas appear to have increased at a faster rate than the reverse movements from many of the larger metropolitan areas and other more prosperous and advantaged centres (Champion, 1987; Stillwell, Rees and Boden, 1992). There remains, however, the research task of assembling migration data for a large number of developed countries on a basis suitable for studying the strength and composition of centrifugal and centripetal flows over the past decade.

D. Explanations of population redistribution trends and their significance

Past research has identified a large number of reasons for the emergence of counter-urbanization as a major dimension of population redistribution in the 1970s. As shown in table 11, they are so numerous that it now seems amazing that the rural population turnaround and metropolitan migration reversal were able to take the experts by surprise. By the same token, however, the new "surprise" produced by the experience of the 1980s requires that a closer analysis of these explanations be made so as to verify whether they really do take account, as their various proponents have maintained, of the powerful forces for population deconcentration. Alternatively, there may have been a temporary change between the 1970s and 1980s in the strength of those deconcentration forces, or perhaps new forces arose during the 1980s that have overridden or counteracted the centrifugal tendencies of earlier times.

At least five sets of reasons can explain the rise and fall of counter-urbanization since the 1960s. The most obvious relates to the availability and mobility of capital for investment in manufacturing. The new spatial division of labour, grounded in the rise of the multinational company and the changing structure of corporate organization, was associated with a major shift in the distribution of manufacturing employment away from major industrial centres towards free-standing cities, towns and more remote regions during the late 1960s and early 1970s. This shift was stimulated by the search for cheap, non-unionized and female labour supplies, by the availability of sites for extensive mass-production plants, and by a high level of government labour subsidies and capital investment grants. The severe cutbacks in investment in manufacturing that took place in many developed countries during the late 1970s and particularly during the 1979-1983 recession accord well with the phasing of population deconcentration.

TABLE 11. EXPLANATIONS FOR THE COUNTER-URBANIZATION PROCESS EXPERIENCED SINCE THE 1960S

1. Expansion of commuting fields around employment centres
2. Emergence of diseconomies of scale and social problems in large cities
3. Concentration of rural population in local urban centres
4. Reduction of the stock of potential out-migrants living in rural areas
5. Availability of government subsidies for rural activities
6. Growth of employment in particular industries like mining, defence and tourism
7. Restructuring of manufacturing industry and the associated growth of branch plants
8. Improvements in transport and communications technology
9. Improvement of education, health and other infrastructure in rural areas
10. Growth of employment in the public sector and personal services
11. Success of explicitly spatial government policies
12. Growth of welfare payments, private pensions and other benefits
13. Acceleration of retirement migration
14. Change in residential preferences of working-age people and entrepreneurs
15. Changes in age structure, household size and composition
16. Effect of economic recession on rural-urban and return migration
17. First round in a new cyclical pattern of capital investment in property and business

Source: *Counterurbanization: The Changing Pace and Nature of Population Deconcentration*, Anthony G. Champion, ed. (London, Edward Arnold, 1989), pp. 236-237.

A second line of explanation can broadly be termed the "rural-resource development" factor. It has been used very effectively in the case of the United States to show how rural areas specializing in particular branches of rurally based economic activity participated in the early 1970s turnaround. Counties specializing in energy resource development, agriculture and forestry were all found to have experienced significant upward shifts in population growth in the early 1970s, as were towns with colleges, places with defence establishments, and other areas with beneficial natural and man-made infrastructures, such as those attracting tourists and retired people (McCarthy and Morrison, 1978). Virtually all these elements of newfound prosperity had taken a heavy blow by the end of the 1970s, because of the downturn in the international demand for raw materials and spending cuts in publicly supported programmes in the areas of farming, forestry and local government services. All these developments have direct implications for employment, as well as negative effects on the general desirability of living in remoter areas or in smaller and less accessible settlements.

Related to the rural resource factor is the role of changing residential preferences. The seminal counter-urbanization work by Berry (1976) saw this as the major force behind the turnaround in the United States. The move away from large cities was portrayed as a return to old "frontier" values or to rural paradise, and it reflected an anti-urban bias which has been allowed to flourish as centripetal pressures have relaxed, easier mobility has increased the range of accessibility to large urban centres, and various developments have reduced the differences in quality of life between metropolitan and rural areas. Retirement migration provides the best documented example of the operation of this factor, in terms not merely of its growth but also of changes in the preferred destinations: from the traditional sea-side resort and spa towns favoured until the 1960s to the countryside in the 1970s. Yet this consumer-led phenomenon has also involved people of working age, particularly those who are self-employed or engaged in footloose types of work. With the rapid developments made in micro-computing and telecommunications during the 1980s, one would expect this factor to have gained importance in recent years, but it may be that the onset of recessionary conditions, combined with attempts by both Government and business to increase productivity and competitiveness, has reaffirmed the importance of contacts and concentration.

A fourth set of considerations builds on the residential-preferences approach by recognizing that different sections of the population hold different preferences and that, as the socio-demographic composition of the population changes, so will the "average preference" of the whole population. Since the early 1960s, developed countries have undergone some more-or-less universal experiences, including the peaking of the birth rate in the 1960s, the resultant surge in the size of the teenage and young adult populations from the mid-1970s onward, the shift of the working population from manual to non-manual occupations, the massive growth of owner-occupied housing, and the general increase in affluence for a large majority of the working population. More specifically, just as deconcentration appears to have gone through a cycle over the last 25-30 years, so too has the demography of most developed countries gone through a cycle—a generational one induced by the inherited imbalance of the age structure and reinforced by recent changes in fertility rates. The peaking of family-rearing activity in the late 1960s and early 1970s correlates with the deconcentration phase, as the subsequent slowdown in urban to rural movements correlates with the increasing size of young adult cohorts that, traditionally are, attracted by "city lights".

Lastly, in some countries, including the United Kingdom, the counter-urbanization experience of the past 25 years can be linked in various ways to changes in the size, nature and distribution of the housing stock. First, the movement of a large number of family-rearing households to peripheral areas during the 1960s and early 1970s can be related to the search for affordable owner-occupied houses. In the early 1980s, however, the emphasis in house-building shifted towards smaller houses and special accommodation for one-person households, one-parent families and the elderly, all needing housing locations easily accessible to work, shops, public services and entertainment facilities. Of equal importance has been the pattern of housing-stock losses, with slum clearance and other demolition accelerating during the 1960s and then slackening markedly during the late 1970s because of the redevelopment of cleared sites, the diminution of public funds for major capital projects, the urban conservation lobby and the introduction of inner-city regeneration policies.

There would seem, therefore, to be ample evidence supporting both the idea that there have been changes since the 1970s in the strength of deconcentration forces and the suggestion that new forces emerged in the 1980s. In relation to the latter, the growth of employment in finance and related business services, within the context of the development of the "information economy", has been highlighted in the recent counter-urbanization literature as a key factor behind the recovery of at least some large cities and core regions over the past decade. As regards fluctuations in the strength of deconcentration forces, attention has been drawn to, among other things, the effects of the 1960s baby boom working its way through the age structure and the role of the economic recession of the late 1970s and early 1980s in restricting the scope for urban-rural shifts. Factors such as lower job turnover, a contraction in primary production, less investment in manufacturing, reduced rates of house-building and greater difficulties faced by retirees in selling their metropolitan homes all served either to increase the rate of out-migration from more rural and remote regions or to slow the pace of out-migration from the larger cities and more urbanized regions.

There are also clear signs that government actions have played an important role both in encouraging population deconcentration and in phasing out counter-urbanization. The 1960s and early 1970s saw the culmination of major investment in infrastructure, notably through the construction of new national highway networks and the setting-up of urban-type facilities related to education and health care in smaller settlements and more remote regions. This was also the period of high spending on defence establishments (usually located at some distance from large population concentrations) and of strong support for agriculture, forestry and rural development in general. Moreover, many countries were at that time implementing urban decentralization policies that included "new town" programmes, while the larger cities began to suffer selective out-migration of residents and businesses and to run into the financial problems associated with rising costs and falling tax revenues.

Subsequent changes in government policy have tended to go hand in hand with the effects of economic downturn. Since the mid-1970s, the focus of government attention has switched away from regional development programmes and rural support towards tackling the problems of urban decline, both intentionally through inner-city development policies and by default as a result of welfare payments to the rising number of urban unemployed. A slower pace of population movement has meant a reduced need for public-service investment in newly expanding towns, while declining national tax revenues have led to large-scale cuts in public expenditure, particularly for infrastructure and other capital projects and also for sectoral support programmes. Government efforts at achieving more rapid economic growth have included progressive deregulation of various elements of economic activity and service provision, which have been aimed at stimulating greater competitiveness and efficiency and may well have renewed the advantages of agglomeration for business.

E. Interpretations of past trends and their implications for the future

Given the variations in recent population distribution trends between countries and over time, and the large number of factors that appear to influence these patterns, there can be little confidence attached to any particular forecast of urban or regional trends for the next decade or two. Certainly, until more accurate and complete information is available on the population changes that took place during the 1980s and on the migration processes underlying them, it is not possible to do more than put forward some alternative interpretations of recent developments. Each will have somewhat different implications for government policy.

One interpretation is that the counter-urbanization of the 1970s was a temporary anomaly, or a "period effect", arising from a chance combination of factors that interrupted the long-term trend towards increasing levels of urbanization and metropolitan concentration, factors such as a once-and-for-all extension of commuting fields due to a major programme of highway construction, the reduction of urban-rural contrasts in service provision (for example, colleges, hospitals), and the attainment of a particular stage in the organization of the manufacturing sector. However, given the number of studies that have portrayed the experience of the 1970s as the logical evolution of the trends experienced during the 1950s and 1960s (Berry, 1976; Hall and Hay, 1980), it might be equally valid to hypothesize that the 1980s constituted the anomaly and represented a short-term hiatus in a new general pattern of deconcentration (Champion, 1988; Champion and Illeris, 1990).

Given the fluctuations in the pace of deconcentration over the past two decades, it is not surprising that speculation has arisen as to whether these period effects are cyclical in nature. This hypothesis is buttressed by the evidence cited regarding changing economic conditions, their association with building cycles and the effects of changes in cohort sizes resulting from past fluctuations in fertility. Klaassen and others (1981) have put forth the idea of cycles in relation to "re-urbanization" in the context of a core-ring model of the evolution of individual metropolitan areas and it is also suggested by the macro-regional trends shown in figure III. Moreover, in discussing Japanese migration, Mera (1988) refers to recent cyclical behaviour in political ideology, citing the cutback in public expenditure in the 1980s and the political moves to liberalize the financial sector. Focusing on the United States experience, Berry (1988 and 1991) has set recent events within the long-term framework provided by the 50-to-60-year Kondratieff cycles that are associated with major phases of innovation and technological development.

By itself, however, a "cycle-perspective" cannot provide a complete answer as to what the long-term future holds in store, since cycles are normally considered to be superimposed on other secular trends rather than on a steady-state system. According to Berry (1988), in the United States each successive Kondratieff wave since the early 1800s has stimulated a progressively lower level of net rural-urban migration, with the intervening troughs moving closer towards a situation of rural-urban migratory balance, until most recently with the trough of the 1970s involving the reversal of net migration so that it flowed from urban to rural areas. A major challenge is to try and identify whether this experience is unique to the United States, or paralleled by that of other countries with a similar settlement history and pattern of economic development, or perhaps is a general phenomenon across the developed world.

In the meantime, a piece of personal speculation can provide a starting point for discussion. Following the lead given by Frey (1987), it is here suggested that two separate processes may be held largely responsible for the trends in population deconcentration observed in developed countries over the past two decades, one being the "counter-urbanization" process *stricto sensu*, involving a shift from larger cities and more heavily urbanized regions to less densely populated areas and down the urban hierarchy; and the other the "regional restructuring" process encompassing shifts in the spatial economy as it adjusts to the new locational requirements of the production and service industries, and normally leading to changes in the balance between regions. Both of these processes are subject to period effects, and there may be certain times when the two are closely aligned in their spatial outcomes and others when they are diametrically opposed or not related at all.

In terms of this approach, the early 1970s becomes a period when the two processes reinforced each other. Very strong counter-urbanization forces were operating at the time, as both residents and jobs moved down the urban hierarchy within macroeconomic regions and in many cases spilled over from regions dominated by large metropolitan centres to less heavily populated regions. This state of affairs coincided with the reorientation of economic activity to more peripheral regions, which was related not only to mineral and tourist resource development but also to what is now seen by some as the "last fling of the Fordist mode of production", with industry actively seeking out cheaper labour and land.

Since the mid-1970s, the factors favourable to counter-urbanization have weakened, partly as a result of difficult economic conditions which have led to cutbacks in investment in housing, manufacturing plants and the public sector, and partly because of demographic shifts. At the same time, a major restructuring of the economic system has taken place, involving a reduction in government support for primary production, the rationalization of the manufacturing industry and the rapid growth of high-level services, particularly in the financial sector. In combination, these changes have resulted in a rather complex regional "mosaic" of growth and decline, as the growth of individual places has been affected by their inherited economic structure, their ability to take advantage of the new sources of growth and their attractiveness as destinations for local urban decentralization (Champion and Illeris, 1990).

If the same mode of thought is applied to anticipating trends over the remainder of this decade and beyond, it is clear that a large number of guesstimates have to be made about the future trajectory of the factors influencing population redistribution trends. In the absence of a full balance sheet of the various factors influencing the counter-urbanization process, it can only be hypothesized that the redistribution of population down the urban hierarchy within macroeconomic regions tends to operate in a cyclical manner and that this process will again accelerate over the next 10-15 years. As regards the process of regional economic restructuring, there are signs that the pace of change will be slower in the next few years than over the past decade, now that manufacturing has already been subjected to a major shakedown and the financial sector appears to be moving into a period of consolidation and rationalization following its extremely rapid growth in the 1980s.

In terms of geographical implications, it can tentatively be suggested that, in the absence of any major exogenous change, some resurgence of population deconcentration will take place. Deconcentration will largely be restricted to the redistribution of residents within regions, involving local decentralization from medium-sized and smaller urban centres as well as from the large cities to less urbanized regions that offer a good quality of life or can take advantage of work in the "information economy" and thus benefit from longer-distance shifts. Most cities that have traditionally specialized in heavy industry and port activities, as well as more remote regions dependent largely on primary production, are likely to experience continuing economic problems and loss of younger people. The large cities that have benefited from the 1980s growth in high-level services are likely to resume their long-term decline as a result of the combination of rationalization in that sector and the resurgence of counter-urbanization.

Inasmuch as both the 1970s and the 1980s have managed to confound expectations, it would be dangerous to attach too much credibility to the above scenario. One factor that is likely to throw this forecast off track is the acceleration of international migration, which has already been rising in most developed countries since 1985. The rise in the number of asylum-seekers and refugees in Europe indicates that, to the extent that global destabilization continues, international migration may well increase. Since most international migrants tend to settle in the larger cities of receiving countries, their presence will have an initial concentrating effect on the urban system. However, over the long run, such a presence may increase the incentive and probably the economic ability for the out-migration of city residents, thus contributing to deconcentration.

F. Consequences and policy implications

Consequences should be considered at a minimum of three levels: at the microlevel (that is, in terms of people, households, firms); at the community level, where communities are not necessarily spatially defined; and at the macrolevel in terms of the national interest. Maybe nowadays a fourth level—the supranational or global level relating to the environment and to the sustainability of development in natural, economic and political terms—should not be omitted. The Governments of modern democratic societies are under a moral obligation, acknowledged to varying degrees, to monitor and review the policy implications of change at all these levels. Unfortunately, our understanding of how the population redistribution scenario described above may affect the relevant levels is still insufficient. Therefore, only some broad generalizations will be made.

At the microlevel, it is normally assumed that the migrants who contribute to population redistribution will gain from their move, whereas those who stay behind will be put at a disadvantage because migration tends to be selective of people with better human capital and thus reduces the assets of the place of origin. This assumption raises questions about the extent to which migration is voluntary as opposed to forced, and about the number of "frustrated migrants" among those who do not move. Thus, opinion polls in the United Kingdom regularly identify considerably more people wishing to migrate from large cities to smaller settlements than actual migrants, while both the business community and central Government have expressed concern about the reluctance of workers to move to areas with job vacancies. Government can play an important role in this context, by assessing the levels of individual mobility and taking action to facilitate movements that are deemed desirable.

As regards the wide range of costs and benefits associated with changes in population distribution and their impact, it is not easy to gauge the overall balance sheet. Thus, although counter-urbanization may involve the renewed growth of certain smaller cities and rural areas that were stagnating or in economic decline, it may also cause greater congestion, environmental degradation and an increase in the cost of living in those areas. Promoting out-migration from the largest cities was once a key goal of government policies aimed at reducing concentration and allowing the redevelopment of older urban areas at lower densities, but the negative effects of selective migration are combining with the effects of economic restructuring and international migration to produce a bipolar social composition comprising a small but powerful elite and a majority of underprivileged people. The impacts of economic restructuring and counter-urbanization have had particularly serious consequences for the older industrial agglomerations, leading in many cases to spiralling decline. Yet, as argued by "liberal" political ideology, it may be against the long-term national interest to prop up those places at the expense of channelling investment to growth areas. Certainly, past government attempts at implementing policies aimed at reversing underlying trends have often had little success and sometimes undesirable consequences.

Given the uncertainty about future trends, the lack of a clear balance sheet of gains or losses, and the reservations about the efficacy of government intervention aimed at modifying population redistribution trends, it is perhaps not surprising that a relatively large number of developed countries consider their present situation "satisfactory" or desire only minor change (United Nations, 1992). Nevertheless, there are a number of actions that Governments should take both to better monitor the process of population redistribution and to improve its outcomes. First, it is important to review the definitions and classifications of “rural” and “urban”, taking into account the major changes that have been occurring in the structure and functioning of urban systems. Of particular relevance is an examination of the concept of the functional urban region and its possible use in different countries in order to improve the comparability of data. Attention should also be given to improving the methods of collecting and collating data on the migration component of population change in order to provide a more accurate and comprehensive record of the role of migration in altering the size and composition of regional and local populations and to furnish a better basis for population projections. In particular, a concerted effort should be made to produce these data for spatial units that provide an appropriate representation of the national settlement system.

In assessing the extent of migration, it is also important to consider to what extent it is the result of push factors and how often potential migrants are frustrated in their intentions to migrate. Such information would allow Governments to assess how far the national interest might be served by actions aimed at allowing greater scope for individuals with respect to realizing their wishes. The identification of factors promoting or retarding mobility, and an assessment of the likely costs and benefits of policy intervention would also be needed.

Special emphasis should be put on assessing the implications of current trends in population distribution and establishing whether they accord with longer-term national objectives relating to the space-economy, social harmony and environmental sustainability. Particular attention should be given to the causes and consequences of socially selective migration, especially in relation to the decline of large metropolitan centres. The wide disparities now existing between and within both urban and rural areas need to be more fully acknowledged and addressed.

Governments should recognize that many of their actions as well as their inaction may have major impacts on population distribution that are often greater than those of explicit spatial policies. Consequently, they should undertake regular assessments of how their programmes and interventions affect population distribution. In particular, Governments should monitor the geographical impact of international migration.

REFERENCES

Beale, Calvin L. (1975). *The Revival of Population Growth in Non-metropolitan America*. Washington, D.C.: U.S. Department of Agriculture, Economic Research Service.

Berry, Brian J. L. (1976). The counterurbanization process: urban America since 1970. In *Urbanization and Counterurbanization*, Brian J. L. Berry, ed. Beverly Hills, California: Sage.

_________ (1988). Migration reversals in perspective: the long-wave evidence. *International Regional Science Review* (Morgantown, Virginia), vol. 11, pp. 245-252.

_________ (1991). *Long-wave Rhythms in Economic Development and Political Behaviour*. Baltimore: The Johns Hopkins University Press.

Champion, Anthony G. (1987). Recent changes in the pace of population deconcentration in Britain. *Geo-forum* (Elmsford, New York and Oxford), vol. 18, pp. 379-401.

_________ (1988). The reversal of the migration turnaround: resumption of traditional trends? *International Regional Science Review* (Morgantown, Virginia), vol. 11, pp. 253-260.

_________, ed. (1989). *Counterurbanization: The Changing Pace and Nature of Population Deconcentration*. London: Edward Arnold.

_________, and Peter D. Congdon (1988). Recent population trends for Greater London. *Population Trends* (London), vol. 53, pp. 11-17.

Champion, Anthony G., and Sven Illeris (1990). Population redistribution trends in Western Europe: a mosaic of dynamics and crisis. In *Unfamiliar Territory: The Reshaping of European Geography*, Michael Hebbert and Jens Christian Hansen, eds. Aldershot, United Kingdom: Avebury.

Cheshire, Paul C., and Dennis G. Hay (1989). *Urban Problems in Western Europe*. London: Unwin Hyman.

Cochrane, Stephen G., and Daniel R. Vining (1986). Recent trends in migration between core and peripheral regions in developed and advanced developing countries. Working Papers in Regional Science and Transportation, No. 108. Philadelphia, Pennsylvania: Regional Science Department, University of Pennsylvania.

_________ (1988). Recent trends in migration between core and peripheral regions in developed and advanced developing countries. *International Regional Science Review* (Morgantown, Virginia), vol. 11, pp. 215-244.

Dogan, Mattei, and John D. Kasarda (1988). *The Metropolis Era*, vol. 1, *A World of Giant Cities*. Beverly Hills, California: Sage.

Fielding, Anthony J. (1982). Counterurbanisation in Western Europe. *Progress in Planning*, vol. 17, pp. 1-52.

_________ (1986). Counterurbanisation in Western Europe. In *West European Population Change*, Allan Findlay and Paul White, eds. London: Croom Helm.

Frey, William H. (1987). Migration and depopulation of the metropolis: regional restructuring or rural renaissance? *American Sociological Review* (Washington, D. C.), vol. 52, pp. 240-257.

_________ (1988). Migration and metropolitan decline in developed countries. *Population and Development Review* (New York), vol. 14, pp. 595-628.

_________ (1989). United States: counterurbanization and metropolis depopulation. In *Counterurbanization: The Changing Pace and Nature of Population Deconcentration*, Anthony G. Champion, ed. London: Edward Arnold.

_________ (1990). Metropolitan America: beyond the transition. *Population Bulletin* (New York), vol. 45, No. 2, pp. 1-49.

_________ (1992). The new urban revival in the United States. Paper presented at the United Kingdom/United States Seminar on Migration in Post-industrial Society, Los Angeles, California, 3 and 4 August.

_________, and Alden Speare (1991). U. S. metropolitan area population growth, 1960-1990: census trends and explanations. Research report, 91-212. Ann Arbor, Michigan: University of Michigan, Population Studies Center.

Fuguitt, Glenn V., C. L. Beale and M. Reibel (1991). Recent trends in metropolitan — nonmetropolitan fertility. Centre for Demography and Ecology Working Paper 91-24. Madison, Wisconsin: University of Wisconsin.

Hall, Peter, and Dennis Hay (1980). *Growth Centres in the European Urban System*. London: Heinemann.

Jones, Philip (1991). The French Census 1990: the southward drift continues. *Geography* (Sheffield, England), vol. 76, pp. 358-361.

Klaassen, L.H., and others (1981). *Dynamics of Urban Development*. Aldershot, United Kingdom: Gower.

Kontuly, Thomas, and Roland Vogelsang (1989). Federal Republic of Germany: the intensification of the migration turnaround. In *Counterurbanization: The Changing Pace and Nature of Population Deconcentration*, Anthony G. Champion, ed. London: Edward Arnold.

Korcelli, Piotr (1984). The turnaround of urbanization in developed countries. In *Population Distribution, Migration and Development: Proceedings of the Expert Group on Population Distribution, Migration and Development, Hammamet (Tunisia), 21-25 March 1983*. Sales No. E.84.XIII.3. New York: United Nations.

McCarthy, Kevin F., and Peter A. Morrison (1978). *The Changing Demographic and Economic Structure of Non-metropolitan Areas in the 1970s*. Santa Monica, California: Rand Corporation.

Mera, Koichi (1988). The emergence of migration cycles? *International Regional Science Review* (Morgantown, Virginia), vol. 11, pp. 269-276.

Salt, John (1988). Highly-skilled international migrants, careers and internal labour migrants. *Geo-forum* (Elmsford, New York and Oxford), vol. 19, pp. 387-399.

_________ (1992). Current and future international migration trends affecting Europe. In *People on the Move: New Migration Flows in Europe*. Strasbourg: Council of Europe.

Stillwell, John, Philip Rees and Peter Boden, eds. (1992). *Migration Processes and Patterns*, vol. 2, *Population Redistribution in the United Kingdom*. London: Belhaven.

United Nations (1992). *World Population Monitoring, 1991: With Special Emphasis on Age Structure*. Population Studies, No. 126. Sales No. E.92.XIII.2.

__________ (1993). *World Urbanization Prospects: The 1992 Revision*. Sales No. E.93.XIII.11.

__________ (1996). *World Population Monitoring, 1993: With a Special Report on Refugees*. Sales No. E.95.XIII.8.

Vining, Daniel R., and Thomas Kontuly (1978). Population dispersal from major metropolitan regions: an international comparison. *International Regional Science Review* (Morgantown, Virginia), vol. 3, pp. 49-73.

Winchester, Hilary P. M., and Philip E. Ogden (1989). France: decentralization and deconcentration in the wake of late urbanization. In *Counterurbanization: The Changing Pace and Nature of Population Deconcentration*, Anthony G. Champion, ed. London: Edward Arnold.

World Bank (1992). *World Development Report, 1992: Development and the Environment*. New York: Oxford University Press.

V. POPULATION DISTRIBUTION IN AFRICA: URBANIZATION UNDER WEAK ECONOMIC CONDITIONS

*Robert A. Obudho**

Until recently, urbanization was not considered a problem in most African countries because it was associated with modernization and industrialization. Both Governments and international donor agencies fostered rural development and agriculturally based strategies without paying attention to the rapid rates of urbanization. Then, national censuses began to show important increases in urbanization. Today, urbanization has been added to the long list of potentially devastating development problems that must be addressed. The fundamental problem is that the urban population is growing very fast while economic growth and the development transformations necessary to support it and enhance the quality of urban life are not occurring as rapidly.

Urbanization estimates indicate that the urban population of Africa is growing four times faster than that of the rest of the developing world. The Governments of most African countries are increasingly aware of the need to ensure viable urban development for sustainable growth. Issues of urban management can only be discussed and properly understood in the wider context of national urbanization policies. The need for active urbanization policies is greater in Africa than it was in developed countries because the African population is growing faster, the rates of economic growth in Africa are low and the role of Government is more pervasive in African countries than in other countries of the world.

Development assistance in Africa has concentrated on rural areas in an effort to alleviate poverty in the continent. In response to limited human and financial resources, international development aid has focused on agricultural production, although Africa's urbanization has been accelerating. The limitations of the agricultural sector with respect to providing adequate economic opportunities to a growing population are now recognized.

The urbanization process is bringing about economic and political changes that affect the ways in which Africans organize spatially and subsist economically. The challenge to Africa's development process is to manage the change that is taking place. The potential contributions of urban centres to national development need to be realized if the success of past gains in agricultural production are to be consolidated. Urbanization provides the production efficiencies that support off-farm employment. Urban centres are the focus of both political and economic decentralization policies and the principal sites for the processing and marketing of agricultural products. The patterns into which the urban population settles today will dictate the standard of living and the possibilities of improving that standard for years to come.

A commitment to an explicit urban policy calls for expanding the activity of international donors in the public, private and informal sectors of urban areas to provide credit, land and infrastructure to low-income families. International donor institutions have an important role to play in supporting the efforts of public and private institutions to manage the urbanization process.

Secondary and small urban centres are at the core of the long-term relationship between urbanization and agricultural development and will play a key role in the region's urbanization. The interdependence of urban and rural populations in Africa is striking. There is a positive correlation between improved urban infrastructure and rising agricultural productivity. Care should be taken in identifying which to support. Secondary and small urban centres should be supported by investment in physical and social infrastructure on the basis of their own growth potential. Public expenditure should be geared to locations where rapid growth is occurring or is very likely to occur soon.

Urban services can only be provided efficiently if there is a partnership between central and local Government,

*Department of Geography, University of Nairobi, Kenya.

with greater responsibilities being given to the latter than has generally been the case. Municipal management must be improved. Decentralization makes better use of local human, financial and physical resources and offers the possibility for more effective resource mobilization, self-sustaining development projects that better reflect local needs, and increased public participation in decision-making.

A. Population distribution and migration: trends and patterns

1. *Population trends*

In 1990, the total population of the world was 5.3 billion of whom 4.1 billion lived in developing countries (United Nations, 1993a). An estimated 43 per cent of the world's population lived in urban areas but the equivalent percentage was 34 per cent in developing countries and 73 per cent in developed countries (United Nations, 1993b). The world's total population was growing at an average annual rate of 1.7 per cent but again there was a large differential in the population growth of developed and developing countries where the annual rates of growth were 0.6 and 2.1 per cent, respectively (United Nations, 1993a). Both differential mortality and fertility contributed to such differences. On average, fertility levels were highest in Africa, where total fertility was estimated at 6.25 children per woman for the period 1985-1990. In the 1980s, fertility remained above 5 children per woman in most African countries (United Nations, 1993a). Such high fertility implied that during 1980-1990 Africa contributed an increasing share of the total number of births in the world. Africa's high birth rates make a major contribution to urban population growth, often higher than that attributable to rural-urban migration or reclassification. In many African countries, 60 per cent or more of urban growth results from the excess of urban births over deaths. Few efforts are thus more important for attaining a more manageable urban population size than to encourage small families by, among other things, improving the status of women and facilitating access to family planning and health services.

In 1990, the world's rural population was 3 billion and is projected to keep on growing to reach a maximum of 3.7 billion persons by the year 2015 when all but 262 million of them will be living in developing countries. In Africa the rural population is expected to keep on growing beyond 2015. Because of such trends, many African countries find themselves in the position of having to absorb large rural populations at the same time as they struggle with rapid urbanization. During 1975-1990, nearly 70 per cent of the total population growth registered in Eastern Africa, 60 per cent of that in Middle Africa, and about half of that in the other African subregions took place in rural areas.

In 1990, Africa accounted for 16 per cent of the rural population of the developing world and is expected to host 24 per cent by 2025 (United Nations, 1993b). It is the only major region projected to increase its share of the rural population worldwide. From 1975 to 1990, 126 million persons were added to the African rural areas and another 289 million persons are expected by 2025. By 2025, the African rural population will be more than twice the size of the entire rural population of developed countries in 1975 (United Nations, 1993b). Because overall rates of population growth are so high in African countries, rural populations are increasing despite substantial rural-urban migration. Population pressure has aggravated the poverty that is endemic in the rural areas of many African countries. Environmental degradation resulting from overcrowding and overgrazing has also contributed to high rates of rural out-migration. Indeed, "push factors" are perhaps as important as "pull factors" in generating rural-urban migration and hence urbanization (United Nations, 1991).

2. *Migration*

Migration in Africa is associated with a series of social, cultural, political and economic factors. Prominent among the social factors is education. Thus, the desire to acquire high levels of education leads people to move to areas where educational facilities are located. In addition, the better educated move to areas offering employment commensurate with their abilities. Other social factors that affect rural-urban migration in Africa include (*a*) the prospective migrants' perceptions of living conditions in urban areas; (*b*) the presence of friends and relatives in urban areas; and (*c*) the existence of expectations among prospective migrants of a rise in social prestige associated with migration.

Political factors have been of increasing importance in causing migration to Africa's urban areas. However, these factors are often interwoven with ethnic and racial considerations. During the pre-colonial era, migration in

Africa was primarily motivated by the desire to conquer. The gradual emergence of a modern economy within the framework of colonial administrative structures changed migration patterns. The colonial urban administration was not under the authority of the African ethnic hierarchy. The political structure was based primarily on non-ethnic considerations. However, although colonial order enhanced the possibility of migration, it also tended to limit it. Most obviously, it stopped movements caused by warfare and conquest.

The most important economic factor usually cited as influencing rural-urban migration in Africa is the urban-rural income differential. Various African Governments concentrated investment in a few urban centres, thus causing regional inequalities which in turn paved the way for rural-urban migration. In such situations, the location of productive activities virtually determines the intensity, pattern and direction of migration. Other economic factors that lead to migration include the existence of high population pressure on land, low rates of investment in agriculture, fragmentation of land ownership, inequalities in the distribution of land and other allocative mechanisms.

Rural-urban migration, while imposing pressures on already strained urban resources, does not necessarily provide relief for rural areas and often increases regional income inequalities. In addition, many African countries are facing serious food shortages requiring the allocation of scarce foreign exchange for imports, although they could well be self-sufficient in terms of food supply, or at least meet a large part of their food needs, if domestic production were supported. Hence, rural development strategies should be pursued to alleviate the negative consequences of rapid urban growth.

B. Consequences of migration and urbanization

Before discussing some of the consequences of the migration and urbanization process in Africa, it is important to highlight some of the human development gains and losses that Africa has recorded since the end of the colonial era. Between 1960-1965 and 1985-1990, infant mortality rates have fallen by 36 per cent and life expectancy has increased from 42 to 52 years (United Nations, 1993a). Adult literacy in Africa had increased by two thirds between 1970 and 1985, but economic growth was slow in the 1980s and the population has been rising at a rate of 3.2 per cent per year. As a result, gross national product (GNP) per capita fell by an average of 2.2 per cent per year in the 1980s (United Nations Development Programme, 1991 and 1992).

Despite improvements, under-five mortality still stands at 178 deaths for every 1,000 live births. More than half of the population has no access to public health services. Almost two-thirds lack safe water. Tropical diseases afflict a high proportion of the population. Unemployment and underemployment is one of the most serious problems. Real wages fell by 30 per cent between 1980 and 1990. Many people have left rural areas, partly because the urban centres offer better prospects. But people have also been driven from the land by population pressure and the degradation of the soil. Family budgets in the region are tight. The net primary enrolment ratio for girls is 44 per cent compared with 54 per cent for boys, and female literacy stands at only 34 per cent compared with 56 per cent for males (United Nations Development Programme, 1991). In several African countries, these problems have been compounded by political violence stemming from cross-border conflicts, ethnic upheavals and civil strife, all of which have created 5.4 million refugees (United Nations High Commissioner for Refugees, 1993).

Displaced persons exist in virtually all the major urban centres in Africa. These include beggars, parking boys and the mentally ill who are unwanted and rejected as participants in urban development (Obudho, 1991a). Rural poverty undeniably affects a greater number of persons and is more severe than urban poverty. However, urban poverty cannot be ignored. In Africa, approximately one in every six urban dwellers is living in poverty. Whatever the share of the urban poor in the urban population, their plight is serious.

The high rate of urbanization in Africa has become a major concern in many countries because of the many sociocultural, economic, political and environmental problems associated with rapid urbanization. Such problems are especially acute because urbanization has resulted in the excessive growth of the major urban centres. Although rural-urban migration is a major contributor to such growth, the contribution of high natural increase is also of key importance (Obudho and Aduwo, 1989). Migration to urban areas is deemed to be the result of unjustified development strategies that favour urban areas, especially in market-oriented economies where investment has been increasingly concentrated in a few rapidly growing urban centres.

1. *Consequences for rural areas*

In rural areas, rural-urban migration affects incomes, rural capital formation and technological change, modes of rural production and the fertility of the rural population. Remittances from urban areas raise rural incomes, increase levels of consumption and, by encouraging technological change, further raise rural incomes. However, the increase depends on the volume of migration, its composition, its destination and the socio-economic status of migrants. Most rural-urban migrants are adults, and generally are better educated than the average rural dweller, hence their movement involves sizeable transfers of human capital from the rural to the urban sector that may adversely affect agricultural productivity and incomes (Hanna and Hanna, 1971).

Where rural out-migration is temporary, the absence of men for extended periods may increase the work burden of the family members left in rural areas and leave the family more vulnerable to unexpected crises. Male rural-rural migrants sometimes replace female workers who will then be restricted to working within the home. Although migrants send remittances to rural areas, in return they often receive help from rural relatives, particularly in the form of food.

A declining labour-to-land ratio often provides an opportunity to change rural production techniques; the transfer of labour to more productive urban activities eventually generates a growing demand for rural output and thus alters the rural-urban terms of trade, raising agricultural prices relative to those of urban goods. The increase in agricultural prices is likely to stimulate agricultural production and raise rural income. Remittances from urban areas are also likely to raise rural incomes and may not only increase levels of consumption but also encourage technological change that further raises rural incomes.

Out-migration from rural areas is likely to push up wage rates and encourage labour-saving technological change or greater work participation by the remaining family members. Technological change would also be stimulated to the extent that out-migrants allocated savings to rural areas in the form of remittances. Remittances may be used for productive investment or better housing and education. Technological change can be attributed to the dynamism of returning migrants, who bring money as well as knowledge and experience of alternative production techniques.

If migration is concentrated among both the fairly rich and the fairly poor, income inequality may tend to grow. However, if the very poor migrate as whole families pushed from rural areas by debt and loss of land, the beneficial effect on wages may reduce rural income inequalities. If migrants from richer households predominate and remittances go to relatively prosperous farmers who can introduce technological innovations to produce higher output and incomes, migration will eventually increase inequality in income and land distribution thus inducing further out-migration.

Rural out-migration tends to be associated with greater reliance on wage labour. The out-migration of young adults changes the age composition of the rural population and often raises the value of the labour of those staying behind to work in family farms, prompting them to hire wage-labour. Migration may also lead to the commercialization of agricultural activity, which is further encouraged by favourable changes in commodity terms of trade and the extension of markets.

Migration affects the distribution of income in rural areas and income distribution is an important determinant of fertility, therefore migration may affect fertility. In addition, the out-migration of young unmarried men of working age may result in severe sex imbalances in rural areas and influence the proportion of persons able to find marriage partners. Furthermore, the large-scale emigration of men in search of urban employment frequently contributes to the dissolution of already existing marriages or delays family formation. Lastly, returning migrants may spread new values and information about family planning and may introduce new family-size ideals.

2. *Consequences for urban areas*

Migrants may lower wage rates in urban areas with this thus leading to the expansion of employment; they may introduce dynamic elements into the urban informal sector thereby providing a catalyst for development; or they may sustain a higher rate of economic growth because of their relatively higher propensity to save, their lower rate of absenteeism and the longer hours they tend to work. Migrants to urban areas may have lower activity rates than locals because of discrimination based on ethnicity or religion and because they are less likely to have the same support of friends and relatives at the place of destination as they had at the place of origin.

The influx of migrants into urban areas increases the demand for infrastructure and community services. Most urban centres are unable to meet the growing demand for these services, since the latter often require large investments. However, migration may also increase the size of an urban centre thus producing economies of scale.

Different groups of migrants will enter the various segments of the urban labour market and hence have different job mobilities. The relatively young and more productive individuals who tend to dominate migration flows will be an asset for the area of destination. In addition, migration of young persons will lower the crude death rate of urban areas and will likely increase the proportion of women in the main child-bearing ages. To the extent that fertility in urban areas is lower than that in rural areas, over the long run urbanization may contribute to the reduction of overall fertility.

While colonial rule led to the establishment of infrastructure and public sector employment, it was not complemented by the expansion of manufacturing and commercial activities in most African urban centres. Indeed the modern sector of most African economies has been heavily dominated by public employment. Such modern sector employment is scarce but it continues to attract the better-educated migrants from rural areas who often remain unemployed. Most of them end up working in the informal sector which is hard put to sustain an ever-increasing population.

Unless positive measures are taken, the rapid increase of the urban population in Africa is likely to exacerbate the prevailing ills associated with the rapid expansion of slum and squatter settlements. Such settlements constitute the living environment for at least one third of the urban population of all African countries and they are growing at a rate that has doubled in the last six years (Obudho, 1987; and Obudho and Mhlanga, 1988). The growing gap in housing needs faced by most major urban centres in Africa has been and will continue to be filled by slum and squatter settlements. In Africa, despite the considerable economic progress made since independence, the general standard of housing is still unsatisfactory for most of the urban population. In many urban areas, there are overcrowding and unauthorized construction of unplanned dwellings built with unsuitable materials. Combined with lack of basic infrastructure such as water supply, sewage and roads, the growing slums and squatter settlements create an unacceptable environment ripe for the spread of disease or the outbreak of fires, violence and other urban disasters. The need for a strategy to build low-income housing has been recognized and attempts have been made towards improving housing for the urban poor, especially through site and service schemes and upgrading efforts (Obudho and Mhlanga, 1988).

Urban centres in Africa have been unable to supply education, health and transport facilities to meet the increasing demand. Although growing demand has put pressure on local authorities, deterioration in the provision of those services continues. Debilitating conditions prevalent in urban Africa include malnutrition and insufficient medical services. Urban transport has also deteriorated as congestion, poor roads and lack of enough terminal facilities prevail. Underlying these problems is the lack of funds available to African authorities. The modern sector is faced with serious constraints including limits on public sector expansion, high domestic resource costs for manufacturing, small national markets and scarcity of credit.

C. Population distribution policies

In 1990, more than 7 out of every 10 countries in Africa considered that their patterns of population distribution required major change, only 13 per cent considered their population distribution to be satisfactory and a further 15 per cent desired minor changes (United Nations, 1992).

In Northern Africa, the Algerian policy is to adjust patterns of population distribution by decentralizing economic activities, promoting small and intermediate urban centres and regrouping rural villages into 1,000 cooperatives. In Eastern Africa, the measures taken by Burundi to redistribute the population have included the creation of rural employment, and the construction of schools, hospitals and health centres as well as the improvement of transportation in rural areas. Ethiopia has sought to adjust population distribution through the resettlement of population from densely populated regions, villagization, rural development and the promotion of growth in urban centres other than Addis Ababa and Asmara. In Malawi, integrated rural development and the creation of rural growth centres are intended to reduce migration to urban areas by improving the level of living in the countryside. In Middle Africa, Angola began setting up a network of agricultural development stations in the mid-1980s in order to provide peasants with inputs, credit, transport and technical support. In Southern Africa, Lesotho encourages the development of regions that are economically deprived, such as the mountain zone and the drought-stricken lowlands of the southern region. To do so, it has undertaken rural road construction, educational programmes, employment creation and water-supply and sanitation programmes. It has also supported the establishment of cottage industries. In Western Africa, Burkina Faso is concerned mainly with rural-rural migration and in Côte d'Ivoire, the major policy goals of the Government are to control north-to-south migration as well as migration to the larger cities and to populate the sparsely settled savannah zone (United Nations, 1992).

Population distribution policy issues are important for many countries, mainly because in countries characterized by strong primacy the spatial distribution of population has generated conditions that conflict with important societal goals such as interpersonal and interregional equity, national security, political stability and integration, improvement in the quality of life, optimal resource exploitation and long-run economic efficiency.

Regarding population policy, there are grounds to suggest that population distribution is the main current and future population issue. However, the basis for formulating population distribution policies remains weak. Countries differ widely in their patterns of population distribution, levels of urbanization, degree of primacy and rates of urban population growth. They also vary in their economic and political systems. Hence, it is hardly surprising that there are major differences in the content and design of the population distribution policies implemented by different countries.

D. The urbanization process in Africa

1. *Trends and prospects*

Urbanization has not been an entirely modern development in Africa, since some of the world's earliest urban centres were located in parts of Western, Northern and Eastern Africa. In Africa, the capitals of some precolonial kingdoms date back to the tenth and eleventh centuries. However, it was not until the end of the nineteenth and the beginning of the twentieth century that many of Africa's major urban centres developed, though most of them remained small for several decades. Since the Second World War, the pace of urbanization in Africa has accelerated markedly and is expected to continue to do so in most African countries for some time to come (Hance, 1970; El-Shakhs and Obudho, 1974; Obudho and El-Shakhs, 1979; O'Connor, 1982 and 1983).

Africa is the least urbanized region of the world, with 32 per cent of its population living in urban areas in 1990. In 1990, 206 million persons lived in the urban centres of Africa, accounting for 15 per cent of all urban dwellers in developing countries. Thirty-eight per cent of the population of Africa will reside in urban areas by the year 2000 and 54 per cent by 2025 (United Nations, 1993b). There is significant variation in the level of urbanization among the different subregions of the continent. In 1990, the proportion urban amounted to only 19 per cent in Eastern Africa, 32 per cent in Middle Africa, 33 per cent in Western Africa, 44 per cent in Northern Africa, and 46 per cent in Southern Africa (table 12). The general direction of these differentials at

Table 12. Percentage of the population of Africa and its subregions living in urban areas, 1975-2025

Subregion	*1975*	*1980*	*1985*	*1990*	*2000*	*2025*
Africa	25.0	27.3	29.6	32.0	37.6	54.1
Eastern	12.3	14.6	16.9	19.1	24.5	41.3
Middle	26.6	28.2	29.8	31.9	37.3	55.5
Northern	38.4	40.2	42.1	43.8	48.4	63.7
Southern	44.1	44.5	45.2	46.2	50.6	66.4
Western	22.7	26.0	29.5	33.2	40.8	59.3

Source: *World Urbanization Prospects: The 1992 Revision* (United Nations publication, Sales No. E.93.XIII.11), table A.1.

the subregional level is expected to be maintained until 2025. Several African countries, including Burkina Faso, Burundi, Ethiopia, Guinea-Bissau, Lesotho, Malawi, the Niger, Rwanda and Uganda, are still barely urbanized, having had less than 20 per cent of their populations living in urban areas in 1990.

Growing at 4.5 per cent per year during 1985-1990, the urban population of Africa has the highest rate of growth in the world. Since 1950-1955, the average annual rate of growth of the urban population in Africa has varied between 4.4 and 4.8 per cent per year (United Nations, 1993b). As of 1990, however, the rate of urban growth in the continent as a whole is expected to decline. Nevertheless, even by 2020-2025, the urban population in Africa is expected to grow at the rate of 3.4 per cent per year, about six times the equivalent rate for the developed countries. Urban growth rates are high for every region of Africa and especially for Eastern and Western Africa where they exceeded 5.5 per cent per year in 1985-1990 and are expected to remain high until the end of this century (table 13).

During 1990-2000, 55 per cent of the population growth in Africa will occur in urban areas, amounting to an additional 117 million persons. During 2000-2025, the equivalent figure will be 535 million or nearly three quarters of total population growth. With its low levels of urbanization, Africa is experiencing a high urbanization rate, estimated at 1.6 per cent per year for 1990-1995. The urbanization rate in Africa is expected to peak at 1.65 per cent in 1995-2000 and then decline to 1.2 per cent by 2020-2025 (United Nations, 1993b).

The urban population in Africa increased from 33 million in 1950 (constituting 15 per cent of the total population) to 164 million (30 per cent) in 1985 and it is likely to reach 857 million (54 per cent) by 2025. Until 1980, Africa had the lowest level of urbanization in the world and is expected to maintain that rank until the end of this century (United Nations, 1993b).

Continental comparisons of urbanization conceal important subregional and national differences. Low levels of urbanization are characteristic of most regions of Africa, though Northern and Southern Africa are considerably more urbanized than the rest. Based on current estimates and projections, rural residents in Eastern Africa are still expected to outnumber urban dwellers by the year 2025. In contrast, already by the year 2000 about 51 per cent of the population of Southern Africa is expected to be living in urban areas. By 2025, at least 60 per cent of the population of Northern and of Southern Africa is expected to live in urban areas, whereas the corresponding proportion will be 59 per cent for Western Africa, 56 per cent in Middle Africa and only 41 per cent for Eastern Africa (table 12). The growth rates of the urban population are highest, however, in Eastern and Western Africa: about 6 per cent and 5 per cent, respectively, during the 1990s (table 13).

In Northern Africa, all countries except the Sudan are highly urbanized. In 1960, about one quarter of the population of the region lived in urban centres of 20,000 inhabitants or more, compared with 13 per cent in the whole of Africa, and the proportion of the population living in urban centres with 100,000 inhabitants or more was 13 per cent against 10 per cent for Africa as a whole. In 1990, the Libyan Arab Jamahiriya was the country with the highest proportion of the population living in urban areas (82 per cent), followed by Western Sahara, Tunisia and Algeria, in order of importance, all with more than 50 per cent of their population urban (United Nations, 1993b). The only country in the region

TABLE 13. AVERAGE ANNUAL RATE OF CHANGE OF THE URBAN POPULATION FOR AFRICA AND ITS SUBREGIONS, 1975-2025

Subregion	*1975-1980*	*1985-1990*	*1990-1995*	*1995-2000*	*2020-2025*
			(Percentage)		
Africa	4.59	4.51	4.53	4.46	3.39
Eastern	6.50	5.55	5.62	5.44	4.22
Middle	3.98	4.39	4.64	4.62	3.83
Northern	3.61	3.42	3.41	3.42	2.41
Southern	2.89	3.01	3.22	3.36	2.26
Western	5.73	5.52	5.32	4.97	3.49

Source: *World Urbanization Prospects: The 1992 Revision* (United Nations publication, Sales No. E.93.XIII.11), table A.5.

with a proportion urban below the African average of 32 per cent was the Sudan (with 22 per cent urban). Morocco exhibits the greatest concentration of the urban population, with 80 per cent living in urban centres of 100,000 inhabitants or more. It is followed by Egypt, the Libyan Arab Jamahiriya, Algeria, Tunisia and the Sudan, in order of importance. In Northern Africa, there are over 213 urban centres with at least 20,000 inhabitants. Almost half of them are located in Egypt. Cairo is the most populous urban centre in the continent (Obudho, 1991b and 1991c).

In Western Africa, the evidence indicates the existence centuries ago of important urban centres, particularly in Abeokuta, Iwo, Oyo, Ibadan and Oshogbo, to mention only a few. However, most urban centres in Western Africa were established during the colonial period. Western Africa exhibits relatively low urbanization levels. Côte d'Ivoire, Liberia and Mauritania are the countries with the highest proportion urban in the region (over 40 per cent in 1990). At the other end of the scale, Burkina Faso, Guinea-Bissau, the Niger and St. Helena all had less than 20 per cent of their populations living in urban areas in 1990 (United Nations, 1993b). There are over 44 urban centres in Western Africa with 100,000 inhabitants or more and over 71 per cent of them are located in Nigeria. Lagos is the largest urban centre in the region. Among urban centres with at least 100,000 inhabitants, those in Côte d'Ivoire, Ghana, Nigeria, Senegal and Sierra Leone grew fastest (Obudho, 1991b and 1991c).

In 1990, Middle Africa exhibited intermediate levels of urbanization, with 32 per cent of its population living in urban areas (table 12). Cameroon, the Central African Republic, the Congo, Gabon, and Sao Tome and Principe all had over 40 per cent of their population living in urban areas (United Nations, 1993b). Only 10 per cent of the population lived in urban centres of 20,000 inhabitants or more, and only seven countries in the region have urban centres with at least 100,000 inhabitants, namely, Angola, Cameroon, the Central African Republic, Chad, the Congo, Gabon and Zaire. Among these, the Congo is the country with the highest proportion of its population in urban centres with more than 100,000 inhabitants, the Central African Republic 10 per cent, Zaire 6 per cent and Cameroon 4.4 per cent, whereas the proportion for the whole of the region is 4 per cent. Among the 18 large urban centres in the region, 11 are located in Zaire and the largest are Kinshasa and Katanga (Obudho, 1991b and 1991c).

Eastern Africa has the lowest level of urbanization despite the fact that it is one of the most densely populated regions in Africa (Obudho, 1987). However, small countries in the region like Djibouti, Réunion and Seychelles have relatively high urbanization levels (near or above 60 per cent in 1990). At the other extreme, Burundi, Ethiopia, Malawi, Rwanda and Uganda had levels of urbanization below 13 per cent in 1990 (United Nations, 1993b). The low urbanization of the region in general can be attributed to the fact that, during the colonial period, Africans were prevented from residing permanently in urban areas. In addition, most of the economies of the Eastern African countries are based on agriculture and there are few large-scale industrial or mining complexes. The urban population is growing very fast in Eastern Africa at an estimated annual rate of 5.6 per cent in 1990-1995. Yet, the percentage of the population living in large urban centres remains low (Obudho, 1993). In those countries with urban centres of 100,000 inhabitants or more, a high proportion of the urban population is concentrated there. In Malawi, for instance, that proportion is 100 per cent, in Kenya 88 per cent, in Somalia 75 per cent, in Ethiopia 74 per cent and in Madagascar 58 per cent. Only in the mainland of the United Republic of Tanzania does less than half of the urban population live in large urban centres. The largest urban centres in the region are Addis Ababa, Dar es Salaam and Nairobi (Obudho, 1991b, 1991c and 1993).

In Southern Africa, a large proportion of the urban population is found in urban centres having at least 20,000 inhabitants. However, the bulk is concentrated in the Republic of South Africa, where nearly 60 per cent of the population lived in urban areas in 1990. Of the 15 urban centres in Southern Africa with populations exceeding 100,000, 11 are in the Republic of South Africa. Johannesburg is one of the eight African urban centres with more than 1 million inhabitants and the Republic also has the other three most populous urban centres in the region, namely, Cape Town, Durban and East Rand. Urbanization in Southern Africa is high because of the large industrial complexes and mining activities in the region.

2. *The problems faced by urban areas in Africa*

Although the level of urbanization in Africa is still relatively low, the continued growth of the urban population poses serious developmental problems. Rapid urban growth is determined by rural-urban migration, high natural increase in urban areas and the reclassification of previously rural areas as urban. There are also non-spatial factors that have significant impacts on the form, nature and extent of urban growth, including fiscal, industrial, defence, equalization, agricultural and immigration policies. Experience has shown that the effects on urban growth of some of those policies may be more significant than those of policies explicitly directed to influence urban growth. Consequently, it is important for Governments to evaluate the impact that non-spatial policies may have on urban growth. Of special importance is the rapid growth of major urban centres whose role has been primate and parasitic in the sense that they have continued to attract development projects and part of the rural population at the expense of other areas.

Given that in Africa major urban centres have been growing at the expense of small and intermediate urban centres, the increasing concentration of population has resulted in practical administrative difficulties related to the provision of public services to a rapidly growing population. There has been a failure to predict and plan for urban growth. Local authorities have been unable to grasp the implications of a population that doubles every nine years. Problems of accessibility to services are compounded by the fact that most urban residents earn low incomes and are not able to pay for the services they need.

Despite the deficiencies in their basic infrastructure and social services, urban centres in Africa are the core of modern economic life and offer opportunities to migrants. Economic growth and economic activity rates in the main urban centres are high, even against a background of poor infrastructure and inefficient urban management. Urban centres are incubators of innovation, new enterprises and are crucial sources of employment. The value added by the urban workforce is generally higher than that added by the rural workforce. In Kenya, for instance, one third of the gross domestic product (GDP) originates in Nairobi. Urbanization is positively correlated with the growth of modern manufacturing in Africa and there are also many instances of positive association with agricultural growth because urban centres are both the distribution points for agricultural inputs and the major markets for agricultural produce. Urban centres are also the financial centres and the source of funds for development. The non-agricultural sector usually pays a larger share of taxes than the agricultural sector in most African countries (Bird, Marsara and Miller, 1985), although agricultural employment as a proportion of the total is high and industrial employment low (De Cola, 1986).

The rate of growth of the urban population of Africa has continued to increase even as the rate of growth of GNP per capita declined from 1.3 per cent in the 1960s to 0.7 per cent in the 1970s and even further in the 1980s. Migration has been fuelled by macroeconomic and spatial policies that have favoured urban dwellers. In a number of African countries, however, the growth rate of the primate urban centre has started to decline in recent years, while the growth rates of secondary urban centres have been exceeding that of the primate city (Baker, 1990; Obudho and Aduwo, 1990; Baker and Pedersen, 1992).

Africa suffers from a severe deficit in terms of the quality, quantity and maintenance of urban infrastructure. In Zaire, for instance, urban infrastructure and services reach less than one half of the urban population. The infrastructure of Kinshasa and secondary urban centres was meant to serve only a fraction of the current population. Furthermore, the existing infrastructure is eroding from lack of maintenance. Urban centres in most African countries face similar problems and, if conditions in the primate urban centres are deficient, secondary urban centres face even more severe deficits (Hamer, 1986). Almost no urban centre in Africa has managed to expand infrastructure to accommodate population growth. Nairobi has achieved more than most. In Nairobi, the problem of maintaining the existing system is the major issue and maintenance has often been neglected. Half of Nairobi's population lacks access to basic urban services (Obudho, 1988).

There are also severe weaknesses in inter-urban and urban-rural linkages. For instance, the condition of roads between Kumasi in Ghana and its surrounding rural areas is so bad that transportation costs are about twice as high as they should be. Many African countries have adequate road networks, but the roads are in poor condition. Thus the "costs" of poor infrastructure affect the African economy in numerous ways. Enterprises that might have come into being have not done so. Those that exist are suboptimal in size, utilize inappropriate technology or face higher-than-necessary operating costs. Clearly, when enterprises themselves have to provide infrastructure their burdens increase. As a consequence, enterprises are not able to expand to the size necessary to take advantage of economies of scale. In addition, lack of infrastructure in small and intermediate urban centres is a major factor inhibiting the realization of urban-rural linkages; but in large urban centres losses stem from an inability to take advantage of so-called economies of agglomeration.

Most African economies are undergoing "structural adjustment", which usually consists of "setting the prices right", better fiscal and monetary discipline, encouraging private savings and investment, and strengthening economic management in general. Part of structural adjustment is likely to change the terms of trade between urban and rural-based economic activities and households. Agriculture will tend to receive more attention under the structural adjustment regimes than in the recent past since one goal is to remove "urban bias" in national and regional development planning.

Kenya's policies, for example, have always been relatively free from urban bias. In Kenya, both Nairobi and small intermediate urban centres have shared urban growth and it is felt that this growth has not been induced by urban price distortions. Urban areas do not receive food subsidies, urban wage increases are controlled, and public transit fares and water tariffs are not subsidized. Kenya has always had a spatial investment strategy whereby resources are allocated on the basis of population. There are, however, cases of subsidy by default; while water fees are not actually subsidized, they are frequently not collected. Similarly, to the extent that property taxes are not collected, urban land and service consumption is subsidized (Obudho, 1988).

The development of Africa's medium-sized urban centres will be the focus of the next phase of urbanization since such urban centres are the spatial centres linking urbanization and agricultural development. The interdependence of urban and rural populations in Africa is striking. An estimated 70 per cent of urban residents maintain strong linkages to the rural sector. In some urban centres, up to 90 per cent of the inhabitants may be linked to the rural sector. Studies of small and intermediate urban centres in Sierra Leone have shown that such urban areas have grown as a direct consequence of the commercialization of agriculture in their surrounding region (Gibb, 1984). Almost all African countries are at a stage of development where the majority of the work-force will be engaged in agricultural activities for many decades. Some secondary urban centres are growing very rapidly because the agricultural areas to which they are linked are prospering. More attention should be given to the role of secondary urban centres in agricultural processing, marketing, storage and distribution.

It is important to understand the nature of urban-rural linkages by considering, *inter alia*, the demand by the rural population for non-food goods, the inputs and services needed by the agricultural sector, and the demand for food by urban dwellers. The first two depend on the demand for agricultural output and the first is highly income-elastic. If policy reforms now under way increase the demand for locally grown food and leave a larger proportion of the export crop surplus in rural areas, then there will be higher rural demand for non-food goods, agricultural inputs and services. In fact, increasing the demand for the last two is necessary for an appropriate supply response on the part of farmers to increases in farm prices.

Increases in rural incomes brought about by improved accessibility to markets and rising agricultural productivity lead to higher levels of activity in secondary urban centres. The rural poor have a high average propensity to consume. Low-income households consume products and services produced locally rather than from distant urban centres. Therefore, increased rural consumption due to increased income will tend to diversify the economic activities of nearby urban centres and create substantial off-farm employment opportunities.

Increasing demand for agricultural inputs and services may be weakened by the removal of subsidies for urban food consumption. Improvements in agricultural productivity will probably owe much to expanding local markets. It is the changing distribution of population between urban and rural areas that will have the most profound impact on agricultural incomes in the long run (Rondinelli, 1983). Over the next 25 years or so, almost all African farms will come under the influence of urban markets. This spread of urban markets should greatly increase the disposable income of farmers. Such an increase will be more rapid than that anticipated for formal or informal sector activities in urban areas. In fact, urban wages are expected to fall in real terms and this has already happened in some urban centres because of the removal of urban subsidies. Hence, the growth of urban centres and the linkages between them and rural areas will support improvements in agricultural productivity and thus help to improve macroeconomic performance. Barriers to the realization of effective urban-rural linkages in Africa include (*a*) overvalued exchange rates and low administered food prices; (*b*) reliance on parastatal companies; (*c*) lack of access to credit; (*d*) lack of transportation networks between and within urban centres; (*e*) lack of market information; (*f*) weak technological support; (*g*) lack of infrastructure; and (*h*) insufficient local institutional strength and ability to generate local revenues (Cohen, 1981; Becker and others, 1985; Hamer, 1986).

Secondary urban centres should be supported by investment in physical and social infrastructure on the basis of their own potential rather than by syphoning off capital to promote "territorial equity". Public expenditure should be geared towards intermediate urban centres where rapid growth is already occurring or is very likely to occur as a result of structural adjustment and changes in macroeconomic and sectoral policies. The best measure of "need" is economic activity levels' being ahead of service provision by the public sector as, for instance, in Kumasi, Ghana. However, the satisfaction of such needs will require all the public sector resources likely to be available in the foreseeable future, given the very high urban population growth rates in many locations. Economic activity levels may be stimulated by public sector investments but in the end local prosperity will depend upon private sector initiative. The urban sector also has a contribution to make to human resource development: effective training programmes are a key to successful decentralization efforts; health is partly an infrastructure issue—not only as a matter of medical services but also of improved water supply and sanitation; high population growth rates are slowed best where family planning programmes are most successful (Obudho, 1983).

The "growth pole" policies advocated during the 1950s and 1960s largely failed to stimulate the growth of secondary urban centres or link them more closely to their surrounding rural areas (Rondinelli, 1983). To the extent that these policies led to the simple transplantation of capital-intensive and export-oriented industries, they failed to benefit the majority of the surrounding population or to strengthen urban-rural linkages. In addition, considerations of "territorial equity" led to investments in infrastructure in areas that lacked real potential and were not attractive to business. Among the most significant forms of support that Governments can give to emerging small and intermediate urban centres and their surrounding areas is that for improvement of their accessibility. An important contributor to the success of Kenya's secondary urban centres is a good national road network.

Development can be characterized according to four stages identified in terms of agricultural and urban activities. The first stage involves a subsistence economy in which almost everyone is engaged in agricultural production and where markets are local, as in Rwanda today. The second stage occurs when local markets emerge and expand, and agricultural productivity begins to rise. Most of the labour force is engaged in agriculture and urban primacy is high. The third stage is characterized by increased agricultural productivity and an increased urban population representing the major market for produce as, for instance, in Zimbabwe. The fourth stage involves increasing industrialization, with most people living in urban centres, as in present-day Mauritius, which is probably the only African country to have reached such a stage. In countries that have reached the fourth stage, the service sector is strong and agricultural and industrial productivity is high, urbanization is high and primacy has been reduced. In order for countries to progress from the first to the second stage, investment in urban infrastructure and building of linkages with rural areas are required. To move into the third stage, small and intermediate urban centres must be promoted as major centres of economic activity linking the urban and rural sectors. The decentralization of governmental functions should be initiated in the third stage and emphasized in the fourth. Donors should then focus on the development of urban centres.

It has been argued that decentralization is not a substitute for the more efficient and equitable use of resources within urban centres (Renaud, 1981). Decentralization policies must encourage the growth of the farm sector as well as the growth of small and intermediate urban centres (Obudho and Aduwo, 1989). The more equal the distribution of income and assets in rural areas, the greater the benefits of decentralization policies. Secondary urban centres can be strengthened through appropriate transport policies, and the creation of industrial estates and of networks joining intermediate urban centres with the regional capital (Obudho, 1983).

E. Africa's mega-cities and large urban centres

In 1950, there were 76 urban centres with over 1 million inhabitants, most of which were in the developed countries. By 1980, 35 per cent of the world's urban population lived in urban centres with more than 1 million inhabitants. The number of such urban centres is expected to increase to 440 by the year 2000 when they will house 43 per cent of the world's urban population. Africa's share of large urban centres is expected to increase from 14 per cent in 1950 to 46 per cent in the year 2000. In that year, Africa will contain the largest number of large urban centres in the world. Although in 1990 only one of the 33 large urban centres was in Africa, urban centres in Africa are growing faster than those in developed countries ever did (Population Crisis Committee, 1990a).

The problems related to the high rates of urban growth in Africa are often further accentuated by the concentration of population in the largest metropolitan areas. To the extent that those large urban centres are the focus of development, they act as a magnet for migrants from both rural and other urban areas. The result of this process is to so increase the concentration of the urban population in one large metropolitan area as to form what has been called a primate urban pattern. The growth of large urban centres and the prospects of their continued expansion rank among the most pressing urban problems in Africa. The size of large urban centres, in addition, magnifies the problems of income and development discrepancies usually inherent in primate urban patterns.

The largest urban centres in Africa are faced with increasing problems as population growth outruns investment in urban infrastructure. A study by the Population Crisis Committee (1990a and 1990b) points out some of the typical problems of the eight largest cities in Africa (table 14). In most African mega-cities, households spend over 40 per cent of their income on food. Housing conditions are typically crowded, given generally high population densities. In Johannesburg and Lagos, for instance, an average of five to six people per room was reported. In Kinshasa and Johannesburg, at least two thirds of the households lacked water or electricity; in Lagos and Cape Town, over 40 per cent. In most African mega-cities, there was less than one telephone for every 10 persons. In Johannesburg and Lagos, at most one third of adolescents aged 14-17 were in school and, although the percentages of school enrolment among that age group were higher in other mega-cities, the highest proportion, recorded by Algiers, amounted to only 67 per cent. As regards infant mortality, Cape Town and Johannesburg have achieved relatively low levels (18 and 22 infant deaths per 1,000 live births, respectively). The two African mega-cities with the highest infant mortality (over 80 infant deaths per 1,000 live births) were Lagos and Kinshasa. Murder rates were high in Alexandria, Cairo and Cape Town (near or above 50 per 100,000 persons annually), but data on infant mortality were generally not available for the African urban centres surveyed. Similarly, measures of air pollution were available only for Cape Town and Johannesburg. Yet in most African mega-cities rush-hour traffic moved slowly (at a speed of less than 25 miles per hour) thus contributing to air pollution.

F. Towards a comprehensive population distribution strategy

Implicit in the above discussion is an awareness that organizational structure in Africa has inhibited the planning process. There are no carefully coordinated strategies focusing on urban problems. There is too much emphasis on the provision of services and too little effort made to involve people and their resources in planning. Secondary urban centres are becoming increasingly important centres of opportunity for migrants and catalytic nodes that link rural areas to the national economy (Rondinelli, 1983; Obudho and Aduwo, 1989). Urban policy should direct its attention to these intermediate urban centres since they have played an important role during the last few decades and will continue to do so. Their rapid rates of growth both of population and of formal employment indicate their dynamism (Obudho, 1983; Baker, 1990). African Governments have stimulated the growth of these urban centres and their impact on their rural surroundings by promoting industrial dispersion and administrative decentralization. Successful policies must be based on the bottom-up approach to planning (Obudho, 1983).

Urban residents have benefited from the urban bias of many government policies. Yet, the reduction or elimination of such biases will be politically difficult, because the urban elite controls political power. Africa has shortages of all the elements needed to formulate and implement policies, including trained manpower, fiscal resources, and executive and administrative organs. For

TABLE 14. SOCIAL AND ENVIRONMENTAL INDICATORS FOR AFRICA'S LARGEST CITIES

Metropolitan area	*Population (thousands)*	*Public safety: murders per 100,000 people*	*Food costs: percentage of income spent on food*	*Living space: persons per room*	*Housing standards: percentage of houses with water/ electricity*	*Communications: telephones per 100 people*	*Education: percentage of children in secondary school*	*Public health: infant deaths per 1,000 live births*	*Peace-and-quiet levels of ambient noise (1-10)*	*Traffic flow: miles per hour in rush hour*	*Clean air/ alternative pollution measurement*	*Urban living standards score*
Cairo, Egypt	11 000	56.4	47	1.5	94	3	53	53	7	12.4	..	35
Johannesburg, South Africa	4 600	19.8	..	5.0	28	6	25	22	3	38.8	1 200 PPM 0_3[a]	46
Lagos, Nigeria	4 000	..	58	5.8	50	1	31	85	7	17.4	..	19
Alexandria, Egypt	3 640	49.3	51	1.5	98	4	47	38	3	16.6	..	42
Kinshasa, Zaire	3 200	..	63	1.7	33	0	60	86	4	18.6	..	29
Casablanca, Morocco	2 900	..	44	2.3	84	3	53	56	5	43.5	..	48
Algiers, Algeria	2 685	..	55	2.5	100	8	67	32	3	18.6	..	48
Cape Town, South Africa	2 425	64.7	34	1.2	58	18	66	18	3	23.0	1 030 PPM 0_3[a]	52

Source: Population Crisis Committee, *Cities — Life in World's 100 Largest Metropolitan Areas* (Washington D.C., 1990), p. 10.
[a]Parts per million ozone.

that reason, programme objectives often cannot be achieved within established time limits and it is difficult to pursue improvements at the national and local levels.

Policies to develop new urban centres have been pursued by some African countries. Political considerations have led to the establishment of new urban centres such as Dodoma in the United Republic of Tanzania and Abuja in Nigeria. Egypt has also embarked on a policy to channel migrants to six new urban centres constructed in the desert areas and to several new development regions in the desert. The development of "growth poles" was aimed at directing urban-bound migrants to small and intermediate urban centres. Such policies assumed that initial government expenditures in land and infrastructure would lead to the self-sustaining economic growth of disadvantaged regions. Although many economic plans in Africa have identified potential growth centres and mechanisms to encourage their development, industrial growth-pole schemes implemented to date have experienced many problems mainly because direct outlays and subsidies for growth-pole ventures have been draining for Governments (Baker, 1990).

"Growth-pole policies" have been aimed at reducing subnational imbalances in per capita income or production and at directing productive investment away from traditional urban centres. So far, however, minimal success has been achieved despite the increasing number of small and intermediate urban centres designated as growth poles. Such failure stems from the tendency to concentrate on the promotion of industrial development in the designated settlements at the expense of industrial development in rural areas. Moreover, some industries have tended to operate below capacity because of difficulty in getting the necessary inputs, power cuts, poor infrastructure and lack of skilled personnel (Renaud, 1981). The encouragement of small-scale industries would be preferable and the improvement of local infrastructure is necessary.

"Growth centres" often provide fewer stimuli to the surrounding areas than expected. Furthermore, such stimuli are generally outweighed by negative effects. Growth-centre strategies have also failed to meet social goals because of confusion in terms of the distinction between social and spatial equity. While government intervention may enhance the role of settlements in national production, it cannot always achieve social and economic benefits for most of the population living there. In most African countries, the basic problem is poverty and it cannot be overcome by such a strategy alone. The desired economic growth has failed to come about because of (*a*) an imprecise diagnosis of existing conditions in growth poles and their zones of influence; (*b*) the simplistic and piecemeal understanding of factors underlying development; (*c*) the lack of integration of settlement policies with microdevelopment policies; (*d*) the inadequate recognition of factors specific to each urban centre; and (*e*) an unrealistic and imprecise estimation of the investments needed to implement proposed policies. A reassessment of the potential role of small and intermediate urban centres therefore seems necessary.

The promotion of intermediate urban centres goes hand in hand with measures to foster rural development. The Government of Kenya hopes to reduce Nairobi's rate of growth by promoting small-scale industries in rural areas and developing alternative growth centres. This strategy also involves the creation of a network of rural service centres throughout the country and the selective allocation of investment using a bottom-up strategy (Obudho, Aduwo and Akatch, 1988). The Government of Botswana is attempting to control migration from rural areas mainly through its national settlement policy which seeks to reduce urban-rural imbalances by offering an adequate allocation of investment, providing infrastructure in rural areas and adopting an integrated rural development programme.

There is a need to foster decentralization in management. Municipal management, which includes the management of urban centres and that of public authorities providing urban services, must be strengthened so that decisions are more sensitive to local needs and opportunities. However, over the past 20 years the tendency in Africa has been towards the centralization of authority. Some devolution of authority has occurred in the case of large urban centres but it has rarely been extended effectively to intermediate urban centres. There is a general recognition, however, that urban services can only be provided efficiently if there is a partnership between central and local Governments, with greater responsibilities being given to the latter than has generally been the case. Decentralization permits better use to be made of local human, financial and physical resources.

Training is the key to effective management for both municipalities and public enterprises. Many African countries have announced decentralization policies that give local authorities more responsibility to provide and maintain local services before they have the necessary human and financial resources to do so. Financial constraints make it difficult for central Government to allocate significant budgets to local authorities. Local Government must be encouraged to raise its own revenue locally and eventually to execute its responsibilities without the supervision of the central Government.

Improved local resource mobilization is important for effective urban management and decentralization. Governments may have to raise more revenue from urban households and businesses when taxes on agricultural output are removed as part of structural adjustment. The share of local Government in the mobilization of public sector resources is low in Africa. There has been more resource mobilization in its capital cities than in its small and intermediate urban centres mainly because the capitals have highly valued taxable property and a working tax collection system. However, municipalities have various sources of local revenue. A major obstacle to improving revenue mobilization is the absence of a relationship between taxes paid and services enjoyed by taxpayers. Thus, revenue collection would be enhanced if taxpayers were informed of the objectives of taxation and if some revenue were earmarked for vital and visible projects. Better local resource mobilization by public enterprises is also important. Urban households and businesses are often willing to pay more for water and power than is assumed by the managers of public utilities.

Land is a vital resource in development that is under increasing pressure as rapid population growth continues. Because land is so important, it constitutes a political issue. Land is an inefficiently used resource in Africa. There is a need to improve land registration, land development, property tax valuation, land tenure and property rights as well as to reduce barriers to the smooth functioning of land markets. Land policy and management issues have been neglected. Although lack of serviced or serviceable land is often identified as a major constraint on efficient urbanization, Governments rarely formulate land policies that deal with this problem. They often lack the information to do so since almost none have the capacity to prepare maps that reflect rapidly changing urbanization, and very few have topographic maps; moreover, hardly anyone knows how much land the public sector owns. Problems arise from the conflict between traditional and modern land tenure systems. A modern economy needs to permit changes in land rights and consequently in land uses as income levels rise and the economy evolves. Urban centres have to be developed at higher densities because the unit costs of supplying infrastructure fall with increasing density.

African Governments attach high priority to the provision of suitable shelter for all. Many are directly engaged in the provision of dwelling units. However, public resource constraints are so severe that Governments will increasingly need to limit their support to only those activities that cannot be carried out by the private sector. Governments could, for instance, acquire land for development, provide assurance of tenure, supply basic infrastructure, establish appropriate planning standards and facilitate the mobilization of domestic private financial resources for housing construction (Obudho and Mhlanga, 1988). Donor agencies should encourage investment in shelter by supporting sites and services projects as well as upgrading projects. Housing is a productive investment. It mobilizes domestic savings, requires few imports and creates substantial direct and indirect employment in the small enterprise sector. The multiplier for housing investment appears to be among the higher multipliers of all industrial sectors. As economies develop, housing development becomes an increasingly important part of annual investment. Squatter upgrading is also important. Africa's urban centres are growing so quickly that settlement sites cannot be prepared and serviced. Squatter settlements dominate many urban centres, accommodating as much as 70 per cent of the population in some cases. Such settlements are characterized by high densities, poor water supply and sanitation, and almost no all-weather roads. Evaluations have shown that squatter upgrading projects involving improved access to water, better sanitation, better roads, the layout of workshop areas, public transportation and loans to improve property have had very high economic rates of return.

G. Inadequacies in public sector intervention

Until the 1960s, the growth of urban centres was such that technical resources, development policies and budgeting were within the capacity of either municipalities or central Government departments. As urban centres in Africa began to grow rapidly, so did the gap between the urban centres' development needs and available resources. By the 1980s, the dominant view in the international community was that the master-plan approach in Africa had failed because (*a*) it was too static in nature; (*b*) it seldom offered guidance on techniques of implementation; (*c*) it seldom evaluated the costs of development; (*d*) it was generally not based on realistic appraisals; (*e*) it rarely provided a compelling rationale for detailed land-use controls; (*f*) it did not involve community leaders; and (*g*) it was updated infrequently. The net effect of such inadequacies is that the urbanization process in Africa has proceeded largely without control.

Financial shortfalls are common in African urban centres, especially for the provision and maintenance of urban services, the acquisition of land for urban expansion and the provision of employment opportunities. Lack of coordination among government agencies often exacerbates the lack of resources. Institutional shortcomings include the lack of trained personnel, especially at the level of local Government, and problems of institutional coordination in development planning and management. The staff often consists of a handful of dedicated and knowledgeable heads of department directing largely unmotivated junior- and medium-level staff. The problems of staff shortages are often compounded by the absence of clear guidelines on areas of responsibility and decision-making powers. Consequently, coordination among the policy-making staff, technicians and those in charge of implementation is ineffective. The filtering down process works poorly and there is no horizontal decision-making at lower levels. Coordination should be a two-way process.

As the accelerating rate of urban growth began to overtake the ability of Governments to provide adequate urban services and guide urban expansion, and as urban unemployment increased and central Governments were unwilling to increase the powers and resources of urban centres to manage themselves, the public became increasingly defensive and the private sector increasingly antagonistic. The private sector has been increasingly unable to rely on the public sector to provide urban services and sees only the negative regulatory aspects of Government. Slowly, Governments are realizing that the "project" approach is only one of the many tools with which to orient urban development and that, in order to cope with urban growth given limited resources, the only realistic option is to find alternatives to direct public sector interventions.

The increasing complexities of urban areas require a re-evaluation of the traditional role of the public sector. Urban-sector interventions characterized by a "project" cannot hope to meet the massive demand for improved services and shelter in many urban centres, let alone contribute significantly to economic growth. As economic conditions have deteriorated in Africa, economic objectives for urban areas, requiring a more integrated approach to urban development, have become more critical. Traditional master planning has paid little attention to the necessary resource allocation needs and financial feasibility of policies. Programmes should be vehicles for institutional development and the strengthening of technical, managerial, financial and administrative capabilities that will sustain development.

H. Challenges for the future

Future policies should have the following objectives: (*a*) to gain control of urban growth; (*b*) to develop alternative methods of providing low-cost urban services; (*c*) to recover investment costs to permit financial replaceability; (*d*) to strengthen national and municipal institutions; (*e*) to stimulate community participation; (*f*) to develop economic and institutional links between the urban and rural sectors; and (*g*) to promote effective urban management strategies. To gain control of urban growth, it is necessary to reduce current levels of national subsidies to the capital and shift the financial burden to those who benefit from it. In order to develop alternative ways of providing urban services, the pricing and distribution of services must be reviewed and the means of recovering an increasing share of the economic cost of the service provided must be devised, probably by eliminating the subsidies enjoyed by urban consumers. The major urban centres are better able to collect the revenue needed for the services they provide because their residents have higher incomes than rural residents. The efficient allocation of scarce national resources is essential if African countries are to improve national incomes in the future. African Governments should re-examine the standard and types of services provided to their growing urban populations. When services aim at maintaining relatively high standards, they are unlikely to reach the majority of urban residents who are primarily low-income. African Governments must experiment with cheaper methods of providing services and improve administrative efficiency.

To recover the investment costs for services provided in urban centres, those using such services must be asked to pay according to their means. The present system whereby services are delivered at very low cost but reach only a small number of people cannot expand unless cost recovery becomes a reality. Implementation of an effective system of municipal taxation would help in this area.

None of the above objectives can be achieved unless African Governments make serious attempts to strengthen the institutions working in the urban centres. In most African countries, the department concerned with urban planning has been understaffed and insufficiently financed to permit the execution of its legal responsibilities. There should be increased financial autonomy for the municipalities, training of staff and a clearer definition of municipal jurisdictions.

Given the financial and institutional weaknesses of the public sector in most African countries, it is imperative that programmes be designed to stimulate greater community participation in the financing and delivery of urban services. Self-help programmes, incentives to the private sector and the use of community organizations or even ethnic and religious organizations must be encouraged in order to generate increasing activity and interest by the urban population.

As part of the economic diversification of urban economies, efforts must be made to develop economic and institutional links between urban centres and rural areas. These links must include the urban production and provision of goods and services needed to support rural development such as the manufacturing of farm implements, developing of repair centres for agricultural machinery and the processing of agricultural products. This approach requires that more resources be devoted to the development of small and intermediate urban centres to perform such functions.

Because policies and programmes to control rural-urban migration and the diffusion of the urban way of life have not been successful, there is an increasing recognition that the growth of urban centres is inevitable and that solutions to the problems of urban centres depend heavily on their effective management. Urban management is based on a holistic concept. It can strengthen the capacity of governmental and non-governmental organizations to identify policy and programme alternatives and to implement these with optimal results. The challenge of urban management is to respond effectively to the problems of particular urban centres so as to enable them to perform their functions effectively. Effective urban management involves (*a*) improving financial management; (*b*) providing urban shelter, services and infrastructure; (*c*) improving urban information systems; (*d*) strengthening the urban informal sector; and (*e*) strengthening the institutional capacities of urban centres.

There has been a heightened public awareness of the need for African Governments to take decisive measures to improve conditions in urban centres and the Governments themselves have not been idle in attempting to articulate the nature of urban problems. The need for a clearly formulated urban policy thus arises precisely because of the importance of ensuring an appropriate role of urban centres in regional and national development in Africa. It is against this background that we have suggested placing productive investments in those urban centres that are the most efficient and have already proved to have high economic potential, whatever their size. A crucial need is to develop links between the economic activities of the mega-cities and national development strategies in Africa. This proposal agrees with the following World Bank recommendations: (*a*) to move beyond isolated projects that emphasize housing and residential infrastructure towards integrated urban-wide efforts that promote urban productivity and reduce constraints on efficiency; (*b*) to increase the demand for labour stressing the generation of jobs for the urban poor; (*c*) to improve access to basic infrastructure and social services emphasizing the needs of women and the poor; (*d*) to address urban environmental problems; and (*e*) to increase our understanding of urban issues through international research. These efforts should promote the role of urban centres as engines of growth for rural areas and hence for the national economy as a whole.

References

Baker, Jonathan (1990). *Small Towns in Africa: Studies in Rural Urban Interaction.* Uppsala, Sweden: Scandinavian Institute of African Studies.

__________, and Paul Ore Pederson (1992). *The Rural-Urban Interface in Africa: Expansion and Adaptation.* Uppsala, Sweden: Scandinavian Institute of African Studies.

Becker, Charles M., and others (1985). *Urban Africa in Macroeconomic and Microeconomic Perspective: Issues and Options.* Washington, D.C.: World Bank.

Bird, R. M., D. Marsara and D. Miller (1985). *Taxation and the Poor in Developing Countries.* Toronto: University of Toronto, Toronto Institute for Policy Analysis.

Cohen, Michael A. (1981). *Urban Sector Strategies for Africa: The Next Twenty Years.* Washington, D.C.: World Bank.

De Cola, Lee (1986). Urban concentration in Africa. *African Urban Quarterly* (Nairobi, Kenya), vol. 1, Nos. 3 and 4, pp. 176-190.

El-Shakhs, Salah, and Robert A. Obudho (1974). *Urbanization, National Development and Regional Planning in Africa.* New York: Praeger.

Gibb, Arthur, Jr. (1984). Tertiary urbanization: agricultural market centre as a consumption related phenomenon. *Regional Development Dialogue* (Nagoya, Japan), vol. 5, No. 1, pp. 110-143.

Hamer, Andrew Marshall (1986). Urban sub-Sahara in macroeconomic perspective: selected issues and options. Discussion Paper No. UDD96. Washington, D.C.: World Bank, Water Supply and Urban Development Department.

Hance, William A. C. (1970). *Population, Migration and Urbanization in Africa.* New York: Columbia University Press.

Hanna, William, and Judith L. Hanna (1971). *Urban Dynamics in Black Africa: An Interdisciplinary Approach.* Chicago, Illinois: Aldine.

Obudho, Robert A. (1983). *Urbanization in Kenya: Bottom-up Approach to Regional Planning.* Washington D.C.: University Press.

__________ (1984). Planning from below: the role of small urban centres in spatial planning in East Africa. In *Equality with Growth: Planning Perspectives for Small Towns in Developing Countries,* H. Detler Kammeier and Peter Swan, eds. Bangkok: Asian Institute of Technology.

__________ (1987). Urbanization and urban policy in East Africa. In *Inequality and Development: Case Studies from Third World,* K. Swindel and others, eds. London: MacMillan.

__________ (1988). The role of metropolitan Nairobi in spatial planning of Kenya. Paper presented at the International Conference on Urban Growth and Spatial Planning of Nairobi, Nairobi, Kenya.

__________ (1991a). Urbanization and street children in Eastern and Southern Africa. In *Street Children in Africa.* Harare, Zimbabwe: Edicesa.

__________ (1991b). The role of megacities in spatial planning. Paper presented at the Commemorative Programme for the Investment. Anniversary of the United Nations Center for Regional Development, Nagoya, Japan, 7-8 November.

__________ (1991c). Gearing urban development in Africa for sustainable growth. The current problems and challenges. Paper presented at the Expert Group Meeting on From Crisis to Sustainable Growth in Africa, Nagoya, Japan, 11-12 November.

__________ (1993). Urbanization and urban development strategies in East Africa. In *Urban Management: Policies and Innovations in Developing Countries,* G. Shabbir Cheema, ed. London: Praeger.

__________, and Gibson O. Aduwo (1989). Policy, strategies and projects for metropolitan areas in Africa. Paper presented at the First Meeting of the Commission on Urban Systems and Urban Development of the International Geographical Union, Paris, France, 22-30 June.

__________, and G. O. Aduwo (1990). The district focus policy for rural development in Kenya: an empirical application of bottom-up concept. *Regional Development Dialogue* (Nagoya, Japan), special issue, pp. 51-68.

__________, Gibson O. Aduwo and S. O. Akatch (1988). The district focus policy for rural development in Kenya: an empirical application of bottom-up concept. *Regional Development Dialogue* (Nagoya, Japan), special issue, pp. 158-188.

__________, and Constance E. Mhlanga (1988). *Slum and Squatter Settlements in Sub-Saharan Africa: Towards a Planning Strategy.* New York: Praeger.

__________, and Salah El-Shakhs (1979). *Development of Urbanization in Africa.* New York: Praeger.

O'Connor, Anthony M. (1982). *Urbanization in Tropical Africa: An Annotated Bibliography.* Boston, Massachusetts: G. R. Hall.

__________ (1983). *The African City.* New York: Africana.

Population Crisis Committee (1990a). *Cities — Life in World's 100 Largest Metropolitan Areas.* Washington D.C.

__________ (1990b). *Cities — Life in the World's 100 Largest Metropolitan Areas: Statistical Appendix.* Washington D.C.

Renaud, B. (1981). *National Urbanization Policy In Developing Countries.* New York: Oxford University Press.

Rondinelli, Dennis A. (1983). *Secondary Cities In Developing Countries: Policies for Diffusing Urbanization.* Beverly Hills, California: Sage.

United Nations (1991). *World Urbanization Prospects, 1990.* Sales No. E.91.XIII.11.

__________ (1992). *World Population Monitoring, 1991: With Special Emphasis in Age Structure.* Population Studies, No. 126. Sales No. E.92 XIII.2.

__________ (1993a). *World Population Prospects: The 1992 Revision.* Sales No. E.93.XIII.7.

__________ (1993b). *World Urbanization Prospects: The 1992 Revision.* Sales No. E.93.XIII.11.

United Nations Development Programme (1991). *Human Development Report, 1991.* New York: Oxford University Press.

__________ (1992). *Human Development Report, 1992.* New York: Oxford University Press.

United Nations High Commissioner for Refugees (1993). *The State of the World's Refugees, 1993.* New York: Penguin Books.

VI. POPULATION DISTRIBUTION IN ASIA: A REGION OF CONTRASTS

*Ernesto M. Pernia**

Although patterns of spatial population distribution have been perceived as a manifestation, if not an integral part, of the development process, concerns about them became prominent only in the 1960s. Indeed, a number of development observers have remarked rather boldly that the population problem is one of distribution, not growth. This observation contributed to the decision of Governments to begin including explicit policies on population distribution in development plans formulated after 1965.

The importance accorded to the issue of the spatial distribution of the population has persisted to the present. According to the *World Population Monitoring, 1991* (United Nations, 1992), of the 38 Asian Governments surveyed by the United Nations, 32 considered the spatial distribution of their population to be unsatisfactory. Twenty Governments out of the 32 felt that their population distribution required a minor change, while the rest considered that a major change was needed.[1] In contrast, when the same 38 Governments were asked about their population growth, 16 regarded their growth rates as being too high, and of these 16, 14 had some form of intervention to lower rates. Several Governments expressed concern over both population distribution and growth.

These results represent perceptions of Governments and it may be argued that they are not necessarily scientifically sound. For instance, the results may vary depending on the individuals answering the questionnaire or on the government agency providing the information, since it is well known that Governments are rarely of one mind. Nevertheless, since the survey of Governments has been conducted repeatedly over a number of years, its results have proved to have some robustness. Clearly, therefore, there is a continuing need to consider the spatial dimensions of development in policies, programmes and projects.

*Economic Research Unit, Asian Development Bank, Manila, Philippines.

The standard theory underlying urbanization and population distribution essentially derives from the relationship between economic growth and structural change. Economic growth involves rising levels of productivity that result in structural changes that are accompanied by spatial shifts. Labour migration occurs in response to changing demands in the agricultural, industrial and service sectors. Since the demand for agricultural output is relatively inelastic while that for non-agricultural products is relatively elastic, the flow of labour is from rural to urban areas, given rising levels of productivity in agriculture. Improvements in productivity are then translated into increases in income which, in turn, result in a changing composition of spending in favour of non-agricultural goods. Apart from this underlying market mechanism, historical and geographical factors and government policies also determine the nature and pace of urbanization and population distribution.

A. Trends and patterns of population distribution

In comparing urbanization trends and patterns between countries, it should be borne in mind that official definitions of the term *urban* vary and that the quality of the data available may be uneven. There are also differences among countries in the definitions of large and small cities or towns. The definitions used here are those established by the countries concerned (a practice followed by the United Nations), since national statistical offices are considered the best judges in this regard.

It is important, moreover, to be clear about the distinction between *urbanization* and *urban population growth*. Following accepted practice, *urbanization* refers to the rise in the proportion of the total population living in urban places. It connotes the changing balance between rural and urban populations brought about by spatial shifts of people from rural to urban areas and by differences in the rates of natural increase of the population in the two areas. *Urban population growth* refers to the growth of the population living in urban places. Hence, urbanization is a measure of a structural phenomenon linked in some way to structural economic change, whereas urban population growth is a measure of population change that refers only to urban places and incorporates no element of concomitant change in rural areas.[2]

1. *Level and tempo of urbanization*

Asia as a whole is not yet highly urbanized by world standards. In 1990, the region was slightly less than one-third urban (31.2 per cent) (table 15), a level similar to that of Africa and far below that of Latin America (71.5 per cent). It was also below the average for the less developed regions (34.3 per cent). However, given the vastness and heterogeneity of Asia, its comparatively low average level of urbanization masks the wide variation existing across subregions and countries.

At the subregional level, Western Asia is the most urbanized subregion, with 62.7 per cent of its population living in urban places, but the range of variation of the level of urbanization at the country level is wide, ranging from 11.0 per cent for Oman to 95.8 per cent for Kuwait. Except for Oman and Yemen, all countries in Western Asia have more than one half of their populations in urban areas and are therefore expected to experience lower rates of urbanization in the future.[3] By the year 2010, most Western Asian countries will be more than two-thirds urbanized.

Southern Asia has the lowest level of urbanization in Asia (26.4 per cent), with the level of urbanization in India, the largest country in the subregion, very close to that figure. Notable deviations from the mean include Bhutan (5.3 per cent), Nepal (10.9 per cent), and the Islamic Republic of Iran (56.9 per cent). Southern Asian countries therefore still have a long way to go and can expect accelerated urbanization in the years to come. This will especially be true of such large countries as Bangladesh, India and Pakistan, although their levels of urbanization will only reach the 30-45 per cent range by 2010. Those levels could be higher if economic growth and structural change speed up as a consequence of the economic reforms that are under way in those countries.

South-eastern Asia's level of urbanization (28.8 per cent) is not much higher than that of Southern Asia, and countries in this subregion also exhibit a wide range of variation, with the level of urbanization ranging from about 12 per cent in the case of Cambodia to 100 per cent for the city-State of Singapore. Many South-eastern Asian countries, mainly those that are members of the Association of South-East Asian Nations (ASEAN), are apt to maintain their rapid urbanization tempos if their economic growth rates remain high and structural change is sustained. The Lao People's Democratic Republic and Viet Nam are likely to follow in the trail of the ASEAN countries, as their economies embark on the change towards a market system. However, by the year 2010 most countries in South-eastern Asia, with the exception of Brunei Darussalam, Malaysia, the Philippines and Singapore, will not have reached the 50 per cent urbanization mark.

The level of urbanization of Eastern Asia (33.2 per cent level) is predominantly a reflection of that of China (26.2 per cent). In contrast, all the other countries in the subregion have levels of urbanization above 50 per cent. Thus, while those countries experience a deceleration of urbanization in the years ahead, China will tend to urbanize at a faster pace, especially if its liberalization and deregulation policies continue. Still, by 2010 China is projected to be only 43 per cent urban.

Although there are many factors involved in the urbanization process, the trends and patterns discussed above suggest that the smaller Asian countries have generally urbanized faster than the larger ones. The salient cases in point are China, India and Indonesia, which will remain more rural than urban through 2010. These trends underscore earlier observations that the critical issue has not been rapid urbanization per se but rather high overall population growth rates occurring in both rural and urban areas (Davis, 1965 and 1972; Pernia, 1977; Fuchs, Jones and Pernia, 1987). A key related problem has been slow economic growth and structural change, particularly prior to 1985 in many Asian countries, including the largest ones.

2. *Urban population growth*

The urban population of Asia as a whole grew at an annual rate of 3.6 per cent during 1970-1990 (table 16). Eastern Asia, the region that encompasses China, experienced a growth rate of 3.1 per cent per year during that period, with China, Mongolia and the Republic of Korea experiencing above-average urban growth rates. Western Asia recorded the highest rate of urban growth within Asia, at 4.9 per cent per year during 1970-1990.

TABLE 15. LEVEL OF URBANIZATION AND ANNUAL PERCENTAGE CHANGE OF THE LEVEL OF URBANIZATION, ASIA, 1950-2010

	Level of urbanization				*Average annual change*		
	1950	*1970*	*1990*	*2010*	*1950-1970*	*1970-1990*	*1990-2010*
Eastern Asia	16.7	24.8	33.2	48.1	2.0	1.5	1.9
China	11.0	17.5	26.2	43.0	2.3	2.1	2.5
Democratic People's Republic of Korea	31.0	53.3	59.8	67.8	2.7	0.6	0.6
Hong Kong	82.5	87.7	94.1	96.6	0.3	0.4	0.1
Japan	50.3	71.2	77.2	81.6	1.8	0.4	0.3
Macau	96.8	97.1	98.7	99.2	0.0	0.1	0.0
Mongolia	18.9	45.1	57.9	69.4	4.4	1.3	0.9
Republic of Korea	21.4	40.7	72.1	86.3	3.3	2.9	0.9
South-eastern Asia	14.8	20.2	28.8	41.8	1.6	1.8	1.9
Brunei Darussalam	27.1	61.5	58.0	64.3	4.2	-0.3	0.5
Cambodia	10.2	11.7	11.6	19.7	0.7	0.0	2.7
East Timor	9.9	10.3	13.1	24.3	0.2	1.2	3.1
Indonesia	12.4	17.1	28.8	44.5	1.6	2.6	2.2
Lao People's Democratic Republic	7.2	9.6	18.6	32.6	1.4	3.4	2.8
Malaysia	20.4	27.0	43.0	58.4	1.4	2.4	1.5
Myanmar	16.2	22.8	24.8	35.4	1.7	0.4	1.8
Philippines	27.1	33.0	42.7	55.7	1.0	1.3	1.3
Singapore	100.0	100.0	100.0	100.0	0.0	0.0	0.0
Thailand	10.5	13.3	22.2	36.6	1.2	2.6	2.5
Viet Nam	11.6	18.3	19.9	27.4	2.3	0.4	1.6
Southern Asia	16.0	19.6	26.4	36.5	1.0	1.5	1.6
Afghanistan	5.8	11.0	18.2	28.2	3.3	2.5	2.2
Bangladesh	4.2	7.6	16.4	30.3	3.0	3.9	3.1
Bhutan	2.0	3.1	5.3	11.4	2.2	2.7	3.9
India	17.3	19.8	25.5	33.8	0.7	1.3	1.4
Iran (Islamic Republic of)	27.0	41.9	56.9	69.7	2.2	1.5	1.0
Maldives	11.0	13.2	29.6	45.4	0.9	4.1	2.2
Nepal	2.3	3.9	10.9	23.2	2.7	5.3	3.8
Pakistan	17.5	24.9	32.0	45.4	1.8	1.3	1.8
Sri Lanka	14.4	21.9	21.4	30.7	2.1	-0.1	1.8
Western Asia	23.9	43.2	62.7	76.3	3.0	1.9	1.0
Bahrain	63.8	78.6	82.9	87.6	1.1	0.3	0.3
Cyprus	29.8	40.8	52.8	65.9	1.6	1.3	1.1
Gaza Strip (Palestine)	50.6	82.0	93.6	95.7	2.4	0.7	0.1
Iraq	35.1	56.2	71.8	81.1	2.4	1.2	0.6
Israel	64.6	84.2	91.6	94.5	1.3	0.4	0.2
Jordan	34.7	50.5	68.0	79.2	1.9	1.5	0.8
Kuwait	59.2	77.8	95.8	98.1	1.4	1.0	0.1
Lebanon	22.7	59.4	83.8	92.1	4.9	1.7	0.5
Oman	2.4	5.2	11.0	21.7	3.9	3.8	3.4
Qatar	64.0	80.2	89.9	94.0	1.1	0.6	0.2
Saudi Arabia	15.9	48.7	77.3	84.7	5.8	2.3	0.5
Syrian Arab Republic	30.6	43.4	50.2	60.7	1.7	0.7	1.0
Turkey	21.3	38.4	60.9	82.4	3.0	2.3	1.5
United Arab Emirates	24.3	57.0	80.9	88.7	4.4	1.8	0.5
Yemen	5.8	13.3	28.9	47.4	4.2	4.0	2.5

TABLE 15 (*continued*)

	Level of urbanization				Average annual change		
	1950	*1970*	*1990*	*2010*	*1950-1970*	*1970-1990*	*1990-2010*
Asia	16.4	22.9	31.2	43.8	1.7	1.6	1.7
World total	29.3	36.6	43.1	52.8	1.1	0.8	1.0
More developed regions	54.3	66.6	72.7	79.1	1.0	0.4	0.4
Less developed regions	17.0	24.7	34.3	46.8	1.9	1.7	1.6

Source: *World Urbanization Prospects: The 1992 Revision* (United Nations publication, Sales No. E.93.XIII.11).

TABLE 16. URBAN POPULATION AND ANNUAL GROWTH RATES OF THE URBAN POPULATION, ASIA, 1950-2010

	Urban population (thousands)				Growth rate (percentage)		
	1950	*1970*	*1990*	*2010*	*1950-1970*	*1970-1990*	*1990-2010*
Eastern Asia	112 406	244 294	448 951	784 052	4.0	3.1	2.8
China	61 024	144 953	302 209	605 995	4.4	3.7	3.5
Democratic People's Republic of Korea	3 015	7 789	13 022	19 676	4.9	2.6	2.1
Hong Kong	1 629	3 458	5 374	6 125	3.8	2.2	0.7
Japan	42 065	74 296	95 332	106 592	2.9	1.3	0.6
Macau	182	238	457	705	1.4	3.3	2.2
Mongolia	144	566	1 269	2 460	7.1	4.1	3.4
Republic of Korea	4 347	12 995	31 288	42 500	5.6	4.5	1.5
South-eastern Asia	26 915	57 889	127 890	255 526	3.9	4.0	3.5
Brunei Darussalam	13	80	149	234	9.5	3.2	2.3
Cambodia	443	812	970	2 557	3.1	0.9	5.0
East Timor	43	62	99	244	1.8	2.4	4.6
Indonesia	9 863	20 534	53 060	109 107	3.7	4.9	3.7
Lao People's Democratic Republic	127	261	782	2 320	3.7	5.6	5.6
Malaysia	1 244	2 929	7 701	15 268	4.4	5.0	3.5
Myanmar	2 881	6 188	10 353	21 798	3.9	2.6	3.8
Philippines	5 695	12 380	26 661	49 800	4.0	3.9	3.2
Singapore	1 022	2 075	2 710	3 158	3.6	1.3	0.8
Thailand	2 097	4 749	12 148	24 459	4.2	4.8	3.6
Viet Nam	3 487	7 820	13 258	26 580	4.1	2.7	3.5
Southern Asia	76 790	147 691	314 690	641 547	3.3	3.9	3.6
Afghanistan	520	1 503	3 021	9 444	5.5	3.6	5.9
Bangladesh	1 774	5 074	18 691	53 757	5.4	6.7	5.4
Bhutan	15	32	82	281	3.9	4.8	6.4
India	61 695	109 616	216 081	401 717	2.9	3.5	3.1
Iran (Islamic Republic of)	4 561	11 911	33 127	72 529	4.9	5.2	4.0
Maldives	9	16	63	169	2.9	7.1	5.1
Nepal	187	450	2 139	7 193	4.5	8.1	6.3
Pakistan	6 923	16 354	37 809	89 830	4.4	4.3	4.4
Sri Lanka	1 106	2 736	3 679	6 627	4.6	1.5	3.0

TABLE 16 (*continued*)

	Urban population (thousands)				*Growth rate (percentage)*		
	1950	*1970*	*1990*	*2010*	*1950-1970*	*1970-1990*	*1990-2010*
Western Asia	10 130	31 854	82 731	164 530	5.9	4.9	3.5
Bahrain	74	173	417	711	4.3	4.5	2.7
Cyprus	147	251	371	543	2.7	2.0	1.9
Gaza Strip (Palestine)	124	287	584	906	4.3	3.6	2.2
Iraq	1 812	5 258	12 989	26 649	5.5	4.6	3.7
Israel	813	2 504	4 269	6 747	5.8	2.7	2.3
Jordan	429	1 162	2 725	5 994	5.1	4.4	4.0
Kuwait	90	579	2 054	2 172	9.8	6.5	0.3
Lebanon	327	1 466	2 296	3 480	7.8	2.3	2.1
Oman	10	34	168	658	6.3	8.3	7.1
Qatar	16	89	384	596	9.0	7.6	2.2
Saudi Arabia	509	2 798	11 495	23 543	8.9	7.3	3.6
Syrian Arab Republic	1 071	2 713	6 202	14 663	4.8	4.2	4.4
Turkey	4 442	13 571	34 114	65 087	5.7	4.7	3.3
United Arab Emirates	17	127	1 286	2 086	10.6	12.3	2.4
Yemen	250	842	3 377	10 692	6.3	7.2	5.9
Asia	226 241	481 728	974 262	1 845 654	3.9	3.6	3.2
World total	737 495	1 352 143	2 282 367	3 778 494	3.1	2.7	2.6
More developed regions	452 081	698 438	880 947	1 060 729	2.2	1.2	0.9
Less developed regions	285 414	653 705	1 401 420	2 717 765	4.2	3.9	3.4

Source: *World Urbanization Prospects: The 1992 Revision* (United Nations publication, Sales No. E.93.XIII.11).

In both Southern and South-eastern Asia, the average annual rate of urban growth increased between 1950-1970 and 1970-1990, reaching 4 per cent and 3.9 per cent per year, respectively. In Southern Asia, some of the least urbanized countries, such as Bangladesh and Nepal, experienced very high rates of urban growth during 1970-1990. India, where one of every five inhabitants lived in an urban area in 1970, and Pakistan, where one in every four inhabitants was an urban-dweller in 1970, experienced moderate rates of urban growth during 1970-1990: 3.5 per cent and 4.3 per cent, respectively. In most countries of Southern Asia, the urban growth rate was expected to remain at moderate levels during 1990-2010.

Despite slowing urban population growth in most Asian countries in the 1990s or thereafter, huge population bases mean that the additional numbers of people expected to live in urban areas will continue to be high for some time to come. Accordingly, greater efforts and more substantial resources will have to be devoted to combating such nagging problems as poverty, unemployment and underemployment, inadequate infrastructure and social services, and environmental degradation.

Internal migration is an important component of urban population growth during the development process. Broadly, in the more developed countries migration and the reclassification of places have accounted for about 60 per cent of urban growth, with natural increase accounting for the rest. In developing countries, including those of Asia, the situation has been the reverse (United Nations, 1984). Indeed, to the extent that economic growth is accompanied by reductions in fertility, the share of migration in urban growth expands relative to that of natural increase. That is the experience of countries that have undergone or are undergoing rapid economic growth and structural transformation, such as the two generations of newly industrializing economies (NIEs), namely, Indonesia, Malaysia, the Republic of Korea, Taiwan Province of China and Thailand. Other Asian countries following in the path of structural reforms to improve economic performance are expected to show the same pattern.

To the extent that economic reforms are complemented by more effective population policy, as in the NIEs, economic growth can be more sustained, and natural increase is likely to play a lesser role in urban growth. In terms of policies and approaches to deal with urban growth, there would, in principle, be an advantage if such growth were caused mostly by an increase in economically active migrants rather than by an increase in dependent children, because workers require fewer social services than children. However, if migrant workers remained unemployed, they would also represent a burden for urban centres.

B. Characteristics of the urban system

1. *Urban primacy and dispersal trends*

A salient characteristic of urban systems in developing countries, not least among those in Asia, has been the concentration of population and economic activity in one or a few large cities, a pattern that is commonly referred to as urban primacy. A convenient and useful measure of urban primacy is the percentage of a country's urban population (or total population) residing in the largest urban agglomeration, which is also known as its primate city. The primate city is typically the national capital. Urban primacy is understood to be a manifestation of economies of agglomeration and scale needed to promote economic efficiency and growth. Its degree is often associated with, *inter alia*, a country's historical factors, geography and physical size, the stage of development, the system of government, and government policies.

The data indicate broadly varying trends in urban primacy among the developing countries of Asia (excluding the special cases of Hong Kong, Singapore and, perhaps, Kuwait). In Eastern Asia, the Republic of Korea, with 23.5 per cent of its urban population in 1950 residing in Seoul, had the highest spatial concentration in that year, followed by the Democratic People's Republic of Korea with 14.9 per cent of its urban population residing in Pyongyang. The urban primacy of Seoul in the Republic of Korea peaked at 40.9 per cent in 1970, thereafter diminishing gradually to 35.7 per cent in 1990, and it is expected to further slide down to 33.6 per cent at the turn of the century (table 17). This development reflects some dispersal of Seoul's population growth and economic activity to the satellite cities of Inchon and Taejon. At the same time, Taegu has served as an intermediate magnet in the corridor between Seoul and Pusan. Nonetheless, the share of the total population living in Seoul rose steadily from 5 per cent in 1950 to 26 per cent in 1990, and will likely surpass 27 per cent by the year 2000.

In contrast, China's urban primacy has been low and declining from about an 8.8 per cent concentration in Shanghai in 1950 to 3.5 per cent in 1990. The share of the urban population in Beijing, the capital and second largest city, has been lower and has also declined over time. China's relative lack of spatial concentration is most likely related to its size and political system. It has several secondary or intermediate urban centres, foremost among which are Guangzhou, Shenyang, Tianjin and Wuhan.

With the exception of the city States in South-eastern Asia, Thailand stands out as having the highest spatial concentration of population in all of Asia, with 56.8 per of its urban population residing in Bangkok as of 1990. It is followed by the Philippines (31.9 per cent in Manila), Myanmar (32.0 per cent in Yangon), Viet Nam (22.2 per cent in Ho Chi Minh City), Malaysia (22.2 per cent in Kuala Lumpur), and Indonesia (16.4 per cent in Jakarta). Bangkok's primacy is probably attributable to its geography and especially to the lack of other adequate harbours to service the economy. However, its high urban concentration appears to have been easing slowly, dropping from a peak of 65.5 per cent in 1970 to a projected 54.7 per cent by the year 2000, as a result of agglomeration diseconomies and dispersal policies, which include the development of the eastern seaboard.

Similar signs of primacy reversal have been apparent with respect to Jakarta in Indonesia beginning in the 1960s, and with respect to Manila in the Philippines and Ho Chi Minh City in Viet Nam starting in the 1980s. Jakarta's share of the urban population, which was 20 per cent to begin with (1950), was down to about 16 per cent by 1990. Even such secondary cities as Bandung and Surabaya have been showing diminishing urban shares, suggesting that yet smaller cities and towns in Indonesia have been absorbing population shifts. However—and there is no parallel case in China or India—Jakarta continues to dominate the country's urban system. Moreover, Jakarta's share of the total population doubled between 1950 and 1990 to reach 5.0 per cent, and is still increasing, as is Manila's share of the total population of the Philippines (table 18). In Malaysia and Myanmar, urban population concentration in their respective capitals, Kuala Lumpur and Yangon, contin-

TABLE 17. PERCENTAGE OF THE URBAN POPULATION OF VARIOUS COUNTRIES IN ASIA RESIDING IN URBAN AGGLOMERATIONS WITH 1 MILLION OR MORE INHABITANTS, 1950-2000

Subregion/country/city	*1950*	*1960*	*1970*	*1980*	*1990*	*2000*
Eastern Asia						
Beijing, China	6.4	5.0	5.6	4.6	2.8	2.3
Shanghai, China	8.8	7.1	7.7	6.0	3.5	2.8
Pyongyang, Democratic People's Republic of Korea	14.9	14.8	12.6	17.3	17.1	16.6
Hong Kong	100.0	94.7	96.1	97.2	98.9	100.0
Tokyo, Japan	16.0	18.2	20.0	19.0	19.1	19.0
Seoul, Republic of Korea	23.5	34.1	40.9	38.2	35.7	33.6
South-eastern Asia						
Jakarta, Indonesia	19.9	19.8	19.1	17.9	16.4	15.9
Kuala Lumpur, Malaysia	16.7	16.7	15.4	19.4	22.2	23.4
Yangon, Myanmar	23.2	23.3	23.1	27.3	32.0	32.0
Metro Manila, Philippines	27.1	27.2	28.6	33.0	31.9	31.2
Singapore	93.2	74.7	75.4	100.0	100.0	100.0
Bangkok, Thailand	64.9	65.1	65.5	58.7	56.8	54.7
Ho Chi Minh City, Viet Nam	24.9	25.9	25.6	26.4	22.2	18.3
Southern Asia						
Kabul, Afghanistan	41.6	41.7	33.6	38.9	51.8	43.9
Dhaka, Bangladesh	23.7	24.5	29.6	33.0	35.0	35.2
Calcutta, India	7.2	6.9	6.3	5.7	5.1	4.7
Delhi, India	2.3	2.9	3.2	3.5	3.8	3.9
Tehran, Iran (Islamic Republic of)	22.2	25.8	28.2	26.4	21.9	19.6
Karachi, Pakistan	14.9	16.7	19.1	20.7	19.6	19.0
Western Asia						
Baghdad, Iraq	32.0	34.7	37.8	38.1	30.0	25.7
Tel Aviv, Israel	51.5	45.4	41.1	41.2	44.7	46.0
Amman, Jordan	21.0	30.1	33.4	36.8	37.6	37.7
Kuwait City	100.0	100.0	96.5	67.0	55.4	53.9
Riyadh, Saudi Arabia	7.7	10.4	14.6	15.9	18.1	20.5
Damascus, Syrian Arab Republic	34.3	34.5	33.7	33.5	32.4	31.0
Ankara, Turkey	12.2	10.7	10.0	9.7	7.6	7.1
Istanbul, Turkey	24.3	21.3	20.5	22.6	19.4	19.3

Source: *World Urbanization Prospects, 1990* (United Nations publication, Sales No. E.91.XIII.11).

ues to rise slowly, and will probably reach 23.4 and 32.0 per cent, respectively, by the year 2000 (table 17). The relatively low primacy of Kuala Lumpur can be ascribed to Malaysia's effective rural development programme and bumiputra policy (Lim, Ogawa and Hodge, 1992).

In Southern Asia, Afghanistan has the highest urban concentration, with 51.8 per cent of the urban population living in Kabul as a result of continued civil strife that has driven many people out of the countryside, followed by Bangladesh with 35.0 per cent of the urban population living in Dhaka, the Islamic Republic of Iran with 21.9 per cent of the urban population living in Tehran, and Pakistan with 19.6 per cent of the urban population living in Karachi. As expected, India has the lowest primacy with about 5 per cent and less than 4 per cent of its urban population residing, respectively, in Calcutta and Delhi (the capital). In Bangladesh, Chittagong and Dhaka switched roles as principal and secondary cities in the early 1960s, with the former experiencing a monotonic reduction in its urban population share and the latter showing the reverse. On the other hand, the primacy of Tehran and that of Karachi peaked in 1970 and 1980, respectively, and each has since been on a downward trend.

In India, as in China and Indonesia, Calcutta's urban primacy has been dropping monotonically, whereas

TABLE 18. PERCENTAGE OF THE TOTAL POPULATION OF VARIOUS COUNTRIES IN ASIA RESIDING IN URBAN AGGLOMERATIONS WITH 1 MILLION OR MORE INHABITANTS, 1950-2000

Subregion/country/city	*1950*	*1960*	*1970*	*1980*	*1990*	*2000*
Eastern Asia						
Beijing, China	0.7	1.0	1.0	0.9	1.0	1.1
Shanghai, China	1.0	1.3	1.3	1.2	1.2	1.3
Pyongyang, Democratic People's Republic of Korea	4.6	6.0	6.7	9.8	10.2	10.5
Hong Kong	88.5	84.4	86.2	89.0	93.1	95.7
Tokyo, Japan	8.1	11.4	14.3	14.5	14.7	14.8
Seoul, Republic of Korea	5.0	9.4	16.6	21.7	25.7	27.4
South-eastern Asia						
Jakarta, Indonesia	2.5	2.9	3.3	4.0	5.0	6.3
Kuala Lumpur, Malaysia	3.4	4.2	4.2	6.7	9.6	12.0
Yangon, Myanmar	3.8	4.5	5.3	6.6	7.9	9.1
Metro Manila, Philippines	7.4	8.3	9.4	12.3	13.6	15.2
Singapore	93.2	74.7	75.4	100.1	100.0	100.0
Bangkok, Thailand	6.8	8.2	8.7	10.2	12.9	16.1
Ho Chi Minh City, Viet Nam	2.9	3.8	4.7	5.1	4.9	5.0
Southern Asia						
Kabul, Afghanistan	2.4	3.3	3.7	6.1	9.5	9.7
Dhaka, Bangladesh	1.0	1.3	2.3	3.7	5.8	8.1
Calcutta, India	1.2	1.2	1.3	1.3	1.4	1.5
Delhi, India	0.4	0.5	0.6	0.8	1.0	1.3
Tehran, Iran (Islamic Republic of)	6.2	8.7	11.6	13.1	12.4	12.4
Karachi, Pakistan	2.6	3.7	4.8	5.8	6.3	7.2
Western Asia						
Baghdad, Iraq	11.2	14.9	21.2	25.2	21.4	19.3
Tel Aviv, Israel	33.3	34.9	34.6	36.5	40.9	43.0
Amman, Jordan	7.3	12.9	16.9	22.1	25.6	27.9
Kuwait City	67.9	86.9	75.1	60.4	52.9	52.4
Riyadh, Saudi Arabia	1.2	3.1	7.1	10.6	14.0	16.8
Damascus, Syrian Arab Republic	10.5	12.7	14.6	15.6	16.4	17.3
Ankara, Turkey	2.6	3.2	3.8	4.3	4.7	5.3
Istanbul, Turkey	5.2	6.3	7.9	9.9	11.9	14.3

Source: *World Urbanization Prospects, 1990* (United Nations publication, Sales No. E.91.XIII.11).

Delhi, though remaining only the third largest city, has been accounting for a continuing larger share of the urban population. India's other big cities, such as Bombay and Madras, have also been on a relative decline, while the intermediate cities of Bangalore, Patna and Surat have been on the upswing. Still other medium-sized urban centres, such as Ahmedabad and Poona, have shown relative stability.

In Western Asia (excluding Kuwait), Tel Aviv appears to be the most primate metropolis, containing 44.7 per cent of Israel's urban population. It is followed by Amman in Jordan (37.6 per cent), Damascus in the Syrian Arab Republic (32.4 per cent), Baghdad in Iraq (30.0 per cent), Istanbul in Turkey (19.4 per cent), and Riyadh in Saudi Arabia (18.1 per cent). While Baghdad, Damascus, Istanbul and Ankara (Turkey's capital) are showing signs of deconcentration, Tel Aviv, Amman and Riyadh exhibit increasing concentration.

2. *Spatial dispersal policies*

The foregoing patterns of spatial concentration indicate that the large countries in the region, China and India, have had comparatively balanced urban structures, as would be expected a priori. As of 1990, the other countries exhibit wide-ranging degrees of primacy, from a low of 16.4 per cent in Indonesia (also a relatively large country) to a high of 56.8 per cent in Thailand.

Owing to official or popular dissatisfaction with spatial imbalances, Governments have for some time been formulating and implementing, in one way or another and with varying degrees of success, different kinds of explicit population distribution policies (Fuchs and Demko, 1981; Fuchs, 1983; United Nations, 1992). Most notable due to their efficacy are the policies that have been adopted by China. These include policies to contain the growth of large cities and foster the development of intermediate cities and small towns via an array of instruments, such as controls on internal migration and family size, land-use planning, development of coastal cities and special economic zones, service provision and employment generation in small towns. In the Republic of Korea, policy has been directed at controlling the growth of Seoul, reducing the imbalance in the urban structure and stemming rural depopulation, by employing such measures as the green belt project, legislation regulating industrial location, fiscal incentives and disincentives, the construction of satellite cities, the promotion of regional growth poles, and rural development through the "new community movement" (*Saemaul Undong*) which operated from 1970 to the early 1980s.

In Indonesia, the Government has tried to lessen the imbalance between Java and the outer islands, as well as to reduce primacy and uneven urban growth through a transmigration programme, rural and urban development in the outer islands, restraints on Jakarta's growth, and industrial dispersal. More recently, the transmigration programme has been curtailed because of budget constraints, and private sector financing for it is being sought. In Malaysia, the goal has been to equalize regional development and the distribution of economic benefits among ethnic groups, with special attention to ethnic Malays (bumiputra policy), and to develop new townships through rural urbanization (by providing infrastructure, credit and extension services), colonization and resettlement (FELDA), the provision of incentives for industrial location in low-income States, and urban development and renewal.

In the Philippines, the objective has been to stem urban concentration in Manila and foster balanced regional development via integrated rural development, the promotion of regional growth centres and medium-sized cities and, more recently, the agrarian reform programme. The Government of Thailand continues to regard limiting the growth of Bangkok as a priority goal and has employed such measures as land-use and zoning regulations, taxation, industrial dispersal, the promotion of regional growth centres and, more recently, the development of the eastern seaboard. The Government of Myanmar is building a new capital city in the forest south of Yangon for a target population of 4 million. In Viet Nam, the Government has tried to slow the tempo of urbanization, especially the growth of Ho Chi Minh City, through the establishment of new economic zones and the opening up of land for cultivation in the south, designed to lead to a large-scale redistribution of the population.

The thrust of India's five-year development plans has been to improve infrastructure and social services in medium-sized urban centres and small towns so that they can function as service centres for rural areas. In Nepal, spatial policy has attempted to promote rural growth centres and resettlement areas in order to curtail migration from the Hills to the Terai (belt of marshy jungle). Pakistan also aims at modifying internal migration flows through rural development programmes such as the electrification of villages, the provision of health-care centres and farm-to-market roads, and the distribution of land to landless households. Sri Lanka has attempted to curtail metropolitan growth and rural-urban migration through resettlement schemes, the establishment of industries outside Colombo, welfare and income transfers, and the provision of social services to address rural-urban disparities.

In Israel, the Government has tried to reduce the acute primacy of Tel Aviv by prohibiting the construction of housing for immigrants and the conversion of agricultural plots to urban uses, as well as by setting up new towns and land colonization schemes. The Government of Saudi Arabia aspires to balanced regional development by fostering growth centres such as the newly completed industrial city of Jubail on the Persian Gulf and the smaller city of Yanbu on the Red Sea. The Syrian Arab Republic is concerned about the housing shortage in Damascus and Aleppo and therefore wishes to reduce rural-urban migration by redirecting investment away from those large urban centres. Similarly, Turkey hopes to redirect rural-urban migration by developing five major urban centres and making small towns more attractive, especially in the southern and south-eastern regions of the country.

There appears to be a common objective in the above spatial policy goals and instruments, which is to slow or reverse urban concentration by means of programmes to spur the development of alternate urban centres, intermediate cities and small towns, as well as to energize rural areas so as to discourage rural out-migration. Apart from China (which in many senses is a special case), Malaysia and Sri Lanka are notable for having contained urban primacy through effective rural development strategies (Abeysekera, 1981; Lim, Ogawa and Hodge, 1992). More recently, several countries have begun to show signs of primacy reversal, as discussed in the preceding section, including Indonesia, the Islamic Republic of Iran, the Philippines, the Republic of Korea, the Democratic People's Republic of Korea, Thailand and Viet Nam. By contrast, other countries are experiencing either persisting or increasing concentration, notably Bangladesh, Israel and Saudi Arabia. To what extent the apparent primacy reversals are attributable to lagged effects of explicit spatial policies, to the natural forces of agglomeration diseconomies, or to macroeconomic policy reforms is an empirical question that must be investigated systematically in each case.

C. Mega-cities

Many primate cities in Asia have grown to become mega-cities, and others are rapidly increasing in size to assume such a status by the end of the century (table 19).[4] This remarkable development is fraught with both favourable and unfavourable implications. In 1960 there were only two mega-cities in the region, namely, Tokyo and Shanghai. In 1970 Beijing was added, and Asia had three of the ten mega-cities in the world, or two of the five in the developing world. By 1980, Bombay, Calcutta, Osaka and Seoul had joined the group, and Asia had 7 of the 15 mega-cities in the world. In 1990, with Delhi, Jakarta, Manila and Tianjin becoming mega-cities, developing Asia contained 9 of the 14 mega-cities in the less developed regions. By the year 2000, 6 more Asian cities are expected to turn into mega-cities, for a total of 15 out of 22 in the developing world, with only 6 left in the more developed regions (the same number as in 1980).

The present mega-cities in Eastern Asia (including Japan) experienced their peak growth rates in the 1950s, long before they became mega-cities. The growth rates during that period ranged from 4.1 per cent per year for Osaka to 8.4 per cent for Seoul (table 20). In the 1990s, their annual growth rates range from as low as

TABLE 19. POPULATION SIZE OF URBAN AGGLOMERATIONS IN ASIA WITH 8 MILLION OR MORE INHABITANTS IN THE YEAR 2000

(*Millions*)

Agglomeration	*Country*	*1950*	*1970*	*1990*	*2000*
Bangalore	India	0.8	1.6	5.0	8.2
Bangkok	Thailand	1.4	3.1	7.2	10.3
Beijing	China	3.9	8.1	10.8	14.0
Bombay	India	2.9	5.8	11.2	15.4
Calcutta	India	4.4	6.9	11.8	15.7
Dhaka	Bangladesh	0.4	1.5	6.6	12.2
Delhi	India	1.4	3.5	8.8	13.2
Istanbul	Turkey	1.1	2.8	6.7	9.5
Jakarta	Indonesia	2.0	3.9	9.3	13.7
Karachi	Pakistan	1.0	3.1	7.7	11.7
Metro Manila	Philippines	1.5	3.5	8.5	11.8
Osaka	Japan	3.8	7.6	8.5	8.6
Seoul	Republic of Korea	1.0	5.3	11.0	12.7
Shanghai	China	5.3	11.2	13.4	17.0
Tehran	Iran (Islamic Republic of)	1.0	3.3	6.8	8.5
Tianjin	China	2.4	5.2	9.4	12.7
Tokyo	Japan	6.7	14.9	18.1	19.0

Source: *World Urbanization Prospects, 1990* (United Nations publication, Sales No. E.91.XIII.11).

TABLE 20. AVERAGE ANNUAL RATE OF CHANGE OF URBAN AGGLOMERATIONS IN ASIA WITH 8 MILLION OR MORE INHABITANTS IN THE YEAR 2000

(*Percentage*)

Agglomeration	*Country*	*1950-1960*	*1960-1970*	*1970-1980*	*1980-1990*	*1990-2000*
Bangalore	India	4.3	3.2	5.5	5.7	5.0
Bangkok	Thailand	4.6	3.7	4.2	4.1	3.6
Beijing	China	4.7	2.5	1.1	1.8	2.6
Bombay	India	3.4	3.6	3.3	3.3	3.2
Calcutta	India	2.1	2.3	2.7	2.7	2.8
Dhaka	Bangladesh	4.3	8.4	7.8	7.0	6.0
Delhi	India	5.0	4.4	4.5	4.6	4.1
Istanbul	Turkey	4.8	4.7	4.6	4.1	3.6
Jakarta	Indonesia	3.4	3.4	4.2	4.4	4.0
Karachi	Pakistan	5.9	5.2	4.6	4.4	4.1
Metro Manila	Philippines	3.9	4.4	5.2	3.5	3.3
Osaka	Japan	4.1	2.8	0.9	0.2	0.1
Seoul	Republic of Korea	8.4	8.1	4.4	2.8	1.5
Shanghai	China	5.1	2.3	0.5	1.3	2.4
Tehran	Iran (Islamic Republic of)	5.9	5.6	4.4	2.9	2.3
Tianjin	China	4.2	3.7	3.3	2.5	3.1
Tokyo	Japan	4.6	3.3	1.3	0.7	0.5

Source: *World Urbanization Prospects, 1990* (United Nations publication, Sales No. E.91.XIII.11).

0.1 per cent for Osaka to 3.1 per cent for Tianjin. In South-eastern Asia, Manila and Bangkok grew fastest in the 1970s when their populations ranged from 3 million to 3.5 million, while Jakarta's growth rate peaked in the 1980s, also before it attained mega-city status. Nevertheless, all three cities continue to expand at a fairly brisk annual rate of 3.3-4 per cent.

Four of the seven emerging mega-cities in Southern Asia are in India. Already among the largest ones in developing Asia, their annual growth rates have remained within the 2.7-5.7 per cent range. Dhaka has been the fastest growing (6-7 per cent per annum during 1980-2000) of all the actual and potential mega-cities in Asia. Karachi's population is growing at an annual rate of over 4.1 per cent, and Tehran's at about 2.3 per cent. Istanbul, the only city to attain mega-city status by the year 2000 in Western Asia, is among the smallest in the region and is growing at less than 4 per cent annually.

The general slowdown in the growth of the largest urban agglomerations in Asia has been closely related to the process of spatial deconcentration and the emergence of regional growth centres and secondary cities in the various countries, as noted earlier. While these trends may be heartening to policy makers and planners concerned with promoting regional development or preoccupied with the familiar problems of large cities, it should be recognized that the momentum for population growth will push those urban agglomerations to higher-scale levels before they stabilize and regional development can become self-sustaining. National urbanization policy must be based on a sound understanding of the key trends and underlying forces in each country, so that suitable instruments can be adopted to help achieve the objectives of regional development without diminishing the important function of the mega-city in the system of supranational cities, especially in the dynamic Asian region. Prudent and efficient urban management can enhance both the national and the supranational roles of the mega-city while dealing with the typical problems of large agglomerations.

D. FACTORS ASSOCIATED WITH URBANIZATION

1. *Urbanization and development*

The relationship between urbanization and economic development is well recognized. If data on levels of urbanization are plotted against the logarithm of per capita incomes, the result is an upward-sloping curve that approximates a logistic function (Renaud, 1981; Pernia, 1988). Countries at higher economic levels tend to be more highly urbanized. However, the level of income alone does not suffice to explain the pattern of urbanization across countries. Historical and geographical factors as well as internal and international market forces, macroeconomic and sectoral policies, and explicit spatial and urban policies, among others, influence directly or indirectly the nature and pace of urbanization. As the effects of these various factors tend to be peculiar to each country's spatial development, an analysis of urbanization should be a country-specific exercise. Suffice it to say that various country studies allude to the aforementioned factors as the principal explanatory variables of urbanization and population distribution (Pernia, 1977; Mills and Song, 1979; Abeysekera, 1980; Mohan, 1984; Becker, Williamson and Mills, 1992).

A number of cross-national studies show consistent results. For instance, data on 66 low- and middle-income countries were analysed to determine the relative significance of factors in explaining the variation in the level of urbanization, the pace of urbanization, and the degree of urban concentration (Tolley and Thomas, 1987). The hypothesized explanatory variables included the per capita income rank of a country, the growth rate of per capita income, the growth rate of total population, and dummy variables denoting the region in which a country is located.

The best-fitting model in terms of proportion of the variation explained was that having as independent variable the level of urbanization. In it, income rank was the most significant independent variable. Total population growth had a significant negative effect, attributable to the negative association between income and population growth. The growth of per capita income exhibited a negative, though less significant, effect, mainly because countries with high levels of income and urbanization usually have relatively low per capita income growth rates, while those experiencing the most rapid growth in per capita income are at intermediate stages of economic development and urbanization. Further, the variables indicating region suggested that the location of countries and their particular regional circumstances also accounted for intercountry differences in urbanization levels.

Another study using a similar approach and focusing on Asian developing countries yielded consistent and complementary results (Pernia, 1982), indicating that (*a*) the degree of industrialization had a significant positive effect on the level of urbanization, while agricultural growth had a negative effect; (*b*) the growth in manufacturing had a positive though not a significant effect on urbanization level, while total population growth had a significant negative effect; (*c*) the initial level of industrialization had a negative effect on the pace of urbanization, implying that urbanization tends to slow down at higher industrialization (or income) levels; (*d*) agricultural growth continued to exert a potent negative effect on the speed of urbanization, while manufacturing had a strong positive effect; (*e*) total population growth had a potent negative effect on the pace of urbanization; (*f*) manufacturing had a significant positive impact on urban primacy while population growth had a weak negative effect; and (*g*) the degree of openness of the economy exerted a potent positive effect on urban primacy.

In sum, while the growth of manufacturing fosters urbanization, population growth and agricultural development slow it. However, population growth does not reduce urban population growth—it only dampens the rise in the level of urbanization. Agricultural development also seems to retard urbanization in that it allows for greater labour absorption by the rural sector. This effect of agricultural growth may, however, only be present at low levels of economic development. It is likely that, at higher levels, agricultural development would have the reverse effect, as the experience of industrialized countries shows.

The observed negative relationship between agricultural growth and urbanization appears to support the view that a buoyant agricultural sector can slow migration to the cities. This interpretation is an alternative to the more conventional notion that agricultural development leads to rural-urban migration. The crux of the matter would seem to be the nature of the prevailing technology used in agriculture and whether it is labour-intensive or labour displacing. For example, farm mechanization, by raising the marginal productivity of labour, results in labour displacement (Squire, 1981). In contrast, irrigation, by raising both land productivity and average and marginal labour productivities, tends to retain labour. Hence, for a given increment in output, irrigation would retain a bigger workforce in agriculture than mechanization (Cho and Bauer, 1987).

It can be inferred from the foregoing analysis that rapid industrialization and manufacturing growth, coupled with agricultural modernization and overall fertility decline, provided a powerful impetus for the urbanization in the Asian NIEs. These processes also appear to contribute to a quicker transition beyond the peak of urban primacy towards a more balanced urban structure.

In comparison, slow economic growth and weak structural transformation, combined with continuing high fertility, as in Southern Asian and other similarly performing countries, have resulted in a generally slower pace of urbanization but a faster growth of both the urban and the rural populations. Rapid urban population growth, stemming from relatively high urban fertility as well as voluminous rural-urban migration, has resulted in concentration in the emerging mega-cities.

In the NIEs, migration appears to have functioned as an effective mechanism to allocate labour between sectors and alleviate poverty and other pressures on rural areas. In other Asian countries, it has tended to compound the problems in the cities and has done little to ease the burden on the countryside. In many cases, migration from the countryside has been a reaction to natural resource depletion, farm density, natural calamities and civil conflict.

All in all, these processes reflect the interrelations among a country's economic growth, structural economic change, demographic transition, population movements and spatial development. In Western Asia, where the economic base and the landscape are unique, those interrelations appear to have been somewhat different, with agriculture and other primary activities playing a minor role, if any, in the economy.

2. *Implicit and explicit spatial policies*

In addition to the factors and processes underlying urbanization, research findings indicate that government policies matter a great deal (Alonso, 1970; Renaud, 1981; Pernia and others, 1983; Fuchs, Jones and Pernia, 1987). Urbanization and spatial concentration appear to have been partly an unintended consequence of macroeconomic and sectoral policies embodied in the import-substitution industrialization strategies that many Asian countries adopted during the post-war period.

There are indications that concentration of population and industrial activity in a country's capital city has been a rational response to the need for nearness to an international gateway and to national agencies that can furnish the required industrial licences, foreign exchange allocations, credit, tax privileges and other incentives. It is also argued that trade policy in industrialized countries and price policy in developing countries exert a strong influence on urban trends and patterns (Kelley and Williamson, 1984). In addition, excessive regulation and centralization of functions by the national Government contribute to spatial concentration (Mills, 1992); and foreign investment has tended to flow to the metropolis (Fuchs and Pernia, 1987; Herrin and Pernia, 1987). On the demand side, import substitutes find their main market among the urban residents of the capital city.

Infrastructure policy has also tended to favour the national capital, while agricultural policy has tended to penalize the rural sector and subsidize urban consumers (David, 1983; Solon, 1992). To the national capital's favoured status as attested by these implicit spatial policies may be added its long-established advantage as the centre of education, culture, communications, modern facilities and services that further enhance its attractiveness.

In reaction to urban primacy and unbalanced spatial development, several countries in Asia began adopting explicit spatial policies in the latter part of the 1960s. Policy measures to pursue decentralization and balanced regional development have generally included investment incentives for lagging regions, controls on industrial location, controls on migration to the metropolitan region, integrated area development, and the creation of export processing zones and regional growth centres. However, there are indications that these measures have been largely ineffective in countering the deep-seated and more potent spatial biases of implicit policies. Moreover, evaluations of the explicit spatial policy instruments suggest that controls on location were fundamentally unsound, investment incentives ill-conceived, the choice of location for export processing zones was based more on political rather than economic considerations, integrated area development projects took more time than anticipated to have a perceptible impact, and sufficient resources may not have been allocated to the designated regional growth centres (Fuchs, Jones and Pernia, 1987; Herrin and Pernia, 1987; Pernia, 1988).

Policy effectiveness varies from country to country and, as discussed earlier, it is possible that the incipient or advanced primacy reversals that have become apparent recently in a number of countries may be partly attributable to the lagged effects of spatial policies. In the Philippines, for instance, there is some evidence that certain regional growth centres are reducing the dominance of Manila with respect to such global factors as foreign direct investment and exports (Solon, 1992). The deconcentration trends observed may also be ascribable to successful economic policy reforms' having the side effect of removing the inherent spatial biases of previous policies. Cases in point are Malaysia, the Republic of Korea, Taiwan Province of China and Thailand.

E. Conclusion

Asia is, by world standards, far from being highly urbanized. However, within this vast and heterogeneous region, the trends and patterns of population distribution and spatial development vary widely. Even within subregions, there is major variation across countries, rendering meaningless any attempt at generalization.

To some extent, the degree and pace of urbanization can be predicted from the level of economic development and rate of economic growth. Thus, the more progressive economies in Eastern and South-eastern Asia, as well as those in Western Asia, have urbanized rapidly and appear to be experiencing a transition from high to low urban population growth and from high concentration to the diffusion of population distribution.

For countries whose economies are lagging, rapid urban population growth has been largely a source of problems, especially because the numbers of people added annually to large urban centres are staggering. Furthermore, the excessive concentration of population in primate cities, a number of which are or are fast becoming mega-cities, further strains their services and infrastructure. High population growth in urban areas further exacerbates the highly charged problems of poverty, unemployment and underemployment, inadequate infrastructure and housing, deficient social services, and environmental degradation (Asian Development Bank, 1987; Asian Development Bank and Economic Development Institute, 1991; Pernia, 1992; Yeung, 1992). These persistent urban ills indicate that the urban population is growing faster than the economic absorptive capacity of cities. In all likelihood, the continuing gap between urban growth and the capacity of cities to absorb population productively largely stems from a market failure that is further compounded by the inadequacy of policy. This problem arises because, although the majority of Governments in Asia have continuously expressed the need for policy intervention to modify the spatial distribution of their populations, difficulties remain in clarifying the specific nature, extent and timing of the interventions needed or most likely to succeed.

Population distribution and urbanization policies are obviously a country-specific matter that cannot be properly dealt with here. Certain general considerations, however, can be laid down. First, it needs to be reiterated that such policies should be formulated and implemented as a functional part of a national development strategy, so that conflicts with other policies are minimized and complementarities optimized. Second, the implicit spatial biases of macroeconomic and sectoral policies, such as fiscal and monetary policies, trade or industrial and agricultural policies, should be sufficiently understood. The policy changes needed to achieve economic restructuring and promote labour-intensive export industries as well as deregulation and decentralization may have the beneficial side effect of reducing biases favouring large cities.

Third, explicit dispersal policies should be thoroughly re-evaluated for their relevance, timeliness and overall effectiveness. Research and experience show that locational controls and fiscal incentives have generally not been effective and are costly. Access to markets, adequate power and water supply, and communication facilities are more important in determining the location of industries. Any new spatial policy schemes, if warranted, should be formulated in consonance with broader economic and other development policies. Fourth, policy tools for urban management that directly tackle such problems as congestion, pollution, transport, housing, and basic social services are strongly recommended as part of the solution to the overall urbanization problem; but perhaps more important is a determined policy to improve basic infrastructure and services in towns and rural areas, because such a policy will raise efficiency and standards of living. Unless such a policy is pursued vigorously, the national development strategy will continue to address merely the symptoms of a broader malaise that is only partly manifested in urban ills. Furthermore, given the important role played by the demographic factor, a strong policy aimed at slowing population growth, especially in countries with lagging economies, should be a key component of any policy package.

Notes

[1]These 12 countries are Cambodia, Cyprus, India, Indonesia, Japan, the Lao People's Democratic Republic, Lebanon, Maldives, Nepal, the Philippines, Thailand and Viet Nam. By way of interregional comparisons, 37 out of the 52 Governments in Africa that responded, and 24 out of the 33 in Latin America, desired a major change in the spatial distribution of their population.

[2]Related terminology used here also needs to be spelled out. The *level of urbanization* refers to the proportion (percentage) of the population that lives in urban areas at a given time, while the *rate of urbanization* is the pace of change in the proportion urban over time. The latter measure should also be distinguished from the *rate of urban growth* which refers to the pace of change of the number of people living in urban places over a period.

[3]The 50 per cent level roughly marks the inflection point of the logistic urbanization curve, with the 40-60 per cent range representing the phase of fastest urbanization.

[4]A mega-city is defined by the United Nations as an urban agglomeration with 8 million inhabitants or more.

References

Abeysekera, Dayalal (1980). Urbanization and the growth of small towns in Sri Lanka, 1901-71. Papers of the East-West Population Institute, No. 67. Honolulu.

__________ (1981). Regional patterns of intercensal and lifetime migration in Sri Lanka. Papers of the East-West Population Institute, No. 75. Honolulu.

Alonso, William (1970). The question of city size and national policy. Institute of Urban and Regional Development, Working Paper, No. 125. Berkeley, California: University of California Press.

Asian Development Bank (1987). Urban policy issues. Proceedings of the Regional Seminar on Major National Urban Policy Issues, Manila, Philippines, 3-7 February.

__________, and Economic Development Institute (1991). *The Urban Poor and Basic Infrastructure Services in Asia and the Pacific.* Manila: Asian Development Bank.

Becker, C. M., J. G. Williamson and E. S. Mills (1992). *Indian Urbanization and Economic Growth since 1960.* Baltimore, Maryland: Johns Hopkins University Press.

Cho, Lee-Jay, and J. G. Bauer (1987). Population growth and urbanization: what does the future hold? In *Urbanization and Urban Policies in Pacific Asia,* Roland J. Fuchs, Gavin W. Jones and Ernesto M. Pernia, eds. Boulder, Colorado, and London: Westview Press.

David, Cristina C. (1983). Economic policies and Philippine agriculture. Philippine Institute for Development Studies, Working Paper Series, No. 83-02. Manila.

Davis, Kingsley (1965). The urbanization of the human population. *Scientific American* (New York), vol. 213 (September), pp. 28-40.

__________ (1972). *World Urbanization, 1950-70,* 2 vols. Population Monograph Series 4 and 9. Berkeley, California: University of California Press.

Fuchs, Roland J. (1983). Population distribution policies in Asia and the Pacific: current status and future prospects. Papers of the East-West Population Institute, No. 83. Honolulu.

__________, and G. J. Demko (1981). Population distribution measures and the redistribution mechanism. In *Population Distribution Policies in Development Planning,* G. J. Demko and R. J. Fuchs, eds. New York: United Nations, pp. 70-84.

__________, Gavin W. Jones and Ernesto M. Pernia, eds. (1987). *Urbanization and Urban Policies in Pacific Asia.* Boulder, Colorado, and London: Westview Press.

__________, and Ernesto M. Pernia (1987). External economic forces and national spatial development: Japanese direct investment in Pacific Asia. In *Urbanization and Urban Policies in Pacific Asia,* Roland J. Fuchs, Gavin W. Jones and Ernesto M. Pernia, eds. Boulder, Colorado, and London: Westview Press.

Herrin, A. N., and E. M. Pernia (1987). Factors influencing the choice of location: local and foreign firms in the Philippines. *Regional Studies* (Cambridge, United Kingdom), vol. 21, No. 6, pp. 531-541.

Kelley, Allen C., and J. G. Williamson (1984). *What Drives Third World City Growth? A Dynamic General Equilibrium Approach.* Princeton, New Jersey: Princeton University Press.

Lim, Lin Lean, N. Ogawa and R. W. Hodge (1992). The impact of an integrated agricultural development program on migration in Malaysia. Nihon University Population Research Institute (NUPRI), Research Paper Series, No. 61 (Tokyo).

Mills, E. S. (1992). Urban efficiency, productivity, and economic development. In *Proceedings of the World Bank Annual Conference on Development Economics* (Washington, D.C.), pp. 221-235.

__________, and B. N. Song (1979). *Korea's Urbanization and Urban Problems.* Cambridge, Massachusetts: Harvard University Press.

Mohan, Rakesh (1984). The effect of population growth, the pattern of demand and of technology on the process of urbanization. *Journal of Urban Economics* (Duluth, Minnesota) vol. 15, No. 2 (March).

Pernia, Ernesto M. (1977). *Urbanization, Population Growth and Economic Development in the Philippines.* Westport, Connecticut, and London: Greenwood Press.

__________ (1982). Asian urbanization and development: a comparative view. *Philippine Review of Economics and Business* (Manila), vol. 19, pp. 383-403.

__________ (1988). Urbanization and spatial development in the Asian and Pacific region. *Asian Development Review* (Manila), vol. 6, No. 1, pp. 86-105.

__________ (1992). Southeast Asia. In *Sustainable Cities: Urbanization and the Environment in International Perspective,* R. Stren, R. White and J. Whitney, eds. Boulder, Colorado: Westview Press.

__________, and others (1983). *The Spatial and Urban Dimensions of Development in the Philippines.* Manila: Philippine Institute for Development Studies.

Renaud, Bertrand (1981). *National Urbanization Policy in Developing Countries.* New York: Oxford University Press.

Solon, Orville (1992). Global influences of recent urbanization trends in the Philippines. Mimeographed.

Squire, Lyn (1981). *Employment Policy in Developing Countries: A Survey of Issues and Evidence.* New York: Oxford University Press.

Tolley, George S., and Vinod Thomas, eds. (1987). *The Economics of Urbanization and Urban Policies in Developing Countries.* Washington, D.C.: World Bank.

United Nations (1984). *Population Distribution, Migration and Development.* Proceedings of the Expert Group on Population Distribution, Migration and Development, Hammamet (Tunisia), 21-25 March 1983. Sales No. E.84.XIII.3.

__________ (1991). *World Urbanization Prospects, 1990.* Sales No. E.91.XIII.11.

__________ (1992). *World Population Monitoring, 1991: With Special Emphasis on Age Structure.* Population Studies No. 126. Sales No. E.92.XIII.2.

__________ (1993). *World Urbanization Prospects: The 1992 Revision.* Sales No. E.93.XIII.11.

Yeung, Yue-man (1992). China and Hong Kong. In *Sustainable Cities,* R. Stren, R. White and J. Whitney, eds. Boulder, Colorado: Westview Press.

VII. POPULATION DISTRIBUTION IN LATIN AMERICA: IS THERE A TREND TOWARDS POPULATION DECONCENTRATION?

*Alfredo Lattes**

In analysing the spatial dimensions of a society, which are the product of various facets of development, spatial processes, such as the economic and socio-demographic ones, must be singled out. Indeed, the highly interrelated dynamics of economic and socio-demographic processes tend to acquire a degree of autonomy that, in conjunction with the tensions between them, lead in time to some of the spatial problems that are a source of concern for policy makers today. Thus, in 1990, 70 per cent of the Governments of Latin American countries reported that the population distribution in their respective countries needed to be changed considerably (United Nations, 1992).

One of the outstanding characteristics of the development of Latin America in the twentieth century has been its rapid urbanization and the concentration of people and activities in a small number of large cities. This process is rooted both in long-standing traditions, exemplified by the fact that some of today's major urban agglomerations are located where pre-Columbian cities used to stand, and in more recent developments, such as the large-scale international migration that resulted in the rapid urbanization of the Southern Cone countries during the late nineteenth and early twentieth centuries. The role of some of these countries as exporters of raw materials led to the rapid growth of port cities. However, the most direct causes of the rapid urbanization of Latin America are more recent and are linked to the two most significant structural transformations that societies in the region underwent as a result of severe international economic crises.

The first phase coincided with the prevalence of a development model based on internalized economic growth that took hold in the region during the 1930s and 1940s and was responsible for, among other things, the rapid industrialization and modernization of most Latin American countries and for unprecedented rural-urban population shifts. During that phase, the region experienced its highest population growth, particularly in urban and metropolitan areas. As that phase began to wane and a new phase began, Latin America experienced the sharp economic slump of the 1980s, which has come to be known as "the lost decade". Indeed, the Economic Commission for Latin America and the Caribbean (ECLAC) (1990b) estimates that in 1989 the gross national product (GNP) per capita in the region was the same as that 13 years earlier. The second phase of urbanization is associated with the adoption of a new development model directed at opening the national economies by changing the orientation of production and promoting structural adjustment (ECLAC, 1990b). So far, the changes that can be observed include de-industrialization, the growth of the informal sector, the increasing precariousness of employment, increases in urban poverty, and a decline in the attraction exerted by large metropolitan areas. Given these social and economic changes, the main trends in population distribution in Latin America will be presented and analysed below, giving particular attention to urbanization. To set the Latin American experience in perspective, observed and expected trends over the period 1925-2025 will be compared with those of other world regions. Then the different experiences of the 22 Latin American countries with at least 2 million inhabitants in 1990 will be analysed. Two key issues will be addressed, namely, a description of current patterns of population distribution in Latin America and whether those patterns indicate a change in the trend towards further population concentration. It must be noted that the data used are based on national definitions of urban areas. Perfect comparability, therefore, cannot be assured. However, even if a consistent definition of urban area were used (as, for instance, places with at least 20,000 inhabitants), the differences between the levels of urbanization of the different countries considered would increase but the main conclusions would remain the same.

*Centro de Estudios de Población (CENEP), Buenos Aires, Argentina.
The author wishes to acknowledge the assistance of Mariano Sana in the preparation of this paper.

A. LATIN AMERICA IN THE GLOBAL CONTEXT

By 1925, the level of urbanization of Latin America was exactly halfway between that of the currently developed world and that of the rest of the developing countries (table 21), but during the next 50 years, Latin America's level of urbanization grew closer to that of the developed regions. That is, although urbanization in Latin America occurred later than in developed countries as a whole, it preceded the urbanization of Africa and Asia. According to the most recent levels of urbanization estimated by the United Nations (United Nations, 1993), 71.5 per cent of the population of Latin America lived in urban areas in 1990, a level similar to that of developed regions but much higher than that of Africa or Asia, where barely one third of the total population lived in urban areas. Projections of the levels of urbanization indicate that towards the end of the twentieth century, Latin America will have reached the same degree of urbanization as Europe, a level only exceeded by Northern America. By 2025, at least 84 per cent of the population of each of the three regions, Northern America, Latin America and Europe, will live in urban areas whereas only 54 per cent of that in Africa and Asia will do so.

The similarity of Latin America's level of urbanization with that of the most developed regions does not imply that other social and economic changes characteristic of development have also been achieved. Although urbanization and the concentration of population in large cities may be a necessary prerequisite of development, it is not a sufficient condition as well. Indeed, the evidence suggests that development in Latin America is slowing down. Thus, recent estimates indicate that urban poverty is increasing. According to ECLAC (1990a), the number of people living in poverty in Latin America remained almost constant until 1977 and thus decreased in relative terms. However, from 1977 to 1986, that number increased from 112 million to 170 million, the share thus rising from 33 to 39 per cent of the population. This increase occurred mostly in urban areas, so that by 1986 the number of urban poor in the region stood at about 94 million.

The rapid urbanization of Latin America occurred during a period of marked population growth. Table 22 shows that during 1925-1975, the population of Latin America grew faster than that of any other region and that the urban population also grew more rapidly than that of almost every other region except for the urban population in Africa during 1950-1975. Consequently, in the course of 50 years, the total population of Latin America more than tripled (increased by a factor of 3.24) and its urban population grew almost eightfold (increased by a factor of 7.96). The resulting net rural-urban redistribution involved 117 million people.

During the 1940s, the average annual growth of the urban population of Latin America reached 5.1 per cent (Lattes, 1990), the highest growth rate observed in any region over a 10-year period[1] with the possible exception

TABLE 21. URBANIZATION LEVELS FOR SELECTED MAJOR WORLD REGIONS, 1925-2025
(*Percentage*)

Major region	*Year* 1925	1950	1975	2000	2025
World	20.5	29.3	37.7	47.6	61.2
More developed regions	40.1	54.3	68.8	75.8	83.8
Less developed regions	9.3	17.0	26.3	40.3	56.7
Europe	47.8	56.2	68.8	76.6	84.5
Northern America	53.8	63.9	73.8	77.7	85.0
Africa	8.0	14.5	25.0	37.6	54.1
Asia	9.5	16.4	24.1	37.1	54.4
Latin America	25.0	41.6	61.2	76.6	84.4

Sources: P. Hauser and R. Gardner, "Urban future: trends and prospects", East-West Population Institute, reprint No. 146 (Honolulu, 1982); and *World Urbanization Prospects, 1990* (United Nations publication, Sales No. E.91.XIII.11).

NOTE: Data for Latin America include both the 22 countries considered in most parts of the present paper and smaller countries, mostly in the Caribbean region, whose inclusion does not change significantly the level of urbanization in the whole region.

TABLE 22. ANNUAL GROWTH RATES FOR THE TOTAL AND URBAN POPULATIONS OF THE MAJOR WORLD REGIONS AND RATE OF URBANIZATION FOR SELECTED PERIODS
(Percentage)

Region	*Period*			
	1925-1950	*1950-1975*	*1975-2000*	*2000-2025*
World				
Total growth rate	1.0	1.9	1.7	1.2
Urban growth rate	2.4	2.9	2.6	2.2
Urbanization rate	1.4	1.0	0.9	1.0
Europe				
Total growth rate	0.6	0.8	0.3	0.1
Urban growth rate	1.3	1.6	0.8	0.5
Urbanization rate	0.7	0.8	0.4	0.4
North America				
Total growth rate	1.1	1.5	1.0	0.7
Urban growth rate	1.8	2.0	1.2	1.0
Urbanization rate	0.7	0.6	0.2	0.4
Africa				
Total growth rate	1.5	2.5	2.9	2.5
Urban growth rate	3.9	4.7	4.5	3.9
Urbanization rate	2.4	2.2	1.6	1.5
Asia				
Total growth rate	1.0	2.1	1.8	1.1
Urban growth rate	3.2	3.7	3.5	2.7
Urbanization rate	2.2	1.5	1.7	1.5
Latin America				
Total growth rate	2.1	2.6	2.0	1.2
Urban growth rate	4.1	4.2	2.9	1.6
Urbanization rate	2.0	1.5	0.9	0.4

Sources: P. Hauser and R. Gardner, "Urban future: trends and prospects", East-West Population Institute, reprint No. 146 (Honolulu, 1982); and *World Urbanization Prospects, 1990* (United Nations publication, Sales No. E.91.XIII.11).

NOTE: Latin America includes all the Caribbean countries.

of that for the urban population of the Union of Soviet Socialist Republics (USSR) during 1930-1939 which, according to Hauser and Gardner (1982), reached 5.4 per cent annually. Since then, the rate of urban growth in Latin America has been declining steadily. Total population growth has also been declining, but the onset of its decline was a decade later (Lattes, 1990). The difference between those two rates, known as the rate of urbanization, has also been falling, passing from 2.8 per cent in the 1940s to 1 per cent in the 1980s. The reduction is mostly due to the sharp decline in urban population growth (table 22).

Since 1925 Latin America has urbanized at a considerably more rapid pace than Northern America and Europe, though somewhat less rapidly than Asia or Africa. Thus, it will have taken 75 years (from 1925 to 2000) for the level of urbanization in Northern America to rise from 53.8 to 77.7 per cent, while Latin America will have covered the same ground in only half the time.

The steady decline in both the growth rate of the urban population and the rate of urbanization in Latin America should not obscure another relevant facet of this phenomenon: the increments to the urban population of the region have continued to increase and will only begin to decline in the next century. As table 23 shows, during 1975-2000, the growth of the urban population in absolute terms is expected to be larger than that of the total population of the region. In Africa, in contrast, the absolute growth of the urban population amounts to only about half of total population growth despite the fact that

TABLE 23. ABSOLUTE INCREASE IN THE URBAN POPULATION OF AFRICA AND LATIN AMERICA AND AS A PERCENTAGE OF TOTAL POPULATION GROWTH, 1950-2025

Period	*Africa*		*Latin America*	
	Urban population growth		*Urban population growth*	
	Number (millions)	*As percentage of total growth*	*Number (millions)*	*As percentage of total growth*
1950-1975	71.6	37.2	127.3	82.1
1975-2000	218.3	49.5	204.5	100.9
2000-2025	534.6	73.6	191.8	107.4

Source: *World Urbanization Prospects: The 1992 Revision* (United Nations publication, Sales No. E.93.XIII.11).

the region is characterized by the highest rate of urban growth among the major world regions.

B. DYNAMICS OF URBANIZATION AMONG LATIN AMERICAN COUNTRIES

There is considerable diversity in the dynamics of urbanization in the countries of the Latin American region, diversity that is closely associated with the level of development of the different countries. Thus, El Salvador, Guatemala, Haiti and Honduras, countries with relatively low levels of urbanization, also have low indices of human development, whereas Argentina, Chile, Uruguay and Venezuela, all highly urbanized, have higher indices of human development (UNDP, 1990).

Figure IV displays schematically the distribution of the level of urbanization of the 22 major countries in the region for different periods. The changing distribution over time corroborates that there has been a significant rise in the median level of urbanization in the region but it also shows that there has been growing diversity. Thus, although the extremes of the distribution have changed relatively little and in the expected direction, the interquartile range (represented by the width of the central box) has been increasing. There has also been a reduction of the upper half of the distribution, indicating that the level of urbanization of the most urbanized countries is converging. However, no such convergence is evident among the less urbanized countries as the constant width of the left side of the distribution indicates.

Table 24 shows the estimated level of urbanization for all major Latin American countries at the beginning of each decade since 1950. The data reveal that, whereas in 1950 only 4 countries in the region had over half of their population living in urban areas, in 1990 there were 16 such countries and by the year 2000 there will be 19. Individual countries have experienced varying changes in rank according to level of urbanization. Thus, Venezuela advanced from fourth place in 1950 to first in 1990; Brazil moved from eleventh to fifth place during the same period; the Dominican Republic, a country that was among the three least urbanized in 1950, experienced the greatest change and by 1990 occupied eleventh place. El Salvador, in contrast, experienced a very small change in its level of urbanization and consequently passed from tenth place in 1950 to nineteenth in 1990. Throughout the period considered, the most urbanized countries remained the same (Argentina, Chile, Uruguay and Venezuela) and countries such as Guatemala, Haiti and Honduras remained (also consistently) among the least urbanized in the region.

Because of the strong concentration of Latin America's population in a few countries, and the positive relation between population size and level of urbanization, the trends observed for the region as a whole reflect mostly those of a small group of countries. Thus, in 1990, 82.3 per cent of the total population and 87.2 per cent of the urban population in Latin America lived in the eight largest countries of the region, namely, Argentina, Brazil, Chile, Colombia, Cuba, Mexico, Peru and Venezuela. Those eight countries plus Uruguay and Puerto Rico are the ten most urbanized. The large flows of international migration that converged on the region in the nineteenth and twentieth centuries have been largely responsible for such a concentration of population, both between and within countries, since immigrants tended to settle in the larger cities. Only eight countries received 95 per cent of the overseas migration that has reached Latin America

Figure IV. Distribution of Latin American countries by level of urbanization, 1950-2000

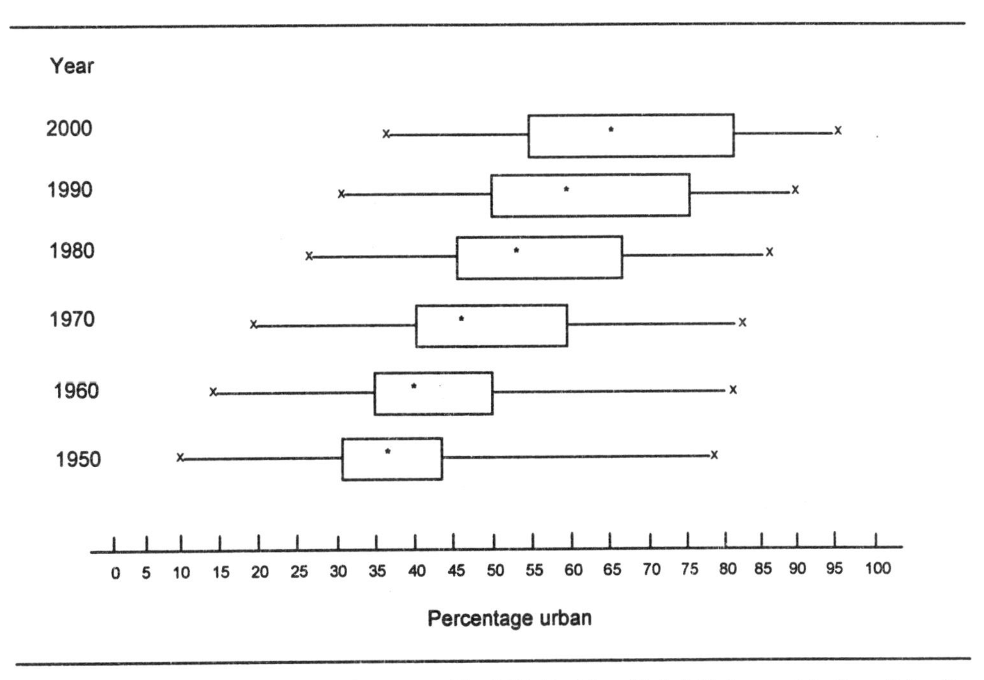

Source: World Urbanization Prospects: The 1992 Revision (United Nations publication, Sales No. E.93.XIII.11).

NOTE: This is a schematic plot of the distribution of 22 Latin American countries by level of urbanization at different times. The box represents the interquartile range, the asterisk within it represents the median and the crosses represent the extremes of the distribution. Half of the countries considered fall within the interquartile range.

since the early nineteenth century (Lattes, 1985). Among them, seven are among the most highly urbanized today. In Mexico and Peru, both of which received relatively low overseas migration, the rural-urban migration of the indigenous population was, since 1800, a major force driving urbanization.

Table 25 shows the countries of Latin America grouped into six geographical subregions and listed according to level of urbanization in 1990. Brazil and Mexico are each considered a subregion because of their size and importance. Both are highly urbanized. Central America is the only subregion where the population is still predominantly rural. The Caribbean, represented here by only the major countries in that subregion, is also characterized by a relatively low level of urbanization and it groups countries with very different levels of urbanization. Similarly, the Andean subregion, despite having a higher average level of urbanization, also includes countries with fairly different levels. The Southern Cone, being the most highly urbanized subregion, shows greater homogeneity, but it also includes moderately urbanized Paraguay.

Analysis of the urbanization trends of individual countries permits the identification of five groups. The first, characterized by an urbanization level of at least 80 per cent in 2000, includes Argentina, Brazil, Chile, Uruguay and Venezuela, whose joint urbanization level in 1990 was 84.2 per cent. The second group, characterized by urbanization levels of between 70 per cent and 80 per cent, includes Colombia, Cuba, Mexico, Peru and Puerto Rico, whose joint urbanization level was 76.8 per cent in 1990. The third group, with levels of urbanization varying between 60 and 70 per cent, includes only three countries, the Dominican Republic, Ecuador and Nicaragua that, as a whole, will have 65.9 of their population living in urban areas by the year 2000. The fourth group, with urbanization levels falling within the 50 to 60 per cent range, includes Bolivia, Costa Rica, Honduras, Jamaica, Panama and Paraguay, with a joint urbanization level of 48.7 per cent. Lastly, El Salvador,

TABLE 24. MAIN COUNTRIES OF LATIN AMERICA ACCORDING TO LEVEL OF URBANIZATION FOR SELECTED YEARS

	Percentage of the population in urban areas										
	1950		*1960*		*1970*		*1980*		*1990*		*2000*
Haiti	12.2	Haiti	15.6	Haiti	19.8	Haiti	23.7	Haiti	28.6	Haiti	34.9
Honduras	17.6	Honduras	22.7	Honduras	28.9	Honduras	36.0	Guatemala	39.4	Guatemala	44.1
Dominican Republic	23.8	Dominican Republic	30.2	Guatemala	35.5	Guatemala	37.4	Honduras	43.7	El Salvador	49.5
Jamaica	26.7	Guatemala	32.4	Paraguay	37.1	El Salvador	41.5	El Salvador	44.4	Honduras	51.6
Ecuador	28.2	Jamaica	33.8	El Salvador	39.4	Paraguay	41.7	Costa Rica	47.1	Costa Rica	52.7
Guatemala	29.5	Ecuador	34.4	Ecuador	39.5	Costa Rica	43.1	Paraguay	47.5	Paraguay	54.0
Costa Rica	33.5	Paraguay	35.6	Costa Rica	39.7	Bolivia	44.2	Bolivia	51.0	Panama	57.2
Paraguay	34.6	Costa Rica	36.6	Dominican Republic	40.3	Jamaica	46.8	Jamaica	52.3	Bolivia	57.9
Nicaragua	34.9	El Salvador	38.3	Bolivia	40.7	Ecuador	47.0	Panama	52.9	Jamaica	58.5
Peru	35.5	Bolivia	39.3	Jamaica	41.5	Panama	49.7	Ecuador	56.3	Ecuador	64.5
Panama	35.7	Nicaragua	39.5	Nicaragua	47.0	Dominican Republic	50.5	Nicaragua	59.8	Nicaragua	65.9
Brazil	36.0	Panama	41.2	Panama	47.6	Nicaragua	53.4	Dominican Republic	60.4	Dominican Republic	68.1
El Salvador	36.5	Puerto Rico	44.5	Brazil	55.8	Colombia	63.9	Peru	69.8	Peru	74.5
Colombia	37.1	Brazil	44.9	Colombia	57.2	Peru	64.6	Colombia	70.0	Colombia	75.2
Bolivia	37.8	Peru	46.3	Peru	57.4	Brazil	66.2	Mexico	72.6	Mexico	77.7
Puerto Rico	40.6	Colombia	48.2	Puerto Rico	58.3	Mexico	66.3	Cuba	73.6	Cuba	78.1
Mexico	42.7	Mexico	50.8	Mexico	59.0	Puerto Rico	67.0	Puerto Rico	73.9	Puerto Rico	78.8
Cuba	49.4	Cuba	54.9	Cuba	60.2	Cuba	68.1	Brazil	75.2	Brazil	81.5
Venezuela	53.2	Venezuela	66.6	Venezuela	72.4	Chile	81.2	Chile	84.6	Chile	87.1
Chile	58.4	Chile	67.8	Chile	75.2	Argentina	82.9	Argentina	86.1	Argentina	88.6
Argentina	65.3	Argentina	73.6	Argentina	78.4	Venezuela	83.3	Uruguay	88.9	Uruguay	91.4
Uruguay	78.0	Uruguay	80.1	Uruguay	82.1	Uruguay	85.2	Venezuela	90.5	Venezuela	94.5
TOTAL	41.7		49.6		57.5		65.2		71.7		76.8
Median	35.8	Median	40.4	Median	47.3	Median	52.0	Median	60.1	Median	67.0
Lower quartile	29.5	Lower quartile	34.4	Lower quartile	39.5	Lower quartile	43.1	Lower quartile	47.5	Lower quartile	54.0
Upper quartile	42.7	Upper quartile	50.8	Upper quartile	59.0	Upper quartile	67.0	Upper quartile	73.9	Upper quartile	78.8
Interquartile range	13.2	Interquartile range	16.3	Interquartile range	19.5	Interquartile range	23.9	Interquartile range	26.5	Interquartile range	24.8

Source: *World Urbanization Prospects: The 1992 Revision* (United Nations publication, Sales No. E.93.XIII.11).

TABLE 25. URBANIZATION LEVELS FOR LATIN AMERICAN GEOGRAPHICAL SUBREGIONS, 1990

Subregion	*Percentage of the population in urban areas in 1990*
Central America	45.7
Nicaragua	59.8
Panama	52.9
Costa Rica	47.1
El Salvador	44.4
Honduras	43.7
Guatemala	39.4
Caribbean	59.1
Puerto Rico	73.9
Cuba	73.6
Dominican Republic	60.4
Jamaica	52.3
Haiti	28.6
Andean region	71.2
Venezuela	90.5
Colombia	70.0
Peru	69.8
Ecuador	56.3
Bolivia	51.0
Mexico	72.6
Brazil	75.2
Southern Cone	82.8
Uruguay	88.9
Argentina	86.1
Chile	84.6
Paraguay	47.5

Source: *World Urbanization Prospects: The 1992 Revision* (United Nations publication, Sales No. E.93.XIII.11).

Guatemala and Haiti, countries whose levels of urbanization are not expected to reach more than 50 per cent by the year 2000, constitute the fifth group with an urbanization level of 37.3 per cent in 1990.

Table 26 shows the growth rates of the total, urban, rural and metropolitan populations of each Latin American country during 1980-1990 in conjunction with the 1990 urbanization levels. As expected, the different growth rates tend to decline as the level of urbanization increases. However, the correlation between each type of growth rate and the level of urbanization is still far from unity. For instance, the correlation between the growth rate of the rural population and the level of urbanization is only 0.73. The cases of countries that deviate markedly from the linear relation between level of urbanization and population growth deserve attention. Venezuela, with the highest level of urbanization in 1990, is still experiencing high rates of total and urban population growth. While Brazil also has a relatively high level of urbanization, its urban and total populations grow fast. Among the countries with low levels of urbanization, Jamaica has very low growth rates among the total, urban and metropolitan populations. El Salvador and Haiti, having very low levels of urbanization, are also experiencing low rates of total population growth, as a result of net international emigration and high mortality.

There is also a high correlation between the rate of growth of the total population and that of the largest metropolitan area in each country (0.67), a fact that supports the view that natural increase is a major determinant of metropolitan growth. There are, however, some deviant cases. Tegucigalpa, for instance, is growing at a considerably lower rate than might be expected from the growth rate of the total population of Honduras. To a lesser extent, the same is true in the case of Panama. In contrast, San Juan, Puerto Rico, is growing at a higher rate than that of the total population of the island.

Since 1950, many countries have experienced major changes in demographic trends, particularly those associated with the demographic transition that involves the passage from high mortality and fertility levels to low ones. Consequently, it is useful to contrast other facets of population dynamics with the urbanization process. Although Argentina, Chile, Uruguay and Venezuela have attained similar levels of urbanization, the growth rates of their urban populations are far from similar (table 26). Thus, if current rates were to remain constant, the urban population of Venezuela would double in 21 years, that of Chile in 33 years, the one in Argentina in 41 years, and that of Uruguay in 69 years. This means that, although their levels of urbanization may be similar, the four countries considered are likely to face very different social and economic conditions associated with urban growth.

Figure V illustrates the differences between the demographic dynamics experienced by Uruguay and Venezuela. The demographic histories of these countries before 1950 are reflected in the different positions they occupy by 1950. Since then, the trends experienced by the two countries have led to a convergence of their urbanization levels although the underlying dynamics of urban and total population growth remain quite distinct. However, both countries evince a convergence of urban and total population growth as the level of urbanization increases. That tendency is also noticeable when the trends experienced by other countries are considered. As

TABLE 26. LEVEL OF URBANIZATION, ANNUAL GROWTH RATES OF THE TOTAL, URBAN, RURAL AND MAJOR-CITY POPULATIONS, AND RATE OF URBANIZATION FOR THE MAIN LATIN AMERICAN COUNTRIES, 1980-1990

(Percentage)

Country	Level of urbanization (1990)	1980-1990 Total rate of growth	Urban rate of growth	Rural rate of growth	Rate of growth of major city	Rate of urbanization
Venezuela	90.5	2.5	3.3	-3.1	3.3	0.8
Uruguay	88.9	0.6	1.0	-2.3	0.1	0.4
Argentina	86.1	1.4	1.7	-0.7	1.5	0.4
Chile	84.6	1.7	2.1	-0.3	2.4	0.4
Brazil	75.2	2.1	3.3	-1.0	3.6	1.3
Puerto Rico	73.9	1.0	1.9	-1.4	2.3	1.0
Cuba	73.6	0.9	1.7	-1.0	0.9	0.8
Mexico	72.6	2.3	3.2	0.3	3.3	0.9
Colombia	70.0	2.0	2.9	0.1	3.2	0.9
Peru	69.8	2.2	3.0	0.6	3.4	0.8
Dominican Republic	60.4	2.3	4.1	0.1	4.5	1.8
Nicaragua	59.8	2.7	3.8	1.3	4.7	1.1
Ecuador	56.3	2.6	4.4	0.7	4.4	1.8
Panama	52.9	2.1	2.7	1.5	1.7	0.6
Jamaica	52.3	1.3	2.4	0.2	1.0	1.1
Bolivia	51.0	2.5	3.9	1.2	4.7	1.4
Paraguay	47.5	3.1	4.4	2.0	5.1	1.3
Costa Rica	47.1	2.8	3.7	2.1	4.1	0.9
El Salvador	44.4	1.3	2.0	0.8	2.9	0.7
Honduras	43.7	3.4	5.3	2.1	3.4	1.9
Guatemala	39.4	2.8	3.4	2.5	4.7	0.5
Haiti	28.6	1.9	3.8	1.3	3.9	1.9
TOTAL	71.7	2.1	3.0	0.0	3.1	1.0

Source: *World Urbanization Prospects: The 1992 Revision* (United Nations publication, Sales No. E.93.XIII.11).
NOTE: Countries are ordered by level of urbanization in 1990.

Figure V. Growth rates for the total and the urban population according to level of urbanization, Venezuela and Uruguay, 1950-2000

(Percentage)

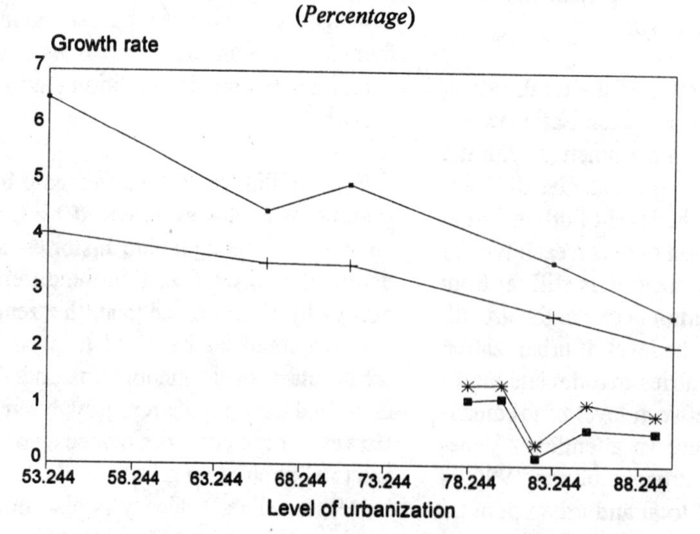

the level of urbanization rises, the difference between the growth rate of the total population and that of the urban population, that is, the rate of urbanization, tends to decline. Nevertheless, the exact path that a country follows towards high urbanization depends on the historical and structural processes that condition its demographic evolution.

Although Latin America is highly urbanized and its rural population is growing very slowly, rural population trends should not be ignored since in several countries the population living in rural areas is still growing at moderate rates. In Central America in particular, the rural population is still growing at 1.9 per cent annually, a rate that is expected to decline. In Haiti as well, the rural population is growing at 1.3 per cent per year, that of Bolivia has an annual growth rate of 1.2 per cent, and in Paraguay the population of the rural areas has an annual growth rate of 2 per cent.

Since the level of urbanization cannot surpass 100 per cent, the rate of urbanization must necessarily decline as the level of urbanization rises. The resulting correlation between those two indices affects comparisons. For that reason, Eldridge (1942) proposed an urbanization index reflecting another dimension of the process and exempt from such problems. The alternative index is the proportion of the rural population that becomes urbanized during a decade and its values for the Latin American countries are shown in table 27. For Latin America as a whole, the urbanization rate of 1.7 per cent estimated for 1950-1960 involved the transfer of 13.5 per cent of the rural population present in 1950 to urban areas during the decade, whereas the urbanization rate for 1980-1990 of 1 per cent is consistent with the transfer of 18.8 per cent of the 1980 rural population to urban areas. According to the urbanization index, Latin America and its most urbanized countries have recently reached a turning point with respect to increases in urbanization since,

TABLE 27. URBANIZATION INDEX FOR SELECTED LATIN AMERICAN COUNTRIES, 1950-2000

	Urbanization index (percentage)				
Country	*1950-1960*	*1960-1970*	*1970-1980*	*1980-1990*	*1990-2000*
Argentina	23.9	18.1	20.7	18.8	18.2
Bolivia	2.4	2.4	5.8	12.2	14.0
Brazil	14.0	19.8	23.6	26.6	25.3
Chile	22.6	23.0	24.3	17.9	16.3
Colombia	17.7	17.4	15.7	16.8	17.4
Costa Rica	4.6	4.9	5.7	7.0	10.6
Cuba	10.8	11.8	19.9	17.3	17.1
Dominican Republic	8.5	14.4	17.1	20.0	19.6
Ecuador	8.6	7.8	12.3	17.6	18.8
El Salvador	2.9	1.8	3.5	4.9	9.1
Guatemala	4.2	4.6	2.9	3.3	7.7
Haiti	3.9	4.9	4.9	6.4	8.8
Honduras	6.3	8.0	9.9	12.0	14.1
Jamaica	9.6	11.7	9.0	10.4	13.0
Mexico	14.1	16.8	17.9	18.5	18.7
Nicaragua	7.1	12.4	12.1	13.6	15.2
Panama	8.5	10.9	4.1	6.3	9.2
Paraguay	1.5	2.4	7.3	9.9	12.5
Peru	16.7	20.7	16.8	14.7	15.6
Puerto Rico	6.6	24.9	20.8	21.0	18.7
Uruguay	9.8	10.0	17.5	24.9	22.9
Venezuela	28.6	17.5	39.3	43.0	42.2
TOTAL	13.5	15.9	18.1	18.8	17.9

Source: *World Urbanization Prospects: The 1992 Revision* (United Nations publication, Sales No. E.93.XIII.11).
NOTE: The urbanization index (*IU*) was obtained using the equation $IU = [PU(t+10) - PU(t)]/[100 - PU(t)]$, where *PU* is the proportion urban and *t* is the length of the intercensal period in years.

beginning in 1990-2000, increases in urbanization will involve tranferring declining proportions of the rural population to urban areas.

C. Structure of the urban population and population dynamics in large cities

Any analysis of urbanization in Latin America would be incomplete if changes in the urban structure were not considered. According to Eldridge (1942), the process of urbanization involves the growth of existing points of concentration as much as the emergence of new ones. Vapnarsky (1969) adds that, until the level of urbanization reaches its asymptotic limit, the quantity, size and density of the points of concentration, that is, their urban structure, will continue to change.

According to United Nations estimates of the population in large cities, the concentration of the Latin American urban population in the largest city of each country reached a maximum sometime during 1945-1955, declined until 1980, and has remained constant since then.[2] Consequently, during 1980-2000, the population in the largest cities and the rest of the urban population will grow at the same rate (table 28). The question thus arises of whether the deconcentration of the largest cities is coming to a halt. To provide an answer, an analysis of urban dynamics at the national level is necessary, since there is a great variety of experiences. In 1990, for

Table 28. Annual growth rate for the major city in each country and for the rest of the urban population, 1950-2000

(Percentage)

	Growth rate of the major city					*Growth rate of the rest of the urban population*				
Country	*1950-1960*	*1960-1970*	*1970-1980*	*1980-1990*	*1990-2000*	*1950-1960*	*1960-1970*	*1970-1980*	*1980-1990*	*1990-2000*
Argentina	2.9	2.2	1.6	1.5	1.1	3.1	2.1	2.6	1.9	1.6
Bolivia	3.3	3.3	4.0	4.7	4.2	2.2	2.4	3.1	3.6	3.3
Brazil	6.6	5.4	4.1	3.6	2.4	5.1	4.9	4.1	3.3	2.3
Chile	4.2	3.3	2.7	2.4	1.7	3.4	3.2	2.1	1.8	1.8
Colombia	6.6	6.0	4.0	3.2	2.8	5.3	4.3	3.1	2.8	2.2
Costa Rica	4.4	4.4	4.4	4.1	3.3	4.7	3.9	2.1	2.9	3.5
Cuba	2.1	2.1	0.9	0.9	1.0	3.3	3.4	3.2	2.0	1.6
Dominican Republic	7.1	6.3	5.1	4.5	3.2	4.5	5.7	4.5	3.6	2.9
Ecuador	5.7	4.4	4.4	4.4	3.6	4.5	4.6	4.8	4.4	3.5
El Salvador	4.5	2.8	2.9	2.9	3.3	2.9	3.9	2.8	1.7	3.2
Guatemala	2.6	2.3	4.4	4.7	4.3	4.8	4.6	2.6	2.4	3.7
Haiti	5.8	5.8	4.2	3.9	3.9	2.8	2.5	2.8	3.7	4.2
Honduras	3.1	3.5	3.2	3.4	3.7	8.6	6.7	6.6	6.0	4.8
Jamaica	2.1	2.3	0.9	1.0	1.6	13.1	6.4	5.3	4.0	2.7
Mexico	5.4	5.5	4.3	3.3	2.4	4.4	4.4	3.9	3.1	2.7
Nicaragua	5.9	6.4	5.2	4.7	4.1	3.5	4.0	3.8	3.2	4.6
Panama	7.7	3.0	1.1	1.7	2.1	0.7	5.8	4.3	3.4	2.8
Paraguay	3.0	4.9	5.5	5.1	4.1	3.0	1.8	2.5	3.2	3.5
Peru	5.5	5.5	4.1	3.4	2.7	5.2	4.7	3.7	2.7	2.6
Puerto Rico	1.5	2.5	4.5	2.3	1.8	1.5	5.6	1.7	1.5	1.2
Uruguay	0.1	0.1	0.1	0.1	0.3	3.8	2.6	1.4	1.8	1.2
Venezuela	6.4	4.7	3.7	3.3	2.4	6.2	4.2	5.3	3.4	2.4
Total	4.3	4.1	3.4	3.1	2.4	4.5	4.2	3.7	3.0	2.4

Source: *World Urbanization Prospects: The 1992 Revision* (United Nations publication, Sales No. E.93.XIII.11).

instance, the proportion of the urban population living in the largest city varied between 15.5 per cent for São Paulo in Brazil to 71.1 per cent for San José in Costa Rica (table 29).

Consideration of trends in the different countries indicates that Latin America is experiencing a reversal of the historical process of concentration in the largest cities. Some countries have experienced that reversal earlier than others, and some have yet to experience it (table 29). However, by 1990 a decline in the urban predominance of the largest city had been observed in eight countries, namely, Argentina, Cuba, Honduras, Jamaica, Panama, Uruguay, Venezuela and possibly Ecuador. In most of these countries, the decline started in the 1970s or earlier. In another five countries, Chile, Costa Rica, Haiti, Mexico and Nicaragua, the reversal was recent (1980-1990) or has just started (1990-2000). In the nine remaining countries, the concentration process in the largest cities continued, although at a markedly slower pace in most of them. Consequently, it is not yet apparent that deconcentration is about to stop.

It is no longer news that the predominance of the largest cities in the region is diminishing. Evidence on that point has been growing since the mid-1970s.[3] The news should be that the recent and profound economic changes that have taken place may have acted as a catalyst facilitating urban decentralization and hence deconcentration, particularly by fostering transfers from the

TABLE 29. POPULATION OF THE MAJOR CITY IN EACH LATIN AMERICAN COUNTRY AS A PERCENTAGE OF THE URBAN POPULATION, 1950-2000

Country	*Population of the major city as a percentage of the urban population*					
	1950	*1960*	*1970*	*1980*	*1990*	*2000*
Argentina	45.0	44.6	44.8	42.4	41.4	40.1
Bolivia	25.4	27.5	29.3	31.2	33.8	35.8
Brazil	12.6	14.4	15.1	15.1	15.5	15.7
Chile	37.5	39.3	39.7	41.1	42.5	42.3
Colombia	15.3	17.0	19.4	20.8	21.5	22.6
Costa Rica	63.3	62.6	63.8	68.7	71.1	70.8
Cuba	39.7	36.9	34.0	29.0	26.9	25.7
Dominican Republic	39.2	45.6	47.1	48.6	50.9	51.4
Ecuador	27.1	29.4	29.0	28.3	28.2	28.4
El Salvador	22.6	25.4	23.3	23.4	25.7	25.8
Guatemala	46.8	41.2	35.9	40.2	46.0	47.5
Haiti	36.3	43.3	51.6	55.1	55.6	54.7
Honduras	56.9	43.2	35.6	28.1	23.2	21.2
Jamaica	90.7	76.4	68.3	58.1	50.6	47.9
Mexico	27.0	29.1	31.6	32.5	32.9	32.1
Nicaragua	28.4	33.5	39.0	42.4	46.1	44.8
Panama	37.6	55.0	48.0	40.1	36.0	34.4
Paraguay	42.8	42.8	50.5	57.9	62.6	63.8
Peru	35.9	36.7	38.7	39.7	41.5	41.6
Puerto Rico	51.9	52.0	44.4	51.2	53.3	54.5
Uruguay	65.5	56.8	50.6	47.5	43.5	41.2
Venezuela	25.3	25.7	26.7	23.6	23.4	23.4
TOTAL	29.0	28.6	28.4	27.8	27.8	27.8

Source: *World Urbanization Prospects: The 1992 Revision* (United Nations publication, Sales No. E.93.XIII.11).

largest city to those of intermediate size. That is, in view of the urban redistribution that has taken place in past decades, the issue is whether the diminishing urban predominance of the largest cities is due to recent economic changes or not.

The most recent census data would indicate that some of the largest cities in Latin America have grown at a slower rate than that estimated by the United Nations. That is the case as regards Buenos Aires and Mexico City. In the case of Buenos Aires, the 1991 census enumerated 11.3 million persons in the metropolitan area (INDEC, n.d.), whereas the 1990 assessment by the United Nations estimated a population of 11.5 million for 1990 (United Nations, 1991). In the case of Mexico City, the 1990 census enumerated 15.1 million persons (INEGI, 1991), a figure considerably lower than the 20.2 million estimated by the United Nations for the same year (United Nations, 1991). Hence, there is reason to believe that population redistribution from the largest cities to other urban areas is more advanced than indicated in table 29.

However, one must point out that the evidence regarding the deconcentration of the largest cities in Latin America during the last decade is scarce or contradictory in a number of cases. Verduzco (1992) shows that the three largest cities of Mexico (Mexico City, Guadalajara and Monterrey) grew at considerably lower rates than expected and suggests that a reversal of the previous tendency towards concentration occurred. Ruiz Chiapetto (1990) maintains that the economic crisis of the 1980s inhibited the continued concentration of the population in the large metropolitan areas of Mexico. Lattes and Recchini de Lattes (1992) show that the predominance of Buenos Aires with respect to the total as well as the urban population of Argentina continued to decline during 1980-1990; at the same time, there was a growing concentration of the urban population in the medium-sized cities of Argentina. Portes (1989), in analysing the cases of Bogotá, Montevideo and Santiago de Chile, finds evidence supporting the occurrence of a reversal of the trend towards concentration by observing that large cities grew at slower rates. However, Portes himself notes that it is too early to tell whether the changes observed represent only a temporary anomaly or a lasting trend. In contrast, Duarte (1991) reports that, during the 1980s, the primacy of Santo Domingo in the Dominican Republic developed through a process of conurbation, and Díaz (1992), analysing the case of Colombia, also underscores the growth of conurbations and satellite cities as well as the settlement of frontier areas in the south of the country and on the Venezuelan border.

Geisse and Sabatini (1988) have suggested that the problems of large cities arise not so much from their size as from their rate of growth. As table 28 indicates, during 1990-2000, the urban population living outside the largest city is expected to grow faster than the population of the largest city in nine Latin American countries, and in another seven, both populations are expected to grow at similar rates. Furthermore, although the respective rates of growth of both components of the urban population are still substantial, ranging from 1.2 per cent for the urban population outside the largest cities in Puerto Rico and Uruguay to 4.8 per cent for that in Honduras, and from 0.3 for the population in the largest city of Uruguay (Montevideo) to 4.3 per cent for that in Guatemala (Guatemala City), they have dropped considerably since 1950.

Indeed, the rate of growth of the largest cities in each country is often surpassed by those of other large cities in the region. Yet, when one considers the largest cities in Latin America, their median rate of growth will have declined from 5.4 to 2.3 per cent between 1950-1960 and 1990-2000 (table 30). Furthermore, whereas in 1950 those cities accounted for 32.4 per cent of the urban population in the region, by 1990 their population constituted 35.9 per cent of the urban population of Latin America. As table 30 shows, the growth rates of the largest cities have shown remarkable convergence towards relatively low levels during the past 50 years. However, such declines should not be interpreted as meaning that large metropolitan areas are experiencing less strain than in the past. Because their base population has increased, lower growth rates do not mean lower absolute population increase. In most cases, the net population gain experienced by large cities is still rising, thus exerting continued pressure on urban infrastructure and services which in any case are under strain because of the decline in financial resources brought about by the economic crisis. In many Latin American countries, a reduction of government spending together with the population's shrinking buying power is eroding urban services and creating problems of an unprecedented magnitude.

For the large and growing proportion of the population living in intermediate urban centres, a major concern is the inability of local authorities to plan and implement

TABLE 30. ANNUAL GROWTH RATE OF THE 22 LARGEST CITIES IN LATIN AMERICA IN 1990, BY DECADE, 1950-2000
(*Percentage*)

		Annual growth rate				
City	*Country*	*1950-1960*	*1960-1970*	*1970-1980*	*1980-1990*	*1990-2000*
Mexico City	Mexico	5.40	5.52	4.33	3.33	2.36
São Paulo	Brazil	6.63	5.40	4.06	3.63	2.40
Buenos Aires	Argentina	2.95	2.17	1.64	1.49	1.13
Río de Janeiro	Brazil	5.40	3.59	2.22	1.98	1.53
Lima	Peru	5.51	5.51	4.14	3.43	2.67
Bogotá	Colombia	6.56	5.99	3.98	3.18	2.80
Santiago	Chile	4.22	3.34	2.70	2.42	1.74
Caracas	Venezuela	6.40	4.68	3.65	3.28	2.45
Belo Horizonte	Brazil	8.61	6.10	4.30	3.87	2.67
Guadalajara	Mexico	7.82	5.41	4.09	3.28	2.53
Porto Alegre	Brazil	8.58	4.19	3.77	3.43	2.43
Monterrey	Mexico	9.05	3.34	4.95	3.87	2.80
Recife	Brazil	6.28	3.91	1.77	1.59	1.44
Salvador	Brazil	6.01	4.69	3.91	3.54	2.53
Brasilia	Brazil	13.24	13.31	7.93	7.09	4.49
Santo Domingo	Dominican Republic	7.11	6.31	5.12	4.55	3.15
Havana	Cuba	2.10	2.10	0.90	0.95	0.96
Fortaleza	Brazil	9.08	4.84	3.71	3.36	2.44
Curitiba	Brazil	12.78	5.03	4.80	4.35	2.99
Goiania	Brazil	11.38	13.42	6.51	5.80	3.80
Guayaquil	Ecuador	5.67	4.42	4.40	4.40	3.59
Campinas	Brazil	5.55	10.10	6.50	5.84	3.82
	TOTAL	5.38	4.44	3.49	3.12	2.32

Source: *World Urbanization Prospects, 1990* (United Nations publication, Sales No. E.91.XIII.11).

new initiatives. According to Hardoy and Satterthwaite (1988), the development of local management capacity is essential to ensure that the potential of smaller urban centres is realized.

D. COMPONENTS OF URBANIZATION, URBAN GROWTH AND THE GROWTH OF LARGER CITIES

Despite efforts made to obtain empirical evidence regarding the components of urbanization, urban growth and the growth of large cities, the information on the subject remains weak, and the studies available lack a comprehensive view and use a variety of analytical methods that compromise their comparability. The need to sort and integrate existing studies, update them, and develop a conceptual framework better able to interrelate urbanization with population dynamics remains to be met.

Although there are differences between countries associated with their stage in the urbanization process, and their level of natural increase and of international migration, in most of them internal migration and natural increase play the most significant roles, though changing ones, in determining the growth of the urban population in general and that of large cities in particular. According to the evidence available, the direct contribution of net rural-urban population transfers[4] to urbanization is so significant that it alone accounts for the magnitude of the rate of urbanization of the region as a whole and that of most of the countries therein (table 31). Since 1950, the net rate of rural-urban population transfer has accounted for between 80 and 125 per cent of the rate of urbanization[5] in most countries of the region. However, the contribution of rural-urban population transfers to the growth of the urban population has decreased. Thus, whereas in 1950-1960 such transfers accounted for 42 per cent of the urban growth in Latin America, by

TABLE 31. RURAL-URBAN MIGRATION AS A COMPONENT OF URBANIZATION AND URBAN GROWTH IN LATIN AMERICAN COUNTRIES, 1950-2000

Country	*Ratio of the rate of net rural-urban transfers to the rate of urbanization*					*Percentage of urban growth attributable to rural-urban migration*				
	1950-1960	*1960-1970*	*1970-1980*	*1980-1990*	*1990-2000*	*1950-1960*	*1960-1970*	*1970-1980*	*1980-1990*	*1990-2000*
Argentina	1.07	1.09	1.13	1.14	1.15	42.4	32.3	28.6	25.1	23.3
Bolivia	0.81	1.01	1.04	1.08	1.11	12.3	13.9	25.4	39.6	39.7
Brazil	1.05	1.11	1.12	1.13	1.13	45.4	49.6	47.6	43.6	39.8
Chile	1.08	1.12	1.11	1.17	1.20	43.6	36.0	36.3	23.0	19.9
Colombia	1.06	1.13	1.12	1.14	1.15	51.7	42.3	38.3	36.2	36.1
Costa Rica	0.87	1.06	1.03	1.05	1.08	17.3	21.0	24.0	25.1	36.4
Cuba	1.04	1.10	1.07	1.07	1.08	38.9	35.3	52.9	49.3	45.9
Dominican Republic	0.98	1.08	1.08	1.10	1.12	43.3	52.9	51.8	48.8	44.8
Ecuador	0.99	1.07	1.08	1.10	1.12	41.3	33.3	40.5	45.8	44.0
El Salvador	0.80	0.92	0.97	1.00	1.05	11.9	7.2	18.2	33.3	35.3
Guatemala	0.87	1.02	0.86	0.82	0.99	21.7	25.2	13.5	13.1	28.4
Haiti	0.89	0.95	0.88	0.88	0.92	55.3	55.9	46.2	43.8	45.8
Honduras	0.90	1.02	1.00	1.03	1.08	41.2	45.8	40.7	38.5	40.4
Jamaica	1.00	1.04	1.04	1.05	1.05	61.9	63.0	49.5	49.4	55.5
Mexico	1.06	1.14	1.16	1.17	1.19	40.4	37.3	34.1	33.0	31.1
Nicaragua	0.99	1.11	1.12	1.15	1.25	29.3	40.0	33.6	33.9	28.1
Panama	1.01	1.10	1.09	1.10	1.11	37.1	37.5	16.6	24.8	34.0
Paraguay	0.50	0.95	1.03	1.06	1.10	4.8	12.4	29.9	32.1	37.2
Peru	1.05	1.11	1.15	1.17	1.19	53.9	49.0	35.2	30.8	29.8
Puerto Rico	1.00	1.05	1.09	1.07	1.08	60.5	69.8	50.1	54.1	45.9
Uruguay	1.05	1.08	1.03	1.05	1.07	18.8	21.1	51.8	43.6	35.4
Venezuela	1.13	1.20	1.20	1.18	1.18	41.6	23.8	35.0	29.6	21.1
TOTAL	1.06	1.12	1.13	1.15	1.16	41.9	40.6	39.3	36.5	33.3

Source: *World Urbanization Prospects: The 1992 Revision* (United Nations publication, Sales No. E.93.XIII.11).

NOTE: Estimates are based on the assumption that natural increase in each country is equal to its growth rate and that the natural increase of the rural population equals that of the urban population in 1950-1960, and is higher than that of the urban population by 5, 10, 15 and 20 per cent, respectively, for subsequent decades.

1990-2000 they have been expected to account for only slightly over 33 per cent of such growth[6] (table 31). At the country level, the contribution of rural-urban population transfers vary greatly, ranging from 20 per cent in Chile to 56 per cent in Jamaica during 1990-2000. The declining importance of rural-urban transfers is related to the growing level of urbanization which leaves smaller proportions of rural populations to be transferred. At the country level, a reduction in the contribution of rural-urban transfers has been noticeable in Argentina, Brazil (since the 1960s) and Chile, but the reduction has not occurred uninterruptedly in Cuba, Uruguay and Venezuela or in other countries with lower levels of urbanization.

There is considerable evidence regarding the declining weight of rural-urban migration among all internal migration movements in countries that are already highly urbanized. In Argentina (Lattes and Mychaszula, 1986), Chile (Raczynski, 1981) and Peru (Aramburu, 1982), for example, rural-urban migration accounts for a quarter or less of all internal migration. For Mexico, the Consejo Nacional de Población (CONAPO) (1991) documents a trend in the same direction, with a growing weight of urban-urban migration, greater diversification of the national urban system, and increased migration from large urban centres to intermediate and small cities. In the Dominican Republic, the proportion of migrants to Santo Domingo who originate in other urban centres has been increasing since 1970 (Duarte, 1991). Cuba constitutes an exception, since 75 per cent of its internal migration still consists of rural-urban or rural-rural movements despite its high level of urbanization. In Cuba, most rural migrants originate in scattered rural areas and move to rural areas with a higher degree of concentration or to towns (Comité Estatal de Estadísticas, 1992).

The direct contribution of net migration, whether from rural or other urban areas, to the growth of the largest Latin American cities during 1960-1970 (table 32) exceeded in many cases that of net migration to urban population growth (table 31). At present, the growth rate of large cities is considerably lower (table 30) and it is likely that natural increase is currently the main component of their growth. Buenos Aires, for instance, grew during most of this century mainly through internal and international immigration but during the last two decades natural increase has accounted for two thirds of its population growth (Lattes and Recchini de Lattes, 1992). Like other large cities in the region, Buenos Aires currently attracts fewer internal and international migrants. Moreover, former migrants are moving from Buenos Aires to their places of origin or to other urban destinations, and a growing number of native city residents have been migrating to the rest of the country. Montevideo's falling population growth is mainly due to substantial international emigration (Aguiar, 1982), which is also significant in the case of Buenos Aires. In the case of Havana, Landstreet and Mundigo (1981) show that low growth resulted from the conjunction of declining internal migration, lower international migration and a sharp reduction of fertility, more pronounced in the capital city than in the rest of the country.

Despite their limitations, these findings suggest that, as the countries of the region become more urbanized, the volume of rural-urban population transfers is declining and, consequently, so is their contribution to the growth of the population of urban areas in general and of large cities in particular. However, migration, especially that from urban to urban areas, is likely to remain a major contributor to the growth of intermediate cities.[7] International migration may also be a factor, particularly for border cities (Arizpe, 1983).

With regard to the contribution of natural increase, recent evidence corroborates that the fertility of large cities tends to be lower than that of other urban areas which is in turn lower than that of rural areas (see table 33, based on World Fertility Survey data). Colombia, Ecuador and Venezuela had particularly high fertility differentials between rural areas and large cities (of the order of 4 children per woman). Colombia, Costa Rica and Ecuador recorded the lowest fertility levels in large cities, whereas the Dominican Republic, Mexico and Venezuela had particularly high fertility levels in rural areas. Data for other regions (not shown) indicate that Latin America, being more urbanized than Africa or Asia, also tended to have higher rural/urban fertility differentials than those regions.

The World Fertility Survey also gathered some evidence regarding mortality differentials between urban, rural and large-city populations (United Nations, 1987a). At the regional level, the proportions of children ever born who were reported as still alive by women aged 40-49 were 94, 90 and 87 per cent, for women living in large cities, those living in other urban areas and those in rural areas, respectively. Such differences in mortality levels contribute to reducing the differential in natural increase between the various population subgroups. Thus, at the regional level, the differential in terms of children ever born between rural areas and large cities was 2.3 children per woman in age group 40-49, whereas that in terms of children surviving amounted to 1.7 children.

Clearly it is necessary to work towards a better integration of the analysis of the demographic transition in terms of fertility and mortality rates with that of the urbanization transition since urbanization itself leads to changes in population dynamics and in the reproductive behaviour of couples and individuals (Villa, 1992).

E. Social and demographic characteristics of migrants in urban areas

It is surprising that in the study of migration and the population dynamics of cities so little attention is paid to specific groups. Thus, although it has been known since the 1950s that female rural-urban migration exceeds that of males in the majority of Latin American countries, research on female migration has been scarce (Recchini de Lattes, 1988; Szasz, 1992). Indeed, if any generalization can be made about migration directed towards urban areas, it is that women predominate among migrants and that urban-bound migrants of both sexes tend to be concentrated in relatively young age groups. Thus, migration makes a significant contribution to the growth of age group 15-29 (table 32).

In one of the few studies available for the region, Peláez and Argüello (1982), having analysed the differences between urban and rural populations with regard

TABLE 32. NET MIGRATION AS A COMPONENT OF THE GROWTH OF LARGE CITIES, 1960-1970

City	Country	Percentage of growth attributable to migration: Total population	Population aged 15-29	Sex ratio of net migration
Belo Horizonte	Brazil	51.9	68.1	89
Bogotá	Colombia	52.9	57.1	79
Buenos Aires	Argentina	43.8	85.2	88
Caracas	Venezuela	40.2	58.0	78
Guadalajara	Mexico	36.6	41.9	92
Lima-Callao	Peru	44.0	58.4	99
Mexico City	Mexico	35.6	54.2	92
Monterrey	Mexico	38.6	51.8	94
Pôrto Alegre	Brazil	48.4	70.6	81
Recife	Brazil	21.6	34.6	66
Rio de Janeiro	Brazil	42.2	71.5	78
Santiago	Chile	47.0	61.3	94
São Paulo	Brazil	59.6	91.9	97

Source: Migration, Population Growth and Employment in Metropolitan Areas of Selected Developing Countries (United Nations publication, ST/ESA/SER.R/57).

TABLE 33. TOTAL FERTILITY RATE FOR WOMEN AGED 15-49, BY PLACE OF RESIDENCE, FOR SELECTED LATIN AMERICAN COUNTRIES DURING THE 1970S

Country	Total fertility: Major city	Other urban areas	Rural	Difference between rural areas and major city
Colombia	2.9	3.9	7.0	4.1
Costa Rica	3.0	3.3	5.2	2.2
Dominican Republic	4.2	4.4	7.4	3.2
Ecuador	3.1	4.9	6.7	3.6
Haiti	4.0	3.4	6.2	2.2
Mexico	4.8	5.7	7.6	2.8
Panama	3.5	3.5	6.2	2.7
Paraguay	3.2	4.0	6.3	3.1
Peru	3.9	5.4	7.2	3.3
Venezuela[a]	3.3	4.3	7.7	4.4

Source: *Fertility Behaviour in the Context of Development: Evidence from the World Fertility Survey*, Population Studies, No. 100 (United Nations publication, Sales No. E.86.XIII.5).

[a]Total fertility refers to women aged 15-44 only.

to their distributions by sex and age, concluded that the proportion of persons aged 15-59 was always greater in large urban centres and would remain so in the future; that the degree of concentration in urban areas of persons in older age groups was related to the degree of urbanization; and that the predominance of women in older ages was virtually universal since, in 16 Latin American countries, women accounted for at least 55 per cent of the population aged 60 years or over in urban areas. Furthermore, other evidence indicates that the proportion of the population aged 60 years or over is considerably higher in large cities than in other urban areas or in rural areas. Thus, in 1980, 14 per cent of the population in the three largest cities of Argentina was aged 60 years or over and women constituted about 59 per cent of that population. In contrast, the proportion of the population aged 60 years or over in Argentina as a whole amounted to 11.8 per cent, of which approximately 56 per cent consisted of women (Recchini de Lattes, 1991).

According to United Nations estimates for 16 of the countries in the region, the sex ratio of net rural-urban migration varied between 34.5 men per 100 women in Uruguay during 1965-1975 and 98.7 men per 100 women in Ecuador during 1972-1982, with an average of 79.8 male migrants per 100 female migrants for all 16 countries (United Nations, 1988, table 79). Recent evidence concerning the Dominican Republic indicates that the majority of the migrants to the capital city of Santo Domingo (73.3 per cent in 1991) originated in other urban areas, that 56.4 per cent of them were women and that a high proportion belonged to the young adult ages (Duarte, 1991). The 1990 census of Mexico corroborates the predominance of women (52 per cent) among migrants to the State of Mexico, the area surrounding Mexico City that is currently attracting major population inflows. Szasz (1992) notes that many other intermediate and smaller urban centres in Mexico are also attracting more female than male migrants.

The stereotyped view of migrant women as passive dependants of men and other simplistic characterizations of migrants have little empirical support. Recent research is beginning to uncover a more varied and complex characterization of internal migration and, particularly, of that directed to urban areas. Yet, much remains to be learned about migration flows, the different components of urban and city growth, and the determinants of those phenomena.

F. TERRITORIAL MOBILITY AND ITS IMPLICATIONS FOR POLICY

In recent years, concern has deepened about the persistence of social and economic structures that perpetuate social inequality, especially in large urban centres (ECLAC, 1989; Coraggio, 1990; UNDP, 1990; and others). It appears that, in Latin America, in contrast with today's developed countries, urbanization and industrialization have not led to a better distribution of wealth but rather have contributed to the polarization and diversification of the social structure, leading to the coexistence of skilled and unskilled workers many of whom are employed in the informal sector (Oliveira and Roberts, 1989). Furthermore, the evidence suggests that the social mobility experienced by rural-urban migrants in the 1950s and 1960s has declined and that their children are unlikely to benefit from the same opportunities open to their parents (Lattes, 1989). At present, Latin American democracies find themselves caught between the traditional centralization of their political and administrative structures and the rising social demands of the population that is mostly concentrated in large cities (Geisse and Sabatini, 1988). Several countries in the region have initiated policies for the decentralization of certain government functions that are expected to influence the distribution of population. In addition, the structural changes that are being implemented also affect territorial mobility, in particular the development and improvement of the transportation and communications systems, the improvement of educational attainment, and the increasing adoption of automation in industry.

These processes of change are modifying the patterns of territorial mobility, yet there is still little evidence on the extent of their impact. Thus, not enough is known about patterns of temporary or semi-permanent population movements. Like other regions, Latin America is experiencing complex movements of persons, whether alone or in family groups, in response to economic opportunities. A recent study documents the various types of movements in the region (PISPAL/CIUDAD/CENEP, 1986), among which those involving a growing number of rural workers settling temporarily in urban areas should be mentioned (Spindel, 1985). The emerging forms of territorial mobility are not resulting as yet in patterns of spatial population redistribution consistent with those studied traditionally. They seem to respond to a different logic and cannot be easily superimposed on the traditional population movements of the past. They have, however, important implications both for the "settled" subpopulation and for that present temporarily in different areas.

Clearly, if planners and policy makers are to devise measures to enhance the benefits of new patterns of population mobility, a comprehensive view of those patterns and their implications is necessary. If the study of mobility is to include both temporary displacements and permanent changes of residence, the issue of how those two types of mobility interrelate must be addressed. A comprehensive analysis of territorial mobility demands the use of novel data-collection systems and of expanded theoretical frameworks that are nevertheless specific enough to reflect the varying processes taking place in different societies. Although the historical-structural approach has been useful in understanding mobility at the macrolevel, it does not give sufficient importance to factors of key significance, such as culture, policy and especially the role of families and households. The influence of such factors needs to be taken explicitly into account in devising better data-collection strategies.

At present, the spatial population dynamics reigning in much of Latin America imply that millions of people are confronted daily with the alternatives of moving or staying, with this state of affairs leading to outcomes that are difficult to apprehend and analyse. As Roberts (1988) points out, spatial mobility is a basic survival strategy in an unstable and unpredictable social and economic environment. The territorial mobility of people affects and is affected by the social and economic policies being implemented. Consequently, its implications should not be ignored when devising policies and actions aimed at achieving a more equitable and sustainable development.

NOTES

[1]The extraordinary rate of growth of both the urban and the total populations of Latin America during the 1940s is reflected in the population projections carried out during the 1960s. Indeed, a comparison of projections made during the 1960s (United Nations, 1969) with those carried out more recently (United Nations, 1993) shows that the former overestimated by considerable margins the urban and the total population of Latin America towards the end of the century. According to the 1960s projections, the urban and total population of the region would be, respectively, 510 million and 638 million compared with 401 million and 523 million, according to current estimates. These major differences are explained by the occurrence of two important changes, namely, a sharp decline in fertility levels since 1960 and a virtual halt of European migration to Latin America.

[2]However, the concentration of the total population in the largest cities continues, albeit at a declining rate. Thus, while in 1950, 12 per cent of the total population in Latin America lived in the largest cities of each country, by 2000 it is expected that 21.3 per cent will do so. Furthermore, as urbanization increases, the relative weights of the population in the largest city with respect to the urban and to the total population tend to converge.

[3]Alberts (1977), among others, concludes that the growth rate of the metropolitan areas of Caracas, Monterrey, San Salvador, Rio de Janeiro and São Paulo has been declining, whereas that of other medium-sized cities with less population has been increasing. Landstreet and Mundigo (1981) discuss the case of Havana. Gatica (1980) and Lattes (1983) suggest that the trend affects the whole Latin American region. Urzúa and others (1981) underscore the declining primacy of Buenos Aires and Montevideo as well as the stagnation of Santiago de Chile. The decline of the growth rate of Buenos Aires in comparison with that of other urban areas in Argentina and certain medium-sized cities is discussed by Lattes and Mychaszula (1986).

[4]Such transfers include net rural-urban migration, the reclassification of places from rural to urban and, in some cases, net international migration that affects differentially rural and urban areas. With respect to internal migration, only its direct effect is being considered, that is, no attempt is being made to estimate the contribution of the migrant population to the natural increase of the area of destination. According to United Nations estimates, the indirect contribution of migration to the growth of 13 of the largest cities in Latin America varied from 3.3 to 7.9 per cent of the growth of the respective city (United Nations, 1985).

[5]The rate of net rural-urban transfers is higher than the rate of urbanization because the rural-urban differential in natural increase either does not contribute to or reduces urbanization, that is, natural increase in rural areas is usually higher than in urban areas. According to United Nations estimates, the average ratio of net rural-urban transfers to the rate of urbanization was 1.22 in the 1950s and 1.14 in the 1960s (United Nations, 1980).

[6]United Nations estimates based on a smaller number of countries and using a more refined estimation procedure show that the contribution of net rural-urban migration to urban growth in Latin America was approximately 39 per cent in the 1950s and 35 per cent in the 1960s.

[7]In Argentina, for instance, the 10 cities experiencing the highest growth during the 1970s had populations in the range of 25,000-250,000 and during the 1980s, in the range of 50,000-500,000.

REFERENCES

Aguiar, C. (1982). *Uruguay, País de Emigración.* Montevideo, Uruguay: Ediciones de la Banda Oriental.

Alberts, Joop (1977). *Migración hacia Areas Metropolitanas de América Latina.* Santiago, Chile: Centro Latinoamericano de Demografía (CELADE).

________, and Miguel S. Villa, eds. (1980). *Redistribución Espacial de la Población en América Latina.* Santiago, Chile: Centro Latinoamericano de Demografía (CELADE).

Aramburu, Carlos (1982). Migraciones internas, proceso social y campesinado en el Perú. Paper presented at the World Congress of Sociology, International Sociological Association, Mexico.

Arizpe, Lourdes (1983). El éxodo rural en México y su relación con la migración a Estados Unidos. *Estudios Sociológicos* (Mexico City), vol. 1, No. 1.

Centro Latinoamericano de Demografía (1983). *Características y Tendencias de la Distribución Espacial de la Población en América Latina.* Santiago, Chile: Instituto Latinoamericano de Planificación Económica y Social (ILPES).

Comité Estatal de Estadísticas (1992). Cuba: el crecimiento urbano y las migraciones internas en el contexto del desarrollo económico y social. Paper presented at the Conference on the Peopling of the Americas, International Union for the Scientific Study of Population, Veracruz, 18-23 May.

Consejo Nacional de Población (CONAPO) (1991). *Sistema de Ciudades y Distribución Espacial de la Población de México*, vol. 1. Mexico City.

Coraggio, José L., ed. (1990). *La Investigación Urbana en América Latina.* Quito, Ecuador: CIUDAD.

Díaz, Luz Marina (1992). La movilidad territorial de la población en Colombia. Paper presented at the Seminar on Territorial Mobility of Populations: New Patterns in Latin America. York University, Centre for Research on Latin America and the Caribbean (CERLAC) and Centro Nacional de Población (CENEP), Toronto, 22-24 October.

Duarte, Isis (1991). Población, migraciones internas y desarrollo en República Dominicana: 1950-1981. Paper presented at the Seminar on Territorial Mobility of Populations: New Patterns in Latin America, the Dominican Case. Organized by CERLAC and CENEP. Santo Domingo, 13-14 December.

Economic Commission for Latin America and the Caribbean (ECLAC) (1989). *La Crisis Urbana en América Latina y el Caribe: Reflexiones sobre Alternativas de Solución.* Santiago, Chile.

__________ (1990a). *Magnitud de la Pobreza en América Latina en los Años Ochenta*. Document No. LC/L533. Santiago, Chile.

__________ (1990b). *Transformación Productiva con Equidad*. Santiago, Chile.

Eldridge, Hope T. (1942). The process of urbanization. *Social Forces* (Chapel Hill, North Carolina), No. 20, pp. 311-316.

Food and Agriculture Organization of the United Nations (FAO) (1992). *Differentials in rural and urban development in selected countries of Latin America*. Santiago, Chile: Regional Office for Latin America and the Caribbean. FAO/RLAC, FPA/RLA/87/P13.

Gatica, Fernando (1980). La urbanización en América Latina: 1950-1970; patrones y áreas críticas. In *Redistribución Espacial de la Población en América Latina*, Joop Alberts and Miguel S. Villa, eds. Santiago, Chile: CELADE.

Geisse, G., and F. Sabatini (1988). Latin American cities and their poor. In *The Metropolis Era*, Mattei Dogan and John D. Kasarda, eds. Newbury Park, California: Sage Publications.

Hardoy, Jorge E., and David Satterthwaite (1988). Small and intermediate urban centers in the Third World: what role for government? *Third World Planning Review* (Liverpool), vol. 10, No. 1, pp. 5-26.

Hauser, P., and Robert Gardner (1982). Urban future: trends and prospects. Honolulu: East-West Population Institute, reprint No. 146.

Instituto Nacional de Estadística y Censos (INDEC) (n.d.). Unpublished tables from the Regional Population and Housing Census 1991. Buenos Aires: INDEC, Cartography Department.

Instituto Nacional de Estadística, Geografía e Informática (INEGI) (1991). *Area Metropolitana de la Ciudad de Mexico, Síntesis de Resultados, IX Censo General de la Población y Vivienda 1990*. Aguascalientes, Mexico.

Landstreet, B., and Axel Mundigo (1981). Internal migration and changing urbanization patterns in Cuba. Paper presented at the Annual Meeting of the Population Association of America, Washington, D.C., 26-28 March.

Lattes, Alfredo E. (1983). Acerca de los patrones recientes de la movilidad territorial de la población en el mundo. *Cuaderno del CENEP* (Buenos Aires), No. 27.

__________ (1984). Territorial mobility and redistribution of the population: recent developments. In *Population Distribution, Migration and Development: Proceedings of the Expert Group on Population Distribution, Migration and Development, Hammamet (Tunisia), 21-25 March 1983*. Sales No. E.84.XIII.3. New York: United Nations.

__________ (1985). Migraciones hacia América Latina y el Caribe desde principios del siglo XIX. *Cuaderno del CENEP* (Buenos Aires), No. 35.

__________ (1989). Emerging patterns of territorial mobility in Latin America: challenges for research and action. In *Proceedings of the International Population Conference, New Delhi 1989*, vol. 2. Liège, Belgium: International Union for the Scientific Study of Population.

__________ (1990). La urbanización y el crecimiento urbano en América Latina desde una perspectiva demográfica. In *La Investigación Urbana en América Latina*, José L. Coraggio, ed. Quito, Ecuador: CIUDAD.

__________, and Sonia Mychaszula (1986). Urbanization, migration and urban deconcentration in Argentina. Buenos Aires: Centro de Estudios de Población. Mimeographed.

__________, and Zulma Recchini de Lattes (1992). Auge y declinación de las migraciones en Buenos Aires. In *Después de Germani*, J. Jorrat and R. Sautu, eds. Buenos Aires: Editorial Paidos.

Oliveira, Orlandina de, and Bryan Roberts (1989). Los antecedentes de la crisis urbana: urbanización y transformación ocupacional en América Latina, 1940-1980. In *Las Ciudades en Conflicto. Una Perspectiva Latinoamericana*, M. Lombardi and D. Vega, eds. Montevideo, Uruguay: CIESU, Ediciones de la Banda Oriental.

Peláez, César, and Oscar Argüello (1982). Envejecimiento de la población de América Latina: tendencias demográficas y situación socioeconómica. *Notas de Población* (Santiago, Chile), vol. 10, No. 30, pp. 9-95.

Programa de Investigaciones Sociales en Población de América Latina (PISPAL)/CIUDAD/CENEP (1986). *Se Fue a Volver. Seminario sobre Migraciones Temporales en America Latina*. Mexico: El Colegio de México.

Portes, Alejandro (1989). La urbanización en América Latina en los años de crisis. In *Las Ciudades en Conflicto. Una Perspectiva Latinoamericana*, M. Lombardi and D. Vega, eds. Montevideo, Uruguay: CIESU, Ediciones de la Banda Oriental.

Raczynski, D. (1981). Naturaleza rural-urbana y patrones geográficos de la migración internal en Chile. *Estudios CIEPLAN* (Economic Research Corporation for Latin America) (Santiago, Chile), No. 5, pp. 85-115.

Recchini de Lattes, Zulma (1988). Las mujeres en las migraciones internas e internacionales, con especial referencia a América Latina. *Cuaderno del CENEP* (Buenos Aires), No. 40.

__________ (1991). Urbanization and demographic ageing: the case of a developing country, Argentina. In *Ageing and Urbanization*. Proceedings of the United Nations International Conference on Ageing Populations in the Context of Urbanization, Sendai (Japan), 12-16 September 1988. Sales No. 1991.XIII.12.

__________, and Sonia Mychaszula (1991). La heterogeneidad de la migración y participación laboral femenina en una ciudad de tamaño intermedio. *Estudios del Trabajo* (Buenos Aires), No. 2, pp. 51-78.

Roberts, Bryan (1988). Ciudades transicionales. In *Repensando la Ciudad de América Latina*, J. Hardoy and R. Morse, eds. Buenos Aires: Grupo Editor Latinoamericano and International Environment and Development Institute.

Ruiz Chiapetto, C. (1990). Distribución de población y crisis económica en los años ochenta: dicotomías y especulaciones. *Revista Mexicana de Sociología* (Mexico City), vol. 52, No. 1, pp. 185-203.

Spindel, C. (1985). Temporary work in Brazilian agriculture: "Boia-Fria". A category under investigation. In *Labour Circulation and the Labour Process*, Guy Standing, ed. London: Croom Helm.

Szasz, Ivonne (1992). La migración femenina en México: tendencias emergentes. Paper presented at the Seminar on Territorial Mobility of Populations: New Patterns in Latin America. York University, CERLAC and CENEP. Toronto, 22-24 October.

United Nations (1969). *Growth of the World's Urban and Rural Population, 1920-2000*. Sales No. E.69.XIII.3.

__________ (1980). *Patterns of Urban and Rural Population Growth*. Population Studies, No. 68. Sales No. E.79.XIII.9 and corrigendum.

__________ (1984). *Population Distribution, Migration and Development: Proceedings of the Expert Group on Population Distribution, Migration and Development, Hammamet (Tunisia), 21-25 March 1983*. Sales No. E.84.XIII.3.

__________ (1985). *Migration, Population Growth and Employment in Metropolitan Areas of Selected Developing Countries*. ST/ESA/SER.R/57.

__________ (1987a). *Fertility Behaviour in the Context of Development: Evidence from the World Fertility Survey*. Population Studies, No. 100. Sales No. E.86.XIII.5.

__________ (1987b). *The Prospects of World Urbanization, Revised as of 1984-85*. Population Studies, No. 101. Sales No. E.87.XIII.3.

__________ (1988). *World Population Trends and Policies: 1987 Monitoring Report. Special topics: fertility and women's life cycle and socio-economic differentials in mortality*. Population Studies, No. 103 (United Nations publication, Sales No. E.88.XIII.3).

__________ (1991). *World Urbanization Prospects, 1990*. Sales No. E.91.XIII.11.

__________ (1992). *World Population Monitoring, 1991: With Special Emphasis on Age Structure*. Population Studies, No. 126. Sales No. E.92.XIII.2.

_________ (1993). *World Urbanization Prospects: The 1992 Revision.* Sales No. E.93.XIII.11.

United Nations Development Programme (UNDP) (1990). *Desarrollo Humano: Informe 1990.* Bogotá, Colombia: Tercer Mundo Editores.

Urzúa, Raúl, and others (1982). Desarrollo regional, migraciones y concentración urbana en América Latina: una investigación comparativa. Santiago, Chile: Centro Latinoamericano de Demografía. Mimeographed.

Vapnarsky, C. A. (1969). *Población Urbana y Población Metropolitana.* Buenos Aires: Editorial del Instituto.

_________ (1981). Aportes teórico-metodológicos para la determinación censal de localidades. In *Investigación e Información Sociodemográfica 2,* Susana Torrado, ed. Buenos Aires: Comisión oblación y Desarrollo de Conselho Latinoamericano de Ciencias Sociais (Latin American Social Sciences Council (CLACSO).

Verduzco, Gustavo (1992). La distribución de la población en México: los efectos de la llamada "década perdida". Paper presented at the Seminar on Territorial Mobility of Populations: New Patterns in Latin America, York University, CERLAC and CENEP. Toronto, 22-24 October.

Villa, Miguel S. (1992). Urbanización y transición demográfica en América Latina: una reseña del período 1930-1990. In *Proceedings of the Conference on the Peopling of the Americas,* vol. 2. Organized by the International Union for the Scientific Study of Population. Veracruz, 18-23 May.

Part Three

SOCIAL, ENVIRONMENTAL AND POLICY ASPECTS OF POPULATION DISTRIBUTION AND INTERNAL MIGRATION

VIII. MIGRATION AS A SURVIVAL STRATEGY: THE FAMILY DIMENSION OF MIGRATION

*Graeme J. Hugo**

In traditional societies the family was not only the basic social unit but also the fundamental unit of economic organization. In their attempts to survive under especially difficult conditions or to improve their life chances in more normal circumstances, families have long developed and employed particular mobility strategies. Such strategies might involve the entire families' moving or, especially if they are operating within a patriarchal system, families may allocate different tasks to their constituent members according to their respective abilities and characteristics. Some of the earliest forms of human organization, such as hunting and gathering or economies based on shifting cultivation, involved population mobility as an integral part of the family's survival strategy. However, in such societies spatial mobility has been restricted to "trodden social space" (Zelinsky, 1971). In modern times, this situation has been substantially transformed. The allocation of family labour to locations other than the hearth area is being used by a broader spectrum of family types in a wider variety of social and economic contexts and, most importantly, the space over which mobility occurs has expanded considerably. The increased mobility of family members has been an important element in the growing scale and complexity of general population mobility as well as in its significance for economic and social development during recent decades.

Before considering the role of the family in fostering territorial mobility, a brief overview of the broad changes that have occurred in the patterns and types of population mobility in developing countries will be presented. This will be followed by a discussion of the processes leading to migration and the ways in which population movement is used as a strategy to ensure and enhance the well-being of the family group, giving special attention to the role that social networks play in sustaining and increasing the level of movement. Lastly, the possible implications of the changes taking place will be explored.

A. Changing patterns and types of population mobility in developing countries

Over the last decade, the territorial mobility of the inhabitants of developing countries has increased dramatically. To move away from the place where home is located, either on a permanent or a temporary basis, has become an option for improving the life chances of a much wider spectrum of the world's population. Earlier in this century, relocation was generally an option for only a part of the population—one that was narrowly defined, whether in terms of socio-economic status, gender, age or ethnic group. Although territorial mobility remains a selective process, it is increasing both in scale and in diversity in terms of the groups involved. In addition to this "shaking loose" of people (Zelinsky, 1971), the movements themselves are more complex in nature and in terms of their spatial patterning, partly because of the improvement of communications.

The permanent relocation of people within countries now involves more persons than ever before, especially because the movement of people from rural to urban areas has gathered momentum over the last decade. The rate of urbanization, defined as the average annual rate of change of the percentage urban, equals the difference between the rate of growth of the urban population and that of the total population. Consequently, it is affected by the redistribution of population from rural to urban areas. According to United Nations estimates, the rate of urbanization of developing countries peaked at 2.5 per cent in 1980-1985 and was 2.4 per cent in 1985-1990, implying that population redistribution from rural to urban areas was intense during the 1980s and will continue into the 1990s (United Nations, 1992). The

*Department of Geography, University of Adelaide, Australia.

extent of rural-urban redistribution has been highest in Asia where the rate of urbanization reached 3.1 per cent per annum during 1985-1990. Rural-urban migration was responsible for a considerable proportion of that redistribution.

Although the increase in permanent migration in developing countries has been substantial, that in non-permanent mobility has been even greater. Such mobility includes a wide array of types of commuting, seasonal movements and circular migration involving periods of absence from the home area ranging from a few days to several months. Temporary mobility is, by its very nature, difficult to quantify and it is usually not detected by the questions normally posed in censuses and national surveys (Hugo, 1982). However, there is growing evidence indicating that the significance of temporary mobility has increased in most developing countries (Standing, 1984; Hugo, 1991a). A major component of temporary mobility is that between rural and urban areas, and it contributes to a blurring of the distinction between rural and urban areas in so far as growing numbers of people who have their permanent residence in rural areas work over lengthy periods in urban areas. Indeed, it could be argued that the accelerating movement of people, information and goods between rural and urban localities in developing countries is leading to a substantial breakdown in the social, economic, cultural and political differences that have long characterized the urban/rural dichotomy in those countries.

Perhaps one of the most unexpected changes in mobility over recent decades has been the increase of international migration originating in developing countries. Indeed, since 1960 the outflow of both permanent and temporary migrants from developing countries to developed countries has shown a tendency to increase, though fluctuations have been experienced by most receiving countries. There has also been an increase in the flows originating and ending in developing countries. These flows have mostly involved migrant labour admitted on a temporary basis by capital-rich developing countries that have experienced labour shortages, such as the oil-rich countries of Western Asia or the newly industrializing economies of South-eastern Asia (United Nations, 1992 and 1996). Flows between developing countries have also involved an increasing number of persons fleeing persecution and being admitted as refugees because of their need for humanitarian assistance (United Nations, 1996). Undocumented migration is yet another important component of international flows, a component that has been rising because of the restrictive international migration policies adopted by most receiving countries. In some cases, such as that of Mexico and the United States of America, undocumented migration has a long history and, though driven mainly by economic disparities, is sustained by the development of social networks.

International migration flows of various types include highly qualified migrants that are considered to be a drain on the society of origin. Developing countries continue to be concerned about the negative effects of such a drain. However, countries such as India, the Philippines and the Republic of Korea argue that the benefits derived from the remittances sent home by skilled personnel largely counterbalance any negative effects of their migration.

An important development regarding population mobility in general is the growing significance of the participation of women in all types of flows (United Nations, 1993 and 1995). Indeed, the increasing opportunities open for the migration of women have important implications for the development of family survival strategies based on territorial mobility. Also of importance is the fact that different types of movements are increasingly linked to one another. Thus, travel for tourism or business favours the development of networks and fosters longer-term mobility of different kinds. An issue that needs to be explored further is the extent to which various types of temporary movements lead to permanent or long-term relocation. Some studies indicate that temporary migration permits the mover to become established gradually at the place of destination and facilitates permanent relocation at a later stage. Others show that circular migration is an entrenched and enduring form of mobility that does not necessarily lead to relocation.

Unfortunately, the methods and procedures to measure and analyse population mobility have not kept pace with the increasing scale and complexity of the phenomenon. Even with respect to the measurement of internal migration of a permanent or long-term nature, many developing countries fail to include questions on place of previous residence or on the timing of migration in their censuses (United Nations, 1970). In addition, even if an attempt is made to measure internal migration, the concepts used are often deficient and compromise the possibility of obtaining a comprehensive picture of the long-term or the permanent relocations of the population within national boundaries. Similarly, national surveys such as labour-force surveys or socio-economic surveys rarely include questions regarding migration. Even demographic surveys, such as the Demographic Health Surveys, have failed to recognize the significance of population mobility as an explanatory variable. The information lacunae with respect to temporary movements are even worse. Although ad hoc surveys carried out during the past 20 years have provided incontrovertible evidence of the scale and significance of the phenomenon, there have simply been no methodological advances with respect to obtaining nationally representative data through well-established data-collection systems. It is necessary to develop adequate concepts and tools to capture the richness and detect the magnitude of different types of temporary population movements.

The situation with respect to the measurement of international migration is similar. Developing countries generally lack adequate systems to record the inflow and outflow of people across their national borders. Yet such information is crucial for realistic social and economic planning. Bangladesh, Pakistan, the Philippines and Sri Lanka are examples of countries where the main source of foreign exchange earnings is the export of people. Some countries have tried to palliate their lack of adequate information on population outflows by asking in censuses or surveys whether a family member is living abroad. However, although that information is better than nothing, it is likely to be subject to important biases that make it a poor basis for policy formulation.

B. Macrostructural forces shaping population mobility in developing countries

Rising population mobility is closely linked to the increasing tempo of the economic and social changes that are being experienced by developing countries. According to Massey (1990), economic development rather than stagnation engenders population mobility, at least in the short-to-medium term. Hence, the modernization of agricultural production in developing countries with large rural populations has been of major significance in prompting large numbers of people to leave rural areas. Land enclosure, the increasing commercialization of agriculture, mechanization and the replacement of labour with capital inputs have all contributed to changing the mode of agricultural production and to release the rural workforce. In Indonesia, for instance, 55 per cent of its 51 million workers were employed in agriculture in 1980 (Biro Pusat Statistik, 1982) but a significant displacement of labour from that sector is expected during the 1990s, as mechanical hulling replaces hand milling in wet rice cultivation, more efficient harvesting methods are introduced, direct seeding of rice is promoted, the use of pesticides and hand tractors reduces the need for manpower, and the use of labour itself becomes increasingly commercialized (Hugo, 1992a and 1993). When these changes are complemented by the impressive gains in educational attainment made since 1970, the potential for retaining a large proportion of the labour force in agriculture becomes small. Most developing countries have similar experiences. With regard to education, for instance, many have achieved universal education, at least at the primary level.

Population mobility has also been influenced by the internationalization of capital. Thus, transnational corporations based in developed countries have been establishing manufacturing operations in developing countries with low labour costs. The improvements in the speed, cost and ease of transport have made the most remote village accessible to the wider world. Similarly, the spread of mass communications has contributed to disseminating information about potentially available opportunities at distant locations. In Indonesia, the proportion of people owning a television increased from less than 1 in 10 in the mid-1970s to 1 in 3 in 1990. Moreover, satellite-based systems ensure the instant transmission of information to the most remote villages. Another factor contributing to the increase of population mobility has been the growth and expansion of a migration industry that seeks to facilitate and profit from various types of internal and international population movements. The industry involves a wide array of recruiters, agents, middlemen, lawyers, transport operators, factory foremen and travel facilitators who have become key actors in many migration systems. In addition, government policies have been instrumental in encouraging population mobility. Thus, certain Governments have actively fostered the emigration of workers by developing markets for their skills and others have promoted the settlement of frontier areas.

Although the macrolevel factors identified above have had important implications for the increase of population mobility, to understand fully the reasons for the changing patterns and scale of territorial mobility it is necessary to consider processes operating at the microlevel, pre-eminent among which is the role of the family in determining who moves, how, when and for what purpose.

C. The family as a unit of analysis of population mobility in developing countries

Classical and neoclassical economic theories see migration as the result of differences in wage rates between one region and another (Lewis, 1954; Ranis and Fei, 1961). Todaro (1976) made the theory more realistic by explaining that migration occurs owing to expected wage differentials between the origin and the potential destination. However, these theories were found to have very limited explanatory power in developing countries and it has been argued that one of the reasons for their failure is that they assume that migrants are "adamistic", that is, that they operate as isolated individuals (Stark, 1991). In fact, most migration occurs as a result of decisions taken by families rather than by individuals acting on their own and consequently it is common for families to allocate family members to different labour markets. In developing countries in particular, the family has traditionally been the unit within which most decisions about production, investment and consumption are taken. With respect to population mobility, the decision makers within the family (usually older men) decide which family members move and under what conditions. Hence decisions about movement may or may not coincide with the wishes or interests of the individual who moves. Case studies corroborate that decisions regarding territorial mobility are often taken at the family rather than at the individual level (Dinerman, 1978; Harbison, 1981; Wood, 1981 and 1982; Schmink, 1984). In addition, by focusing on the family rather than on the individual it is easier to consider broader macrostructural forces that impinge on the decision-making process. That is, consideration of intermediate units, such as families, households, communities or kin groups, provides a means of bridging the gap between the social and the individual level of analysis.

When the family is the unit of production, the family, usually through its patriarch, allocates its labour resources over the range of tasks that are necessary to achieve a desired level of production. Since family members have different capacities, characteristics and skills, a division of labour is instituted so as to make the best possible use of the human capital available to the family. In rural areas, this division involves allocating family members to off-farm and on-farm tasks where off-farm tasks would be carried out at nearby locations. The allocation of family labour should achieve two objectives: (*a*) the maximization of production and income; and (*b*) the minimization of risk. The latter is especially important since many rural families live at the subsistence level and lack the surplus needed to palliate failures. An important strategy of risk minimization is the diversification of sources of income or of the types of production so that failure in one area will not necessarily result in a total disaster for the family. In such a context, families tend to adopt survival strategies that involve the allocation of family labour to a range of tasks to be carried out at a variety of locations. The allocation adopted will generally represent a trade-off between maximizing total production and income and minimizing risk. Population mobility is an important part of this strategy because it not only provides the family with access to new and additional sources of income at new locations but also diversifies the family's portfolio in terms of sources of income and thus minimizes risk.

Rahmato (1987) has suggested that coping with risk is a central part of the peasant economy and that peasants develop particular "anticipatory strategies" to insure themselves against dearth and hardship. Hence, survival strategies involving some form of population mobility can take the form of either a response to the onset of a crisis, such as a crop failure or the destruction of a major source of family income, or an "insurance policy" against the possible future failure of an important source of family income.

In some cases, the family makes decisions about population mobility only for certain family members. Thus, among poor families in the Philippines, men were generally found to make relatively autonomous decisions, whereas young women were highly dependent on family decisions (Torres, 1992). Usually, girls who are not attending school are expected to stay indoors or work in the company of their parents and other elders. Even to leave the house, young women require permission from their elders or brothers. If they go out, they usually do so with relatives. Hence, they are unlikely to acquire first-hand information on job opportunities in Metro Manila. Their families, being vested with normative authority to decide for them, must take the initiative. Filial piety and respect for the elders are a strong Filipino value. Young women are therefore expected to abide by the decisions of their kin group, especially with regard to working away from home.

The kinship system often determines who in the family will migrate (Janet Rodenburg, 1993). This can be determined on the basis of age, birth order and gender as well as of the particular skills and attributes of the different individuals making up the family group. In some situations, families adopt deliberate strategies whereby older children are kept away from school to work on the family farm while younger children are given schooling in preparation for migration (Connell and others, 1976).

The adaptive strategies outlined above, or some version of them, have operated in most traditional societies for a long time. To be effective, such strategies depend on norms establishing inter-generational relations whereby children owe absolute loyalty to their parents and must give their parents part of their income (Caldwell, 1976). Because of the changing contexts in developing countries, the area over which the family can allocate its labour resources has extended considerably. Whereas the bulk of the mobility associated with the family's deployment of its labour tended to remain within a short distance of the locality of origin, today many families have the option of allocating their labour to faraway locations, be they distant cities, newly settled frontiers or overseas destinations.

The geographical expansion of labour markets that has accelerated particularly over the last decade implies that, whereas most labour markets tended to be local or regional, they now cover the whole nation or large parts of it, as is especially the case for labour markets centred on major metropolitan areas, which typically draw workers from most of the country. Indeed, a growing number of labour markets overlap national boundaries and for some highly specialized occupations labour markets are practically global. This is the case for managers, professionals and other highly skilled personnel that move frequently between countries as they perform their duties and acquire new skills; but even less well-qualified personnel is tending to move within international labour markets. Thus, agricultural workers from Mexico and the Caribbean operate within a labour market that encompasses the United States. Similarly, Philippine nurses work in a number of countries and Philippine women working in domestic service are found throughout Asia and in a number of European countries (United Nations, 1995).

The sum of individual migration decisions does not add up to a family migration strategy, since the needs of the family may dictate that a family member is to be allocated to a task and type of movement different from those that the individual himself or herself would have chosen. Families weigh carefully their options regarding the allocation of family labour. Rural Mexican families, for instance, send their better-educated children to Mexico City where they can get white-collar jobs and the less well educated ones to the United States, where they are likely to work in agriculture or in blue-collar occupations (Massey, 1990). Such a strategy not only involves a realistic matching of the strengths of family members to available opportunities but also permits a minimization of risk through the diversification of destinations. The reduction of risk may even reduce pressure on the family farm so that experimentation with innovative crops or agricultural techniques can be carried out (Stark, 1991). The importance of minimizing risk through migration implies that wage differentials between areas of origin and destination are not a necessary prerequisite of the occurrence of migration, since the aim of mobility may mostly be to diversify the portfolio of the family's income opportunities.

The expansion of temporary mobility in developing countries cannot be explained in terms of conventional classical and neoclassical economic theory, but it is readily understood from the family perspective. The temporary deployment of family members at distant locations allows the family to take advantage of periodic lulls in labour demand at home or of temporary upswings in demand at the potential destination. Because the family has an interest in maintaining ties with the migrant and in using his or her labour at home when demand is high, circulatory movements are likely to be fostered. Circulation permits the family to maintain control over the migrant and the income that he or she generates. Indeed, the evidence suggests that remittances tend to diminish the longer the migrant is away from the place of origin. Patriarchal authority over the migrant is also likely to wane the longer he or she remains away from the family. In addition, the costs of the temporary migration of a single family member are likely to be lower than those involved in the long-term relocation of several family members or of the whole family group. Besides, earning in the city while spending in the village ensures that earnings go much further, since the cost of food and housing is usually lower in the village.

The mechanics of how the decision to migrate is made within the family are likely to vary considerably according to context. However, it cannot be assumed that the decision to migrate is made in a democratic fashion that involves consideration of the views of all or, at least, of the most important family members, including, of course, the prospective migrant. Migration-related decisions reached by true consensus within the family may not be the rule in certain contexts. Rather, decisions are likely to be taken by those wielding power within the family. In patriarchal societies, the male head of household or the family patriarch is most likely to decide which migration strategy best suits the interests of the family group. However, recent research suggests that the hitherto widespread belief that women play a minor role in the making of the decision to migrate is wrong. In many contexts, women are active and key players in the decision-making processes related to migration, especially when they themselves are potential migrants (Riley and Gardner, 1993). It is therefore incorrect to characterize women, even when they migrate in the company of family members, as passive followers of male migrants (United Nations, 1993).

When migration is used by the family as a form of survival or adaptive strategy, it may occur in response to a specific crisis in which the deployment of family members to different locations may be absolutely necessary for the survival of the family group. Thus, families living in places experiencing periodic problems in obtaining stable and adequate access to food develop strategies to minimize the impact of food shortages (Corbett, 1988). A model depicting the strategies used by families to cope with famine in Africa is presented in table 34. According to it, migration responses may occur in two of the three stages of reaction to food scarcity. In the first stage, characterized by the onset of declining food supplies, families will respond in ways that do not involve the disposal of their key productive assets. One such response is to deploy family labour elsewhere to earn income and even to secure food from areas not affected by food shortages. Such responses can be interpreted as forms of self-insurance or inter-household insurance which are likely to have evolved as coping mechanisms to deal with predictable and not very severe risks. Using these strategies, families may be able to support themselves for long periods, but if food shortages persist, there will come a time when families will have to dispose of their key productive assets, thus entering the second stage of the response to crises. In the third stage, families will be virtually destitute and consequently their ability to generate both current and future income will be severely curtailed. Even the sale of family labour may no longer be possible, especially if household members are weak because of hunger or disease. At this stage, distress migration of the whole family group is likely to occur. Asian and African case studies confirm the use of this response, which may be the only option remaining at this point (Corbett, 1988).

Even when the survival of the family group is not at stake, families may deploy their labour in different locations in order to adapt to shifts in the pattern of

TABLE 34. FAMILY STRATEGIES TO COPE WITH FOOD SCARCITY

Stage 1. Declining food supplies

* Increased petty commodity production and trading
* Dispersed grazing; change in cropping and planting practices
* Migration to towns
* Collection of wild foods; inter-household transfers and loans; credit from merchants and moneylenders
* Migration to rural areas; rationing of food consumption

Stage 2. Sale or loss of productive assets

* Sale of productive household assets
* Access to food distributed by relief programmes
* Sale of possessions

Stage 3. Response to full-fledged crisis

* Break-up of household
* Distress migration

Source: Jane Corbett, "Famine and household coping strategies", *World Development* (Great Britain), vol. 16, No. 9 (1988), pp. 1099-1112.

NOTE: An asterisk (*) indicates strategies often involving mobility.

available opportunities. In such contexts, the temporary mobility of family members is used not so much to ensure survival as to maintain or improve the family's standard of living. Clearly, one key factor for the success of such strategies is the ability of the family or in many instances the patriarch to maintain sufficient control over migrant family members so as to ensure that their earnings are remitted to the family of origin. That is, those moving away from the family must continue to abide by the patriarchal system even when they are not co-resident with the authority figures in the system. This form of social control is common in many societies. In Northern Sumatra, Indonesia, for instance, a strong mutual moral obligation is felt between family members living in the village of origin and those who have moved elsewhere (Janet Rodenburg, 1993). Circular migration, by maintaining the ties between the migrant and the family left behind, is more likely to ensure that the family maintains control over the resources generated by migrant family members.

In sum, a comprehensive explanation of the increased scale and complexity of contemporary population mobility in developing countries needs to go beyond an understanding of the macroeconomic, social and political changes occurring in those countries. It must encompass an appreciation of how family strategies operate in their changing environment. One important element in shaping those strategies is the operation of social networks.

D. Social networks in the migration process

Migrants in developing countries are stereotypically characterized as wide-eyed, naive individuals overwhelmed by the alien nature of their destination and having no contacts there. That image could not be further from the truth. The evidence shows that the majority of persons who move do so along well-trodden paths with which they themselves, and their relatives or friends are familiar. In many instances, those relocating move in the company of friends or relatives and have a range of contacts at the place of destination. Consequently, social networks linking places of origin with those of destination play a key role in generating and sustaining population flows between them. In many contexts, social networks inject a self-perpetuating dynamism into migration flows that allows them to continue long after the original economic, social or political reasons for migration have disappeared.

In the tightly knit communities characteristic of many developing countries, whenever a person migrates, every individual that knows him or her acquires social capital in the form of a contact at the mover's destination. The networks established by earlier waves of movers act as conduits through which later waves to those destinations are channeled in an atmosphere of certainty. In many instances, previous waves of movers not only supply valuable information and encouragement to facilitate the move of others, but also pay the costs of relocation. At the place of destination, social networks provide valuable assistance in the adjustment process, especially by securing housing and employment for the newly arrived migrant. The fundamental role of networks is to reduce the risks associated with migration.

The role of networks based on kin or friendship ties in facilitating population mobility in developing countries can scarcely be exaggerated. While the generation of networks may depend on fortuitous events, once the pioneer migrants are established, a flow may develop rapidly. In the case of international migration, pioneers may gain access through formal labour recruitment processes, as immigrants or by being admitted into particular categories, such as those of students, trainees or refugees. If the receiving country allows family reunification, a more substantial flow may develop in the short run. In the case of undocumented migration, social networks play an even more crucial role in making possible the clandestine movement of people, and those networks need not be composed mainly of migrants. In the United States, for instance, local church groups played a key role in facilitating the migration of asylum-seekers from Central America. It is also possible for pioneer migration to occur without leading to the development of a dynamic network, as in cases where the flow of refugee migration is stopped by stringent controls imposed by the country of origin. This constitutes just one example, but others doubtlessly exist. There is still only a modicum of information about how networks are generated and maintained (Gurak and Caces, 1992).

The important point is that, once created, networks change the environment in which subsequent relocation occurs. Networks link individuals not only with friends and relatives at a range of destinations but possibly also with employers or labour recruiters. Frequently, a patron-client relationship characterized by mutual dependence develops between an employer and a family or groups of families from a particular origin. This relationship guarantees employment to new waves of migrants and assures the employer of a regular and trusted supply of labour. In such cases, it is inadequate to characterize networks as purely "social" since they involve more than blood and friendship ties. In some contexts, networks involve not only migrants, their relatives, friends and employers, but also other actors involved in the migration process. Thus, networks may encompass a variety of intermediaries, including agents, recruiters, lawyers, travel agents and transport operators, all of whom facilitate the flow of movers. In the case of undocumented migration from a village in East Java to the plantations of peninsular Malaysia, a prospective migrant is likely to pass from the local *calo*, to a middleman in Surabaya and to another one in Jakarta who will make contact with a broker in Riau; the broker will in turn arrange matters with the border police and passage with a boat captain. The captain will transport the migrant and put him in contact with a *toke* in Malaysia who will eventually lead him to an employer. This example illustrates the complexity of the networks of intermediaries that make migration possible, networks which are highly institutionalized in some contexts. In the Philippines, a complex network of underground travel agents, talent promoters and job brokers is known to expedite the illegal entry of female entertainers into Japan (Torres, 1992).

One of the most important features of migration networks is their capacity to sustain migration independently of the economic conditions prevailing at origin and destination. Because networks operate largely outside the sphere of influence of policy makers, their impact on migration is often beyond policy control. Almost all attempts to prevent the types of migration movements that rely heavily on the functioning of networks have failed. Networks appear to be resistant even to the most vigorous actions by Governments. However, there is much about the functioning of networks that is not yet known, particularly regarding those that are no longer capable of eliciting new flows. It has been suggested that networks follow a specific pattern of expansion and attrition, so that migration flows may grow according to a logistic curve with movement initially of a few persons, followed by a rapid increase in movement, and then a tapering off when a certain threshold is reached. It may also be that certain flows never expand. More information regarding those cases is urgently needed.

The migration linkages established between origin and destination not only create channels for the further movement of people but also operate as important conduits of information, money and goods in both directions. Remittances to the place of origin are of major significance. In the 1970s, the conventional wisdom about remittances was that they only had a small impact on development because they were relatively low and were spent mostly on consumption and non-productive investment (Lipton, 1980). Today, a new assessment of the impact of remittances has been made, noting that past measurement of their magnitude has been at best partial and has severely underestimated their scale. In times of crisis, remittances are essential for facing emergencies and have a major impact on the place of origin. Indeed, the very fact that support will be available if needed can have a significant effect on development activity in the area. The argument that remittances have been used for non-productive investment fails to take into account the multiplier effects of investments in land and housing. In the case of international migration, remittances are having not only a very significant impact in the communities of origin of migrants but also important macroeconomic impacts at the country level.

The significance of networks in sustaining and developing particular migration flows has been overlooked in many attempts to explain migration in developing countries. The development of networks is facilitated in cases where the extended family is the rule. In such systems, the patriarch or the person wielding power within the family (sometimes a matriarch) is able to deploy family members to particular areas and retain control over their earnings. The existence of networks explains why neighbouring communities, which are apparently identical with respect to their economic and social circumstances, may experience very different levels and patterns of population mobility. Those communities having better linkages with other areas will tend to experience higher population mobility than communities lacking those linkages. Networks can both facilitate the relocation of family members and aid families in maintaining control over migrants. The experience of Toba Batak traders moving from rural areas in Indonesia to cities illustrates this point. Upon their first arrival in town, most traders stay with friends or relatives, who provide not only shelter but also funds for the initiation of trade. By staying with co-villagers, the migrants can send messages or money back home when one of them returns. Wives who stay behind feel that co-villagers exert some social control on their husbands' behaviour, especially because any infringement of rules would be known in the village (Adriana N. Rodenburg, 1993).

Networks often facilitate the migration of groups that are normally not allowed to move by the family. For example, Torres (1992) maintains that the existence of networks has been crucial in convincing Filipino families to send their educated single daughters to work in Manila. The increasing participation of Filipino women in migration is evident in both internal and international migration flows. These moves are encouraged both by the demands of the labour market and by the expectations fostered by the social network. Women who are still young and single are encouraged to leave their towns or villages to work in Manila not only because there are few or no jobs for them in rural areas but also because the jobs that they will secure in Manila will generate better incomes. According to the migrants themselves, a sizeable proportion of their salaries is used for their families' sustenance back home. In this context, the woman's family usually makes the decision about her migration.

E. Conclusion

Population mobility in all its forms has become a major adaptive and survival strategy across a broad spectrum of groups in developing countries. The stereotypical view according to which most residents of developing countries live in rural communities from which they seldom leave has never been fully accurate but is certainly very far from today's reality. Population movements within and between countries have greatly expanded and have substantial economic, social and demographic consequences for the communities involved. Yet our understanding of the changes taking place is partial at best. Indeed, in most countries even the scale of population mobility is not accurately known, much less its composition, causes and implications. There is a pressing need to arrive at a better conceptualization, measurement and analysis of patterns of population mobility, particularly that taking place in developing countries. In doing so, it is important to consider all types of movement, since many of those that are not detected by conventional data-collection systems have considerable economic and social significance. That is the case for many types of temporary migration and of undocumented international migration.

It is anticipated that population mobility will increase during the 1990s because the forces that have led to the recent increases in the scale and complexity of territorial mobility are expected to persist. Thus, the substitution of capital for labour in agriculture is accelerating in many developing countries and the commercialization of agricultural production is increasing. The move towards the privatization and consolidation of land, increased emphasis on markets, and mechanization persists and will lead more people to leave rural areas. Increased mobility will also be facilitated by the rising levels of educational attainment to which most Governments are strongly committed. The influence of the mass media will not only persist but become more pervasive. Furthermore, the costs of transportation will continue to decline and both national and international transport systems will improve their accessibility and speed. The internationalization of capital will continue to foster international population mobility although the barriers to international migration will be maintained or even reinforced.

Neoclassical economists have argued that interregional differences in wage rates are the fundamental cause of migration and, although such differences provide only a partial explanation of current population mobility, there is no indication that international or intranational differences in the supply of or the demand for labour will diminish during the 1990s. Indeed, at the international level those differences are likely to increase. On the demand side, there are major differences in the level of economic growth and overall prosperity of different countries and their differentiation in terms of the growth of their labour force is increasing. Those countries that began to experience substantial fertility reductions in the 1960s and early 1970s will begin to experience a reduction in the size of the cohorts entering the labour force during the 1990s. Already Hong Kong, Japan, Singapore and Taiwan Province of China are experiencing labour shortages (Hugo, 1991b and 1992b), and they are located near countries that still have large labour surpluses. In Indonesia, for example, the population of working age is increasing annually by 2.4 million persons while 40 per cent of the existing labour force is underemployed (Hugo, 1993). China is estimated to have an excess of 200 million workers (Hugo, 1992b). Hence the potential for substantial international labour migration during the 1990s is large. In addition, inequalities within developing countries are also increasing, especially between large metropolitan areas and their hinterlands. Consequently, there are few grounds for expecting a reduction of population mobility within developing countries during the 1990s.

Any comprehensive explanation of the changes in population mobility in developing countries must incorporate not only macroeconomic, political and social forces but also factors operating at the individual, family and community levels. The family in particular plays a crucial role in mediating the influence of macrostructural forces on individual migrants. To a large extent, population mobility in developing countries has resulted from decisions made at the family level regarding the best strategies to follow to ensure family survival and well-being. A number of survival and adaptive strategies adopted by families involve the allocation of family labour across a variety of income-earning activities in a number of locations. Such strategies are designed to maximize family income while minimizing risk by diversifying the range of income sources. They have resulted in a wide array of population movements, ranging from the permanent to the temporary and involving different lengths of absence when they are temporary. Movers and migrants have included both men and women, and persons from a variety of socio-economic strata and ethnic groups. Both internal and international mobility has been used. The changing context has facilitated the expansion of territorial mobility, in terms of both the number of persons involved and the geographical area covered.

Networks have played a crucial role in fostering population mobility in developing countries, a role that is only beginning to be understood. Networks can be conceptualized as the set of interpersonal ties that link migrants and non-migrants in areas of origin and destination, often through the bonds of kinship, friendship and shared origin (Massey, 1988). Once established, networks set in motion a process of cumulative causation that tends to perpetuate migration even independently of the forces that first gave rise to it. By providing information and assistance, networks increase the likelihood of migration and reduce the costs of relocation. To be effective, policies intending to modify patterns of population mobility must take account of the functioning of networks. Thus, Governments seeking to encourage mobility should consider policies that reinforce the development of networks. In cases where networks are well established, policy interventions that seek to reduce migration will not be effective unless the mode of operation of existing networks is taken into account. Clearly, a better understanding of the operation of networks is necessary to devise appropriate policy interventions. This is particularly the case regarding the networks of intermediaries that facilitate migration, including agents, recruiters, lawyers, migration advisors and marriage brokers.

Lastly, it is also important to consider the effects that current changes are having on families in developing countries. Traditional extended families in which patriarchal authority is supreme are being eroded as nuclear families become predominant. Although men may still exercise considerable power within nuclear families, the latter tend to foster more equitable relationships between spouses as well as between parents and children. In addition, continued declines in fertility will reduce the labour available within the family. Consequently, it is likely that, in future, the family may wield less power in allocating the labour of its members to different locations or in controlling their earnings. This does not mean, however, that migrants will become the ideal "adamistic" individuals posited by economic theory. Even in countries where the nuclear family predominates, the migration decisions of individuals are not made in isolation from the family group. Furthermore, networks also play an important role, though they may be less dependent on kinship and friendship ties than on other institutionalized linkages between individuals.

References

Biro Pusat Statistik (1982). *Population of Indonesia, Results of the Sub-Sample of the 1980 Population Census*. Jakarta, Indonesia: Biro Pusat Statistik.

Caldwell, John C. (1976). Toward a restatement of demographic transition theory. *Population and Development Review* (New York), vol. 2, Nos. 3-4, pp. 321-366.

Connell, John, and others (1976). *Migration from Rural Areas: The Evidence from Village Studies*. Delhi: Oxford University Press.

Corbett, Jane (1988). Famine and household coping strategies. *World Development* (Great Britain), vol. 16, No. 9, pp. 1099-1112.

Dinerman, I. (1978). Patterns of adaptation among households of U.S.-bound migrants from Michoacan, Mexico. *International Migration Review* (Staten Island, New York), vol. 12, No. 4, pp. 485-501.

Gurak, Douglas T., and Fe Caces (1992). Migration networks and the shaping of migration systems. In *International Migration Systems*, Mary M. Kritz, Lin L. Lim and Hania Zlotnik, eds. Oxford: Clarendon Press.

Harbison, S. F. (1981). Family structure and family strategy in migration decision making. In *Migration Decision Making. Multidisciplinary Approaches to Microlevel Studies in Developed and Developing Countries*, Gordon F. De Jong and Robert W. Gardner, eds. New York: Pergamon.

Hugo, Graeme J. (1982). Circular Migration in Indonesia. *Population and Development Review* (New York), vol. 8, No. 1, pp. 59-84.

__________ (1991a). Population movements in Indonesia: recent developments and their implications. Paper presented at the International Conference on Migration, National University of Singapore, Singapore, 7-9 February.

_________ (1991b). Recent international migration trends in Asia: some implications for Australia. In *Immigration Population and Sustainable Environments*, J. Smith, ed. Adelaide, Australia: The Flinders Press.

_________ (1992a). Indonesian labour migration to Malaysia: trends and policy implications. Paper presented at the International Colloquium on Migration, Development and Gender in the ASEAN Region, University of Malaya, Population Studies Unit, Kuantan, Malaysia, 28-31 October.

_________ (1992b). Asia on the move: the transformation of international migration in Asia in the 1990s and its implications for Australia. Inaugural Lecture, University of Adelaide, Adelaide, Australia, 23 October.

_________ (1993). *Manpower and Employment Situation in Indonesia 1992*. Jakarta: Department of Manpower.

Lewis, W. A. (1954). Economic development with unlimited supplies of labour. Manchester School of Economic Studies, Manchester, United Kingdom, 22 May. Mimeographed.

Lipton, M. (1980). Migration from rural areas of poor countries: the impact on rural productivity and income distribution. *World Development* (Great Britain), vol. 8, pp. 1-24.

Massey, Douglas S. (1988). Economic development and international migration in a comparative perspective. *Population and Development Review* (New York), vol. 14, No. 3, pp. 383-413.

_________ (1990). Understanding Mexican migration to the United States. University of Pennsylvania, Population Studies Center, Department of Sociology. Mimeographed.

Rahmato, Dessalegn (1987). Famine and survival strategies: a case study from northeast Ethiopia. *Food and Famine Monograph Series*, No. 1 (May). Institute of Development Research, Addis Ababa, Ethiopia: Addis Ababa University.

Ranis, G., and J. G. H. Fei (1961). A theory of economic development. *American Economic Review* (Madison, Wisconsin), vol. 51, pp. 533-565.

Riley, Nancy E., and Robert W. Gardner (1993). Migration decisions: the role of gender. In *Internal Migration of Women in Developing Countries*. Sales No. E.94.XIII.3. New York: United Nations.

Rodenburg, Adriana N. (1993). *Staying Behind: Rural Women and Migration in North Tapanuli, Indonesia*. Amsterdam: University of Amsterdam.

Rodenburg, Janet (1993). Emancipation of subordination? Consequences of female migration for migrants and their families. In *Internal Migration of Women in Developing Countries*. Sales No. E.94.XIII.3. New York: United Nations.

Schmink, M. (1984). Household economic strategies: review and research agenda. *Latin American Research Review* (Texas), vol. 19, No. 3, pp. 87-101.

Standing, Guy, ed. (1984). *Labour Circulation and the Labour Process*. London: Croom Helm.

Stark, Oded (1991). *The Migration of Labour*. Cambridge, United Kingdom: Basil Blackwell.

Todaro, Michael P. (1976). *International Migration in Developing Countries*. Geneva: International Labour Office.

Torres, Amaryllis T. (1992). Features of the migration of men and women in the Philippines. Paper presented at the International Colloquium on Migration, Development and Gender in the ASEAN Region, Population Studies Unit of the University of Malaya, Kuantan, Malaysia, 28-31 October.

United Nations (1970). *Manual VI: Methods of Measuring Internal Migration*. Population Studies, No. 47. Sales No. 70.XIII.3.

_________ (1992). *World Population Monitoring, 1991: With Special Emphasis on Age Structure*. Population Studies, No. 126. Sales No. E.92.XIII.2.

_________ (1993). *Internal Migration of Women in Developing Countries*. Sales No. E.94.XIII.3.

_________ (1995). *International Migration Policies and the Status of Female Migrants*. Sales No. E.95.XIII.10.

_________ (1996). *World Population Monitoring, 1993: With a Special Report on Refugees*. Sales No. E.95.XIII.8.

Wood, Charles H. (1981). Structural changes and household strategies: a conceptual framework for the study of rural migration. *Human Organization* (Kentucky), vol. 40, pp. 338-344.

_________ (1982). Equilibrium and historical-structural perspectives on migration. *International Migration Review* (Staten Island, New York), vol. 16, No. 2 (Summer), pp. 298-319.

Zelinsky, Wilbur (1971). The hypothesis of the mobility transition. *Geographical Review* (New York), vol. 41, No. 2, pp. 219-249.

IX. HEALTH AND ENVIRONMENTAL PROBLEMS IN THE CITIES OF DEVELOPING COUNTRIES

*David Satterthwaite**

In recent decades, most developing countries have experienced a very rapid growth in their urban populations which has not been accompanied by the needed expansion in the provision of urban infrastructure and services or in the availability of safe shelter of adequate quality. In most urban centres—from large cities and metropolitan areas to regional centres and small market towns—much of the population lives in shelters and neighbourhoods with little or no access to the infrastructure and services that are essential for the maintenance of human health in an urban environment, including a piped and clean water supply; sanitation and drainage; paved roads; regular garbage collection; health care and emergency services; and schools, social centres and sites for play or recreation.

It is estimated that, between 1950 and 1990, the urban population of the developing world grew from 286 million to more than 1.5 billion persons (United Nations, 1991), so that it is now larger than the total population of the developed world. In Africa, the urban population grew sevenfold during that period, and that of Asia grew more than sixfold. Although the urban population of Latin America and the Caribbean grew less rapidly, by 1990 more than 70 per cent of that region's population lived in urban areas (United Nations, 1991).

In most cities of the developing world, a large proportion of families live in one or two rooms in cramped and overcrowded tenements, cheap boarding houses or shelters built on illegally occupied or subdivided land. In many cities, there is a smaller proportion of persons who live and sleep in public or semi-public spaces and thus have great difficulties in having access to water, sanitation and security. In most cases, neither the Government nor the formal private sector has contributed much to expanding the housing stock that is affordable to the low-income majority. During the last few decades, most new housing for that segment of the population has been planned and built at the margin of the law. Large numbers of urban dwellers have no choice but to build, buy or rent an illegal dwelling (or a legal dwelling illegally subdivided) because they cannot afford the cheapest "legal" house or apartment. It is common for 30-60 per cent of an entire city's population to live in homes and neighbourhoods that have been developed illegally. In the majority of cities in the developing world, 70-95 per cent of all new housing is built illegally. An estimated 600 million urban dwellers in Africa, Asia and Latin America are in circumstances that threaten life and health because of inadequate housing, lack of adequate infrastructure and services, and the use of sites prone to flooding or some other natural disaster (Cairncross, Hardoy and Satterthwaite, 1990).

(*a*) Water: Several hundred million urban dwellers have no alternative but to use contaminated water or water whose quality is not guaranteed. A small minority have water piped to their homes; a larger number have to collect water from a standpipe nearby; but hundreds of millions "are obliged to use water from streams or other surface sources which in urban areas are often little more than open sewers, or to purchase water from insanitary vendors. It is little wonder that their children suffer frequently, often fatally, from diarrhoeal diseases" (Cairncross, 1990).

(*b*) Sanitation: It is estimated that about two thirds of the urban population in developing countries have no hygienic means of disposing of excreta and an even greater number lack adequate means of disposing of waste waters (Sinnatamby, 1990). Most cities in Africa and many in Asia have no sewers at all, and some of them are major cities with over 1 million inhabitants. Untreated human excreta and waste waters generally end up in rivers, streams, canals, gullies and ditches. In cities with sewers, the sewerage system rarely serves more than a small proportion of the population, typically those dwelling in the richer residential, government and commercial areas. The majority of dwellers in major

*Human Settlements Programme, International Institute for the Environment and Development, London, United Kingdom.

cities, such as Accra, Calcutta, Dar es Salaam, Jakarta, Kampala, Khartoum and Manila, lack adequate sanitation. In India, where one third of the urban population, or over 50 million people, have no latrine of any kind and another third rely on bucket latrines, defecating in the open is common practice. One third may use latrines connected to sewers but only 10 per cent have sewerage connections to their homes (Centre for Science and Environment, 1983).

(c) Garbage collection and disposal services: It is estimated that between one third and one half of the solid wastes generated in urban centres of the developing world remain uncollected (Cointreau, 1982). Poorer areas in cities usually have the least adequate garbage collection services or lack services altogether. Most poor families live in quarters with very limited space (especially in tenements and high-density illegal settlements) and have difficulty storing and disposing of waste. Only a small proportion is able to transport garbage to a supervised dump site (Cointreau, 1982).

(d) Health care: The data available do not provide an adequate estimate of the number of people lacking ready access to health-care services in urban areas, but WHO (1991a) estimates that around 1.6 billion people in developing countries lack access to primary health care, a figure that probably includes several hundred million living in urban areas. These figures underestimate, however, the number of people lacking access to effective primary health care since many clinics are understaffed relative to the demand for their services and lack essential drugs. UNICEF (1991) documented a decline in health spending per person in more than three quarters of the countries of Africa and Latin America in recent years and noted that hundreds of health clinics had been closed and those remaining open were often understaffed and lacked essential supplies.

(e) Overcrowding: The average number of persons per room in most urban areas of the developing world is three to five times that in urban areas of the developed world. Many health problems affecting poorer groups are associated with overcrowding, including household accidents, acute respiratory infections, tuberculosis and other airborne infections (WHO, 1992a). In the predominantly low-income residential areas of cities in developing countries, there is often an average of four or more persons per room and in many instances the average floor space per person is less than one square metre (Aina, 1989; Murphy, 1990). Diseases such as tuberculosis, influenza and meningitis are thus easily transmitted from one person to another. Their spread is aided by low resistance among co-dwellers because of malnutrition and by the frequent contact between infective and susceptible persons.

A. ADVANTAGES AND DISADVANTAGES OF POPULATION CONCENTRATION

Although much of the literature on urbanization associates rapid urban growth with contamination of the environment and unhealthy living conditions, the concentration of population and production in urban centres provides many opportunities and cost advantages with respect to maintaining a cleaner environment, better environmental health and comprehensive health-care coverage. The concentration of population greatly reduces the unit costs of providing each building with piped water, sanitation, garbage collection, paved roads, electricity and drains. It also reduces the unit costs for the provision of health services, schools, pre-school centres and child development centres.[1] Even in most squatter settlements, population densities are not too high to pose problems for the cost-effective provision of such infrastructure and services.

Industrial concentration in cities reduces the cost of enforcing regulations on environmental and occupational health and those on pollution control. It also lowers the cost of many specialized services and waste-handling facilities, including those that reduce waste levels or recover materials from waste streams for reuse or recycling. The concentration of households and enterprises in cities makes it easier for public authorities to collect taxes and charges for public services, while in prosperous cities there is a larger revenue base, higher demand and increased capacity to pay. Population concentration can also facilitate people's involvement in taking action or implementing decisions within their own district or neighbourhood.

Environmental problems are most severe where there is a lack of effective governance, including weaknesses in the institutional means to ensure the control of pollution and the provision of infrastructure and services. Good governance is particularly critical for cities to ensure both that their potential environmental and health advantages are fully exploited and that their potential disadvantages are avoided. The extent to which adequate environmental quality in urban areas is achieved and maintained is one of the most revealing indicators of the competence and capacity of city and municipal authorities, as is the extent to which their policies respond to the needs and priorities of the city's population.

B. Failure of Government with respect to urban environmental issues

In countries characterized by the rapid growth of the urban population, net rural-urban migration usually contributes significantly to urban population growth. In most developing countries, the rapid growth of the urban population has overwhelmed the capacity of city and municipal authorities to provide basic infrastructure and services and to ensure a rapid expansion of affordable housing of good quality. However, since some of the most rapidly growing cities in the developing world have coped relatively well with rapid population growth and have better social and environmental indicators than urban centres that have grown slowly, the relation between the rate of population growth and the extent of social and environmental problems is not straightforward or unidirectional.

Countries that have experienced high levels of rural-urban migration in recent decades have also tended to record rapid growth in gross national product (GNP) and generally perform better than countries experiencing slow or negative economic growth in terms of social and health indicators. For countries in Asia, Latin America and Northern Africa, there is a strong positive correlation between the rate of economic growth in recent decades and the extent of net rural-urban migration (Hardoy and Satterthwaite, 1989). In much of sub-Saharan Africa, that correlation was weaker during the 1960s and 1970s, mostly because of the unique social and political circumstances underlying the process of urbanization during the post-independence period, including the dismantling of colonial controls on the African population's right to live and work in urban areas. Most countries that have experienced high levels of rural-urban migration have also had important gains in life expectancy and reduction of infant and child mortality. This relation is not surprising since better economic performance and the increasing prosperity it implies can both provide a higher proportion of the population with adequate incomes and permit greater government expenditure on social services. However, there are countries where major improvements in health have been achieved without rapid economic growth and others where improvements in health indicators have not kept in step with rapid economic growth. Nevertheless, since high levels of net rural-urban migration are strongly associated with rapid economic growth, it is counterproductive to consider that rural-urban migration per se is an economic drain or a social problem.

It is well known that, among migrants to urban centres, those who are young and seek work are overrepresented. They are often prepared to work for low wages in enterprises where employers save money by not abiding by regulations on occupational health and social security. Ironically, the very people who contribute much to a city's economic base by providing cheap labour or cheap goods and services and whose movements in and out of the city largely correspond to the changing demand for labour are the ones most often characterized as a drain on the economy. There is a long tradition of ascribing cities' problems to recent migrants (see, for example, Portes, 1979) but, at a macrolevel, such a conclusion cannot be supported if the scale and direction of net rural-urban migration correspond to changing economic circumstances (as has been shown in national, regional and local studies[2]) and if migration movements are logical responses by individuals or households to changing economic circumstances (as has been shown by most detailed field studies of migrants in areas of origin and destination).

Migrants might be considered a serious social problem if government expenditures on social services for migrants constituted a large part of the social welfare budget. Yet, in most urban areas of developing countries, the Government spends little on social services, whether for migrants or for poor city-born persons. Indeed, in most urban centres, low-income households, be they composed of city-born or migrant members, contribute significantly to building the city's housing stock through their own self-help and mutual help efforts, usually with little or no support from Government. Furthermore, the informal or illegal processes by which such housing is built bring many advantages to both the Government and city businesses. Illegal construction costs neither of these groups any money (except the opportunity costs of land occupied illegally) and it provides housing for a high proportion of the urban population. There are numerous case studies describing the development of different illegal or informal settlements in cities of the developing world and it is in those settlements that most new urban housing has been developed over the past 30 or 40 years (John F. C.

Turner, 1976; Bertha Turner, 1988; Hardoy and Satterthwaite, 1989). Case studies reveal the ingenuity with which new settlements have been built at low monetary cost and with scarce resources, thus demonstrating the capacity that poor people have to organize and plan (John F. C. Turner, 1976; Sobreira de Moura, 1987; Bertha Turner, 1988; Hardoy and Satterthwaite, 1989; Peattie, 1990; Cuenya and others, 1990).

It is clear that most social and environmental costs, in terms of ill health, disablement and premature death, are borne by people living in informal or illegal settlements in cities of the developing world. Whatever their individual ingenuity and collective organization, they can rarely address the problems related to having no paved roads, drains, sewers, piped water or garbage collection. Lack of access to health care implies that injuries and disease often go untreated. Even where community organization seeks a solution to some of these problems, it is often undermined by the hostility of public authorities.

Poor migrants and poor individuals in general contribute little to the other environmental problems associated with cities in the developing world. With regard to air pollution, for instance, the main contributors are industrial processes, thermal power stations and motor vehicles (Lee, 1985). Only in cities where the poorer groups use biomass fuels or coal for cooking or heating are they likely to make a significant contribution to ambient air pollution. Poorer households contribute hardly at all to the creation of hazardous chemical wastes and can hardly be held accountable for the existence of industrial effluents, since they are able to afford so few of the goods produced by those industries. On a per capita basis, poorer groups draw much less than their more affluent counterparts on the finite capacities of local or regional ecosystems for the provision of fresh water and biomass. Much of the housing in which poorer groups live makes widespread use of recycled or reclaimed materials, but little use of cement and other materials requiring a high energy input. In addition, low-income households have too few capital goods to be regarded as accounting for much use of metals and other non-renewable resources, and they both generate low volumes of household waste and tend to reuse or recycle a number of materials, including metal, glass, paper and clothes, among others. Most poor people walk, bicycle or rely on public transport, and thus have a low average oil consumption per person. Low-income groups generally occupy only a small proportion of a city's surface, relative to their proportion of the total urban population, and hence cannot be considered a major factor in respect of the use of valuable land. Thus, a recent estimate for Manila suggested that close to half of the population lives in illegal settlements that take up just 5.3 per cent of the land area of the city (ANAWIM, 1990). The contribution of poor urban dwellers to greenhouse gas emissions and to the release of chemicals responsible for the depletion of stratospheric ozone also remains small.

Although there are cities with very serious environmental problems that have experienced high in-migration levels, the association is not necessarily causal. Where environmental problems have centred on chemical contamination from industries (as in Cubatao, Brazil, during the 1970s and much of the 1980s), the problem did not stem from in-migration (although the city grew rapidly because of it), but rather from the lack of pollution controls among the industries located in the city and the failure of Government to enforce such controls.

When a city's environmental problems are largely the result of the inadequate provision of piped and treated water, sanitation, drainage, garbage collection and health services, they may have been exacerbated by rapid population growth to which net in-migration might have made a major contribution. However, the real cause of such environmental problems is the incapacity or unwillingness of Government to cope with the social and environmental implications of rapid urban growth. It is also difficult to implicate high in-migration as the cause of deficiencies in the public provision of basic infrastructure and services when there are many cities with low or even negative net migration in which environmental problems are also very serious (for instance, Calcutta, Mexico City[3] and Buenos Aires during the past 10 or 20 years). There are also numerous urban centres that have been experiencing net out-migration and that have serious deficiencies in the provision of water, sanitation, drainage and garbage collection. In a sense, this is not surprising because net out-migration from a city is often a symptom of economic decline which is in turn associated with decreasing purchasing power and increasing poverty for a considerable proportion of the city's inhabitants. This association contradicts the suggestion that in-migration to a city is necessarily associated with an increase of social and environmental problems.

An analysis of the environmental problems of the cities of developing countries and of their underlying causes shows that it is the failure of Government to adapt to economic change that underlies the most serious environmental problems. The main areas where deficiencies in Government action are evident include:

(*a*) Failure to ensure adequate provision of water and solid and liquid waste collection and treatment systems, and failure to ensure adequate provision of health care that not only treats environment-related diseases but also implements preventive measures to limit their incidence and severity. Provision of such services can only be undertaken in a cost-effective manner either by the Government itself or with its support. Lack of action in this area is partly due to the refusal by central Government to grant city or municipal authorities the fund-raising powers needed to provide the necessary services effectively;

(*b*) Failure to implement pollution controls, mostly stemming from Government's reluctance to implement existing legislation rather than from a lack of appropriate regulations or legislation;

(*c*) Failure to promote good practices regarding occupational health and safety standards;

(*d*) Failure to make available at affordable prices legal land sites for housing developments on a sufficient scale to put a downward pressure on land prices and within a planning framework that guarantees sufficient open space and minimizes infrastructure costs. Governments have also failed to collect the taxes and user charges from those who benefit from publicly funded infrastructure and services. The costs of these failures fall most heavily on poorer groups.

C. Environmental hazards and their impact on health

Box 1 lists seven kinds of health hazard that are common in urban environments. Four—biological pathogens; chemical pollutants; a shortage of (or lack of access to) particular natural resources; and physical

Box 1. Hazards to health within the urban environment

1. **Biological pathogens or pollutants within the human environment that impair human health:** These include pathogenic agents and their vectors and reservoirs, such as the pathogenic micro-organisms in human excreta, and disease vectors such as the anopheline mosquitoes that transmit malaria and airborne pathogens such as those causing acute respiratory infections and tuberculosis.
2. **Chemical pollutants within the human environment:** These include chemicals added to the environment by human activities, such as industrial wastes or particulate matter remaining in the air after fossil fuel combustion, and chemical agents present in the environment that do not result from human activities.
3. **Shortage of (or lack of access to) the natural resources on which human health depends in terms of their availability, cost and quality:** These include food, water and fuel.
4. **Physical hazards:** For instance, the risk of flooding in houses and settlements built on flood plains or the risk of mud slides or landslides affecting houses built on slopes.
5. **Aspects of the built environment having negative consequences on psychosocial health:** For instance, inadequate protection against noise; lack of security from eviction, robbery or violence; inadequate provision of infrastructure or services; lack of common areas for recreation.
6. **Natural resource degradation:** For instance, decline in soil or water quality resulting from the disposal of gaseous, liquid or solid wastes.
7. **National or global environmental degradation:** This includes the depletion of non-renewable resources. Also, wastes from human activities may threaten the functioning and stability of global cycles and systems. Examples of such global environmental degradation include changes in climatic conditions caused by greenhouse gases and the depletion of the stratospheric ozone layer.

hazards—have a direct bearing on health. In terms of their impact on health, particularly that of infants and children, these are the four most pressing urban environmental problems in Africa and much of Asia and Latin America. The other three, which also influence health albeit less directly, are aspects of the built environment that have negative consequences on psychosocial health; natural resource degradation; and national or global environmental degradation, including a rising concentration of greenhouse gases in the atmosphere and the depletion of the stratospheric ozone layer. It is these last three that tend to dominate international discussions on the environment, including those at the 1992 United Nations Conference on Environment and Development (Earth Summit) in Rio de Janeiro, although their health impact is only indirect.

1. *Biological pathogens within the human environment*

In urban areas of developing countries, biological pathogens in the human environment—in water, food, air or soil—are the single most serious environmental problem in terms of their toll on human health (WHO, 1992a). According to the medium through which human infection takes place, biological pathogens can in turn be subdivided into three categories: food-borne; airborne; and water-related. A number of biological pathogens fall into more than one category, since they can be ingested both through contaminated food or through contaminated water.

Box 2 presents examples of the main water-borne, water-washed, water-based or in other ways water-related diseases with estimates of morbidity, mortality and population at risk (where these are available). Water-borne diseases (which are the single largest category of communicable diseases contributing to infant mortality worldwide) account for more than 4 million infant deaths per year (WHO, 1992a). These diseases, which are second only to tuberculosis in contributing to adult mortality, account for 1 million adult deaths per year. In contrast, very few fatal cases of waterborne diseases are now recorded in developed countries. Among the many water-related and water-based diseases common in the urban areas of developing countries, filariasis and intestinal worms (especially ascariasis or roundworm parasitosis) cause the progressive debilitation of millions of people, although only a small proportion die from them (WHO, 1991b). Case studies in urban areas of developing countries have shown that a high proportion of the population in certain low-income settlements or districts suffer from intestinal worms, as in Allahabad (Misra, 1990) and Kuala Lumpur (Bundey, Kan and Rose, 1988).

The quality of housing has an important influence on the incidence and severity of respiratory infections, especially through overcrowding, poor ventilation, dampness and indoor air pollution from coal or biomass combustion for cooking or heating. Acute respiratory infections remain one of the main causes of infant and child mortality. Between 4 million and 5 million infants and children die each year from these infections (mostly from measles or pneumonia) and a child who contracts bronchitis or pneumonia in a developing country is 50 times more likely to die than a child with that disease in a developed country (Pio, 1986). WHO (1992a) notes that acute respiratory infections tend to be endemic rather than epidemic. They tend to affect children and are more prevalent in urban than in rural areas. The frequency of contact, the density of the population and the concentration and proximity of infective and susceptible individuals in urban areas promote their transmission. Poorer groups are at higher risk of contracting those diseases because they tend to be younger, have limited access to health services (especially to vaccines and antibacterial drugs), and live in overcrowded households. The influx of migrants susceptible to infection or who are carriers of new strains of infective agents also contributes to the prevalence of respiratory infections (WHO, 1992a).

Overcrowding also contributes to increasing the incidence of tuberculosis in urban areas. The highest incidence of the disease tends to be found among people living in the poorest areas, experiencing high levels of overcrowding and having a high number of social contacts. A combination of overcrowding and poor ventilation often means that if a family member becomes infected, tuberculosis spreads quickly to the rest of the family (Cauthen, Pio and ten Dam, 1988; WHO, 1991b). Each year, tuberculosis alone is responsible for about 3 million deaths and is the single largest cause of adult mortality (WHO, 1992a). Overcrowding also facilitates the transmission of rheumatic fever and meningococcal meningitis (Sapir, 1990; WHO, 1992a), and there are links between communicable respiratory diseases and indoor air pollution caused by coal-burning or wood-burning stoves, but their nature remains poorly understood (Bradley and others, 1991) and under-researched (Stephens and Harpham, 1992).

BOX 2. EXAMPLES OF THE MAIN WATER-RELATED INFECTIONS WITH ESTIMATES OF MORBIDITY, MORTALITY AND POPULATION AT RISK

Disease		*Morbidity*	*Mortality*	*Population at risk*
(Common name)	*(Name)*			
1. Water-borne (and water-washed); also foodborne (designated by an asterisk (*))				
Cholera	Cholera*	More than 300,000	More than 3,000	
Diarrhoeal diseases	This group includes salmonellosis*, shigellosis, campylobacter*, *Escherichia coli*, rotavirus, amoebiasis* and giardiasis*	700 million or more infected each year	More than 4 million	More than 2,000 million
Enteric fevers	Paratyphoid and typhoid	500,000 cases; 1 million infections (1977-1978)	25,000	
Infective jaundice	Hepatitis A*			
Pinworm	Enterobiasis			
Polio	Poliomyelitis	204,000 (1990)	25,000	
Roundworm	Ascariasis	800 million-1,000 million cases; 1 million cases of disease	20,000	
Leptospirosis				
Whipworm	Trichuriasis			
2. Water-washed				
(*a*) Skin/eye infections				
Scabies				
School sores	Scabies			
Trachoma	Impetigo			
Leishmaniasis	Trachoma	6 million-9 million people blind		500 million
	Leishmaniasis	12 million infected; 400,000 new infections per year		350 million
(*b*) Other				
Relapsing fever	Relapsing fever			
Typhus	Rickettsial diseases			
3. Water-based				
(*a*) Penetrating skin				
Bilharzia	Schistosomiasis	200 million	Over 200,000	500 million-600 million
(*b*) Ingested				
Guinea worm	Dracunculiasis	Over 10 million		Over 100 million

BOX 2 (*continued*)

Disease (Common name)	(Name)	Morbidity	Mortality	Population at risk
4. Water-related insect vector				
(*a*) Biting near water				
Sleeping sickness	African trypanosomiasis	20,000 new cases annually (thought to be an underestimate)		50 million
(*b*) Breeding in water				
Filaria	Filariasis (lymphatic)	90 million		900 million
Malaria	Malaria	267 million (107 million clinical cases)	1 million-2 million (three quarters are children under age 5)	2,100 million
River blindness	Onchocerciasis	18 million (over 300,000 blind)		85 million-90 million
Yellow fever	Yellow fever	10,000-25,000		
Breakbone fever	Dengue fever	30 million-60 million infected every year		

Sources: World Health Organization, *Our Planet, Our Health: Report of the WHO Commission on Health and Environment* (Geneva, 1992), derived from Sandy Cairncross and Richard G. Feachem, *Environmental Health Engineering in the Tropics: An Introductory Text* (Chichester, United Kingdom, John Wiley and Sons, 1983); and G. F. White, J. Bradley and A. V. White, *Drawers of Water: Domestic Water Use in East Africa* (Chicago, Illinois, University of Chicago Press, 1972). Figures for morbidity, mortality and population at risk from World Health Organization, *Global Estimates for Health Situation Assessments and Projections, 1990* (WHO/HST/90.2) (Geneva, WHO, Division of Epidemiological Surveillance and Health Situation and Trend Analysis, 1991).

Crowded and cramped conditions, inadequate water supply and inadequate facilities for preparing and storing food greatly exacerbate the risk of food contamination. McGranahan (1991) notes that contaminated food contributes to a high incidence of acute diarrhoea in developing countries and of food-borne diseases such as cholera, botulism, typhoid fever and different parasitoses. Microbial activity generally contributes to food spoilage and unsafe chemicals may deliberately be added to food to retard or disguise spoilage. Food contamination is closely linked to the sanitary conditions of food preparation, processing and even production. In addition, there are likely to be numerous links within the home among water, sanitation, the presence of flies, animal and personal hygiene, and food storage and preparation that contribute to the transmission of diarrhoea (Esrey and Feachem, 1989 as cited in WHO, 1992a).

A wide variety of disease vectors live, breed or feed within or around houses and settlements. The diseases they cause or carry include some of the major causes of ill health and premature death in many cities of the developing world, especially malaria (anopheles mosquitoes) and diarrhoeal diseases (cockroaches, blowflies and houseflies). Other diseases caused or carried by insects, spiders or mites include bancroftian filariasis (*Culex* mosquitoes), Chagas' disease (triatomid bugs), dengue fever (*Aedes* mosquitoes), hepatitis A (houseflies, cockroaches), leishmaniasis (sandfly), plague (certain fleas), relapsing fever (body lice and soft ticks), scabies (scabies mites), trachoma (face flies), typhus (body lice and fleas), yaws (face flies), and yellow fever (*Aedes* mosquitoes) (Schofield and others, 1990; WHO, 1992a). Leptospirosis outbreaks have been associated with flooding in São Paulo and Rio de Janeiro, where the disease has passed to humans through water contaminated with the urine of infected rats or certain domestic animals (Sapir, 1990).

Some of the diseases transmitted by insect vectors have long been urban problems in the developing world. Malaria, for instance, is among the most common causes of infant mortality in many low-income urban settlements. Other diseases remain concentrated in rural areas but have become serious urban problems as infected individuals move to urban areas, disease vectors adapt to urban environments or the expansion of urban areas produces changes in the local ecology that favour the emergence or multiplication of certain disease vectors. That is the case of the increase in lymphatic filariasis and malaria in urban populations (WHO, 1992a). Chagas' disease, which infects an estimated 18 million persons in Latin America, affects primarily poor rural inhabitants but has increasingly become an urban problem through both the migration of infected persons to urban areas (there is no effective treatment for the disease) and the expansion of peri-urban informal settlements where the insect vectors can reproduce (Gomes Pereira, 1984; Briceño-León, 1990).

2. *Chemical pollutants in the urban environment*

Box 3 lists some of the chemical pollutants commonly found in urban areas that affect human health or about which there is concern in regard to human health, even if their precise health impact is not well established. Concern with regard to health has tended to concentrate on the effects of lead (in food, water and air); indoor air pollutants resulting from fuel combustion; toxic or hazardous wastes; and ambient air pollution.

Among the heavy metals, lead raises particular concerns, especially in terms of its effects on children, since there is increasing evidence that relatively low concentrations of lead in the blood may have a damaging effect on their mental development that will persist into adulthood (Needleman and others, 1991). Exposure to lead may also contribute significantly to higher risks of heart attack and stroke among adults. The four major sources of lead are exhausts from motor vehicles (except those using lead-free fuel); lead water pipes (especially if the water supply is acidic); industrial emissions; and paint. A study of the lead levels in the blood of adult volunteers in 10 cities, undertaken by the World Health Organization (WHO) and the United Nations Environment Programme (UNEP) between 1979 and 1981, found the highest lead concentrations among residents of Mexico City where lead levels in the blood were above the WHO guidelines and two to four times higher than those of people residing in cities where low-lead or lead-free gasoline was used (UNEP and WHO, 1988). In Mexico City and Bangalore (one of India's major metropolitan centres), 10 per cent of the sampled population had lead concentrations in the blood well above the level at which biochemical changes therein begin to occur (UNEP and WHO, 1987). A 1988 study conducted in Mexico City found that over one quarter of newborn infants had lead levels in their blood high enough to impair neurological and motor-physical development (Rothenburg and others, 1989). A study in Bangkok that sought to rank urban environmental problems on the basis of the health risks they posed suggested that lead should be ranked with airborne particulates and biological pathogens (primarily intestinal worms and those causing acute diarrhoea, dengue fever and dysentery) among the environmental problems posing the highest risk (USAID, 1990).

Among the agents causing indoor air pollution, fumes from coal, wood and other biomass fuels are likely to be the main health hazard. Indoor air pollution from fuel combustion affects rural dwellers more than urban dwellers, since rural households are more likely to be exposed regularly to potentially harmful emissions from open fires or poorly designed stoves lacking adequate ventilation. Estimates suggest that the number of rural inhabitants suffering from ill health as a result of such exposure is of the order of hundreds of millions whereas that for their urban counterparts is probably of the order of tens of millions. Yet, in many cities, coal and biomass fuels are widely used, especially among poorer households. The most serious health risks are likely to result from burns and smoke inhalation (WHO, 1992a). The long-term effects of smoke inhalation include chronic inflammation of the respiratory tract which reduces resistance to acute respiratory infections which in turn enhances the susceptibility of the individual to bronchial inflammation. These processes may lead to emphysema and chronic obstructive pulmonary disease that may cause heart failure (ibid.). Yet, although some studies have suggested that indoor air pollution has significant effects on health, others have been unable to demonstrate a clear link (McGranahan, 1991). "It is not clear whether this variation reflects methodological differences in discerning important real effects that do exist, spurious results in the few cases where significant health effects have been detected or underlying differences in the relative importance of indoor air pollution in different areas" (ibid., p. 23). The extent to which health problems in developing countries result from the use of building materials that are increasingly recognized as having a significant health impact in developed countries, such as those that emit formaldehyde, chloroform or radon, is not known.

BOX 3. EXAMPLES OF HAZARDOUS CHEMICAL POLLUTANTS WITHIN THE HUMAN ENVIRONMENT

Chemicals that can be found in food and water

Lead: found in food and drinking water, especially when lead water pipes are used with acidic water
Food contaminants introduced deliberately or accidentally
Nitrates in drinking water which become nitrites in the body
Aflatoxins or other natural food toxins
Trace pollutants in the water supply, many derived from agrochemicals such as halogenated organic chemicals
Minerals (high content) in drinking water
Aluminium in food and drinking water
Arsenic and mercury

Chemicals found in the indoor environment of home or workplace

Carbon monoxide resulting from the incomplete combustion of fossil fuels
Lead from paint, for instance
Tobacco smoke
Asbestos which is carcinogenic
Smoke from combustion of coal, wood or other biomass fuels
Potentially dangerous chemicals used without health and safety safeguards
Formaldehyde from insulation or some wood preservatives or adhesives

Chemicals found outdoors

Lead from exhaust from automobiles in which gasoline with a lead additive is used, or external paint
Sulphur dioxide and oxides of nitrogen from industries, power stations, auto emissions and so forth
Ozone and photochemical smog
Carbon monoxide resulting from the incomplete combustion of fuels in internal combustion engines
Cadmium, mercury compounds and other heavy metals
Dioxins, polychlorinated bipheryls (PCBs) and organochlorine pesticides
Other chemical wastes from industries

Chemicals found both indoors and outdoors

Micropollutant mixtures at trace level that have additive effects
Pollens and organic dusts

Source: G. Matthews, Health and the environment, (World Health Organization, Regional Office for Europe, July 1990 (ICP/GEH/211/9)), mimeographed.

In many cities of the developing world, the concentrations of pollutants in ambient air are already high enough to cause illness in the more susceptible individuals and premature death among the elderly, especially those with respiratory problems (WHO, 1992a). Current levels of air pollution may also be impairing the health of people in far greater numbers than have been ascertained by available studies. The limited data available on air pollution in cities of developing countries also suggest that pollution is generally getting worse. Box 4 outlines the effects on health of some of the air pollutants most commonly found in urban areas.

BOX 4. SOME URBAN AIR POLLUTANTS AND THEIR EFFECTS ON HEALTH

Pollutant	Absorption / action	Health effects
1. Traditional pollutants from the combustion of coal or heavy oil		
Smoke or suspended particulates (some contribution also from diesel use)	Can penetrate to lungs; some retained with possible long-term effects. May also irritate bronchi.	**London smog complex**: *Short-term effects*: sudden increases in deaths, hospital admissions and illness among bronchitic patients. Temporary reductions in lung function among both weak and some normal individuals.
Sulphur dioxide	Readily absorbed on inhalation. Irritates bronchi, may cause bronchospasm.	
Sulphuric acid (mainly a secondary pollutant formed from sulphur dioxide in air)	Hygroscopic. Highly irritant to upper respiratory tract. If absorbed with other fine particles may penetrate further and promote bronchospasm.	*Long-term effects*: increased frequency of respiratory infections (children). Increased prevalence of respiratory symptoms (adults and children). Higher death rates from bronchitis in polluted areas.
Polycyclic aromatic hydrocarbons (some contribution also from traffic)	Mainly absorbed with smoke; can penetrate with it to lungs.	*Possible carcinogenic effects*: may increase incidence of lung cancer in urban areas.
2. Photochemical pollutants from traffic and other hydrocarbon emissions		
Hydrocarbons (volatile: petrol and so on)	Non-toxic at moderate concentrations.	**Los Angeles smog complex**: *Short-term effects*: Primarily eye irritation. Reduced athletic performance. Possibly small increases in deaths and hospital admissions.
Nitric oxide	Capable of combining with blood haemoglobin but no apparent effect in humans.	
Nitrogen dioxide and ozone (mainly secondary pollutants formed in photochemical reactions)	Neither gas is very soluble: some irritation of bronchi but at high concentrations can reach lungs and cause oedema. Concentrations in urban areas too low for such effects, but evidence of reduced resistance to infections in animals.	*Long-term effects*: Increased prevalence of respiratory illness (children), increased asthma attacks (adults). No clear indication of increased bronchitis.
Aldehydes, other partial-oxidation products, peroxyacetyl nitrate	Eye irritation, odour.	
3. Others from traffic		
Carbon monoxide (smoking and other sources also contribute)	Combines with blood haemoglobin, reducing oxygen-carrying capacity.	Possible effects on central nervous system (reversible unless concentrations are very high). Some evidence of effects on perception and performance of finely detailed tasks at moderate concentrations.
Lead (some industrial sources also contribute; human intake often dominated by lead in food and water)	Transported by blood to soft tissues and some to bone.	Possible effects on central nervous system (longer timescale than in the case of carbon monoxide and not necessarily reversible). Indications of neuropsycho-logical effects on children within overall environmental exposure range, but role of traffic lead uncertain.

Source: R. E. Waller, "Field investigations in air", in *Oxford Textbook of Public Health,* vol. 2, W. W. Holland and others, eds. (Oxford, United Kingdom, Oxford University Press, 1991).

An estimated 1.4 billion urban residents worldwide are exposed to annual averages of suspended particulate matter or sulphur dioxide (or both) that are higher than the minimum standards recommended by the World Health Organization (UNEP and WHO, 1988) and the partial evidence available for developing countries suggests that such concentrations are increasing (UNEP, 1991). Comparable estimates regarding exposure to nitrogen oxides and carbon monoxide are not available, although some studies in particular cities or city districts indicate that there are ambient air pollution levels that can impair health. There are also concerns about the health impacts of such secondary pollutants resulting from reactions between primary pollutants and the air as, for instance, acid sulphates and ozone (see box 4).

In certain industrial centres, air pollution levels can be so high as to cause demonstrable health impairment. In Cubatao, Brazil, for instance, air pollution levels have been linked to reduced lung function in children (Hofmaier, 1991, as cited in WHO, 1992a). Non-ferrous metal smelters are often major contributors to air pollution and although no well-documented example of their health impact in developing countries could be found, a recent study of the Katowice district in Upper Silesia, Poland, showed that the high output of lead and cadmium of four non-ferrous metal industrial plants was responsible for elevated lead and cadmium concentrations in the blood of 20 per cent of children in the region. Some of those tested (especially children) already exhibited detectable symptoms of lead poisoning (Jarzebski, 1992). Links between health problems and air pollution levels have also been suggested by comparisons between the health of people living in highly polluted areas within cities and those living in less polluted areas. Such comparisons have shown a strong association between the incidence of respiratory infections and pollution levels. In addition, in cities where high concentrations of air pollution occur at particular times (for instance, when high emissions coincide with particular meteorological conditions), higher mortality among particularly vulnerable groups is common (WHO, 1992a).

In Latin America, recent studies suggest that air pollution levels in major cities—such as Belo Horizonte, Rio de Janeiro and São Paulo in Brazil, Santiago in Chile, Bogotá in Colombia, Guadalajara, Mexico City and Monterrey in Mexico, Lima in Peru, and Caracas in Venezuela—are sufficiently high for priority to be given to their control. One estimate suggests that over 2 million children suffer from chronic cough as a result of urban air pollution and that air pollution causes an excess of 24,300 deaths a year in Latin America (Romieu, Weitzenfeld and Finkelman, 1990). The same source estimates that about 65 million person-days of work are lost each year because of respiratory tract-related problems caused by air pollution. While these are rough estimates, they give an idea of the order of magnitude of the problem.

Regarding industrial and commercial wastes, in most cities of the developing world, toxic or hazardous materials are dumped in water bodies or land sites without being so treated as to render them less damaging and without ensuring that they remain isolated from the rest of the environment. Generally, there are few incentives for industrial or commercial enterprises to reduce polluting emissions or waste since polluters are rarely penalized for dumping and the penalties imposed are generally too small to have a deterrent effect. Reports of severe health problems arising from human contact with toxic or hazardous wastes in cities of developing countries are common (Hardoy, Mitlin and Satterthwaite, 1992). Occupational exposure to chemical pollutants and exposure as a result of accidents have also become common. Environmental hazards are evident in work-places ranging from large factories and commercial institutions to small "backstreet" workshops or worksites at home. They include dangerous concentrations of toxic chemicals and dust; inadequate lighting, ventilation and space; and inadequate protection of workers from machinery and noise. Many case studies indicate that the health of a high proportion of the workers in particular industries or industrial plants has been affected by exposure to environmental hazards at the workplace. Thus, a recent study of an Egyptian pesticide factory found that "about 40 per cent of the workers had problems related to pesticide poisoning, ranging from asthma to enlarged livers" (Pepall, 1992, p. 15). In most countries, the incidence of occupational injuries and disease is considerably under-reported. The Mexican Social Security Institute, for instance, reported an average of 2,000-3,000 cases of work-related illnesses in the country in 1988, whereas a study of just one steel mill found 4,000-5,000 such cases and documented that more than 80 per cent of the steelworkers were exposed to extreme heat, loud noise and toxic dust (Castonguay, 1992). Similarly, a study of Bangkok's environmental problems noted that numerous Thai workers were exposed to poor working environments but that the number suffering from occupational diseases was small.

It concluded, however, that such an outcome was probably a reflection not of satisfactory working conditions but rather of the difficulty of linking disease to working conditions (Phantumvanit and Liengcharernsit, 1989). Other studies have documented the serious health effects of workers' exposure to toxic chemicals: benzene poisoning among leather workers in Turkey (Askoy, Erdem and Dincol, 1976); and lead poisoning among persons working in lead-acid battery repair shops in Kingston, Jamaica (Matte and others, 1989).

Large-scale accidents involving chemical hazards have attracted considerable attention because of the immediacy of their detrimental effects. Thus, the accidental release of methyl isocyanate at Bhopal, India, caused over 3,000 deaths and injured about 100,000 persons; the natural gas explosions in Mexico City in 1984 caused over 1,000 deaths and, more recently, explosions of gases that had accumulated in the sewers and drains of downtown Guadalajara resulted in numerous deaths and property damage. It appears that the illegal disposal of industrial wastes in Guadalajara's sewer system was the main cause of that accident. Such accidents are often cited as indicating the significance of the health impact of environmental hazards on urban populations. The attempts by European and North American industries to dispose of their toxic wastes in developing countries have also received much publicity. However, the impact on health associated with occupational exposure to chemical hazards or from exposure to indoor and outdoor air pollution is likely to be larger and more widespread. Coming decades will probably show that the health effects of air and water pollution in developing countries have been underestimated.

3. *Availability, cost and quality of natural resources*

The availability to any individual or household of such resources as food, fuel and fresh water is clearly central to health. The environmental dimension is prominent in that the ecosystem defines the limits for the availability of fresh water, fertile soils and forests but social, economic and political factors tend to be dominant in influencing who has access to them or to the land and water sources from which they can be drawn. While access to land and forests for food and fuel is normally considered a rural issue, there is increasing evidence that access to land on which food can be grown is of considerable importance to poorer households in many cities, especially those in poor and less urbanized developing countries. Thus, a study of Nairobi, Kenya, noted the importance for most households of the food that they had grown or produced themselves, thus underscoring their need to have access to land (Lee-Smith and others, 1987).

In regard to water, only rarely does an overall shortage of fresh water explain why so many urban households lack access to safe and sufficient water supplies. More commonly, lack of access to fresh water is the result of society's refusal to assure an adequate supply for all urban inhabitants. Thus, water supplies are usually inadequate in squatter settlements and it is common for their inhabitants to pay private water vendors between 4 and 100 times as much per unit of water as middle- and upper-income groups pay for publicly provided piped water (see box 5). That is to say, it is neither a lack of water nor a lack of the willingness to pay that prevents piped water from reaching poorer areas but rather the refusal of Government to extend this service to those areas.

BOX 5. RATIO OF THE PRICE OF WATER CHARGED BY WATER VENDORS TO THE PRICE CHARGED BY PUBLIC UTILITIES IN SELECTED CITIES OF DEVELOPING COUNTRIES, MID-1970S TO EARLY 1980S

City	*Ratio of price of water from private vendors to price charged by public utilities*
Abidjan	5:1
Dhaka	12:1 to 25:1
Istanbul	10:1
Kampala	4:1 to 9:1
Karachi	28:1 to 83:1
Lagos	4:1 to 10:1
Lima	17:1
Lomé	7:1 to 10:1
Nairobi	7:1 to 11:1
Port-au-Prince	17:1 to 100:1
Surabaya	20:1 to 60:1
Tegucigalpa	16:1 to 34:1

Source: World Bank, *World Development Report, 1988* (New York, Oxford University Press, 1988), p. 146.

4. *Physical hazards*

The extent of accidental injuries is often grossly underestimated. In 1982, a study of accidents among children in 10 developing countries found that they were the main cause of death among children aged 5-9 and 10-14 (Manciaux and Romer, 1986). Furthermore, for every accidental death, there were several hundred accidental injuries. Road accidents account for about half a million deaths a year; the figure for injuries caused by such accidents is several times that for deaths (WHO, 1992a). Accidents linked to the quality of housing and its surroundings are among the most common causes of death or injury. Thus, burns, scalds and household fires are more common in overcrowded shelters, where space limitations prevent the protection of occupants, especially children, from open fires, stoves or kerosene heaters. The use of flammable building materials, such as wood, cardboard, plastic, canvas and straw, further increases the risk of fires in poor settlements. In addition, overcrowded dwellings make it difficult for adults to safely store medicines and dangerous household chemicals (such as bleach) safely out of the reach of children. Here, as with many environmental problems, the level of risk is compounded by social factors, including the lack of adult supervision caused by the need to work. The effect of accidents on general health is further amplified by lack of access to health services and emergency treatment, not to mention to long-term treatment and care (Goldstein, 1990).

Physical hazards stemming from the land sites on which housing has been developed are also of importance. In virtually all cities of the developing world, there are sizeable clusters of illegal housing built on dangerous sites, such as steep hillsides, flood plains and desert land, or housing that has been built on or near polluted sites, including solid waste dumps, open drains and sewers, or in industrial areas with high levels of air pollution. Some housing developments are on sites that, being close to major highways or airports, are subject to high noise levels. Such developments arise not because of a lack of land (since most of the major cities and metropolitan areas in developing countries have large pieces of valuable land that are not prone to being affected by natural or man-made hazards but that are nevertheless left undeveloped or are developed only partially (Sarin, 1983; Hardoy and Satterthwaite, 1989)) but rather from the fact that poorer groups have no means of securing access to adequate land sites.

D. Aspects of the built environment with negative consequences on psychosocial health

Several psychosocial disorders are associated with poor-quality housing and inadequate living environments, as well as with such non-environmental factors as insecurity with respect to tenure of the shelter and the constant fear of eviction. The most serious psychosocial health problems and phenomena include depression; drug and alcohol abuse; suicide; and different kinds of violence (child mistreatment and abuse; spouse mistreatment; target violence; and assault and rape). Psychosocial and chronic diseases are becoming a major cause of death and morbidity among adolescents and young adults in many urban areas of developing countries, especially in certain districts within urban areas. In cities as diverse as Shenyang, China, and Rio de Janeiro, Brazil, these are the main causes of death among adolescents (Bradley and others, 1991). In 1986, homicides were responsible for 5 per cent of all deaths in São Paulo (Leitmann, 1991a). Housing of poor quality and overcrowded living environments contribute to an increase in the stress that underlies many such diseases.

A number of the physical characteristics of the housing and living environment can increase stress, including noise, overcrowding, inappropriate design, poor sanitation, lack of garbage collection services and inadequate maintenance (WHO, 1992a). Housing of good quality and an adequate living environment can reduce or eliminate stress by providing sufficient space to individuals, facilitating access to services, providing areas for recreation, maintaining a low noise level, and ensuring that personal hazards are few (Schaeffer, 1990; Ekblad and others, 1991). Within the wider neighbourhood in which a house is located, a sense of security, adequate physical infrastructure in terms of roads, pavements, drains and street lights, adequate services in terms of sanitation, emergency care, education and health, and the provision of amenities can reduce stress and contribute to good mental health.

Ekblad and others (1991) suggest the need to consider three aspects of the physical environment in relation to its possible impact on people's psychosocial health: (*a*) the subjective experience of the dweller, that is, his or her level of satisfaction with the house, the neighbourhood in which it is located and its general location within the urban area; (*b*) the dwelling's physical structure, that is, the amount of space, state of repair and facilities, which may influence the level of privacy, the possibility of meeting relatives and friends, and child-rearing practices; and (*c*) the characteristics of the neighbourhood, including the quality of its services and facilities as well as its level of security.

Many characteristics of urban neighbourhoods that are not easily identified or defined may have important influences on the individual's level of satisfaction and on the incidence of crime, vandalism and interpersonal violence. These are factors more fully explored with regard to the cities of developed countries as, for instance, in the critique of urban planning by Jacobs (1961) which focuses on the characteristics of cities and city neighbourhoods and streets that make them pleasant, safe and valued so that urban degradation is avoided. The work of Newman on what he termed "defensible space" provides another example of an exploration of such factors: he showed how the particular form of open space within a neighbourhood, including the extent to which it was subject to informal supervision and the extent to which there was a clear visual definition as to who had the right to use it and who was responsible for its maintenance could be linked to levels of crime and vandalism (Newman, 1972).

The extent to which individuals or households have the possibility of modifying or changing their housing environment and working with others in the locality to effect change in the wider neighbourhood is an important determinant of psychosocial health. Many critiques of public housing and urban planning in cities of the developing world, especially of measures such as slum and squatter clearance and redevelopment, have centred on the loss of individual, household and community control that such measures imply (Turner and Fichter, 1971; John F. C. Turner, 1976). Work on the topic documents the hardships caused to households and, occasionally, the negative health consequences of such measures, but there is generally no examination of the social pathologies that may ensue. The literature on the importance of being able to command events that affect one's life in order to maintain both physical and mental health is large and varied (Duhl, 1990) and medical doctors and psychiatrists are increasingly recognizing the importance of such a link between command of events and health (Duhl, 1990; WHO, 1992a).

The precise linkages between different elements of the physical environment and each psychosocial disorder are difficult to ascertain and to separate from the effects of the other variables that can promote or prevent a process that might lead to disease such as, for instance, the existence of a social support network which may mitigate the effect of an inadequate physical environment (Kagan and Levi, 1975 as quoted in Ekblad and others, 1991; WHO, 1992a). Indeed, WHO (1992a) notes that the strong social networks and sense of community organization in many run-down inner-city districts and squatter settlements might help to explain the remarkably low level of the psychosocial problems detected in them. The importance of such networks can also be inferred from the increase in physical and mental health problems among populations that have undergone relocation from inner-city tenements or illegal settlements to "better-quality" housing, a change that disrupts the functioning of existing networks (John F. C. Turner, 1976).

E. Local, regional, national and global natural resource degradation

In addition to the direct links between health and particular pollutants or pathogens in the urban environment, there are many processes that, by affecting the environment, can affect human health indirectly. Thus, production or waste generation processes taking place in cities can lead to natural resource degradation which in turn affects human health; for example, when acid rain compromises the quality of the land or water used by urban households for their livelihood or damages food production in rural areas, this leads to reductions in food supplies which are likely to affect poor urban households most. A similar process is in operation when local fisheries are damaged or fish and seafood cannot be consumed by humans because of their contamination by industrial wastes. The depletion of fish stocks in rivers, estuaries and coastal waters has been associated with the generation of wastes in urban areas. Some urban populations suffer from water pollution caused by city-based activities degrading particular rivers or other water bodies upstream.

Recently, attention has been focused on two problems of global environmental degradation: the depletion of the stratospheric ozone layer; and the increasing concentrations of greenhouse gases in the earth's atmosphere which may lead to global warming. Both problems are expected to affect human health. The main effect of the depletion of stratospheric ozone will be an increased incidence of skin cancer. Atmospheric warming will have direct health effects such as heat stress from higher temperatures, and indirect effects, by forcing changes in agricultural production, producing a rise in sea levels, disrupting the fresh-water resources available for city use and producing an expansion of the areas in which tropical disease vectors can survive and breed. In general, the indirect effects are likely to have a more pronounced impact on human health and well-being than the direct ones.

F. WHO BEARS THE ENVIRONMENTAL COSTS?

1. *Vulnerability*

Among urban dwellers, certain groups suffer more than others from environmental problems and it is crucial for adequate policy formulation to identify both the groups subject to greatest risks and the reasons for their vulnerability. It has been noted that the risks of being exposed to the most serious environmental hazards of the urban environment—biological pathogens, chemical pollutants, physical hazards and inadequate access to water—are closely associated with the quality of housing, its location and the services available (access to water, sanitation, drainage, garbage collection and health care). The extent to which safe practices are followed in the work environment is also an important factor determining risk. Consequently, the people living in housing of the worst quality, lacking adequate infrastructure and services, and working in the most dangerous occupations are those subject to the greatest risks.

However, the presence of a pathogen, pollutant, physical hazard or cause of stress does not necessarily imply that it will harm someone. Its effect depends on the characteristics of the individual, household or social group exposed to it. Factors that influence the level of risk to which a particular individual is subject or the severity of the likely effect include that person's (*a*) age, sex and other demographic characteristics; (*b*) health characteristics including physical capability, genetic make-up, health status and acquired immunity to certain diseases; (*c*) socially determined roles, especially those related to gender; and (*d*) access to means that may permit the avoidance of certain hazards or the possibility of combating their detrimental effects (via access to health care, for instance).

These factors can be divided into two categories: (*a*) those that influence whether ill health or injury can be avoided; and (*b*) those that influence how the individual, household or community can cope with the illness or injury. Among the first set of factors, a person's age is a key one, since both young children and the elderly have weaker body defence mechanisms for fighting biological pathogens and they are more likely to suffer from physical hazards because of physical incapacity and mobility problems. The disabled are also particularly vulnerable to physical hazards because of limitations on their mobility and persons suffering from certain ailments, such as asthma, are particularly sensitive to chemical hazards. A person's nutritional status also affects the body's capacity to fight biological pathogens or to withstand chemical exposure. Acquired immunity, whether because of previous exposure or through vaccination, also influences the ability of the individual to withstand different biological pathogens. Genetic factors may also play a role in determining the relative sensitivity of individuals to certain pollutants.

The second set of factors that influence how individuals are able to cope with injury or illness resulting from environmental hazards include the extent of public, private and community provision of health care and emergency services; the individual's ability to afford such services, to purchase medicines and to take time off to recuperate when sick or injured; and the extent to which an individual or household can devise new or make use of existing coping mechanisms once sickness or injury has occurred, including knowing what to do, who to rely on, which services to use and so forth. The people who are most vulnerable to environmental hazards are those least able to avoid them or to cope with the illness or injury caused by those hazards.

2. *Low income*

According to Bradley and others (1991), most studies on morbidity and mortality due to communicable diseases indicate that those most vulnerable are predominantly the poor, be they children, adults living in crowded and unhygienic conditions, or workers in certain occupations. Low-income groups are the least able to afford the type of housing that offers some protection against environmental hazards and they are also considerably more likely than higher-income groups to have dangerous jobs or work in hazardous environments.

Low-income individuals and households generally have the greatest difficulties in terms of obtaining treatment for any injury or illness, including emergency services in the case of serious accidents and treatment from a health centre or hospital (Goldstein, 1990). They are also the most lacking in the means to afford medicines and, generally experience the smallest possibility of taking time off to recover, since any loss of income would press heavily on their ability to survive and they are unable either to afford health insurance or to secure jobs offering health insurance as a fringe benefit.

Low-income individuals and households generally have the least access to adequate water supplies and are therefore at greater risk of acquiring water-related diseases. In cases where water consumption per person in cities averages 20-40 litres per person per day (well below the minimum required for good health), the consumption of poorer groups is often less than half the average (Hardoy, Mitlin and Satterthwaite, 1992). In this respect, most cities in the developing world are comparable with cities in the developed world many decades ago, before it became the accepted responsibility of city or municipal authorities to ensure that all citizens had access to piped water, sewers, drains and garbage collection.

If there were sufficient information to construct a map of each city in the developing world showing the level of risk from all environmental hazards in each neighbourhood, the areas with the highest risks would coincide with the areas where low-income groups live. The correlation between income levels and environmental hazards would be particularly strong for such factors as quality and quantity of water, sanitation, solid waste collection and risk from floods, landslides and other natural disasters. This outcome is expected because poorer groups are generally priced out of safe, well-located, well-serviced housing and land sites. In many cities, a strong correlation between indoor air quality and income will also be noticeable because poorer groups have no choice but to use more-polluting fuels and less-efficient stoves (or open fires). The fact that poorer groups live in overcrowded conditions exacerbates environmental problems and facilitates the transmission of infectious diseases. A high proportion of the poor live in shacks made of flammable materials, thus increasing their risk of suffering accidental fires. Poorer groups will generally have the least access to playgrounds, parks and other open spaces managed for public use. With regard to ambient air pollution, the correlation between income level and level of air pollution will be weaker. However, poor groups are more likely than the better-off to live near certain "hot spots" such as quarries, cement works and highly polluting industries.

Studies of the differentials in health status or mortality rates between city districts (boroughs or municipalities) indicate that conditions in poorer areas are worse than both the city average and those in wealthier areas (Harpham, Lusty and Vaughan, 1988; Jacobi, 1990; Bradley and others, 1991). Infant mortality rates in poorer areas are often at least four times higher than those in richer areas, with larger differentials being apparent if the poorest district is compared with the richest. Large differentials between rich and poor districts are also common in respect of the incidence of many environment-related diseases, such as tuberculosis and typhoid, as well as in respect of the mortality rates associated with diarrhoeal diseases and acute respiratory infections. Since most of the people living in districts or settlements subject to high risks of floods or landslides are poor, they are also subject to higher risks of injury or death from natural disasters. Box 6 gives examples of differentials between different areas within a city in terms of pollution levels, access to public services and health indicators.

Other differentials between high- and low-income groups are likely to be found with regard to maternal mortality rates, since low-income women have less access to basic health care and prenatal services. Available statistics show maternal mortality differentials only at the level of whole countries, among which maternal mortality rates in the poorest countries are at least 100 times higher than those in richer countries (WHO, 1992a). Large differentials between low- and high-income areas within cities are also likely with regard to the proportion of people who are disabled or chronically ill.

G. Why are low-income groups vulnerable?

To a large extent, capacity to pay for housing defines the scale of the environmental hazards present in the housing and living environment, including whether or not there is safe and sufficient water supply, sanitation, garbage collection and drainage. In most cities of developing countries, low-income individuals and house-

Box 6. Intra-city differentials in environmental hazards, access to public services and health indicators

(*a*) Differentials in environmental hazards:

Mexico City (Mexico): The highest concentrations of particulate matter in the air are found in the south-east and the north-east areas which are predominantly low-income areas.

Bombay (India): In 1977, a study compared the health of residents in two districts with heavy industrial concentration with that of residents in a district with little industry. In the industrialized districts, there was a considerably higher incidence of diseases such as bronchitis, tuberculosis, skin allergies, anaemia and eye irritation and a notable increase in deaths from cancer was detected in one of the industrialized districts.

Caracas (Venezuela): An estimated 574,000 people live in illegal settlements on slopes with a significant risk of landslides. Most of the areas continuously affected by slope failures are sites of low-income settlements.

(*b*) Differentials in access to public services:

Buenos Aires (Argentina): Of the 11 million or so inhabitants of the Buenos Aires metropolitan area, only 57 per cent have running water and only 45 per cent have connections to the sewer system. Virtually all of those in the Capital Federal (the centre of the metropolitan area with some 3 million inhabitants) have piped water and connection to sewers. It is in the poorer, peripheral municipalities outside the Capital Federal, that a high proportion of the population lack either running water or connection to sewers. Thus, in the municipalities of General Sarmiento, Merlo and Moreno, less than 5 per cent of the population have running water and less than 4 per cent have connections to sewers. In three other municipalities, less than 12 per cent of their population have running water and less than 10 per cent have connections to sewers.

Accra (Ghana): In the high-class residential areas with water piped to the home and water closets for sanitation, water consumption per capita is likely to be well in excess of the recommended figure of 200 litres per person per day. In slum neighbourhoods such as Nima-Maamobi and Ashiaman, where residents commonly buy water from vendors, the water consumption is about 60 litres per capita per day.

Surabaya (Indonesia): The wealthiest 20 per cent of the population are reported to consume 80 per cent of the public services, including those essential to a healthy environment.

Mexico City (Mexico): Residents in the high-income area of Lomas de Chapultepec consume, on average, some 450 litres of water per person per day. The average in the slum area of Nezahualcóyotl is only 50 litres per day. Overall, 9 per cent of consumers account for 75 per cent of the water consumption while 2 million people have very inadequate access to safe water supplies.

Santiago (Chile): In some of the richest districts, per capita water consumption is between 300 and 450 litres per day. In many of the poorer districts, it is 100 litres a day or less (for instance, it is just 80 litres a day in La Pintana, a district on the periphery).

(*c*) Differentials in health indicators:

Jakarta (Indonesia): Official estimates suggest that the infant mortality rate for the whole city is 33 deaths per 1,000 live births, while estimates for some of the poorer areas are four to five times the city average.

Karachi (Pakistan): In three low-income areas, between 95 and 152 children out of every 1,000 born alive die before age 1. In a middle-class area, only 32 children out of every 1,000 births die before that age.

BOX 6 (*continued*)

Manila (Philippines): In Tondo (one of the largest squatter settlements in the city), the infant mortality rate was 210 per 1,000 live births in the mid-1970s compared with 76 per 1,000 in non-squatter areas in Manila. The proportion of people with tuberculosis in Tondo was nine times higher than the average for non-squatter areas; typhoid was four times as common.

Port-au-Prince (Haiti): In the "slums", 1 in 5 infants die before their first birthday while another 1 in 10 die between their first and second birthdays. These estimates are almost three times as high as mortality rates in rural areas and many times those in richer areas of Port-au-Prince where infant and child mortality have levels similar to those in urban areas of the United States.

Pôrto Alegre (Brazil): Infant mortality rates among residents of shanty towns were three times as high as those among residents living outside shanty towns. Neonatal mortality rates were twice as high and post-neonatal mortality was more than five times as high. Mortality from pneumonia and influenza was six times higher, and from septicaemia eight times higher, in shanty towns.

São Paulo (Brazil): Infant mortality rates can vary by a factor of four, depending on the district. In the core area, 42 children out of every 1,000 born alive died before age 1, while in one of the predominantly poor peri-urban municipalities, infant mortality was 175 per 1,000 live births. In the city's periphery, infant death rates from enteritis, diarrhoea and pneumonia were twice as high as those in the core.

Tianjin (China): A study of environment-related morbidity and mortality in the different subdistricts of the city found large variations. Thus, the average for subdistricts' infant mortality rates was 13 but in one district it was 31. The prevalence of tuberculosis in the subdistricts averaged 172 per 100,000 people with the highest subdistrict figure being 347 and the lowest just 54. There were also major differences in mortality rates from lung and cervical cancer.

Sources: Accra: Jacob Songsore, "Review of household environmental problems in the Accra metropolitan area, Ghana", working paper (Stockholm, Stockholm Environment Institute, 1992); Bombay: Centre for Science and Environment, *The State of India's Environment: A Citizen's Report* (Delhi, 1983); Buenos Aires: Silvia Zorrilla and Maria Elena Guaresti, *Sector Agua Potable y Saneamiento: Lineamientos para una Estrategia Nacional* (Buenos Aires, United Nations Development Programme, 1986). Caracas: Virginia Jiménez Díaz, "Landslides in the squatter settlements of Caracas: towards a better understanding of causative factors", *Environment and Urbanization* (London), vol. 4, No. 2 (October 1992), pp. 80-89; Jakarta: Trudy Harpham, Paul Garner and Charles Surjadi, "Planning for child health in a poor urban environment: the case of Jakarta, Indonesia", *Environment and Urbanization* (London), vol. 2, No. 2 (October 1990), pp. 77-82; Karachi: Aga Khan University students, unpublished and quoted in *In the Shadow of the City: Community Health and the Urban Poor*, Trudy Harpham, Tim Lusty and Patrick Vaughan, eds. (Oxford, Oxford University Press, 1986); Manila: Samir S. Basta, "Nutrition and health in low-income urban areas of the Third World", *Ecology of Food and Nutrition* (United Kingdom), vol. 6 (1977), pp. 113-124; Mexico City: Martha Schteingart, "The environmental problems associated with urban development in Mexico City", *Environment and Urbanization* (London), vol. 1, No. 1 (April 1989); Port-au-Prince: J. E. Rohde, "Why the other half dies: the science and politics of child mortality in the Third World", *Assignment Children*, vols. 61/62 (1983); Pôrto Alegre: J. J. Guimaraes and A. Fischmann, "Inequalities in 1980 infant mortality among shanty town residents and non-shanty town residents in the municipality of Pôrto Alegre, Rio Grande do Sul, Brazil", *Bulletin of the Pan American Health Organization* (Washington, D.C.), vol. 19 (1985), pp. 235-251; Santiago: Vicente Espinoza, *Para una Historia de los Pobres de la Ciudad* (Santiago, Ediciones SUR, 1988); São Paulo: World Bank, *Staff Appraisal Report: Brazil Second Health Report* (Washington, D.C., World Bank, Population, Health and Nutrition Department, 1984); Surabaya: Johan Silas quoted in Mike Douglass, "The political economy of urban poverty and environmental management in Asia: access, empowerment and community-based alternatives", *Environment and Urbanization* (London), vol. 4, No. 2 (October 1992); Tianjin: Alain Bertaud and M. Young, "Geographical pattern of environmental health in Tianjin, China", research/sector paper (Washington, D.C., World Bank, 1990), as quoted in Josef Leitmann, *Tianjin Urban Environmental Profile*, Urban Management and the Environment, Discussion Paper Series (Washington, D.C., UNDP/World Bank/United Nations Centre for Human Settlements (Habitat), April 1991).

holds have very little chance of obtaining healthy legal accommodation within a neighbourhood where environmental risks are low, that is, one with sufficient space, security of tenure, services and facilities, and on a site not prone to flooding, waterlogging or landslides. Many low-income groups live in fear of eviction from their homes because they are tenants, temporary boarders in cheap rooming houses, settlers of illegal housing developments or land renters. The insecurity and environmental hazards evident within their homes and neighbourhoods result from a combination of three factors: their low incomes; the refusal or inability of Government to guarantee their access to shelters that are not so dangerous or to the resources that would permit them to build those shelters themselves; and the refusal or inability of Government to provide community-based health care and emergency services that can prevent illness or injury or reduce their negative impacts.

If an individual or household cannot afford an accommodation satisfying minimum standards, there is no choice but to opt for a less-than-adequate one. Options available usually involve worse environmental quality entailing higher health risks and much inconvenience. However, the poor cannot give priority to the reduction of those risks since they must devote most of their resources to assuring survival, especially to buying food, paying for children's education or buying the necessary tools or materials to engage in some productive activity. The preferred trade-offs of low-income individuals or households will vary, and in some cases will involve the size of the accommodation, in others the terms under which it is occupied, and in yet others the suitability of its location, its quality, its access to infrastructure and services, and so on. To reduce housing costs, a household of several members may, for instance, sacrifice space and live in one room, or sacrifice secure tenure and access to piped water and live in a house built by themselves on illegally occupied land. To understand the possibilities for improving the housing environment of such people, one must understand their diverse needs and priorities. Complex questions have to be explored, including legality of site or house occupation, legality of the housing structure, and the terms under which the occupants live there (whether as guests, legal tenants, illegal tenants, subtenants or owners). Furthermore, the housing needs of households change over time, so that any government action seeking to improve shelter conditions must promote a range of housing solutions that match the diversity of needs of poorer groups.

Ironically, dangerous or polluted sites located near areas where poorer groups find income-earning opportunities often serve those groups well, for their access to such opportunities is possible mostly because environmental hazards make those sites unattractive to other potential users. It was thus that the high concentration of low-income residents around the Union Carbide factory in Bhopal, India, resulted in several thousand deaths and over 50,000 serious injuries (Centre for Science and Environment, 1985). In Manila, some 20,000 people live around a garbage dump known as Smokey Mountain where the decomposition of organic wastes produces a permanent haze and a powerful rank smell affecting the whole area. Some of these people have lived there 40 years or more. Moving to a cleaner and safer location is beyond their means and many of them make a living scavenging on the dump, generally sorting garbage with their bare hands (Jimenez and Velasquez, 1989).

H. Differentials within low-income groups

There are considerable differences in the scale and nature of the environmental hazards to which different low-income groups are exposed and consequently in the severity of the illnesses or injuries to which such exposure contributes. Thus, workers in certain industries or occupations face greater risks from environmental hazards because of the nature of their work and because of the ineffectiveness of Government in promoting occupational health and safety. In most cities, there are also considerable differences among the various districts where low-income groups predominate because of variations in the presence of physical hazards or service provision. Furthermore, there also are likely to be differences within squatter settlements or tenements depending on the status of different dwellers. Thus, in some low-income settlements in Delhi, where child mortality was estimated at 221 deaths per 1,000 live births, the level of such mortality was nearly twice as high among the poorer castes living within those settlements (Basta, 1977). Within any city, the poorest groups are likely to live in the most dangerous housing or, being homeless, to sleep in open spaces, parks, graveyards, railway stations, bus stations, and so forth.[4]

There is also the differentiation within low-income groups caused by demographic, health or social characteristics. Thus, infants and young children are particularly susceptible to infection or poisoning by toxic substances. Elderly persons whose mobility, reactions and strength are impaired face particular risks from certain environmental hazards. Educational attainment is also associated with differences among low-income groups, mainly because illiteracy and low educational attainment limit the income-earning capacity of individuals and reduce their access to information, thus preventing them from developing adequate coping mechanisms. The health both of adult individuals and of their children is influenced by their knowledge, capacity and motivation to act on the basis of sound information. Thus, the children of better-educated women generally experience lower mortality risks than those of women with low educational levels (Caldwell and Caldwell, 1985).

The gender of an individual is also associated with differential exposure to certain environmental hazards and to differences in susceptibility to them. Some of these differences are biological in nature: women, for instance, are particularly vulnerable to certain environmental hazards during pregnancy and childbirth. However, most are related to the particular social and economic roles that men and women are expected to play (UNICEF, 1992b), as will be discussed below.

1. *Infants and children*

The unborn child is particularly vulnerable to the mother's exposure to toxic chemicals in the workplace (UNICEF, 1992b) and is also likely to be affected by environmental hazards that influence the health and nutritional status of the mother as, for instance, malarial infection that leads to low birth weights and increased infant mortality. Because of their immature immune systems, infants and young children are at greater risk of dying from many environment-related diseases, including diarrhoeal diseases, malaria, pneumonia and measles (UNICEF, 1992b; WHO, 1992a). The time when infants stop relying exclusively on maternal milk for nutrients is particularly hazardous in environments where ready access to clean water is lacking and the facilities needed for the hygienic preparation and storage of food are non-existent. Infections and parasitosis contracted through the ingestion of contaminated food or water contribute to the undernutrition of children which in turn retards their growth and lowers their immunity.

Infants and children are more at risk than adults from various chemical pollutants, especially from lead contamination and high nitrate concentrations in water. They are also particularly at risk from hazards commonly found in low-income settlements, including those stemming from the use of flammable materials to build shelters combined with overcrowding and the widespread use of open fires, stoves or kerosene heaters and cookers that increase the risk of burns and accidental fires. It is common for open spaces where children play to be contaminated with faecal matter and household wastes that attract rats and other disease vectors. Furthermore, the lack of safe play sites often leads to situations where children play on roads, in garbage dumps and in other hazardous places. The increasing mobility of infants and children as they learn to crawl and then walk coupled with their natural curiosity can also expose them to a number of environmental hazards, especially in overcrowded dwellings where dangerous substances cannot be kept out of their reach.

From the time an infant begins to crawl through childhood and adolescence, the size and quality of the home and its surroundings exert a strong influence on the level of risk to which the child is subject. Thus, a survey of 599 slum children in Rio de Janeiro found that accidents accounted for 19 per cent of all health problems. Among children having had an accident, 66 per cent had experienced falls, 17 per cent cuts and 10 per cent burns (Reichenheim and Harpham, 1989). The age of the child was an important determinant of accidents, with the peak number occurring between the second and fifth years of life. Both the hazardous physical environment in which children lived and the lack of parental care and supervision were factors contributing to the high level of accidents detected.

A poor physical environment can also inhibit or damage a child's physical and mental development (Myers, 1991), especially when children lack opportunities to play and learn informally from their peers. In most poor districts of urban areas, there are few open spaces or facilities for the use of children. Yet it is increasingly recognized that safe and stimulating play is crucial in promoting the evolution of a child's motor, communication and problem-solving skills, as well as a child's logical thinking, emotional development and positive social behaviour (Hughes, 1990).

Infants and children are also affected by the psychosocial disorders that the deficiencies in their physical environment promote in their parents or caregivers. To develop normally, children need to interact with other persons who can provide stimulation, they need to be able to explore and discover, and they need consistency and predictability in their caregiving environment (Myers, 1991). A poor physical environment makes these conditions difficult to meet, especially since adults in low-income households usually work long hours and therefore have little time to care for or supervise children, let alone interact positively with them.

Child labour is still common in many developing countries. Certain environmental risks are associated with occupations that children and adolescents are likely to engage in including picking through garbage or working in particularly hazardous industries (Furedy, 1992). In Asia and Latin America, industries that employ child labour involve high risks related to the use of dangerous machinery, toxic chemicals, exposure to dust and heat (Lee-Wright, 1990).

Street children are particularly vulnerable to environmental hazards, especially if they lack family ties with adults and have to live on the streets. The United Nations Children's Fund (UNICEF) distinguishes between children who work on the streets but have strong family connections, may attend school and, in most cases, return to some kind of family shelter at the end of the day, and those who both live and work on the streets. The latter group may or may not have adult relatives. Those who are on their own are normally the most vulnerable. Those living on the street have difficulties finding shelter for sleeping, places for washing or defecating, and access to health services if they are sick or injured (Patel, 1990). They often work in dangerous situations, involving dodging traffic, for instance, as they sell goods to passing motorists, and are exposed to child abuse, especially since child prostitution is one of the more dependable ways of ensuring sufficient income for survival.

2. *Women*

Women are particularly vulnerable to a range of environmental hazards partly because of their reproductive role and partly because of the tasks that they undertake according to socially determined roles. Thus, pregnant women and their foetuses are particularly sensitive to environmental hazards. Every stage of the reproductive process can be disrupted by external environmental agents, leading to increased risks of abortion, birth defects, fetal growth retardation and perinatal death (WHO, 1992b). A woman's increased vulnerability during pregnancy, childbirth and the period just following it, although biologically determined, is nevertheless exacerbated by the fact that Governments and aid agencies have accorded low priority to its reduction. It is estimated that about half a million women die each year of causes related to pregnancy and childbirth, leaving about 1 million children without mothers (UNICEF, 1992a).

The exposure of women to certain environmental hazards is largely determined by their roles in child-rearing, household management and income generation (Moser, 1985 and 1987). In low-income settlements lacking basic infrastructure and services, women's exposure to environmental hazards is greater than that of men because of the tasks they undertake, including caring for the sick, washing soiled clothes where water supplies and sanitation facilities are inadequate, cooking in poorly ventilated rooms, and so forth (Sapir, 1990). In most societies, women's capacity to earn an income is reduced because of their child-rearing and caring responsibilities, their role as homemakers and their role as caregivers for sick or injured family members. Low-income women in particular have no alternative but to bear those responsibilities themselves. In poor households, the persons who are responsible for water collection and its use for laundry, cooking and domestic hygiene suffer most if supplies are contaminated and difficult to obtain and those persons are generally women. Low-income urban households tend to use biomass fuels or coal for cooking or heating over open fires or in poorly ventilated stoves, and it is generally women or girls who, being responsible for tending the fire and cooking, inhale larger concentrations of pollutants over longer periods (WHO, 1992a).

Women's vulnerability to the environmental hazards linked to the inadequate provision of water, sanitation, drainage and garbage collection is increased because their practical needs as caregivers and homemakers are rarely given the priority they deserve in the planning related to public services or housing programmes. Even when their needs are considered, the women themselves are rarely consulted about the most appropriate design and provision of services (Moser, 1987; Oruwari, 1991). Among health services, prenatal and post-natal care are rarely accorded the priority they deserve given their cost-effectiveness in reducing health burdens. Generally, health-care services do not provide the needed focus on women's reproductive health (including advice on family planning) that can do so much to reduce maternal mortality and severe health problems among women (Germain and Ordway, 1989; WHO, 1992b). Often, publicly funded health services are not easily accessible to women because of their location or their opening hours. The same is true for crèches, childcare centres and other services for children that could facilitate women's participation in the labour force (Moser, 1987). The provision of public transport is generally scheduled to serve primary income earners but it does not serve the needs of secondary income earners (often women) or of those responsible for shopping, and taking children to and from school, and for visits to health centres (Moser, 1987). Regulations on residential zoning often prohibit the kinds of informal productive activities in which women commonly engage in their homes (Moser, 1987; Oruwari, 1991). Public provision of housing, housing finance and skill-training tends to exclude women, whether implicity or explicitly, even women who are heads of household.

Among "low-income households", those headed by women usually face particular problems. In many low-income settlements, 30 per cent or more of households are headed by women either because a male partner is temporarily absent or because of separation or death (Moser, 1987). Women who are heads of household are usually the sole income earners in the household and have to combine income-earning with child-rearing and household management responsibilities, thus facing all the problems noted above in regard to inadequate access to infrastructure and services. They also face constraints when the kinds of income-earning activities in which they can engage are prohibited or when they are denied access to government programmes. Female heads of household often have special housing needs because their need to be at the same time income earners, child-rearers and homemakers reduces their ability to take part in "self-help" housing schemes. Yet, rarely do government housing programmes make special allowance for their needs (Moser, 1987; Falu and Curutchet, 1991).

I. Migrants

A key question is whether migrants constitute a vulnerable group within urban areas of developing countries. To answer it, one must ascertain whether urban migrants are distinct from non-migrants in terms of access to social services, unmet housing needs, and the pressure that they put on urban services and the environment. In terms of access to social services and unmet housing needs, it is difficult to distinguish the particular social or environmental problems faced by recent migrants from those faced by people who have lived in the city for long periods or who were born there. Where recent migrants are found to face particularly serious housing, health or environmental problems or restricted access to social services in comparison with non-migrants, the issue is whether it is their status as recent migrants or some other characteristic such as, for instance, their low income, that is the main cause of such differences. Thus, although a particular squatter settlement with a high proportion of recent migrants and particularly serious environmental and health problems might be taken as furnishing evidence of the problems faced by migrants, it might also be regarded as providing evidence of the problems faced by those with inadequate incomes, a category that may encompass a high proportion of both recent migrants and long-term city dwellers.

There are, however, certain environmental problems faced by recent migrants that are clearly related to their status as migrants. One is their possible lack of immunity to particular diseases that are common within the urban area to which they have moved. Sapir (1990) notes that the possible susceptibility of recent migrants to a disease that is endemic in a given urban centre can trigger an epidemic. In addition, when diseases such as tuberculosis or malaria are not common in their area of origin, migrants may be at higher risk of contracting them. However, the reverse effect may be in operation when migrants introduce new diseases into the urban environment (Sapir, 1990).

Recent migrants to a city may also face higher environmental risks because of their inability to find suitable accommodation and a source of income. However, assumptions that this will be the case have often been proved to be incorrect. Thus, reports of women interviewed in an inner-city bustee in Khulna, Bangladesh, showed that 72 per cent of the households with whom the women had migrated received assistance from friends and relatives in finding accommodation and 68 per cent received assistance in finding employment (Pryer, 1992). Over half had stayed in the homes of friends and relatives when they arrived (ibid.). Similarly, reports from female slum dwellers in Bangkok also showed that nearly all those who had migrated had a place to stay for an initial period while they found work and a place to live (Thorbek, 1991). Except where natural disasters, wars or other catastrophes force rural people to suddenly flee from their homes into urban areas, most migrants moving into a city have family, kin or friends to whom they can turn for help when they arrive.

A certain proportion of migrants may be at greater risk of exposure to environmental hazards than city-born or long-term urban residents because of the work they secure, especially if the migrants are poor or lack education. In general, all other things being equal, those who already live in a city have advantages over newcomers in terms of enjoying easier access to information on income-earning opportunities and being more likely to have relevant working experience. It is therefore noteworthy, that in some cities, a high proportion of recent migrants secure jobs in the formal sector. Thus, in Bangkok, migrants tend to enter directly into low-paying jobs in small and large firms: male migrants tend to fill unskilled wage positions while women tend to be employed in large-scale firms (often foreign-owned) or as domestic workers (Douglass, 1981). In Bangladesh, many of the young and single women who migrate from rural areas secure jobs in the garment industry in Dhaka and Chittagong (Kabeer, 1991). In Bangkok, few recent migrants become hawkers or vendors in the informal sector since those occupations are highly organized and have access rules and protective barriers against casual entry (Douglass, 1981). The problem may not be that migrants cannot find paid work but that the work they find is so poorly paid or dangerous. Since, in most instances, both migrants and long-term city residents have no source of income unless they find paid work, they are forced to take whatever is available to ensure their own survival. Initially, therefore, they may be prepared to work in industries where working conditions and pay levels are so poor that few long-term city residents choose to work there.

A high proportion of migrants may be disadvantaged, in comparison with long-term city-dwellers, in terms of the quality of accommodation that they can find or of its quality relative to its cost. A comparative study of the shelter strategies of low-income groups in Bamako and La Paz found some differences between recent migrants and longer-term city dwellers in the housing submarkets that they used (van Lindert and van Western, 1991). City-born households in La Paz were more successful than migrant households in avoiding settlement in residential neighbourhoods that were further from the location of income sources, especially the inhospitable El Alto area (ibid.). However, there is no simple association between recent migrants and slums; the level of income available to a household is a more important determinant of the quality of its housing. There are many squatter settlements and inner-city tenement districts where most of the population consists of long-term city residents who live there because they are poor (Cuenya and others, 1990; Pryer, 1992).

There are strong associations between poor housing conditions and particular survival or shelter strategies adopted by certain individuals or households. For instance, many low-income households, including both migrants and long-term city residents, choose to live in poor conditions in illegal settlements if there are possibilities that they will eventually receive legal tenure and basic infrastructure and services. That is, they choose to forgo environmental quality in the short term in return for the possibility of developing their own owner-occupied house and the prospect of legal tenure and services in the future. Many single people (or household members who migrate alone to a city) will tolerate poor housing conditions (in cheap boarding or rooming houses) to minimize the amount spent on accommodation. That strategy is most commonly used by temporary migrants. The level of risk associated with a particular shelter depends on the particular trade-off that an individual or household chooses to make in terms of location, quality of accommodation, tenure and cost, within the limited options available. The chosen trade-off usually changes for any given person or household through time, depending on age, household composition and so on

Among migrants to urban areas, some groups may be especially vulnerable. One such group consists of people forced to move to urban areas as a result of war, famine, drought or some other sudden catastrophe. In such cases, careful plans can rarely be made by the migrants themselves and strategies to minimize risk can hardly be formulated. Those migrants may have few if any social contacts already living in the urban centre to which they move. The city of Khartoum, for instance, is one of a number of African cities whose population has grown rapidly in recent years because of an influx of people fleeing war and natural disaster, whose health conditions are particularly poor (Omer, 1990).

Another vulnerable group consists of foreign immigrants, especially if illegal, since they often face discrimination in the housing and job markets and may have no protection under the law from exploitation by landlords or employers. A third vulnerable group comprises persons who have been evicted from or forced out of their homes in circumstances where there is little or no provision for rehousing them (Asian Coalition for Housing Rights, 1989). In addition, gender inequality will often make female migrants more vulnerable than male migrants to environmental risks. Since women face discrimination in the job and housing markets, and have less access to education and to land than men, the extent of their geographical mobility will be affected as will the likelihood of their earning decent incomes or of avoiding high-risk jobs in urban areas (Chant, 1992).

Although there are too few studies through which to be able to gauge the extent to which recent migrants suffer more than other groups within a city from social or environmental problems, there are enough studies to show that many common generalizations about migration are inaccurate.[5] It has already been noted that migrants do not tend to live predominantly in slums or squatter settlements and that a high proportion of migrants do not tend to be unemployed. In addition, detailed micro-studies and regional studies carried out over the past 20 years have shown that most migrants who move to cities are not poorly informed about their prospects there.

However, one must be careful not to suggest or imply other generalizations in regard to migration, given the fact that "migrants" into any particular city are usually a very diverse group, varying in terms, among others, of age, level of education, extent to which the move to the city is considered permanent or temporary, and extent to which the move is part of a complex household survival strategy. Recent studies have highlighted the extent to which migration patterns are also differentiated by gender (Chant and Radcliffe, 1992; Hugo, 1992; Pryer, 1992). Furthermore, in most urban centres with significant migrant inflows, there are usually major outflows too (Nelson, 1976). The scale of in-migration and out-migration from a city and the form each takes are constantly changing, reflecting, among other things, changes in the city's economic base as well as social, economic and political changes within the region or the country as a whole.

Lastly, there are about 30,000 urban centres in the developing world and each has its own unique pattern of in-migration and out-migration conditioned not only by changes in its own economy and employment pattern but also by such factors as crop prices, land-owning structures and changes in agricultural technologies and crop mixes in surrounding areas and distant regions. Each detailed study of migrants in urban settings and of conditions in areas of out-migration reveals a long list of factors that influence migration, ranging from those relating to individuals, and to household structures and gender-relations within households, passing on to local social, economic and cultural factors, and ending in regional or national social and economic changes related to international factors. In each location, the relative importance of the different factors is subject to constant change. This observation thus cautions against seeking to make too many generalizations and general recommendations in regard to rural-urban migration.

NOTES

[1]In general, the costs per household of installing most forms of infrastructure and supplying most kinds of services fall with increasing population density. Higher capital expenditures per person in infrastructure and service provision in urban areas is associated with increases in the quality of services provided rather than with higher population density per se. However, increases in capital expenditure per person reflect a public expenditure bias in favour of urban areas only if the beneficiaries of such expenditures fail to pay the full costs involved. On the other hand, increasing population density can also demand more expensive infrastructure. Thus, although low-cost ventilated and improved pit latrines can provide hygienic and convenient forms of sanitation in rural settlements and in urban areas with low population densities, more expensive systems are needed in higher-density settlements. The costs of infrastructure and services may also rise with city size, if the costs of acquiring land for their provision is a significant part of the total cost. The need for more complex and sophisticated pollution controls may also rise with increasing population density. For instance, it may not be necessary to provide a treatment of the effluent from the sewers and storm drains of a small urban centre that is as complex and as expensive as treatment carried out in larger cities. The costs to public authorities of formulating and implementing environmental legislation may also rise with city size (Linn, 1982; World Bank, 1991).

[2]See, for instance, Chant and Radcliffe (1992) for a recent summary; Saint and Goldsmith (1980) for a study of household decision-making; and Manzanal and Vapnarsky (1986) for an example of how migration flows respond to changing economic circumstances within a region.

[3]Using preliminary data on Mexico City from the 1990 census of Mexico and the adjustments to the 1980 figure suggested by Garza (1991).

[4]In some cities, however, certain groups may live permanently in particularly dangerous locations that pose even greater health risks than those to which people sleeping outdoors or in public places may be subject.

[5]An example of a myth about migration is implied in the text of the World Population Plan of Action (United Nations, 1975), according to which "urbanization in most countries is characterized by a number of adverse factors: drain from rural areas through migration of individuals who cannot be absorbed by productive employment in urban areas ..." (para. 44) and "internal migration policies should include the provision of information to the rural population concerning economic and social conditions in the urban areas, including information on the availability of employment opportunities" (para. 47).

REFERENCES

Aga Khan University students (1986). Unpublished paper quoted in *In the Shadow of the City: Community Health and the Urban Poor*, Trudy Harpham, Tim Lusty and Patrick Vaughan, eds. Oxford: Oxford University Press.

Aina, Tade Akin (1989). *Health, Habitat and Underdevelopment - with Special Reference to a Low-Income Settlement in Metropolitan Lagos*. London: International Institute for Environment and Development Technical Report.

ANAWIM (1990). (Philippines), vol. IV, No. 4. Published by The Share and Care Apostolate for Poor Settlers.

Arrossi, Silvina, and others (forthcoming). *Funding Community Level Initiatives*. London: Earthscan Publications.

Asian Coalition for Housing Rights (1989). Evictions in Seoul, South Korea. *Environment and Urbanization* (London), vol. 1, No. 1 (April), pp. 89-94.

Askoy, M., S. Erdem and G. Dincol (1976). Types of leukaemia in a chronic benzene poisoning. A study of thirty-four patients. *Acta haematologica* (Basel, Switzerland), vol. 55, pp. 65-72.

Basta, Samir S. (1977). Nutrition and health in low income urban areas of the Third World. *Ecology of Food and Nutrition* (United Kingdom), vol. 6, pp. 113-124.

Bertaud, Alain, and M. Young (1990). Geographical pattern of environmental health in Tianjin, China. Research/Sector paper, ASTIN. Washington, D.C.: World Bank.

Bradley, David, and others (1991). *A Review of Environmental Health Impacts in Developing Country Cities*. Urban Management Program Discussion Paper No. 6. Washington D.C.: World Bank, UNDP and United Nations Centre Human Settlements (Habitat).

Briceño-León, Roberto (1990). *La Casa Enferma: Sociologia de la Enfermedad de Chagas*. Caracas: Consorcio de Ediciones, Capriles C.A.

Bundey, D., S. Kan and R. Rose (1988). Age-related prevalence, intensity and frequency distribution of gastrointestinal helminth infection in urban slum children from Kuala Lumpur, Malaysia. *Transactions of the Royal Society of Tropical Medicine and Hygiene* (London), vol. 82, pp. 289-294.

Cairncross, Sandy (1990). Water supply and the urban poor. In *The Poor Die Young: Housing and Health in Third World Cities*, J. Hardoy, S. Cairncross and D. Satterthwaite, eds. London: Earthscan Publications.

__________, and Richard G. Feachem (1983). *Environmental Health Engineering in the Tropics: An Introductory Text*. Chichester, United Kingdom: John Wiley and Sons.

Cairncross, Sandy, Jorge E. Hardoy and David Satterthwaite (1990). The urban context. In *The Poor Die Young: Housing and Health in Third World Cities*, J. Hardoy, S. Cairncross and D. Satterthwaite, eds. London: Earthscan Publications.

Caldwell, John C., and Pat Caldwell (1985). Education and literacy as factors in health. In *Good Health at Low Cost*, S. B. Halstead, J. A. Walsh and K. S. Warren, eds. New York: Rockefeller Foundation.

Castonguay, Gilles (1992). Steeling themselves with knowledge report on the work of Cristina Laurell. *IDRC Reports* (Ottawa), vol. 20, No. 1 (April), pp. 10-12.

Cauthen, G. M., A. Pio and H. G. ten Dam (1988). *Annual Risk of Tuberculosis Infection*. Geneva: World Health Organization.

Centre for Science and Environment (1983). *The State of India's Environment: A Citizen's Report*. Delhi: Centre for Science and Environment.

_________ (1985). *The State of India's Environment: A Second Citizens' Report*. Delhi: Centre for Science and Environment.

Chant, Sylvia (1992). Conclusion: towards a framework for the analysis of gender-selective migration. In *Gender and Migration in Developing Countries*, S. Chant, ed. London: Belhaven Press.

__________, and Sarah A. Radcliffe (1992). Migration and development: the importance of gender. In *Gender and Migration in Developing Countries*, S. Chant, ed. London: Belhaven Press.

Cointreau, Sandra (1982). *Environmental Management of Urban Solid Waste in Developing Countries*. Urban Development Technical Paper, No. 5. Washington, D. C.: World Bank.

Cuenya, Beatriz, and others (1990). Housing and health problems in Buenos Aires: the case of Barrio San Martin. In *The Poor Die Young: Housing and Health in Third World Cities*, J. Hardoy, S. Cairncross and D. Satterthwaite, eds. London: Earthscan Publications.

Douglass, Mike (1981). Thailand: territorial dissolution and alternative regional development for the Central Plains. In *Development from Above or Below*, W. B. Stohr and D. R. Fraser Taylor, eds. Chichester, United Kingdom: John Wiley and Sons.

___________ (1992). The political economy of urban poverty and environmental management in Asia: access, empowerment and community-based alternatives. *Environment and Urbanization* (London), vol. 4, No. 2 (October), pp. 3-32.

Duhl, Leonard J. (1990). *The Social Entrepreneurship of Change*. New York: Pace University Press.

Ekblad, Solvig, and others (1991). *Stressors, Chinese City Dwellings and Quality of Life*. Stockholm: Swedish Council for Building Research. Document D12.

Espinoza, Vicente (1988). *Para una Historia de los Pobres de la Ciudad*. Santiago: Ediciones SUR.

Esrey, S. A., and R. G. Feachem (1989). Interventions for the control of diarrhoeal disease: promotion of food hygiene. Geneva: World Health Organization. WHO/CDD/89.30.

Falu, Ana, and Mirina Curutchet (1991). Rehousing the urban poor: looking at women first. *Environment and Urbanization* (London), vol. 3, No. 2 (October), pp. 23-38.

Furedy, Christine (1992). Garbage: exploring non-conventional options in Asian cities. *Environment and Urbanization* (London), vol. 4, No. 2 (October), pp. 42-54.

Garza, Gustavo (1991). The metropolitan character of urbanization in Mexico, 1900-1988. Mimeograped (April).

Germain, Adrienne, and Jane Ordway (1989). *Population Control and Women's Health: Balancing the Scale*. New York: International Women's Health Coalition in cooperation with the Overseas Development Council.

Goldstein, Greg (1990). Access to life saving services in urban areas. In *Housing and Health in Third World Cities*, J. Hardoy, S. Cairncross and D. Satterthwaite, eds. London: Earthscan Publications.

Gomes Pereira, M. (1984). Characteristics of urban mortality from Chagas' disease in Brazil's Federal District. *Bulletin of the Pan American Health Organization* (Washington, D.C.), vol. 18, No. 1, pp. 1-9.

Guimaraes, J. J., and A. Fischmann (1985). Inequalities in 1980 infant mortality among shanty town residents and non-shanty town residents in the municipality of Porto Alegre, Rio Grande do Sul, Brazil. *Bulletin of the Pan American Health Organization* (Washington, D.C.), vol. 19, pp. 235-251.

Hardoy, Jorge E., and David Satterthwaite (1989). *Squatter Citizen: Life in the Urban Third World*. London: Earthscan Publications.

Hardoy, Jorge E., D. Mitlin and D. Satterthwaite, eds. (1992). *Environmental Problems in Third World Cities*. London: Earthscan Publications.

Harpham, Trudy, Paul Garner and Charles Surjadi (1990). Planning for child health in a poor urban environment: the case of Jakarta, Indonesia. *Environment and Urbanization* (London), vol. 2, No. 2 (October), pp. 77-82.

Harpham, Trudy, Tim Lusty, and Patrick Vaughan, eds. (1988). *In the Shadow of the City: Community Health and the Urban Poor*. Oxford: Oxford University Press.

Hofmaier, V. A. (1991). *Efeitos da Poluicão do ar sobre a Funcão Pulmonar: un Estudo de Cohorte em Criancas de Cubatao*. São Paulo, Brazil: São Paulo School of Public Health. Doctoral thesis.

Hughes, Bob (1990). Children's play - a forgotten right. *Environment and Urbanization* (London), vol. 2, No. 2 (October), pp. 58-64.

Hugo, Graeme (1992). Women on the move: changing patterns of population movement of women in Indonesia. In *Gender and Migration in Developing Countries*, Sylvia Chant, ed. London: Belhaven Press.

Jacobi, Pedro (1990). Habitat and health in the municipality of São Paulo. *Environment and Urbanization* (London), vol. 2, No. 2 (October), pp. 33-45.

Jacobs, Jane (1961). *The Death and Life of Great American Cities*. London: Pelican Books.

Jarzebski, L. S. (1992). Case study of the environmental impact of the non-ferrous metals industry in the Upper Silesian area. In World Health Organization Commission on Health and Environment. *Report of the Panel on Industry*, annex II. Geneva: World Health Organization.

Jimenez Diaz, Virignia (1992). Landslides in the squatter settlements of Caracas: towards a better understanding of causative factors. *Environment and Urbanization* (London), vol. 4, No. 2 (October), pp. 80-89.

Jimenez, Rosario D., and Sister Aida Velasquez (1989). Metropolitan Manila: a framework for its sustained development. *Environment and Urbanization* (London), vol. 4, No. 2 (October), pp. 51-58.

Kabeer, Naila (1991). Cultural dopes or rational fools? Women and labour supply in the Bangladesh garment industry. *European Journal of Development Research*, vol. 3, No. 1, pp. 133-160.

Kagan, A. R., and L. Levi (1975). Health and environment: psycho-social stimuli - a review. In *Society, Stress and Disease: Childhood and Adolescence*, L. Levi, ed. Oxford: Oxford University Press.

Lee, James A. (1985). *The Environment, Public Health and Human Ecology*. Baltimore, Maryland and London: Johns Hopkins University Press.

Lee-Smith, Diana, and others (1987). *Urban Food Production and the Cooking Fuel Situation in Urban Kenya*. Nairobi: Mazingira Institute.

Lee-Wright, Peter (1990). *Child Slaves*. London: Earthscan Publications.

Leitmann, Josef (1991a). *Environmental Profile of São Paulo*. Urban Management and the Environment Discussion Paper Series. Washington, D.C.: UNDP/World

__________ (1991b). *Tianjin Urban Environmental Profile*. Urban Management and the Environment Discussion Paper Series. Washington, D.C.: UNDP/World Bank/United Nations Centre for Human Settlements (Habitat). April.

Linn, Johannes F. (1982). The costs of urbanization in developing countries. *Economic Development and Cultural Change* (Chicago), vol. 30, No. 3, pp. 625-648.

Manciaux, M., and C. J. Romer (1986). Accidents in children, adolescents and young adults: a major public health problem. *World Health Statistics Quarterly* (Geneva), vol. 39, No. 3, pp. 227-231.

Matte, T. D., and others (1989). Lead poisoning among household members exposed to lead-acid battery repair shops in Kingston, Jamaica (West Indies). *International Journal of Epidemiology* (Oxford), vol. 18, pp. 874-881.

Matthews, G. (1990). Health and the environment. World Health Organization, Regional Office for Europe (July). ICP/GEH/211/9. Mimeographed.

Mazingira Institute (1987). *Urban Food Production and the Cooking Fuel Situation in Urban Kenya - National Report: Results of a 1985 National Survey*. Nairobi, Kenya.

McGranahan, Gordon (1991). *Environmental Problems and the Urban Household in Third World Countries*. Stockholm: Stockholm Environment Institute.

Misra, Harikesh (1990). Housing and health problems in three squatter settlements in Allahabad, India. In *Housing and Health in Third World Cities*, J. Hardoy, S. Cairncross and D. Satterthwaite, eds. London: Earthscan Publications.

Moser, Caroline O. N. (1985). *Housing Policy and Women: Towards a Gender Aware Approach*. London: Development Planning Unit (DPU) Gender and Planning Working Paper, No. 7. London: University College.

__________ (1987). Women, human settlements and housing: a conceptual framework for analysis and policy-making. In *Women, Housing and Human Settlements*, C. Moser and L. Peake, eds. London and New York: Tavistock Publications.

Murphy, Denis (1990). *A Decent Place to Live: Urban Poor in Asia*. Bangkok: Asian Coalition for Housing Rights.

Myers, Robert (1991). *The Twelve Who Survive: Strengthening Programmes of Early Child Development in the Third World*. London and New York: Routledge.

Needleman, Herbert L., and others (1991). The long-term effects of exposure to low doses of lead in childhood: an eleven-year follow-up report. *New England Journal of Medicine* (Boston), vol. 322, No. 2 (January), pp. 83-88.

Nelson, Joan M. (1976). Sojourners versus new urbanites: causes and consequences of temporary versus permanent cityward migration in developing countries. *Economic Development and Cultural Change* (Chicago), vol. 24, No. 4 (July), pp. 721-757.

Newman, Oscar (1972). *Defensible Space: Crime Prevention through Urban Design*. New York and London: Macmillan.

Omer, Mohamed I. A. (1990). Child health in the spontaneous settlements around Khartoum. *Environment and Urbanization* (London), vol. 2, No. 3 (October), pp. 65-70.

Oruwari, Yomi (1991). The changing role of women in families and their housing needs: a case study of Port Harcourt, Nigeria. *Environment and Urbanization* (London), vol. 3, No. 2 (October), pp. 6-12.

Patel, Sheela (1990). Street children, hostel boys and children of pavement dwellers and construction workers in Bombay: how they meet their daily needs. *Environment and Urbanization* (London), vol. 2, No. 2 (October), pp. 9-26.

Peattie, Lisa (1990). Participation: a case study of how invaders organize, negotiate and interact with government in Lima, Peru. *Environment and Urbanization* (London), vol. 2, No. 1 (April), pp. 31-39.

Pepall, Jennifer (1992). Occupational poisoning reporting on the work of Mohamad M. Amr. *IDRC Reports* (Ottawa), vol. 20, No. 1 (April), p. 15.

Phantumvanit, Dhira, and Wanai Liengcharernsit (1989). Coming to terms with Bangkok's environmental problems. *Environment and Urbanization* (London), vol.1, No. 1 (April), pp. 31-39.

Pio, A. (1986). Acute respiratory infections in children in developing countries: an international point of view. *The Pediatric Infectious Disease Journal* (Baltimore, Maryland), vol. 5, No. 2, pp. 179-183.

Portes, Alejandro (1979). Housing policy, urban poverty and the state: the favelas of Rio de Janeiro. *Latin American Research Review*, No. 14 (summer), pp. 3-24.

Pryer, Jane (1992). Purdah, patriarchy and population movement: perspectives from Bangladesh. In *Gender and Migration in Developing Countries*, Sylvia Chant, ed. London: Belhaven Press.

Reichenheim, M., and T. Harpham (1989). Child accidents and associated risk factors in a Brazilian squatter settlement. *Health Policy and Planning* (London), vol. 4, No. 2, pp. 162-167.

Rohde, J. E. (1983). Why the other half dies: the science and politics of child mortality in the Third World. *Assignment Children*, vols. 61/62.

Romieu, Isabelle, H. Weitzenfeld and J. Finkelman (1990). Urban air pollution in Latin America and the Caribbean: health perspectives. *World Health Statistics Quarterly* (Geneva), vol. 23, No. 2, pp. 153-167.

Rothenburg, Stephen J., and others (1989). Evaluación del riesgo potencial de la exposición perinatal al plomo en el Valle de México. *Perinatologia y Reproducción Humana*, vol. 3, No. 1, pp. 49-56.

Saint, William S., and William D. Goldsmith (1980). Cropping systems, structural change and rural-urban migration in Brazil. *World Development* (Boston), vol. 8, pp. 259-272.

Sapir, D. (1990). *Infectious Disease Epidemics and Urbanization: A Critical Review of the Issues*. Paper prepared for the World Health Organization Commission on Health and Environment. Geneva: WHO, Division of Environmental Health.

Sarin, Mahdu (1983). The rich, the poor and the land question. In *Land for Housing the Poor*, S. Angel, R. Archer, S. Tanphiphat and E. Wegelin, eds. Singapore: Select Books.

Schaeffer, B. (1990). Home and health - on solid foundations? *World Health Forum* (Geneva), vol. 11, pp. 38-45.

Schofield, C., and others (1990). The role of house design in limiting vector-borne disease. In *The Poor Die Young: Housing and Health in Third World Cities*, J. E. Hardoy, S. Cairncross and D. Satterthwaite, eds. London: Earthscan Publications.

Schteingart, Martha (1989). The environmental problems associated with urban development in Mexico City. *Environment and Urbanization* (London), vol. 1, No. 1 (April), pp. 40-49.

Sinnatamby, Gehan (1990). Low cost sanitation. In *The Poor Die Young: Housing and Health in Third World Cities*, J. E. Hardoy, S. Cairncross and D. Satterthwaite, eds. London: Earthscan Publications.

Sobreira de Moura, Alexandrina (1987). Brasilia Teimosa - the organization of a low-income settlement in Recife, Brazil. *Development Dialogue* (Stockholm), vol. 18, pp. 152-169.

Songsore, Jacob (1992). Review of household environmental problems in the Accra metropolitan area, Ghana. Working paper. Stockholm: Stockholm Environment Institute.

Stephens, Carolyn, and Trudy Harpham (1992). The measurement of health in household environmental studies in urban areas of developing countries: factors to be considered in the design of surveys. London: Urban Health Programme, London School of Hygiene and Tropical Medicine. Mimeographed.

Thorbek, Susanne (1991). Gender in two slum cultures. *Environment and Urbanization* (London), vol. 3, No. 2 (October), pp. 71-81.

Turner, Bertha, ed. (1988). *Building Community: A Third World Case Book.* London: Habitat International Coalition.

Turner, John F. C. (1976). *Housing by People: Towards Autonomy in Building Environments.* London: Ideas in progress, Marion Boyars.

__________, and Robert Fichter, eds. (1971). *Freedom to Build.* New York and London: Macmillan.

United Nations (1975). *Report of the United Nations World Population Conference, 1974, Bucharest, 19-30 August 1974.* Sales No. E.75.XIII.3. Chapter I.

__________ (1991). *World Urbanization Prospects, 1990.* Sales No. E.91.XIII.11.

United Nations Children's Fund (UNICEF) (1991). *The State of the World's Children, 1990.* Oxford: Oxford University Press.

__________ (1992a). *The State of the World's Children, 1991.* Oxford: Oxford University Press.

__________ (1992b). *Environment, Development and the Child.* New York: UNICEF, Environment Section, Programme Division.

United Nations Environment Programme (UNEP) (1991). *Environmental Data Report, 1991-2.* Global Environment Monitoring System (GEMS) Monitoring and Assessment Research Centre, Oxford, United Kingdom: Blackwell.

__________, and World Health Organization (WHO) (1987). Global Pollution and Health: Results of Health-related Environmental Monitoring. Global Environment Monitoring System (GEMS).

United Nations Environment Programme (UNEP), and World Health Organization (WHO) (1988). *Assessment of Urban Air Quality.* London: Global Environment Monitoring System (GEMS), Monitoring and Assessment Research Centre (MARC).

United States Agency for International Development (USAID) (1990). *Ranking Environmental Health Risks in Bangkok.* Washington, D. C.: Office of Housing and Urban Programs.

van Lindert, Paul, and August van Western (1991). Household shelter strategies in comparative perspective: evidence from low-income groups in Bamako and La Paz. *World Development* (Boston, Massachusetts), vol. 19, No. 8, pp. 1007-1028.

Waller, Robert E. (1991). Field investigations in air. In *Oxford Textbook of Public Health,* vol. 2, W. W. Holland and others, eds. Oxford: Oxford University Press.

White, G. F., D. J. Bradley and A. V. White (1972). *Drawers of Water: Domestic Water Use in East Africa.* Chicago: University of Chicago Press.

World Bank (1984). *Staff Appraisal Report: Brazil Second Health Project.* São Paulo Basic Health Care and National Health Policy Studies. Washington, D.C.: World Bank, Population, Health and Nutrition Department.

__________ (1988). *World Development Report, 1988.* New York: Oxford University Press.

__________ (1991). *Urban Policy and Economic Development: An Agenda for the 1990s.* Washington, D.C.: World Bank.

World Health Organization (WHO) (1989). *Urbanization and its Implications for Child Health: Potential for Action.* Geneva: WHO.

__________ (1991a). *Global Estimates for Health Situation Assessments and Projections, 1990.* Geneva: WHO, Division of Epidemiological Surveillance and Health Situation and Trend Analysis. WHO/HST/90.2.

__________ (1991b). Health trends and emerging issues in the 1990s and the twenty-first century. Geneva: World Health Organization, Monitoring, Evaluation and Projection Methodology Unit, Division of Epidemiological Surveillance and Health Situations and Trend Assessment.

__________ (1992a). *Our Planet, Our Health.* Report of the WHO Commission on Health and Environment. Geneva: WHO.

__________ (1992b). *Reproductive Health: A Key to a Brighter Future.* J. Khanna, P. F. A. Van Look and P. D. Griffin, eds. Biennial Report 1990-1991. Special 20th Anniversary Issue. Geneva: WHO, Special Programme of Research Development and Research Training in Human Reproduction.

Zorrilla, Silvia, and Maria Elena Guaresti (1986). *Sector Agua Potable y Saneamiento: Lineamientos para una Estrategia Nacional.* Buenos Aires: United Nations Development Programme.

X. THE MODERATE EFFICIENCY OF POPULATION DISTRIBUTION POLICIES IN DEVELOPING COUNTRIES

*Carlos Antonio de Mattos**

After the Depression that began in 1929 and particularly after the Second World War, a growing number of countries of the capitalist periphery began to shape their public administration processes in accordance with an economic and social development strategy influenced by the Keynesian theories that were popular at the time in the core countries. This strategy was based on the conviction that it was possible to correct the problems that economic growth itself was generating by means of greater and more rational State intervention. Consequently, faith in State planning became widespread together with a strong confidence in its feasibility and potential.

Such ideas gained currency in the periphery under the assumption that they offered a sure means of combating underdevelopment. Wherever their implementation was possible, the policies adopted brought about major structural changes, accelerating industrialization and strengthening the integration of economic and spatial processes. The evolution of such processes led, among other things, to a gradual intensification of urbanization and internal migration, which in turn gave rise to major changes in the spatial distribution of both productive activities and population. In many cases, such changes were regarded as the negative effects of industrialization.

In accordance with the conviction that planned State intervention would make it possible to correct or eradicate such negative effects, specific types of policies aimed at influencing the spatial distribution of productive activities and population were designed and implemented. The main goal of those policies was to achieve a more balanced spatial distribution of productive activities, employment and population by ensuring the diffusion of growth from the more to the less developed regions of a country (Stöhr and Todtling, 1978). After trying to implement those policies for several decades, confidence in their effectiveness waned. Consequently, towards the mid-1970s, Keynesian thinking began to be displaced by a new theoretical and ideological approach based on the assumptions of neoclassical economics which became the basis for public policy formulation in the capitalist world. The emergence of this new paradigm brought about major changes in the orientation and content of the prevailing development model and, consequently, in the prevailing mode of regulation of development processes.

The present paper analyses the characteristics, scope and effectiveness of spatial distribution policies during both the Keynesian period and the period that followed. Since the object of the analysis is a particular type of public policy, that is, a component of the public administration process, it is necessary to analyse its relevance, compatibility and function in relation to the development model prevailing during each period. Consequently, we will consider the spatial distribution policies adopted by developing countries under the influence of the development model prevalent after the Second World War, giving due regard to the modes of regulation that accompanied such policies. Attention will mostly be focused on Latin American countries, though examples from other regions will be cited as appropriate. However, given the growing diversity among developing countries, any generalization is hazardous. Even within Latin America, the conclusions derived from an analysis of the experience of countries like Argentina, Brazil, Chile, Colombia, Mexico and Venezuela, all of which have relatively high levels of industrialization, cannot be

*Instituto de Estudios Urbanos, Pontificia Universidad Católica de Chile, Santiago.

automatically extended to the cases of less industrialized countries in the region.

A. Population distribution policies during the Keynesian period

Under the impetus of the economic dynamics that began to take shape in the global capitalist economy during the years that followed the 1929 crisis—subject to a new international division of labour, which has gradually become established since then—an increasing number of countries began to frame their modernization strategies according to Keynesian thinking. This phase of the development process had as basic reference the fordist mode of production that had been taking hold in the core countries after the Depression. From the point of view of its organizational tenets, the fordist system could be characterized as a regime of intensive accumulation based on (*a*) taylorist principles concerning the organization of production, including the division of labour, the separation of management from property ownership, and the managerial control over production, in conjunction with mechanization to increase productivity; (*b*) a stable regulation of the relation between labour and capital as a means of ensuring the reproduction of labour and strengthening the internal market; and (*c*) the consolidation of large manufacturing enterprises capable of producing standardized goods for mass consumption and of becoming the dynamic core of the economic system. The aim of reproducing that regime in developing countries was espoused by various political forces that were gaining support as proponents of modernization.

Furthermore, in the formulation of national development projects, a key role was played by Keynesian theories of economic and social development, which, associated with such names as Hirschmann, Myrdal, Nurkse, Robinson and Rosenstein-Rodan, gained widespread popularity in developing countries. In Latin America, this approach found its clearest and most influential expression in the normative model proposed by the Economic Commission for Latin America and the Caribbean (ECLAC) which influenced significantly the formulation of economic policy during the 1950s and 1960s.

The appearance of new social forces in the socio-political arena of developing countries, such as the urban middle and working classes and the industrial bourgeoisie, provided a political base that supported the viability of modernization initiatives through the pursuance of new interests and claims. These changes led to the establishment of a strategy of accumulation and growth which was accompanied by a serious concern about ensuring the compatibility of economic growth and social justice.

Three basic principles constituted the core of the national development strategies adopted during the period: (*a*) the promotion of inward-oriented economic growth, so that internal markets would become the main support of productive activities; (*b*) the promotion of industrialization for import substitution as the dynamic core of economic growth; and (*c*) the establishment of a new mode of regulation based on growing State intervention which embodied the new theory on development planning and on a set of institutional arrangements aimed at stabilizing the relations between capital and labour.

This strategy began to be tried with different degrees of consistency and intensity by many developing countries after the world crisis of 1929 and with greater determination after the Second World War. Accordingly, it constituted the basis for the populist political projects that prevailed during that period in the developing world and particularly in Latin America. Consequently, the aim of maintaining economic growth while ensuring social justice became one of the basic dilemmas confronting Governments and public administrators in developing countries. Indeed, given that the scarcity of resources for capital formation slowed accumulation, the maximization of economic growth proved difficult to reconcile with the improvement of social equity.

Faced with this quandary, policy makers tended to give priority to the objectives and policies that they regarded as having greater and more direct impact on the stimulation of accumulation and economic growth, and hence on industrialization. Consequently, the national and sectoral policies necessary for the attainment of such goals were those implemented most consistently and attracting the strongest political backing, whereas the policies dealing with issues considered to have less priority were applied less consistently or with insufficient political commitment. The effectiveness of these policies was therefore impaired and their objectives, when attained, were mostly the unintended result of the national and sectoral policies that had been given priority.

The adoption of such development policies led to the replication of various features of fordism in the many developing countries that embarked on industrialization, including intensive accumulation accompanied by growing mechanization, the application of taylorist principles of production, the relative growth of mass consumption of durable goods, and advances in the regulation of wages. However, these changes occurred only fragmentarily and the expression "peripheral fordism" was coined to describe the emerging situation (Lipietz, 1986). The production system thus established, which advanced to varying degrees in different developing countries, shaped the global socio-economic framework within which public administrators had to circumscribe their efforts so as to handle the "territorial problems" to whose accentuation peripheral fordism had itself contributed.

1. *Territorial dynamics of peripheral fordism*

As already noted, the development policies pursued during the post-war period contributed greatly to strengthening a capitalist mode of economic and territorial integration, and led to the increased industrialization of the countries in which they were effectively applied. Despite the various objections that have been raised against the mode of industrialization pursued in Latin America by describing it, for instance, as "incomplete industrialization" (Fajnzylber, 1983), the rate of industrialization achieved was high in most countries of the region. Thus, between 1950 and 1977, the industrial growth rate in Latin America averaged 6.7 per cent per annum, a figure higher than the average annual growth rates for the world as a whole (5.9 per cent), for Canada and the United States of America (3.6 per cent) and for Western Europe (5.2 per cent). Only the annual industrialization rates of the Eastern-bloc countries of Europe (10.2 per cent) and Japan (12.7 per cent annually) surpassed that of Latin America (Comisión Económica para América Latina (CEPAL, 1981).

These rates of industrialization, combined with the concomitant increase in economic and territorial integration, led to the progressive introduction of capitalist modes of production in rural areas, accelerated the transition from an economy predominantly rurally-based to one based predominantly in urban areas, and increased both internal migration and urbanization. Thus, between 1960 and 1990, the level of urbanization in Latin America increased from 49 to 72 per cent (Lattes, 1989).

The changes that took place during this period in the developing countries that industrialized the fastest led to a specific pattern of human settlements whose basic features, which had already been discernible in earlier periods, became more profoundly marked and consolidated. The main characteristics of the resulting pattern, which was the territorial expression of the peculiar socio-economic dynamics set in motion by the development policies in place, are detailed below.

When industrialization for import substitution began, the urban centres that had greater advantages, in terms of size and accessibility to markets, availability of manpower and infrastructure, including transportation and communication systems, and access to the nation's political power, were the ones that most strongly attracted both the new productive activities and new settlers. Indeed, these characteristics of the major cities of each country set the initial conditions that impelled and sustained the processes of territorial concentration. The very fact that industries were being concentrated in the major cities generated feedback mechanisms between the external economies of agglomeration and industrialization.[1] The characteristics of large cities together with the externalities emerging from the territorial concentration of productive activities gave rise to a form of capitalist socialization of the forces of production (Topalov, 1979) which, from the standpoint of the strategy of capital profitability, was perceived as providing comparative advantages superior to those offered by smaller urban centres. As the capitalist process evolved, the strategies and policies adopted to ensure the profitability of capital became a key element with respect to setting in motion and sustaining the process of concentration and the uneven territorial distribution of both capital and people. Given those dynamics, underdevelopment merely became one facet of inequitable accumulation, and the hyperconcentration of capital in locations that ensured high profitability, the other (Topalov, 1979).

However, beyond certain thresholds, the increasing levels of industrialization and urbanization that were closely associated with the growth of major cities began to generate diseconomies of agglomeration, prompting new industries, and many of those already established in major cities, to consider alternative locations. In most cases, the new locations involved were urban centres located near the major cities, so that a process of suburbanization began that generally took the shape of a tentacle-like expansion of the central city, which in turn gave rise to the formation of extensive and complex central subsystems.[2]

The territorial concentration of industry in steadily expanding agglomerations, by stimulating a rise in production, the adoption of new forms of managerial organization, and the continuous incorporation of technological change, generated conditions favourable to the setting in motion of feedback mechanisms between the concentration of capital and the territorial concentration of people. Essentially, territorial concentration proved favourable to the intensification of economic concentration[3] and the latter, in turn, stimulated the trend towards territorial concentration through the effects triggered by the horizontal and vertical transmission of externalities. As Singer (1973) pointed out, in capitalism the concentration of capital and the territorial concentration of economic and social activities have common causes.

As the process of economic and territorial integration proceeded, it entailed the gradual but inexorable propagation of the capitalist mode of production in rural areas. Consequently, the new social organization of labour led to the proletarianization of large contingents of the rural labour force, the introduction of capital-intensive agricultural production and the expansion of land owned by rural capitalist enterprises. These changes brought about substantial rises in migration to the cities (Singer, 1973) and, among the urban hierarchy, central subsystems exerted the greatest attraction because they offered better employment prospects than smaller or medium-sized cities.

As these processes evolved, central subsystems became the main centres of capitalistic accumulation within countries, thus helping to articulate the rest of the national system around their own activities and shaping the process of economic and territorial integration as a function of their needs and interests.[4] Clearly, such dynamics could not but result in a territorial distribution characterized by unequal accumulation.

2. *Beliefs and opinions about territorial "problems"*

The processes described above led to patterns of population distribution considered to be the source of problems that had to be solved by government intervention. However, there arises the question of what the basis is for regarding a certain phenomenon as a territorial "problem". From a public administration perspective, a territorial problem arises when some aspect of the pattern of population distribution is perceived as undesirable by policy makers who consequently decide to correct or erradicate it. The identification of any aspect of population distribution as a problem stems, therefore, from the implicit or explicit comparison that policy makers make between their perception of the current population distribution pattern and an ideal pattern used as reference. This means that, in the final analysis, the perception and description of population distribution problems are based on the values and beliefs of decision makers.[5] Consequently, a certain "ideological bias" underlies most population distribution policies, thus justifying the assertion that the basis for population distribution policies is riddled with extreme value judgements, ideological positions and inconsistent assertions, many of which are little more than myths (Richardson, 1984).

During the post-war period, the most common ideal reference used to determine the existence of population distribution problems was the utopia of balanced development. Not surprisingly, an ideology favouring a more equitable distribution of the benefits of development buttressed the ideal of a balanced distribution of production, employment and population, thus echoing the ongoing debate regarding equity versus efficiency. In cases where the position favouring redistribution prevailed, it was common for it to be based on assumptions that were incompatible with the structural conditions inherent in capitalist dynamics. There was thus a tendency to accept, in a rather simplistic manner, the position that a better territorial distribution of the population was a means of ensuring greater social equity and that appropriate methods for achieving such population distribution were available. To validate such positions, it was often necessary to disregard the restrictions imposed by the prevailing development model and the fact that a better interregional distribution of income did not ensure a better interpersonal distribution of it.

What, then, were the territorial phenomena identified as negative and therefore as possible targets for government action? The plans, strategies and policies formulated during the post-war period indicate that two phenomena inherent in the dynamics of peripheral fordism were usually considered to be problems: (*a*) the trend towards progressive territorial concentration of production, employment and population with the consequent increase in the primacy of cities; and (*b*) the "imbalance" in the territorial distribution of the forces of production, their level of development, the rates of accumulation and growth, and the standards of living of the subpopulations concerned.

In general, assessments regarding population distribution trends and their consequences were made under the influence of certain theoretical approaches that were very influential among the intellectual and political elites of developing countries during the late 1950s and early 1960s. One such theory was proposed by Myrdal (1957) who, rejecting neoclassical tenets, asserted that the free play of market forces led to self-sustaining processes that stimulated development in the core and underdevelopment in the periphery. Similarly, the conclusion of Perroux (1955), namely that the expansion of modern oligopolistic large-scale industry and the industrial complexes that grew up around it led to the formation of growth poles in the urban hierarchy whose attraction further contributed to the maintenance of territorial inequality, gained considerable influence. Consequently, studies carried out in developing countries usually characterized the trend towards concentration as unfavourable for the attainment of equitable economic and social development. Garza (1980), for instance, concluded that, in the case of Mexico, the disadvantages of a high urban concentration were evident in so far as it led to the poor utilization of natural resources and manpower, produced great inequalities between regions, laying the foundation for political and social conflict, implied investments with high opportunity costs in the main urban agglomeration of the country, entailed high social costs, and led to the proliferation of social problems in urban areas. Such negative effects of spatial concentration could not but retard the socio-economic development of the country.

However, arguments of this type were and continue to be the subject of controversy. Many authors maintain that, given the circumstances under which the processes of accumulation and growth in developing countries must take place, the externalities associated with the main urban agglomerations constitute, at least during the initial phases of industrialization, a factor favourable to the attainment of high productivity.[6] In fact, this was the position taken by most Governments when they framed their strategies of substitutive industrialization.

From an economic perspective, perhaps the strongest arguments in favour of policies for the redistribution of production, employment and population are those based on the "costs of urbanization". Analysing the cases of four developing countries (Bangladesh, Egypt, Indonesia and Pakistan), Richardson (1987) concluded that the marginal cost of absorbing an additional person was approximately three times as high in an urban as in a rural area and that this ratio rose as a city's degree of primacy increased. This finding suggested that a change in population distribution could lead to a more efficient use of scarce resources (Hansen, 1990).

3. *The modest effects of population distribution policies*

Irrespective of the arguments used in favour of or against population distribution policies, the fact is that certain aspects of population distribution were identified as problems and, according to the Keynesian approach, government intervention was called for to correct the imbalances brought about by development. Consequently, it was considered feasible to formulate and implement population distribution policies even as the macroeconomic priorities set by the chosen mode of development were being pursued.

Since the goal of population distribution policies was to achieve better territorial distribution of production, employment and population, they were based on the idea of spreading economic growth from the core to the periphery. To that end, efforts were made to promote "concentrated deconcentration", that is, to generate external economies at new locations in order to attract flows of capital and thus start the processes of accumulation and growth within them. As Stöhr and Todtling (1978) noted, since migration was beyond direct political control in most mixed and market economies, population distribution policies had to operate mostly by changing the flows of capital and technology, and by creating greater external economies in peripheral areas through public investment in infrastructure. Thus, population distribution policies usually promoted the establishment of growth poles or industrial parks in selected medium-sized cities of the periphery and, in some cases, they also involved frontier development programmes, promoting both the settlement of new agricultural areas and integrated rural development.

Given the restrictions imposed by capitalism, in order to generate external economies in peripheral areas and unleash the process of accumulation and growth, a number of instruments were used that treated locations differently in order to direct private capital flows to them. These included a variety of subsidies and incentives to private investment, investment in public infrastructure, restrictions and controls on industrial location, differential tariffs on land development, and subsidies on employment (Stöhr and Todtling, 1978; Richardson, 1984).

Assessments of the impact of these policies tended to conclude that they were ineffective or that the results obtained were modest at best.[7] Thus, policies aimed at promoting the establishment of industry in peripheral urban centres did not, in general, achieve the desired effect. In some cases, most of the enterprises that relocated moved to medium-sized cities within the same region from which they originated and not to poorer regions in the periphery. Consequently, the policies in place ended up strengthening the trend towards the growth and expansion of central subsystems by fostering suburbanization.[8] In other cases, a significant proportion of the industries that relocated to peripheral areas were subsidiaries of multiregional and multinational enterprises and, being capital-intensive, did not have a major impact on local or regional employment levels.

Since population distribution policies did not succeed in reversing the persistent feedback mechanisms favouring the concentration of private capital in central subsystems, a significant transfer of capital from the periphery to the core continued. In many cases, the profits of industries established in peripheral regions were used as sources of investment in the core, where profitability was higher.

Policies promoting the expansion of transport and communications infrastructure were designed, in principle, to favour a greater territorial dispersion of productive activities and population. Nevertheless, to the extent that they were not accompanied by higher growth in production and employment in peripheral regions, they ended up promoting migration, including that to central subsystems.

Policies fostering the industrialization of peripheral regions, while helping to reduce interregional disparities in income in some cases, were more frequently the cause of rising intraregional disparities, especially in peripheral regions where growing industrialization was based on the use of abundant and cheap labour. In such regions, the benefits of economic growth did not spread to all the population, but remained concentrated among the urban middle and upper classes, thus intensifying interpersonal disparities in the distribution of income within the region and especially between urban and rural areas.[9]

Policies favouring the modernization of agriculture contributed to the intensifying of pre-existing trends regarding the expansion of capitalist relationships into the rural environment and stimulated rural-urban migration (Martine, 1987), although in many cases it was asserted that their purpose was to improve the living conditions of the rural population and thereby provide incentives for that population to remain in the countryside.[10] In addition, policies favouring the settlement and expansion of frontier areas tended, according to capitalist logic, to foster exclusion and concentration. Even programmes aimed at regional agricultural development, which were successful in achieving significant increases in farm production, ended up by causing a greater concentration of income by relying on traditional capitalist mechanisms in relation to marketing, access to credit or land. Thus, higher-income groups were those most likely to profit from their privileged access to capital markets, education and new investment opportunities (Barkin, 1972).

These developments support the conclusion that the population distribution policies implemented during the post-war period by developing countries did not succeed in improving the territorial distribution of production, employment and population. On the contrary, in almost all countries, internal migration to the major cities grew and the trends towards population concentration and urban primacy were strengthened, especially between 1950 and 1970 (Portes, 1989; Lattes, 1989).

4. *Reasons for the modest impact of population distribution policies*

Why did population distribution policies fail to achieve better results? An answer to this question must take into account the way in which decision makers assigned priorities to different objectives and the response of private investors to public measures. As already noted, during the post-war period, decision makers showed a marked preference for macroeconomic or sectoral policies that would promote economic growth and the accumulation of capital through industrialization. Those policies, by favouring industrial growth and increases in productivity, had as secondary effect the concentration of capital in places that offered comparative advantages. In addition, policies aimed at increasing territorial integration also contributed to concentration.

In contrast, the redistribution of population and productive activities was usually not given priority, and concern about emerging population distribution patterns was expressed mostly for political reasons. Lacking priority in the eyes of policy makers, population distribution goals were largely excluded from the basic agenda of public administration and were therefore not pursued in terms of effective public policy.

Furthermore, in countries where population distribution policies were implemented, they were largely incapable of changing the perception of private investors regarding the profitability of investing in different locations, especially in growth poles or centres of growth in peripheral regions. Consequently, investment continued to flow mostly to central subsystems, where existing externalities were perceived by private investors as a more satisfactory alternative for the utilization of capital. Indeed, it must be stressed that in capitalist economies, public policies have a limited capacity to determine the decisions made by the private sector. In societies having a high structural dependency of capital, the effectiveness of any instrument of public policy is conditioned by its capacity to induce private strategies compatible with the purposes established by the public sector (Przeworski and Wallerstein, 1986).

In most cases, the capitalist socialization of the forces of production offered by central subsystems continued to be perceived by private investors as the environment ensuring the best profitability of capital; consequently, they tended to discard the alternatives presented by public policy. Population distribution policies had, therefore, a minor impact and whatever changes in population distribution actually occurred were mostly the result of the macroeconomic and sectoral policies adopted to promote overall development. In most cases where peripheral regions did increase their power to attract population, major investments were made in those regions by the public sector itself. This was the case for the creation of a growth pole in Ciudad Guayana in Venezuela and for Brasilia in Brazil.

It can be concluded, therefore, that the most relevant changes in territorial distribution during the Keynesian period were caused by the dynamics of economic growth set in motion by the macroeconomic and sectoral policies implemented according to the development model of the time.[11] Some of those policies seem to have contributed to strengthening population distribution trends that had begun to take shape spontaneously as the process of capitalist integration of developing countries proceeded. Attempts to change those trends through population distribution policies that were not accorded priority were doomed to fail. As Richardson (1984) noted, policies tend to be effective when they anticipate or are consistent with market forces but they are unlikely to succeed if they run against those forces.

B. A NEW STRATEGY FOR PUBLIC ADMINISTRATION

In both developed and developing countries, the Keynesian model began to show unmistakable signs of exhaustion in the 1960s, but its collapse became evident only during the 1970s when a profound structural crisis manifested itself in a decline in productivity, an increase in production costs, a decrease in profitability, a consequent reduction of the rate of accumulation, and a sustained rise in unemployment (Lipietz, 1986). In a world economy that was becoming more interdependent, this crisis spread rapidly to developing countries, thus aggravating the progressive deterioration of peripheral fordism. Because these developments were soon perceived as a failure of the Keynesian model itself, they led to the conviction that countries that continued to follow that model would find it difficult to devise satisfactory solutions to the complex economic and social problems confronting them. Thus, many Governments had to accept the necessity of restructuring by adopting so-called structural adjustment policies.

These policies are based on a new theoretical and ideological approach dominated by neoliberal ideas that propound a strategy aimed at (re)establishing conditions favourable to the profitability of private capital as a means of injecting new dynamism into the processes of accumulation and growth. As a justification for this strategy, it is alleged that the changes introduced under the influence of Keynesian thinking have reduced the rate of accumulation, thereby interfering with economic growth and reducing the development prospects of the countries concerned. In particular, the expansion of the State, especially the welfare State, has led to an unacceptable increase in the fiscal deficit and therefore to rising taxes that have tended to discourage investment, accumulation and growth.

Given this diagnosis, the new strategy propounds fiscal austerity and balanced budgets as the means of overcoming the State's fiscal crisis and achieving adequate rates of accumulation. As a logical corollary, the State must be reformed to become smaller and more efficient, even if less powerful. This is the cornerstone of the new neoclassical approach. The aim is to free the State from anything that interferes with its actions so as to make it a more effective regulator of the functioning of the market and an arbitrator of conflict between social groups.

Basically, the new approach gives priority to market forces as regulators of the general functioning of each system, thus increasing the prominence of private capital in the processes of accumulation and growth, injecting flexibility into the regulation of relations between labour and capital, reducing the size of the State and its direct intervention in the economy, especially by restricting its role as investor or producer, and fostering an open economy that has strong links with the world economy and where exports are the main engine of growth.

Given the growing globalization of the world economy and the technological changes that facilitate transport and communication, the adoption and implementation of such policies have become a necessary condition for the modernization of each national economy, so that growth based on its competitive incorporation into the new international division of labour can take place. The processes set in motion by structural adjustment policies have been strengthened by an equally important set of private policies and are producing profound transformations in the world economy, at the international, national, regional, and local levels, particularly with regard to the organization of production and management.

1. *New criteria for addressing population distribution problems*

Under the new theoretical and ideological principles established by the neoclassical approach, the free play of market forces is assumed to be a necessary condition and—to a certain extent, a sufficient one—for achieving a more balanced territorial distribution of productive activities and population. Consequently, it is believed that urban primacy is a phenomenon inherent in the initial stages of development and that it is beneficial because it maximizes the economies of scale (Hansen, 1990). This perspective has been buttressed by hypotheses arising mainly from empirical research on the evolution of developed countries. Thus, it has become popular to argue that with time per capita income will rise and regional inequalities will tend to diminish, following an inverted U-shaped curve as proposed by Williamson (1965). Similarly, Berry's hypothesis that, as development proceeds, a more balanced distribution of cities in accordance with their size will be attained has been resurrected (Berry, 1961). These observations suggest that, once the conditions for the free play of market forces are restored, population distribution policies will no longer be needed. That is to say, in contrast with the Keynesian approach, it is now explicitly acknowledged that macroeconomic policies, by regulating market forces, will ensure a better territorial distribution of production, employment and population and that, in general, there is no need for explicit population distribution policies which, in any case, should be compatible with the workings of the free market.

This new approach to public administration implies the use of a new set of instruments which differ in important ways from those used in the preceding period. Such instruments include incentives, but incentives subject to selective criteria; the improvement of transport and technological infrastructure, with special emphasis on communication and information networks; the provision of services to enterprises so as to allow them to increase their competitiveness within the global market; and sustained support for small and medium-sized enterprises (del Río Gómez and Cuadrado Roura, 1991). The new approach is therefore oriented towards the integration of social actors and, as such, it requires a high degree of coordination to avoid wasting efforts and resources.

In what way have these changes in strategy and public policy affected the territorial dynamics of population distribution? Undoubtedly, the policies that have had the greatest impact on the territorial mobility of capital are those that modified the relative importance of public versus private investment and those that reduced the importance of the internal market in relation to the external market. To the extent that these policies have affected the most dynamic urban centres of each system, they have established conditions for the modification of the territorial pattern of accumulation and growth.

The territorial distribution of private capital

Given that one of the basic tenets of the new development strategy is to minimize public investment and reduce the entrepreneurial role of the State, there has been a growing acceptance of the belief that, in investment matters, the State should play only a supplementary role, chiefly for the purpose of providing infrastructure. Consequently, the main responsibility for regional investment has been left to the private sector. Under the new scenario, private investment strategies have tended to increase the territorial mobility of capital. Two processes have contributed to such increased mobility: the modernization and expansion of financial systems, which have facilitated and stimulated the circulation of capital in monetary form within national territories in accordance with sectoral or territorial differences in profitability; and the concomitant modernization and territorial diffusion of new information technologies, which have made possible the real-time management of the different components of the financial network, irrespective of the national or international location of its key nodes.

The combined effects of these two processes have enabled major enterprises and economic groups to expand their operations within the national territory by relocating certain production processes or subprocesses, whether carried out by the enterprise itself or by subcontractors. In addition, stimulated by the new export-led orientation, productive systems have tended to adopt more flexible types of organization and administration, characterized by greater vertical segmention which has often been associated with a growing spread of productive enterprises over the territory.

These changes have been accompanied by the diffusion of modes of entrepreneurial organization and management that combine centralization and decentralization. Thus, while the determination and management of a global strategy for the utilization of capital that includes planning and control are carried out centrally by major enterprises, to the extent that such a strategy involves relocating investment and production processes, it is accompanied by the decentralization of management and decision-making to the relocated subsidiaries or plant divisions.

These new forms of organization and management are being adopted by the developing countries that have advanced the most in their restructuring. Thus, just as during the preceding period some aspects of the newly introduced fordist model coexisted with earlier economic strategies (Gatto, 1990), so today some post-fordist practices are coexisting with those that developed during the Keynesian period.

With the intensification of the territorial mobility of capital, there has been progressive sectoral and territorial diversification of a growing number of major enterprises, which has transformed them into vast conglomerates of a multisectoral and multiregional nature. In developing countries, these conglomerates, which consist mainly of the most profitable activities in increasingly dynamic export-led productive processes, have become the core of the processes of accumulation and growth, at both national and local levels. Such changes, plus those brought about by the growing importance of the vertical segmentation of large enterprises and of the small and medium-sized enterprises that are linked to export-led productive processes, are exerting the greatest impact on the rest of the national productive system.

However, entrepreneurial conglomerates, being profit-oriented, give scant regard to considerations related to the development of the national, regional or local communities in which they operate. As soon as those communities begin to be too small to ensure the continued growth of the conglomerates concerned, the latter relocate. Consequently, the flexibility gained through the sectoral and territorial diversification of conglomerates has meant that financial capital is no longer bound by any kind of border, be it national, regional or provincial. Capital has thus been losing its roots, be they sectoral, regional or local (de Mattos, 1990). Such behaviour is found even in the most dynamic regional or local enterprises, whose growth strategies compel them to diversify their investments sectorally and territorially in order to ensure their continuing expansion. In Mexico, for instance, entrepreneurial groups traditionally based in the northern city of Monterrey have expanded their investments to other parts of Mexico and even to the United States. In Chile, not only are major conglomerates diversifying investment within Chilean territory, but they have also started investing in Argentina, Brazil and Peru. These developments have led to a weakening of the traditional type of large-scale fordist enterprise, which had predominated during the Keynesian period and been closely associated with the expansion of certain urban centres.

Although the changes that have taken place do not imply that entrepreneurs are no longer potential agents of development for specific locations, it must be recognized that most of the regional or local accumulation and growth depend on large conglomerates which lack territorial roots. Given that the economic system gravitates around the activities of such conglomerates, their strategies and decisions regarding vertical quasi-integration[12] or the horizontal transmission of externalities lead to processes of accumulation and growth that affect greatly the territorial distribution of all economic activities. Thus, although autonomous small and medium-sized enterprises continue to play a fundamental role in the regional or local economic system, only under exceptional circumstances do they become major promoters of regional or local growth. Indeed, small and medium-sized enterprises are themselves dependent on the activities of large corporations or conglomerates.

Territorial distribution in an open market

As developing economies become more open, the increasing importance of the external market and the profitability of export activities have tended to overshadow the profitability of activities oriented towards satisfying the internal market. Exports have therefore become the major force underlying economic dynamics, both in sectoral and in territorial terms. The empirical evidence available suggests that export-oriented production has already begun to change the territorial pattern of accumulation and growth prevalent during the earlier period (Castells, 1986). Thus, the activation of export bases in locations that were not exploited before has unleashed new processes of accumulation and growth within them and numerous medium-sized cities linked to areas that produce for export have experienced rapid economic and population growth, indicating a greater territorial expansion of production. There is also evidence of reductions in the rate of growth of primate cities which nevertheless continue to maintain their territorial pre-eminence, and of stagnation in urban centres dependent on industries producing mostly for the internal market.

During the Keynesian period, when production was oriented mostly towards the internal market, both public and private entrepreneurial strategies tended to locate the most dynamic productive activities in the proximity of the main markets of each country, thus stimulating the concentration of both production and population in a small number of large urban agglomerations. This process is beginning to change as national economies open up and export activities are established in locations offering comparative advantages in terms of access to foreign markets. The increasing sectoral and territorial diversification of major conglomerates has been oriented towards the promotion of export-oriented activities considered to be more profitable. Pozas (1991) documents such strategies in the case of the entrepreneurial conglomerates of Monterrey, Mexico, as the Mexican economy began to open up during the 1980s. The flexibility inherent in the new forms of management and the mobility of capital have made possible and promoted the territorial spread of the different plants or divisions of a conglomerate, which have been better able to exploit the comparative advantages offered by different locations in terms of access to natural resources, cheap labour or external markets. The modernization and territorial expansion of new information technologies, by making possible the real-time management at various locations of the production processes carried out by a single conglomerate, have facilitated the changes taking place. Consequently, as Lipietz and Leborgne (1987) suggest, a neo-taylorist development model has taken over, characterized both by the concentration of financial activities and high-level services to enterprises in the core of large cities and by the territorial dispersion of specialized plants located in areas where wages are low.

In countries such as Chile, which are still predominantly exporters of natural resources and whose products have relatively low value added, new investment has tended to be located in the proximity of natural resources or in places where products can be exported (de Mattos, 1991). In countries where manufactured goods constitute a significant component of exports, the location of production plants has been determined by the existence of entrepreneurial skills, a favourable labour market and adequate availability of cheap labour, access to communication facilities, and adequate transportation services for products to markets, as is the case in some Mexican States bordering the United States (Quintanilla, 1991). In addition, the exploitation of new exports in areas that had little productive activity has promoted the growth of certain medium-sized cities. In Ecuador, for instance, the growth of medium-sized cities during the past 25 years has been shaped largely by the impact of national export cycles, rather than by government policies (Lowder, 1991). The transformations noted are therefore the consequence of ongoing processes of economic transformation that are influenced significantly by the new development paradigm.

However, some settlements that were growing in the past are experiencing stagnation, often of an irreversible nature, largely because their growth was based on industries, such as the steel and coal industries, that have been suffering from declining world demand or were highly protected under the import-substitution regime. Indeed, integration into the world market often implies the demise of economic activities that fail to reach the needed degree of competitiveness.

In any case, none of the processes outlined above has the necessary strength to change the pre-eminence of the major urban agglomerations within each country. Indeed, central subsystems have become consolidated and their growth, though slowing down, is unlikely to become negative, mainly because many activities that cannot be easily relocated are concentrated in primate cities, including highly specialized financial services and the top-level management of the main conglomerates. In the case of the Republic of Korea, for instance, the concentration of private decision-making functions in Seoul is associated with the emergence of major conglomerates during the 1970s. In 1981, 46 of the 50 largest business enterprises in the Republic of Korea had their central headquarters in Seoul (Kim, 1988, p. 69). Moreover, major agglomerations are usually the headquarters of national political power and of most of the public administration apparatus, a fact that contributes to the persistence of their attraction.

Although in recent years the population growth rates of certain urban agglomerations have been slowing down and there has been a reduction in the indices of urban primacy (Portes, 1989; Lattes, 1990), these trends do not necessarily signal a reversal of urban concentration or polarization. Thus, Campolina Diniz (1991) argues that the deconcentration allegedly taking place in Brazil can best be characterized as the expansion of the zone of agglomeration, since most of the new economic activities have been captured by a limited number of new growth poles located within the state of São Paulo itself or near it. The physical expansion of the industrial area around central subsystems is occurring in other countries through a process by which the area of influence of an urban agglomeration incorporates ever-growing areas or urban centres located close to it. Thus, enterprises located in those adjacent urban centres can escape the location-based disadvantages of the large urban agglomeration without renouncing its advantages (Azzoni, 1989).

It must be added that in those developing countries in which restructuring has brought about a certain degree of economic recovery and where there has been a reactivation of the internal market with the concomitant increase in demand for food, clothing, textiles, construction materials and other goods produced nationally, industrial plants have found it advantageous to stay near their main markets, that is, within the main urban agglomerations and in the medium-sized cities adjacent to them.

The growth of the medium-sized cities located near urban agglomerations has also been promoted by the increasing number of persons migrating from the core agglomeration proper to cities nearby in order the improve their quality of life. Indeed, the continuing growth of the urban core of large agglomerations under a free market regime where controls are minimal has had various negative consequences, including an increase in environmental deterioration and in insecurity and delinquency. In fact, social inequity, in which affluence coexists with the growing poverty of large sectors of the resident population, has become even more marked in the major urban agglomerations of many developing countries (de Oliveira and Roberts, 1989). Furthermore, the recent years of economic stringency have witnessed a marked polarization of the labour markets of developing countries. There has been a reduction of industrial urban employment as a consequence of the contraction of the formal sector, a large increase of open urban unemployment and informal industrial employment, a sharp decline in wages, a growing participation of women in the labour force and a steady increase in the concentration of the income distribution (Portes, 1989).

The various transformations affecting the population distribution pattern indicate that, despite a possible greater territorial expansion of production, a tendency towards uneven territorial growth remains and a new form of polarization is arising. As long as the new development model remains in use, those areas or regions within a country that lack the competitive advantages to enter the world market seem doomed to remain the weak links in the pattern of territorial accumulation and will continue to be bypassed by the process of development.

C. Dangers of post-fordist mythology

Given that the current development model propounds a subsidiary role for the State and the free play of market forces, it seems unlikely that, as long as it remains in force, effective population distribution policies will be formulated or implemented. However, one may hope that the dynamics of export-led growth combined with the new forms of entrepreneurial organization may continue to promote the territorial expansion of production. Indeed, since in the more industrialized developing countries there is a growing diffusion of flexible forms of production based on the use of new technologies in which the processes of vertical segmentation and quasi-integration provide small and medium-sized enterprises with favourable conditions for their multiplication and growth, this diffusion can be the basis for promoting changes in the territorial distribution of employment and population by strengthening medium-sized cities (Hansen, 1990).

The evolution of post-fordist processes of flexible accumulation seems to be generating greater possibilities than those offered by the earlier model for the development of new production processes in peripheral regions endowed with the conditions for producing goods for export competitively. In such circumstances, the policy instruments available might be applied to stimulating the development of processes of territorially integrated vertical quasi-integration (Lipietz and Leborgne, 1987), based on the use of subcontracting and the strengthening of small and medium-sized enterprises, particularly those located in medium-sized cities. Such efforts may be buttressed by strengthening regional and local Government as part of the decentralization policies being instituted as a basic component of State reform. Local authorities may then be better able to create the conditions necessary for attracting capital and population, so that their medium-sized cities may become the recipients of new productive activities and especially of "footloose" industries.

However, it seems important to consider cautiously and critically any normative model based on the success of the experiences of a few regions or localities in developed countries (since such models have become fashionable in recent years), for it is precisely the propensity to work uncritically from imported models, transforming them into the foundation of norms for action, that in the past has been at the root of the failure of population distribution policies and programmes in developing countries. Unfortunately, this tendency continues, as developing countries attempt to produce the scientific and technological parks that have been proliferating all over the world (Benko, 1991), the supposed success of Italy's Marshal industrial districts, or the political and administrative decentralization that is supposed to promote local development "from below" (Stöhr, 1981) and in which the multiplication of small and medium-sized enterprises would be the main propelling force.

On the basis of such experiences, the literature promoting with enthusiasm the "new orthodoxy" of local endogenous development has been increasing (Piore and Sabel, 1984; Scott and Storper, 1987; Vázquez Barquero, 1988; Pecqueur, 1989). However, the results of various assessments conducted recently on the experiences constituting the basis for the new orthodoxy do not seem to justify such enthusiasm. For instance, the current crisis of the "Third Italy" has exposed some of the less idyllic aspects of the accumulation process within the industrial districts of the country which remained hidden during the period of prosperity, including the existence of a labour market based on a sexist division of labour and the informal recruitment of migrants. Similarly, an evaluation of the 20 most important French technological poles, which include Sophia-Antipolis and Meylan ZIRST (Grenoble), has confirmed that they have had little economic impact on surrounding areas except in terms of the financial flows associated with the wages of their workers (Bruhat, 1990). Moreover, the analysis of wage structures in other technological centres, such as Silicon Valley, has revealed an extremely polarized distribution where a large proportion of the workers are women or members of ethnic minorities who are paid rather low wages.

Furthermore, in many cases, the experiences considered are far from attesting to the attainment of any long-awaited paradise for the expansion of small and medium-sized enterprises. Thus, the productive structure of most technological poles generally consists of a basic core comprising mostly the subsidiaries of multinational enterprises which are essential for the survival of such poles and have major influence in their success. Indeed, in a recent interview, René Roy, founding director of Meylan ZIRST (Grenoble), one of the paradigmatic French technological poles, confessed that industrialization had been carried out mostly by large conglomerates and that no small and medium-sized industries had yet been spawned by the research at *Zone pour l'innovation et les réalisations scientifiques et techniques* (ZIRST), (*L'Express*, (Paris), No. 2153, 9 October 1992). Consequently, the synergistic effects of cross-fertilization and spin-off, which have constituted the theoretical basis for the technological pole paradigm, do not seem to have manifested themselves in most of the experiences evaluated.

In addition, although there has been a great deal of talk about the footloose nature of high-technology industries, it should be borne in mind that because of the diversity of conditions prevailing in peripheral urban centres, the activities of those industries are most likely to find all the elements required for their development only in the major urban agglomerations. Consequently, common conjectures about the extraordinary possibilities for territorial diffusion of high-technology industries are little more than wishful thinking. Even in the case of the Republic of Korea, where the level of industrialization is higher than in other developing countries, Kim (1988) considers that high-technology industries and research and development activities are more likely to be established close to Seoul than in other locations since they require easy access to technical information, the availability of skilled workers and technicians, the services of highly developed industries and a good residential environment.

There seems, therefore, to be some justification for the doubts expressed by some of the critics of the new course of action. As Hadjimichalis and Papamichos (1990) argue, the approximations to local development based on the rationalization of cases studied evince, both in theory and in practice, the same limitations as two decades ago. In fact, the new development model is similar to the old one in that it implies a process of unequal capitalist development and the relations between local society and social change are envisioned in the same manner, emphasizing facts that are devoid of content.

Furthermore, the sudden enthusiasm for things small and local arises just as a protracted process of mergers and acquisitions evidences a clear and powerful trend towards conglomeration and globalization of the productive apparatus. As Amin and Robins (1991) suggest, the debate over entrepreneurial artisans and the segmentation of enterprises should not hide the growing power and influence of global industrial and financial capital.

From this standpoint, as we re-examine the foundations and proposals of the new orthodoxy, there is the lingering impression that we are being tricked, since they generally omit or include only in a superficial manner an analysis of the implications of the fact that, whether we like it or not, we are living in societies in which the structural dependence of capital is becoming persistently and progressively accentuated (Przeworski and Wallerstein, 1986). Consequently, the analysis is leaving out precisely the territorial logic of capital and the strategies or policies for its private utilization. That is to say, in discussing new paths for the policies of accumulation, growth and industrialization in developing countries, there is a need to consider explicitly the structural conditions imposed by the new technical and economic model that has been taking shape during the past decade. It is necessary to recognize, as Storper (1989) suggests, that development strategies in the context of post-fordism must necessarily be different from those that were used to implement the fordist methods of production, and to identify both the restrictions and possibilities imposed by the new context, so that the policies required to confront population distribution problems can be formulated on a realistic basis and truly contribute to improving the quality of life in developing countries.

NOTES

[1] Studying territorial concentration in the case of Mexico, Hernández Laos (1980) suggests that the industrialization strategy adopted by the Mexican Government, together with the policy of creating infrastructure and providing subsidies in major urban centres, generated a process that tended to consolidate and make self-sustaining the regional concentration of manufacturing. A point worth emphasizing is that the process becomes cumulative to the extent that it generates mechanisms that make it self-perpetuating as, for instance, through the expansion of favourable circumstances for the growth of industrial centres resulting from a substantive reduction in unit costs associated with greater internal and mostly external economies.

[2] Although having different magnitudes, the processes of suburbanization and tentacular expansion around major cities can be observed both in countries with large territories, such as Argentina, Brazil and Mexico, and in those of intermediate size, such as Chile and Venezuela. In fact, even in small countries, such as Costa Rica and Uruguay, similar patterns of population distribution are discernible. Studies conducted in the cases of São Paulo, Brazil (Azzoni, 1982, 1986 and 1989) and Mexico City (Unikel, 1976) illustrate the form of central subsystems. Such patterns of population distribution have shown their persistence even among countries having advanced levels of industrialization, such as the Republic of Korea where, notwithstanding the policies of population redistribution that are

being promoted by the Government, the process of suburbanization of Seoul has continued (Kim, 1988; Choi, 1990).

[3]This feedback mechanism had already been emphasized by Perroux in his classic study of growth poles. Perroux pointed out the importance of the effects of the economic intensification that territorial agglomeration stimulated in a complex of industries, promoting their rapid growth and at the same time favouring the processes of industrial oligopolization (Perroux, 1955).

[4]Cano studied this phenomenon in the case of São Paulo, Brazil, pointing out that since the 1930s the accumulation of capital controlled by São Paulo residents had contributed to the integration of the national market and had thereby made that market complementary to the needs imposed by the process of capital accumulation in São Paulo itself (Cano, 1985).

[5]In regard to migration, one may recall the observation by Martine (1979) that channelling efforts to study migration in order to formulate policies is the product of a value judgement that something is wrong with the direction, intensity and characteristics of the migratory flows themselves. He further asserts that, in the social sciences, experts do not believe in the possibility of objective research, since any hypothesis or thesis is based implicitly on a series of values and prejudices.

[6]According to Mera (1973), available empirical analysis shows that large cities are more productive and that larger cities are likely to be more productive in relation to smaller cities particularly in the less developed countries. Consequently, countries should not embark on a policy of decentralization of investment and redistribution of the population, especially if the goal is to maximize the rate of economic growth.

[7]Despite some differences in emphasis, most of the assessments of the territorial policies applied during this period agree that the results obtained were modest. See, for instance, Stöhr and Todtling (1978) on mixed economies and market economies in general; Helmsing and Uribe-Echeverría (1981) and de Mattos (1989) on Latin America; Jatobá and others (1980) and Cano (1985) on Brazil; Unikel (1975) and Lavell, Pirez and Unikel (1978) on Mexico; and Kim (1988) and Choi (1990) for the Republic of Korea.

[8]An assessment of population distribution policies in the Republic of Korea shows that one of their effects was to promote a process of suburbanization involving the rapid growth of six satellite cities in the vicinity of Seoul during the 1970s and early 1980s (Kim, 1988). In the case of Brazil, Azzoni (1986) shows that the cities that experienced greater industrial growth were those located in regions already having large urban agglomerations and thus having the greatest potential for concentration. The same phenomenon has been observed around most of the major urban agglomerations in Latin America.

[9]The development policies pursued in north-eastern Brazil provide a classic example because, although they helped to unleash major industrialization processes in the regional capitals (Fortaleza, Recife, Salvador), they maintained and may even have aggravated the level of poverty in other parts of the region (Baer and Geiger, 1978).

[10]In the case of Brazil, Jatobá and others (1980) show that the modernization of rural areas, the rising use of agricultural technology, the promotion of livestock raising, and the ensuing proletarianization of the rural labour force have contributed to increase migration from rural areas and expand monopolistic land ownership. Similar experiences have been documented in other countries.

[11]A similar assessment has been made regarding other population policies. Thus, Miró (1992) asserts that the major demographic changes that have taken place in Latin America and the Caribbean during the past 30 year are mostly the result of economic, social and political transformations rather than of the isolated population policies or programmes adopted during the period, including family planning.

[12]According to Lipietz and Leborgne (1987), vertical quasi-integration occurs when an enterprise benefits simultaneously from the advantages of integration and from those of vertical segmentation through a process of redefinition of economic relations based on subcontracting.

References

Amin, Ash, and Kevin Robins (1991). Distritos industriales y desarrollo regional: límites y posibilidades. *Sociología del Trabajo* (Madrid), special issue.

Azzoni, Carlos Roberto (1982). *Teoria da Localização: Una Análise Crítica. A Experiência de Empresas Instaladas no Estado de São Paulo.* São Paulo: Instituto de Pesquisas Econômicas.

_________ (1986). A lógica da dispersão da indústria no Estado de São Paulo. *Estudos Econômicos* (São Paulo), vol. 16 (special issue), pp. 45-67.

_________ (1989). La nueva dirección de la industria en San Pablo: ¿reversión de la polarización en Brasil? In *Revolución Tecnológica y Reestructuración Productiva: Impactos y Desafíos Territoriales,* Francisco Alburquerque and others, eds. Buenos Aires: Grupo Editor Latinoamericano.

Baer, Werner, and Pedro Pinchas Geiger (1978). Industrialização, urbanização e a persistência das desigualdades regionais no Brasil. In *Dimensões do Desenvolvimento Brasileiro,* Werner Baer and others, eds. Rio de Janeiro: Campus.

Barkin, David (1972). ¿Quiénes son los beneficiarios del desarrollo regional? In *Los Beneficiarios del Desarrollo Regional.* Mexico City: SepSetentas.

Benko, Georges (1991). *Géographie des Technopôles.* Paris: Masson.

Berry, Bryan (1961). City size distribution and economic development. *Economic Development and Cultural Change* (Chicago), vol. 9 (July), pp. 573-587.

Bruhat, Thierry (1990). *Vingt technopôles, un premier bilan.* Paris: La Documentation Française.

Campolina Diniz, Clelio (1991). Desenvolvimento poligonal no Brasil. Nem desconcentração nem contínua polarização. Paper presented at the Primer Encuentro Iberoamericano de Estudios Regionales, Santa Cruz de la Sierra, Bolivia, 28 October-1 November.

Cano, Wilson (1985). *Desequilibrios Regionais e Concentração Industrial no Brasil: 1930-1970.* São Paulo: Global Editora.

Castells, Manuel (1986). Mudança tecnológica, reestruturação econômica e a nova divisão espacial do trabalho. *Espaço & Debates* (São Paulo), vol. 6, No. 17, pp. 5-23.

Choi, Jin-Ho (1990). Patterns of urbanization and population distribution policies in the Republic of Korea. *Regional Development Dialogue* (Nagoya, Japan), vol. 11, No. 1 (spring), pp. 130-151.

Comisión Económica para América Latina y el Caribe (CEPAL) (1981). *La Industrialización de América Latina y la Cooperación Internacional.* Santiago, Chile.

del Río Gómez, Clemente, and Juan Ramón Cuadrado Roura (1991). El papel de los servicios a la producción en la nueva política regional. Paper presented at the Primer Encuentro Iberoamericano de Estudios Regionales, Santa Cruz de la Sierra, Bolivia, 28 October-1 November.

de Mattos, Carlos A. (1989). Mito y realidad de la planificación regional y urbana en los países capitalistas latinoamericanos. In *La Investiga-ción Urbana en América Latina: Viejos y Nuevos Temas,* M. Unda, ed. Quito, Ecuador: CIUDAD.

_________ (1990). Reestructuración social, grupos económicos y desterritorialización del capital. El caso de los países del Cono Sur. In *Revolución Tecnológica y Reestructuración Productiva: Impactos y Desafíos Territoriales,* Francisco Alburquerque and others, eds. Buenos Aires: Grupo Editor Latinoamericano.

_________ (1991). Modernización neocapitalista y reestructuración productiva y territorial en Chile, 1973-1990. *Estudios Territoriales* (Madrid), No. 37 (September-December), pp. 121-138.

de Oliveira, Orlandina, and Bryan Roberts (1989). Los antecedentes de la crisis urbana: urbanización y transformación ocupacional en América Latina, 1940-1980. In *Las Ciudades en Conflicto: Una Perspectiva Latinoamericana,* Mario Lombardi and Danilo Veiga, eds. Montevideo: Ediciones de la Banda Oriental.

Fajnzylber, Fernando (1983). *La Industrialización Trunca de América Latina*. Mexico City: Editorial Nueva Imagen.

Garza, Gustavo (1980). *Industrialización de las Principales Ciudades de México*. Mexico City: El Colegio de México.

Gatto, Francisco (1990). Cambio tecnológico neofordista y reorganización productiva. Primeras reflexiones sobre sus implicaciones territoriales. In *Revolución Tecnológica y Reestructuración Productiva: Impactos y Desafíos Territoriales*, Francisco Alburquerque and others, eds. Buenos Aires: Grupo Editor Latinoamericano.

Hadjimichalis, Costis, and Nicos Papamichos (1980). Desarrollo local en el sur de Europa: hacia una nueva mitología. *Estudios Regionales* (Málaga, Spain), No. 26, pp. 113-144.

Hansen, Niles (1990). Impacts of small and intermediate-sized cities on population distribution: issues and responses. *Regional Development Dialogue* (Nagoya, Japan), vol. 11, No. 1 (spring), pp. 60-76.

Helmsing, Bert, and Francisco Uribe-Echeverría (1981). La planificación regional en América Latina: ¿teoría o práctica? In *Experiencias de Planificación Regional en América Latina*, Sergio Boisier and others, eds. Santiago, Chile: CEPAL/Instituto Lationamericano y del Caribe de Planificación Económica y Social (ILPES)/Sociedad Interamericana de Planificación (SIAP).

Hernández Laos, Enrique (1980). Economías externas y el proceso de concentración regional de la industria en México. *El Trimestre Económico* (Mexico City). vol. 47, No. 185 (January-March), pp. 119-157.

Jatobá, Jorge, and others (1980). Expansão capitalista: o papel do Estado e o desenvolvimento regional recente. *Pesquisa e Planejamento Econômico* (Rio de Janeiro), vol. 10, No. 2 (April), pp. 273-318.

Kim, Won Bae (1988). Population redistribution policy in Korea: a review. *Population Research and Policy Review* (Dordrecht, Netherlands), No. 7, pp. 49-77.

Lattes, Alfredo E. (1989). La urbanización y el crecimiento urbano en América Latina, desde una perspectiva demográfica. In *La Investigación Urbana en América Latina. Caminos Recorridos y por Recorrer*, J. L. Coraggio, ed. Quito, Ecuador: CIUDAD.

Lavell, Allan, Pedro Pirez and Luis Unikel (1978). *La Planificación del Desarrollo y Redistribución Espacial de la Población: El Caso de México, 1940-78*. Santiago, Chile: Centro Latinoamericano de Demografía (CELADE).

Lipietz, Alain (1986). *Mirages et miracles. Problèmes de l'industrialisation dans le tiers monde*. Paris: Editions La Découverte.

________, and Danièle Leborgne (1987). *L'après-fordisme et son espace*. Paris: Couverture orange (CEPREMAP), No. 8807.

Lowder, Stella (1991). El papel de las ciudades intermedias en el desarrollo regional: una comparación de cuatro ciudades de Ecuador. *Revista Interamericana de Planificación* (Guatemala), vol. 24, No. 93 (January-March), pp. 45-60.

Martine, George (1979). Migraciones internas, ¿investigación para qué? *Notas de Población. Revista Latinoamericana de Demografía* (Santiago de Chile), vol. 7, No. 19 (April), pp. 2-38.

________ (1987). Exodo rural, concentração urbana e fronteira agrícola. In *Os Impactos Sociais da Modernização Agrícola*, George Martine and Ronaldo Coutinho Garcia, eds. São Paulo: Editora Caetes.

Mera, Koichi (1973). On the urban agglomeration and economic efficiency. *Economic Development and Cultural Change* (Chicago), vol. 21, pp. 309-324.

Miró, Carmen (1992). *Políticas de Población: Reflexiones sobre el Pasado y Perspectivas Futuras*. Santiago, Chile: CELADE.

Myrdal, Gunnar (1957). *Economic Theory and Underdeveloped Regions*. London: Gerald Duckworth and Company.

Pecqueur, Bernard (1989). *Le développement local*. Paris: Syros Alternatives.

Perroux, François (1955). Note sur la notion de pôle de croissance. *Economie appliquée* (Paris), Nos. 1-2, pp. 307-320.

Piore, Michael, and Charles Sabel (1984). *The Second Industrial Divide*. New York: Basic Books.

Portes, Alejandro (1989). La urbanización de América Latina en los años de crisis. In *Las Ciudades en Conflicto: Una Perspectiva Latinoamericana*, Mario Lombardi and Danilo Veiga, eds. Montevideo: Ediciones de la Banda Oriental.

Pozas, María de los Angeles (1991). Estrategias empresariales ante la apertura externa. *Ciudades* (Mexico City), vol. 3, No. 9 (January-March), pp. 26-38.

Przeworski, Adam, and Michael Wallerstein (1986). Soberanía popular, autonomía estatal y propiedad privada. In *Crisis y Regulación Estatal: Dilemas de Política en América Latina y Europa*. Buenos Aires: Grupo Editor Latinoamericano.

Quintanilla, Ernesto R. (1991). Tendencias recientes de la localización en la industria maquiladora. *Comercio Exterior* (Mexico City), vol. 41, No. 9 (September).

Richardson, Harry W. (1984). Population distribution policies. In *Population Distribution, Migration and Development: Proceedings of the Expert Group on Population Distribution, Migration and Development, Hammamet (Tunisia), 21-25 March 1983*. Sales No. E.83.XIII.3. New York: United Nations.

________ (1987). The costs of urbanization: a four-country comparison. *Economic Development and Cultural Change* (Chicago), vol. 35, No. 3 (April), pp. 561-580.

Scott, Alan, and Michael Storper (1987). High technology industry and regional development: a theoretical critique and reconstruction. *International Social Science Journal*, No. 112, pp. 215-232.

Singer, Paul (1973). *Economia Política da Urbanização*. São Paulo: Brasiliense.

Stöhr, Walter (1981). Development from below: the bottom-up and periphery-inward development paradigm. In *Development from Above or Below? The Dialectics of Regional Planning in Developing Countries*, Walter Stöhr and D. R. F. Taylor, eds. London: J. Wiley.

________, and Franz Todtling (1978). Una evaluación de las políticas regionales. Experiencias en economías de mercado y en economías mixtas. *Revista Interamericana de Planificación* (Mexico City), vol. 12, No. 45, pp. 5-31.

Storper, Michael (1989). La industrialización y el desarrollo regional en el Tercer Mundo, con especial referencia al caso de Brasil. *Estudios Demográficos y Urbanos* (Mexico City), vol. 4, No. 2 (May-August), pp. 313-342.

Topalov, Christian (1979). *La Urbanización Dependiente. Algunos Elementos para su Análisis*. Mexico City: Editorial Edicol.

Unikel, Luis (1975). Políticas de desarrollo regional en México. *Demografía y Economía* (Mexico City), vol. 11, No. 2, pp. 143-181.

________ (1976). *El Desarrollo Urbano de México. Diagnóstico e Implicancias Futuras*. Mexico City: El Colegio de México.

Vázquez Barquero, Antonio (1988). *Desarrollo Local. Una Estrategia de Creación de Empleo*. Madrid: Ediciones Pirámide.

Williamson, J. G. (1965). Regional inequality and the process of national development: a description of the patterns. *Economic Development and Cultural Change* (Chicago), vol. 13, pp. 3-45.

Part Four

INTERNATIONAL MIGRATION TRENDS AND PROSPECTS

XI. SOUTH-TO-NORTH MIGRATION

*Philip Muus**

The 1960s mark an important turning point in international migration. For the first time in this century, international migration from developing countries (the South) towards developed countries (the North) was allowed to increase and the result has been that a growing percentage of the migration flows directed to developed countries is now accounted for by migrants from the developing world (Zlotnik, 1991). Until 1980, the changes observed had largely been the result of policies adopted by the main receiving countries in the developed world. More recently, however, migration from developing countries to certain developed countries has occurred despite the anti-migration stance of the receiving countries. Consequently, particularly in Europe, there is considerable concern about the potential for yet larger inflows of migrants from developing countries, and a search for policies and other control measures to stop migration is well under way. The issue, therefore, is whether migration from the South will continue to increase despite the increasingly restrictive admission policies prevalent in most of the North.

To assess the likelihood of any future development of migration trends, a better understanding of the causes of international migration would be needed. Yet, the theoretical basis in the field of international migration is still fairly weak. Two promising approaches have been proposed in recent years. The first, based on the concept of international migration systems, posits that international migration is yet another of the processes linking countries with one another and, therefore, that it does not arise in isolation from other processes creating ties between countries (Kritz, Lim and Zlotnik, 1992). Although the systems approach is a useful tool for analysing migration flows systematically at the macrolevel, it does not answer the question of why people migrate. The second approach is better oriented towards explaining how migration flows expand and are maintained through the operation of social networks (Boyd, 1989; Fawcett, 1989). It has the potential of linking the macrolevel and the microlevel analyses of migration, but it fails in respect of determining how migration flows get started in the first place.

The fact that international borders act as barriers to the movement of people distinguishes international migration from that occurring within countries (Zolberg, 1989), as does the fact that international migration policies generally have an effect on migration flows, though it may not necessarily be the intended one. Thus, if international migration policies are unrealistic and make no allowance for the kind of migration fostered by economic processes, migration will occur at the margin of policy. That is to say, international migrants may enter a country by the main gate or through the back door, depending on the opportunities that policy opens up to them (Zolberg, 1992).

Existing theories of international migration tend to be partial in that they focus only on certain types of migration. The first general theory of territorial mobility was perhaps that proposed by Zelinsky (1971) who combined the theory of demographic transitions with the so-called laws of migration. Zelinsky noted that there had been definite, patterned regularities in the growth of territorial mobility during recent history and that those regularities were related to the modernization process. Consequently, he argued that in the early stages of the vital transition (when mortality began to decline), a country would experience a rapid rate of urbanization and major outflows of emigrants to available and attractive foreign destinations. At later stages of the vital transition, when fertility reduction became rapid, the rate of urbanization would slow down and emigration would decline or even stop. Lastly, when both fertility and mortality reached low levels, migration would occur largely between urban agglomerations or would originate in developing countries which would be the source of unskilled or semiskilled workers.

*Centre for Migration Studies, Department of Human Geography, University of Amsterdam, Netherlands.

Amersfoort, Muus and Penninx (1984) have proposed a scheme of analysis at the macrolevel that encompasses the major factors at play in international migration and gives prominence to the role of international migration policies (figure VI). The basic premise is that differences between the socio-economic conditions in countries of origin and those in countries of destination are a major factor in mobilizing potential migrants. This assumption is valid for all types of migrants. Admission policies establish the framework for the legal migration of persons for specific periods. Migration, however, is not just the result of a schematic macroeconomic rule adjusted by admission policy. When a critical number of migrants from a certain country settle in the country of destination, migration networks may develop and add inertia to the flow of people. Because of network dynamics, migration may continue even when the circumstances originally leading to it have drastically changed (Muus and Penninx, 1991).

The factors included in the analytical scheme are meant to cover a wide array of subfactors. Thus, labour-market conditions will depend on demographic trends determining the size of the labour force and on social trends determining, among other things, the extent of women's participation in the labour force or age at retirement. Respect of human rights depends on the degree of political freedom, freedom of speech or freedom of assembly, the absence of conflict or persecution and so on. Migration policies include a wide array of measures aimed at controlling or shaping migration, including the admission of the descendants of former emigrants or members of a certain ethnic or nationality group.

The characteristics of the migrant population include the legal status of migrants which determines, among other things, the possibility of family reunification. The demographic characteristics of the migrant population are also important for determining the possibility of further migration through family reunification and the potential for the creation of migration networks.

A. Migration to Europe

Before the Second World War, most European countries were countries of emigration. Migration from developing countries to Europe took place within the context of colonialism, as migration from the colonies to the colonizing country. France, for instance, recruited workers from the Maghreb during the First World War. North African soldiers fought on the side of the French during the Second World War. The United Kingdom of Great Britain and Northern Ireland and the Netherlands admitted small numbers of students from their colonies. Since Germany did not possess colonies in the developing world during most of the twentieth century, it tended to look for labour supplies in Central and Eastern Europe. The Mediterranean countries and most Central European countries were generally countries of emigration.

As a consequence of the Second World War, Europe experienced a major refugee crises. During the 1950s and early 1960s many Europeans emigrated to overseas countries but, by the mid-1960s emigration from the Northern and Western European countries had slowed down. Indeed, several countries in those regions had become labour importers as their economies recovered. The first sources of foreign workers were Southern European countries but, as competition for foreign workers grew, recruitment followed colonial ties, as was the case for France and the United Kingdom, or proximity, as was the case for Belgium, the Netherlands, Sweden and Switzerland. Germany re-established its eastern ties with Greece, Turkey and Yugoslavia. In addition, the end of the era of colonization led to important migration flows from the former colonies or territories to the European colonizing countries. Thus, Algerians moved to France, Surinamese to the Netherlands, and persons from India, Pakistan and the West Indies to the United Kingdom.

The rise in oil prices in 1973 caused a major economic crisis in Europe that led to the discontinuation of labour recruitment policies in the labour-importing countries. However, migrant workers already present in those countries were allowed to stay and, under certain conditions, to bring in their close relatives. Owing to family reunification, the labour-importing countries of Europe continued to receive sizeable inflows of migrants during the late 1970s and early 1980s, although they registered relatively low or even negative net migration gains because important return flows also occurred (table 35).

Return flows were particularly strong among migrants from Southern European countries, which began experiencing net gains through migration in the late 1970s for the first time in decades. In addition, as the economies of countries like Greece, Italy and Spain began gaining strength, they in turn became magnets for migrants from developing countries. Most of these migrants entered or

Figure VI. Schematic approach to the analysis of international migration

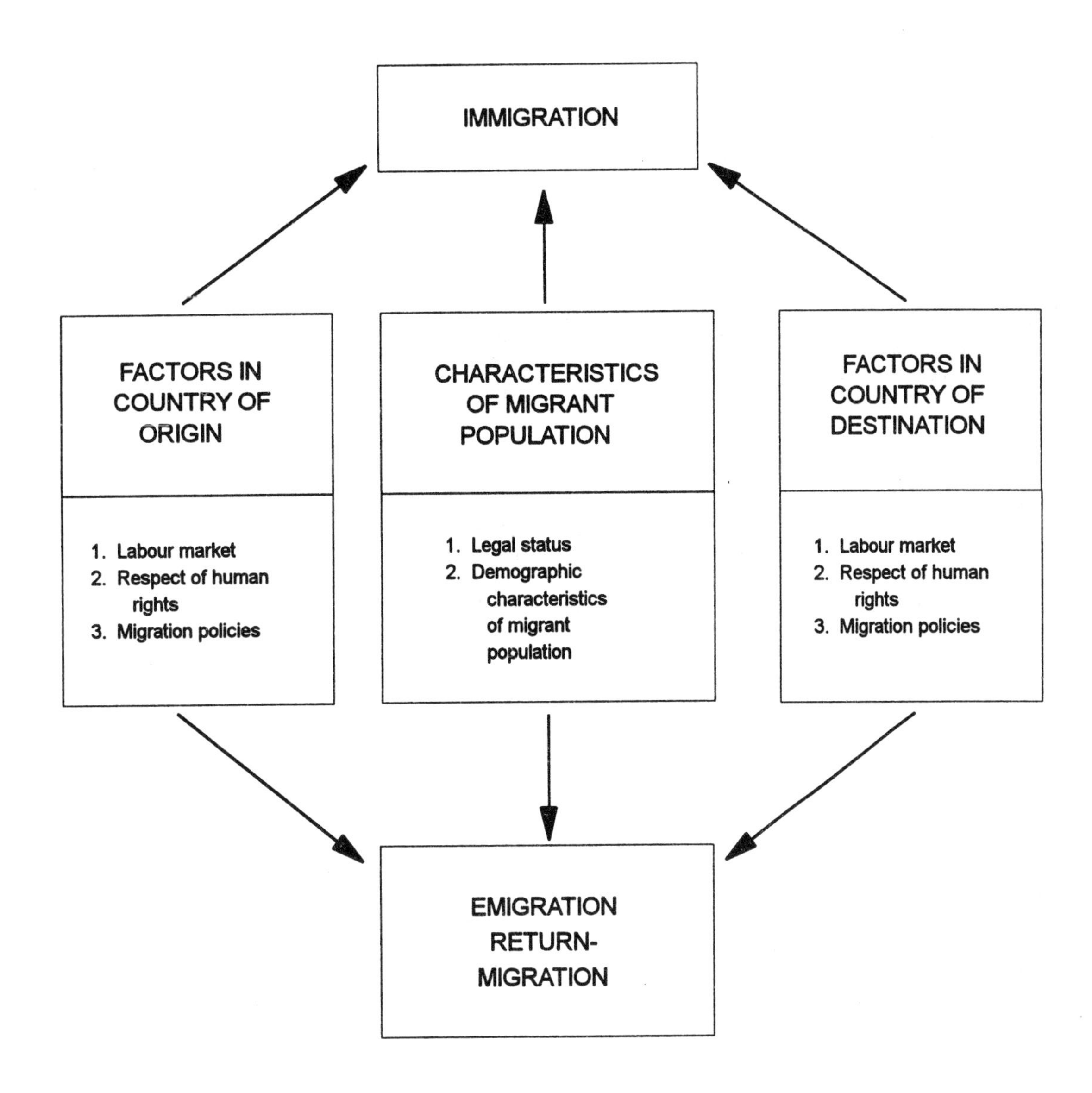

worked illegally in the new receiving countries, largely because those countries lacked explicit policies on the admission of unskilled migrant workers. To gain control over such undocumented migration, Italy and Spain adopted immigration laws and resorted to regularization campaigns. In Spain, 44,000 applications for regularization were filed after the introduction of the Law of July 1985 (United Nations, 1992) and in 1991 a second regularization programme yielded another 133,000 applications (OECD, 1992). In Italy, the status of about 119,000 undocumented migrants was regularized under the Acts of 31 December 1986 and 16 March 1988 (United Nations, 1992) and a third regularization campaign carried out in 1990 led to the legalization of a further 216,000 migrants (OECD, 1991). In both Italy and Spain, the majority of the migrants regularized originated in developing countries.

An important feature of the migration directed to the Western-bloc countries of Europe has been its having

TABLE 35. AVERAGE ANNUAL NUMBER OF IMMIGRANTS ORIGINATING IN DEVELOPING AND DEVELOPED COUNTRIES AND AVERAGE ANNUAL NET MIGRATION, BY REGION OF DESTINATION AND PERIOD, 1960-1991

	1960-1964	*1965-1969*	*1970-1974*	*1975-1979*	*1980-1984*	*1985-1989*	*1990-1991*
				A. *Developing countries*			
Emigrants to							
Northern America	128 650	236 959	372 864	605 159	782 155	704 306	545 972
Oceania	8 035	14 038	27 239	29 064	36 444	57 718	69 645
France	40 869	48 245	79 159	32 474	43 841	22 623	..
Western Europe	73 555	226 028	362 266	334 121	292 996	339 053	..
Total	251 109	525 270	841 527	1 000 817	1 155 436	1 123 700	615 617
Net migration							
Oceania	7 334	12 702	25 783	27 726	34 995	56 560	68 915
Western Europe	53 164	93 944	176 852	97 591	29 073	156 127	..
Maximum net emigration to developed countries	230 017	391 850	654 658	762 950	890 064	939 616	614 887
				B. *Developed countries*			
Emigrants to							
Northern America	243 161	303 965	208 199	153 364	133 090	125 854	134 086
Oceania	138 278	164 177	146 063	55 244	69 248	56 767	51 813
France	133 416	155 490	113 426	37 924	35 048	14 037	..
Western Europe	655 598	874 469	935 176	577 898	549 873	890 435	..
Total	1 170 453	1 498 101	1 402 864	824 429	787 258	1 087 094	185 899
Net migration							
Oceania	115 771	121 124	103 670	24 022	42 016	47 305	46 493
Western Europe	214 575	88 565	146 857	-44 297	-38 830	303 315	..
Maximum net emigration to developed countries	706 923	669 144	572 153	171 013	171 324	490 511	180 579
				C. *All regions*			
Emigrants to							
Northern America	371 811	540 924	581 063	758 523	915 244	830 160	680 058
Oceania	146 313	178 215	173 302	84 307	105 692	114 485	121 458
France	174 285	203 735	192 585	70 398	78 889	36 660	..
Western Europe	729 153	1 100 497	1 297 442	912 018	842 868	1 229 489	..
Total	1 421 562	2 023 370	2 244 392	1 825 247	1 942 694	2 210 794	801 516
Net migration							
Oceania	123 105	133 826	129 453	51 748	77 011	103 865	115 408
Western Europe	267 738	182 509	323 710	53 294	-9 757	459 442	..
Maximum net emigration to developed countries	936 939	1 060 994	1 226 811	933 964	1 061 388	1 430 127	795 466

Source: Hania Zlotnik, "International migration: causes and effects", in *Beyond the Numbers. A Reader on Population, Consumption and the Environment*, Laurie Ann Mazur, ed. (Washington, D.C., Island Press, 1994), table 1, pp. 362-363.

NOTE: The data for Canada and the United States (Northern America) and Australia and New Zealand (Oceania) are classified by place of birth; the data for Belgium, France and the Netherlands are classified by country of citizenship, and those for Western Germany, Sweden and the United Kingdom are classified by place of last or next residence. The total for Belgium, Germany, the Netherlands, Sweden and the United Kingdom is presented under the label Western Europe. Data for the United Kingdom are available only as of 1965.

The data for the United States include information on the number of migrants legalizing their status as a result of the Immigration Reform and Control Act (IRCA) of 1986. Those legalizing their status as a result of the general amnesty programme were redistributed over the period 1972-1982 according to a linearly increasing trend. Those legalizing their status under the Special Agricultural Workers Programme were redistributed uniformly over the period 1982-1986.

tended to originate in only a few countries (Zlotnik, 1992a). As already noted, former colonies were usually the main sources of foreign workers and hence of migrants. Thus, as of 1990, out of the 11.6 million foreigners residing in Belgium, France, Germany, the Netherlands, Sweden and Switzerland, about 5.3 million originated in developing countries and of the 5.3 million, 4 million were citizens of only four countries: Algeria, Morocco, Tunisia and Turkey (United Nations, 1996). That is to say, citizens of those four countries accounted for 76 per cent of the nationals from developing countries present in six of the major migrant receiving countries of continental Europe. Ten years earlier, the corresponding proportion had been 86 per cent. This means that a trend towards the diversification of migrants' origins had taken hold during the 1980s, although its impact was still moderate.

One of the major changes in migration directed to the Western-bloc countries of Europe during the 1980s was the rapid increase in the number of asylum-seekers, a high proportion of whom originated in developing countries. Between 1983 and 1989, of the over 1.3 million applications for asylum that were lodged in European countries, 70 per cent were filed by citizens of developing countries (Zlotnik, 1991). In contrast with previous waves of migrants from the South, whose selection was generally controlled through recruitment programmes, the arrival of asylum-seekers was largely spontaneous and outside the direct control of the receiving countries. Hence, the variety of origins was considerably greater.

With the collapse of communism in Eastern-bloc countries, controls on the migration of citizens were reduced and a growing number of persons found their way to Western countries. Following the practices established during the cold war era, when citizens of Eastern-bloc countries had generally been admitted as refugees by Western-bloc countries, persons leaving Eastern and Central European countries during the 1980s sought asylum in the West. Their numbers grew as the decade progressed and, whereas in 1983 they were responsible for the filing of about 19,000 applications for asylum in Europe, by 1989 they accounted for 128,000. The weight of these persons in relative terms (that is, as a share of all applications) also increased, passing from about one quarter in 1983-1984 to nearly 37 per cent in 1989 (Zlotnik, 1991). By 1992, as the conflict in the former Yugoslavia produced yet larger numbers of persons in need of protection, over half of all asylum applications lodged in Europe were from citizens of former Eastern-bloc countries (UNHCR, 1993).

Another important flow of migrants from Eastern- to Western-bloc countries is that constituted by ethnic minorities with homelands abroad, such as ethnic Germans and Jews. Both groups have been allowed to emigrate from former Eastern-bloc countries, particularly the Union of Soviet Socialist Republics (USSR). Germany and Israel, respectively, have been the main destinations of those migrants. The emigration of ethnic Germans from Eastern-bloc countries to Germany accelerated during the late 1980s. Thus, whereas between 1968 and 1984, a total of 653,000 ethnic Germans (*Aussiedler*) were admitted by the then Federal Republic of Germany, between 1985 and 1992 the number of admissions rose to 1.6 million (Germany, 1993). These numbers exclude the further inflow of Germans from the former German Democratic Republic to the Federal Republic of Germany; before reunification, those Germans were considered international migrants and their numbers are reflected in the statistics provided in table 35.

During the cold war era, Eastern-bloc countries kept their borders closed both to the movement of their own citizens and to the inflow of foreigners. However, some countries imported workers from developing countries with socialist regimes, such as Angola, Cuba and Viet Nam. Statistics on the number of migrants involved have come to light only recently and show both that their numbers were relatively small and that they have been declining rapidly, as migrant workers are repatriated even before their official contracts expire.

B. The traditional countries of immigration

As former British colonies, Australia, Canada and the United States of America were settled mainly by Europeans. Only the United States also received a significant number of African people, brought to the country as slaves. Even during most of the twentieth century, it was Europeans who dominated immigration flows to the countries of permanent immigration. However, towards the mid-1960s and 1970s, with the liberalization of admission policies and the adoption of universalist criteria for the selection of immigrants, the proportion of immigrants originating in developing countries rose rapidly. In the United States, immigrants from developing countries began to outnumber those from developed countries in 1966, in Canada that turning point was reached in 1975, and in Australia it was reached during fiscal year 1984/85.

As table 36 shows, the share of migration from developing regions to the United States was high even during the late 1960s, when European migration still accounted for 83 per cent of the immigrants to Australia and for 68 per cent of those to Canada. Indeed, by the late 1980s, immigrants from developing countries accounted for over 85 per cent of all immigrants to the United States and these figures do not take into account the large number of migrants who regularized their status in the United States during the late 1980s. Changes in the composition of immigration were reflected in the migrant stock. Thus, according to United States census data, the number of persons born in Europe declined from 5.7 million in 1970 to 5.1 million in 1990. As expected, the number of persons born in the main countries or regions of origin in the developing world increased markedly.

TABLE 36. TOTAL NUMBER OF ADMISSIONS OF PERMANENT IMMIGRANTS AND PROPORTION OF IMMIGRANTS ORIGINATING IN DIFFERENT REGIONS: AUSTRALIA, CANADA AND THE UNITED STATES, 1965-1969 TO 1985-1989

Region of origin and receiving country	*1965-1969*	*1970-1974*	*1975-1979*	*1980-1984*	*1985-1989*[a]
			Total number (thousands)		
World					
Australia	781.0	620.0	344.8	468.1	616.2
Canada	909.9	794.3	650.6	570.3	689.5
United States	1 794.7	1 923.4	2 412.6	2 825.0	3 028.4
			(Percentage)		
Africa					
Australia	2.8	3.5	4.7	4.7	5.6
Canada	3.2	4.9	5.7	4.5	6.1
United States	1.2	1.8	2.1	2.6	3.0
Asia					
Australia	6.6	12.4	31.2	32.7	41.4
Canada	11.5	22.5	32.4	44.1	48.1
United States	14.4	29.9	38.1	47.7	44.1
Europe					
Australia	83.3	69.8	42.3	44.6	30.7
Canada	67.5	41.7	33.7	29.9	24.0
United States	35.1	24.8	15.3	11.7	10.7
The Americas					
Australia	3.1	9.1	7.1	5.0	5.9
Canada	15.5	28.8	26.4	20.2	21.0
United States	48.7	42.8	43.7	37.3	41.5
Oceania					
Australia	3.5	4.6	14.3	13.0	16.4
Canada	2.1	1.6	1.7	1.2	0.8
United States	0.6	0.8	0.9	0.7	0.7

Source: *World Population Monitoring, 1993: With a Special Report on Refugees* (United Nations publication, Sales No. E.95.XIII.8), table 59.

[a]Data for the United States excluding persons who legalized their status in 1989 under the Immigration Reform and Control Act of 1986.

The figure for the members of the population born in Mexico increased from nearly 800,000 in 1970 to 4.3 million in 1990; that for those born in the Caribbean rose from nearly 700,000 to about 2 million; the figure for those from South America increased from about a quarter of a million to 1.1 million; and that for those born in developing Asia (excluding Japan) increased from 700,000 to over 5.4 million (United States Bureau of the Census, 1973 and 1993). By 1990, the major developing countries of origin were Mexico (4.3 million), the Philippines (912,000), Cuba (737,000), the Republic of Korea (568,000), Viet Nam (543,000) and China (530,000).

In Canada, the proportion of immigrants originating in the developing world tended to rise steadily after 1970, though the total number of immigrants declined until 1985. Of particular importance was the rise in the proportion of immigrants originating in Asia, which accounted for 48 per cent of all immigrants by 1985-1989. In contrast with the United States, Canada still attracted nearly one quarter of its immigrants from developed countries by the late 1980s.

In comparison with the United States and Canada, Australian immigration was dominated by migrants from developed countries for a longer period. Yet, the elimination of the White Australia policy during the late 1970s opened the door to greater immigration from Australia's immediate vicinity: the countries of South-eastern Asia. The share of Asian migration to Australia increased rapidly. From scarcely 12 per cent in 1970-1974, it jumped to over 30 per cent in 1975-1979 and reached 41 per cent by 1985-1989. As a result of these changes, the stock of persons born in Asia increased fivefold, rising from 167,000 in 1971 to over 898,000 in 1991, whereas those born in Europe increased by less than 10 per cent, passing from 2.2 million in 1971 to 2.4 million in 1991 (McMahon, 1992). In 1991, the major countries of origin of Asian-born migrants in Australia were Viet Nam (133,000), Malaysia and Brunei Darussalem (84,000), the Philippines (74,000), Hong Kong and Macau (73,000), China (69,000) and India (65,000).

C. Japan

Japan has traditionally been a closed society, although it fostered the emigration of its citizens early in the twentieth century and received the inflow of significant numbers of foreign workers during the Second World War (most were imported as forced labour). During the 1980s, however, Japan began to experience increased inflows of foreigners, most of whom entered Japan as tourists or for short stays in relation to business. Yet, a significant component of the inflow has consisted of foreigners intending to stay for longer periods, whether legally (as businessmen, temporary workers, trainees, descendants of former Japanese emigrants, and so on) or illegally, as undocumented workers.

Japanese statistics indicate that the number of foreigners residing in Japan has increased significantly, although the sizeable Korean minority, which comprised both the workers who had been forcefully moved to Japan during the Second World War and their descendants, remained the main component of the foreign population in the country. Migrants from such countries as China (including Taiwan Province of China), Brazil, the Philippines, the United States and Peru, among others, had increased significantly in number between 1980 and 1991 (table 37). It must be noted that migrants from Brazil and Peru are mostly the descendants of Japanese emigrants and are therefore ethnic Japanese. Japanese law grants them the right of residence in Japan. In that respect, Japan resembles Germany, since the latter also has provisions granting residence and citizenship rights to ethnic Germans.

Until 1980, Japan had been nearly unique among developed countries in not having resorted to labour migration. Indeed, the country had made an effort to restructure its industry and to export labour-intensive manufacturing to other Asian countries in order to avoid the need of importing workers. However, like every other developed country, Japan has found that there are some jobs that it cannot export and that are not sufficiently attractive for the Japanese labour force, particularly those in construction and a variety of the low-paying jobs in the service sector. Lack of local manpower has prompted Japanese employers to hire migrants, even if the latter lack official permission to work. Consequently, undocumented migration to Japan has been rising.

D. Undocumented migration

Undocumented migration can be said to arise because of a failure of migration policy, since in an increasing number of cases the policies in place to control migration

TABLE 37. FOREIGNERS REGISTERED IN JAPAN, BY COUNTRY OF CITIZENSHIP, 1980-1991
(*Thousands*)

	1980	1985	1990	1991
Australia	1.1	1.8	4.0	5.4
Brazil	1.5	2.0	56.4	119.3
Canada	1.7	2.4	4.9	5.9
China (including Taiwan Province of China)	52.9	74.9	150.3	171.1
Germany	2.7	3.0	3.6	3.8
Indonesia	1.4	1.7	3.6	4.6
Malaysia	0.7	1.8	4.7	5.6
Pakistan	0.4	1.0	2.1	3.7
Peru	0.3	0.5	10.3	26.3
Philippines	5.5	12.3	49.1	61.8
Republic of Korea	664.5	683.3	687.9	693.1
Thailand	1.3	2.6	6.7	8.9
United Kingdom	5.0	6.8	10.2	11.8
United States	22.4	29.0	38.4	42.5
Viet Nam	2.7	4.1	6.2	6.4
Other	18.9	23.4	36.9	48.7
Developed countries	32.9	43.0	61.1	69.4
Developing countries	731.2	784.2	977.3	1 100.8
Others	18.9	23.4	36.9	48.7
TOTAL	783.0	850.6	1 075.3	1 218.9

Source: Japan, Ministry of Justice, Immigration Bureau.

are unrealistic and fail to allow the use of foreign manpower that the labour market demands. Consequently, economic forces operate against migration policies and the result is a growing lack of control over migration, particularly when foreigners can gain legal entry to a country's territory and then become undocumented migrants by working illegally.

One of the problems faced in the study of undocumented migration is such migration's inclusion of a wide array of situations. Furthermore, an undocumented status need not last forever. Thus, undocumented migrants may become regularized, or regular migrants who fulfil all the requirements for admission set by the host country and to reside and work in that country may become undocumented because of changes in regulations or administrative inefficiencies. Hence, the undocumented population is liable to change over time, even in the absence of major fluctuations in migration flows.

The measurement of undocumented migration is fraught with difficulties. Some indication of the level of undocumented migration can be obtained from the results of regularization programmes. To date, the largest legalization drive has been that carried out in the United States as a result of the Immigration Reform and Control Act (IRCA) of 1986. More than 3 million people applied for legalization under two different programmes: 1.8 million under the normal regularization programme and another 1.3 million under the Special Agricultural Workers programme. Most of those applying for legalization were Mexican citizens (United States Department of Justice, 1992). By the end of fiscal year 1991, 2.5 million persons—of whom 75 per cent had been born in Mexico and a further 10 per cent in El Salvador, Guatemala or Haiti—had legalized their status under IRCA (Zlotnik, 1992b).

As already mentioned, several regularization drives were carried out in Spain and Italy during the late 1980s and early 1990s. In addition, about 124,000 persons were regularized in France in 1981-1982 (OECD, 1990). The largest proportion of regularized migrants in each of those three countries originated in Morocco. In France, there were in addition sizeable proportions of Algerian, Portuguese, Tunisian and Turkish migrants regularized. In Italy, important sources of undocumented migrants other than Morocco included China, the Philippines, Sri Lanka, Senegal, Tunisia and Yugoslavia (OECD, 1991). In Spain, significant numbers of migrants from Argentina, the Gambia, Portugal, Senegal and the United Kingdom were regularized. Although in all three countries a relatively large proportion of the migrants regularized originated in countries within their immediate vicinity, the number of migrants coming from countries further away was also of importance, indicating the expanding diversification of migration flows.

Japan has not as yet carried out any regularization drive and consequently lacks comparable data on undocumented migration. However, information on the number of illegal migrant workers apprehended indicates that their number has been increasing steadily (table 38). However, there has been a significant change in the distribution by sex of the migrants apprehended (a reduction in the proportion of women) and important changes in their distribution by nationality. Thus, whereas in 1985 Philippine and Thai migrants predominated, by 1989 equally large numbers originated in

TABLE 38. APPREHENDED UNDOCUMENTED MIGRANT WORKERS, BY NATIONALITY AND SEX, JAPAN, 1985, 1987 AND 1989

Nationality		*Year* 1985	1987	1989
Bangladesh	Male ...	1	437	2 275
	Female ..	0	1	2
	Total ...	1	438	2 277
China	Male ...	126	210	316
	Female ..	301	284	272
	Total ...	427	494	588
Pakistan	Male ...	36	905	3 168
	Female ..	0	0	2
	Total ...	36	905	3 170
Philippines	Male ...	349	2 253	1 289
	Female ..	3 578	5 774	2 451
	Total ...	3 927	8 027	3 740
Republic of Korea	Male ...	35	109	2 209
	Female ..	41	99	920
	Total ...	76	208	3 129
Thailand	Male ...	120	290	369
	Female ..	953	777	775
	Total ...	1 073	1 067	1 144
Other	Male ...	20	85	2 165
	Female ..	69	83	395
	Total ...	89	168	2 560
Total	Male ...	687	4 289	11 791
	Female ..	4 942	7 018	4 817
	Total ...	5 629	11 307	16 608

Source: K. Morita, "Japan and the problem of foreign workers", paper presented at the Expert Group Meeting on Cross-national Labour Migration in the Asian Region: Implications for Local and Regional Development, United Nations Centre for Regional Development (Nagoya, Japan, 5-8 November 1990).

Bangladesh, Pakistan and the Republic of Korea. Other data suggest that most undocumented male migrants are finding work in factories, whereas female migrants tend to be concentrated in personal services. Relatively few women declare their work in the sex industry but at any event, this is a sector where no male undocumented migrants are found.

Although a number of countries have carried out regularization drives with the purpose of controlling undocumented migration, it is apparent that these drives have not stopped the further inflow of undocumented migrants. As OECD (1990) notes, regularization on its own cannot be expected to solve the problem of illegal migration. Tighter controls at the borders and within a country, as well as the use of effective sanctions against employers of and traffickers in undocumented migrants, are necessary to control their flow.

However, although all developed countries have sought to control immigration across their borders by adopting laws and instituting systems or agencies to enforce bans on illegal migration, no country has been totally successful in its efforts (North, 1993). In the United States, for instance, the average number of apprehensions of undocumented migrants per Border Patrol agent decreased only temporarily after IRCA was passed and then increased again to reach pre-IRCA levels.

According to North (1993), the problem is not that the control of migration is impossible. Examples abound of countries that have successfully sealed their borders (Eastern-bloc countries during the cold war, for instance) and of countries that have expelled large numbers of undocumented and even documented migrants (Ghana, Nigeria and Uganda, among others). The problem lies in the fact that Western democracies find it difficult to impose the draconian measures needed to control the movement of people. This does not mean, however, that they are powerless. Indeed, given the supposedly enormous pressures for emigration that exist in many developing countries, the surprising fact is that so few people have moved, whether legally or illegally, to developed countries. Migration policies and the various means of control at the disposal of Governments do make a difference. In addition, the economic situation matters greatly. The need for unskilled labour is limited. Furthermore, if that labour cannot be obtained at bargain prices, other means may be found to carry out the needed tasks. Here lies the importance of regularization drives which, by providing undocumented migrants with guarantees regarding their legal situation, are the main tool for combating the unfair exploitation of foreign labour.

E. INTERNATIONAL MIGRATION POLICIES

What has been the impact of international migration policies in developed countries on migration originating in developing countries? The answer to this question must take account of the different policy approaches that characterize European countries, the traditional countries of immigration, and Japan. In particular, each set of countries has different goals regarding international migration. The European countries in general do not consider themselves "countries of immigration" and would therefore like to keep international migration at a minimum. If Western-bloc countries in Europe have become de facto countries of immigration, this change has occurred largely against their will. It is not surprising, therefore, to find that their policies regarding international migration are often at odds with reality.

Thus, when the labour-importing countries of Europe decided to stop the recruitment of labour and allow family reunification under restricted conditions, insufficient attention was given to the effect that family reunification would have in the long run. Indeed, labour-importing countries appeared to equate the closure of borders to foreign workers with a general closure of borders. Furthermore, family members were seen mostly as dependants and not as future workers. It is therefore not surprising to find that even the restricted family reunification regulations now in place in most European countries are being considered too generous.

During the 1980s, two developments called into question the soundness of international migration policies in the West. The first was the growing number of asylum-seekers arriving directly from developing countries. The second was the growing number of migrants originating in Eastern-bloc countries for whom there were two major paths of entry: one involved requesting asylum and, in a way, competing with migrants from developing countries. The other was open only to ethnic minorities with homelands abroad. Ethnic Germans thus came to constitute the largest flow of East-West migrants in Europe during the late 1980s and early 1990s, a flow that was made possible almost exclusively by the citizenship laws of the Federal Republic of Germany.

The asylum policies of Western-bloc countries had been developed to respond to the needs of the cold war era, when few migrants managed to leave Eastern-bloc countries and refugees from developing countries were resettled on a quota basis. The asylum system was not prepared for the spontaneous arrival of large numbers of people seeking asylum and having to undergo a lengthy adjudication procedure. A country like Germany, where the right to seek asylum was enshrined in the constitution, had a particularly difficult time coping with the growing number of asylum applications filed. The need to introduce greater controls was soon felt and it has already led to the adoption of more stringent criteria on the granting of asylum and even on the basis for allowing an asylum application to be filed.

As the European Union (EU) moves towards making the single European Economic Space a reality, the free movement of EU citizens within that space is materializing. In addition, member States are moving towards the harmonization of migration policies with respect to non-EU citizens. The Schengen countries (which include all the EU member States except Denmark, Ireland and the United Kingdom) and the TREVI Group comprising the 12 European ministers of justice have been preparing blueprints and treaties on the harmonization of migration policies for the 9 Schengen countries and the 12 EU member States, respectively. Late in 1992, Germany agreed to change its liberal asylum policy and adopt a more restrictive one compatible with European standards while other countries agreed to move towards a more restrictive family reunification policy similar to that already in force in Germany. Common employment policies, mainly regarding temporary migrant workers, are being considered by the Schengen countries. The effect of this move towards harmonization of international migration policies will be to further restrict the possibilities for migration from developing countries. If labour-market needs subsist under such restrictive policies, undocumented migration will likely fill the gap.

In contrast with European countries, the traditional countries of immigration favour the admission of immigrants. However, especially in Australia and Canada, there has been growing concern about matching the intake of immigrants with economic needs. Therefore, a series of tools has been used to select immigrants whose skill profiles match those needed by the labour market and to adjust the level of immigrant intake to the economic situation in general. Consequently, the number of immigrants admitted annually has fluctuated considerably during the past 20 years as economic conditions have changed. However, despite the emphasis put on the admission of immigrants with needed skills or capital, immigrant flows to both Australia and Canada are still dominated by persons admitted on the basis of their family ties with other immigrants or with Australian or Canadian citizens, respectively. Thus, during 1981-1987, 43 per cent of the immigrants to Canada were admitted under the Family Class and a further 9 per cent under the Assisted Relative Class. In Australia, admissions in the Family Class amounted to 23 per cent during

1986-1989 and those in the Concessional Class, which is similar to the Assisted Relatives Class in Canada, accounted for an additional 25 per cent. Furthermore, in both Australia and Canada, persons admitted as refugees or for humanitarian reasons accounted for 10-15 per cent of the total immigrant intake (United Nations, 1992).

In the United States, the importance of family ties as the basis for admission has been even greater. During 1985-1989, 7 out of every 10 immigrants were admitted by virtue of being related either to United States citizens or to permanent resident aliens and an additional 2 out of every 10 immigrants were admitted as refugees (United Nations, 1992). Furthermore, because immediate relatives of United States citizens were not subject to numerical limitations, the number of immigrants to the United States had been growing steadily since 1965. Concern about these developments were at the root of the debate leading to a change in international migration policy in the United States in 1990. Thus, the Immigration Act of 1990 established increased intakes of employment-based immigrants and a "hard cap" of 714,000 annual admissions of immigrants during 1992-1994. That figure included 465,000 places allocated to family immigrants; 55,000 to immediate relatives of persons regularized under IRCA; 40,000 for nationals of countries under-represented in previous years; 12,000 for Hong Kong nationals who were high-level employees of large United States multinationals having subsidiaries there; and 140,000 for immigrants with needed skills (Zlotnik, 1992b). Starting in 1995, a soft global cap of 675,000 was established, but that cap can be pierced to assure that at least 226,000 places are available for family immigrants who are relatives of permanent residents or United States citizens. In effect, therefore, the 1990 law establishes for the first time a cap on the number of admissions of immediate relatives of United States citizens (Zlotnik, 1992b).

Japan's immigration policy is closer to that of European countries than to that of the traditional immigration countries. In 1989, the Government of Japan amended the Immigration Control and Refugee Recognition Act in order to grant long-term residence status to migrants with Japanese ancestry and to increase the categories of migrants that could be admitted on a temporary basis (Zlotnik, 1994b). However, none of the categories added includes unskilled workers. That is to say, to the extent that Japanese authorities are willing to admit foreigners, they are interested only in those possessing needed skills.

Despite their differences, the migration policies of developed countries show considerable convergence in two respects: there is an interest in attracting skilled manpower that can enhance the competitiveness of receiving countries in an ever more global and interconnected economic environment; and there is a common concern regarding the unwanted inflow of migrants, be they asylum-seekers or undocumented workers. The second concern is leading to growing cooperation in regard to the control and regulation of international migration. The establishment of the Inter-governmental Consultations on Asylum, Refugee and Migration Policies in Europe, North America and Australia, for instance, is indicative of the perceived need for the harmonization of international migration policies, particularly those regarding asylum, as is the increasing attention being paid to trafficking in undocumented migrants. The success of such cooperation is likely to lead to further restriction on migration from developing countries, especially from those countries whose human capital does not allow them to export highly skilled personnel. However, developed countries are also increasingly aware of the many forces that are fueling migration from developing countries and there is growing recognition that some migration may need to be accommodated. Traditional countries of immigration can do so through their normal immigration programmes. European countries and Japan will have to devise appropriate mechanisms if and when they decide to avoid relegating developing-country migration to the back door.

F. International migration, population and development

One of the important issues facing the developed world is the extent to which development will lead to reductions in international migration. As part of the implementation of IRCA, the United States set up a Commission for the Study of International Migration and Cooperative Economic Development to study the issue. The report of the Commission concluded that, in the short run, economic development tends to stimulate migration by raising people's expectations and enhancing their ability to migrate. It may take many years, perhaps generations, for development to result in low migration. Consequently, efforts to enhance the development process in today's developing countries are not a solution to undocumented migration (United States Commission for the Study of International Migration and Cooperative Economic Development, 1990).

The Commission further concluded that the United States could best assist developing countries by reducing its own fiscal deficit and by providing debt relief to developing countries whose economic growth was being compromised by excessive debt burdens. Another major strategy to promote the economic growth of developing countries was the expansion of trade between them and the industrialized countries. For the latter, the choice was clear: either accept the goods and services produced by developing countries and bear their impact on domestic industries and labour markets, or keep on countenancing unauthorized migration. Development aid on its own, even if substantially increased, would not have the same impact on economic growth as expanded trade.

Despite these conclusions, the liberalization of trade between industrialized countries and the developing world has been slow in coming. Although the successful conclusion of the Uruguay Round of multilateral trade negotiations is encouraging, the products that are most competitively produced by developing countries are still far from having free access to the markets of developed countries. Indeed, the move towards the creation of free trade areas, such as that encompassing the European Union and the European Free Trade Association (EFTA) countries (without Switzerland), indicates a tendency to foster free trade between industrialized countries at the expense of barriers to goods produced abroad.

Given the reluctance of developed countries to permit free trade and the current world recession that is draining the resources available for development aid, more attention is being focused on other concomitants of migration. When the low population growth rates experienced by European countries are juxtaposed with the high growth rates experienced by the developing countries in their vicinity, it is tempting to reach the conclusion that international migration from the latter to the former will increase if only because economic growth in developing countries will be incapable of keeping pace with the increasing labour force (Golini, Righi and Bonifazi, 1993). In the past, however, there has been a low correlation between population growth and South-to-North migration flows (Zlotnik, 1994a). Indeed, since a large proportion of the migration flows directed to European countries originate in Europe itself, the correlation between migration flows and population growth has been negative, with the slowest-growing regions being those sending the largest numbers of migrants. Although the experience of the past 30 years may not repeat itself during the next decades, it is nevertheless likely that population growth will not be the main factor fuelling international migration to developed countries. Economic and political factors will, as in the past, play more important roles.

References

Amersfoort, H. van, Philip Muus and Rinnus Penninx (1984). International migration, the economic crisis and the State: an analysis of Mediterranean migration to Western Europe. *Ethnic and Racial Studies*, vol. 7, No. 2 (April), pp. 239-286.

Boyd, Monica (1989). Family and personal networks in international migration: recent developments and new agendas. *International Migration Review* (Staten Island, New York), vol. 23, No. 3 (fall), pp. 638-670.

Fawcett, James T. (1989). Networks, linkages, and migration systems. *International Migration Review* (Staten Island, New York), vol. 23, No. 3 (fall), pp. 671-680.

Germany, Statistisches Bundesamt (1993). *Statistisches Jahrbuch, 1993*. Wiesbaden: Metzler Poeschel.

Golini, Antonio, Alessandra Righi and Corrado Bonifazi (1993). Population vitality and decline: the North-South contrast. In *Changing the Course of International Migration*. Paris: Organisation for Economic Cooperation and Development.

Kritz, Mary M., Lin L. Lim and Hania Zlotnik, eds. (1992). *International Migration Systems. A Global Approach*. Oxford: Clarendon Press.

McMahon, Vincent (1992). Australia: 1993 Continuous Reporting System on Migration (SOPEMI) report. Canberra, Australia: Department of Immigration. Mimeographed.

Morita, K. (1990). Japan and the problem of foreign workers. Paper presented at the Expert Group Meeting on Cross-National Labour Migration in the Asian Region: Implications for Local and Regional Development, United Nations Centre for Regional Development (UNCRD), Nagoya, Japan, 5 to 8 November.

Muus, Philip, and Rinnus Penninx (1991). *Immigratie van Turken en Marokkanen in Nederland*. Amsterdam: University of Amsterdam, Centre for Migration Research, Department of Human Geography.

North, David S. (1993). Why democratic Governments cannot cope with illegal immigration. In *Changing the Course of International Migration*. Paris: Organisation for Economic Cooperation and Development.

Organisation for Economic Cooperation and Development (OECD) (1990). *Continuous Reporting System on Migration. SOPEMI 1989*. Paris: OECD.

__________ (1991). *Continuous Reporting System on Migration. SOPEMI 1990*. Paris: OECD.

__________ (1992). *Trends in International Migration: Continuous Reporting System on Migration. SOPEMI*. Paris: OECD.

Takayama, T., and Y. Iguchi (1992). Japan: 1992 SOPEMI report. Tokyo: Ministry of Labour. Mimeographed.

United Nations (1992). *World Population Monitoring, 1991: With Special Emphasis on Age Structure*. Population Studies, No. 126. Sales No. E.92.XIII.2.

__________ (1996). *World Population Monitoring, 1993: With a Special Report on Refugees*. Sales No. E.95.XIII.8.

United Nations High Commissioner for Refugees (UNHCR) (1993). *The State of the World's Refugees: The Challenge of Protection*. New York: Penguin Books.

United States Bureau of the Census (1973). *1970 Census of Population*, vol. 1, *Characteristics of the Population. Part 1: United States Summary, Section 2*. Washington, D.C.: United States Government Printing Office.

__________ (1993). *1990 Census of Population. The Foreign-born Population in the United States*. Washington, D.C.: United States Government Printing Office.

United States Commission for the Study of International Migration and Cooperative Economic Development (1990). *Unauthorized Migration: An Economic Development Response*. Washington, D.C.: United States Government Printing Office.

United States Department of Justice (1992). Immigration Reform and Control Act. Report on the Legalized Alien Population. Washington, D.C.: U.S. Department of Justice, Immigration and Naturalization Service.

Zelinsky, Wilbur (1971). The hypothesis of the mobility transition. *The Geographical Review* (American Geographical Society), vol. 61, pp. 219-249.

Zlotnik, Hania (1991). South-to-North migration since 1960: the view from the North. *Population Bulletin of the United Nations*, Nos. 31/32, pp. 17-37. Sales No. E.91.XIII.18. New York: United Nations.

__________ (1992a). Empirical identification of international migration systems. In *International Migration Systems. A Global Approach*, Mary M. Kritz, Lin L. Lim and Hania Zlotnik, eds. Oxford: Clarendon Press.

__________ (1992b). Who is moving and why? A comparative overview of policies and migration trends in the North American system. Paper presented at the Conference on Migration, Human Rights and Economic Integration, North York, Ontario, Canada, 19-22 November.

__________ (1994a). International migration: causes and effects. In *Beyond the Numbers. A Reader on Population, Consumption and the Environment*, Laurie Ann Mazur, ed. Washington, D.C.: Island Press.

__________ (1994b). Comparing migration to Japan, the European Community and the United States. *Population and Environment: A Journal of Interdisciplinary Studies* (New York), vol. 15, No. 3 (January), pp. 173-186.

Zolberg, Aristide R. (1989). The next waves: migration theory for a changing world. *International Migration Review* (Staten Island, New York), vol. 23, No. 3 (fall), pp. 403-430.

__________ (1992). Reforming the back door: perspectives historiques sur la réforme de la politique Américaine d'immigration. In *Logiques d'états et immigrations*, J. Costa-Lascoux and Patrick Weil, eds. Paris: Editions Kimé.

XII. MIGRATION BETWEEN ASIAN COUNTRIES

*Nasra M. Shah**

The significance of migration between Asian countries has increased during the last decade, as a result of both labour migration and refugee movements. The present paper outlines recent trends in intra-Asian migration in terms of its volume, origin and destination, describes the characteristics of such movements, and discusses their major socio-economic consequences. Migration may take at least three distinct forms, namely, permanent migration, short-term labour migration, and refugee movements. Within the first two, one can make a further distinction between legal and clandestine movements. This paper focuses mostly on labour migration and refugee movements, mainly because most Asian countries discourage the permanent resettlement of foreigners in their territories.

A. Labour migration to Western Asia

A general problem one faces in dealing with international migration within Asia is the lack of reliability and the varying quality of the data available. Both the International Labour Organization (ILO) and the Economic and Social Commission for Western Asia (ESCWA) have published data on migration to and from various countries in the region. In addition, Birks and Sinclair (1979) and Birks, Seccombe and Sinclair (1988) have published estimates of the number of foreign workers in Western Asian countries. However, a comparison of data sources leaves one bewildered and confused. An example of such differences is given in table 39. The profiles of foreign workers in Saudi Arabia provided by ESCWA for 1984 (ESCWA, 1985) and by Birks, Seccombe and Sinclair (1988) for 1985 are quite different. One hardly knows which profile to accept. Conclusions about numbers and especially about workers' characteristics therefore remain highly tentative.

Table 39. Distribution of the economically active foreign population in Saudi Arabia by sector of economic activity

(Percentage)

Sector of economic activity	*ESCWA for 1984*	*Birks, Seccombe and Sinclair for 1985*
Agriculture and fishing	2.5	11.2
Mining and quarrying	2.5	1.0
Manufacturing	20.6	10.2
Electricity, gas and water	1.4	4.1
Construction	28.6	29.0
Trade	20.3	13.3
Transport and communication	3.6	5.0
Finance and business	4.1	1.8
Community and personal services	16.4	24.4
Total	100.0	100.0
Number of persons	1 323 858	3 522 700

Sources: Figures for 1984 are estimates presented in ESCWA, Demographic and related socio-economic data sheets (Baghdad, 1985), p. 147; those for 1985 appear in J. S. Birks, I. J. Seccombe and C. A. Sinclair, "Labour migration in the Arab Gulf States: patterns, trends, and prospects", *International Migration* (Geneva), vol. 26, No. 4 (1988), p. 275.

Labour migration to the oil-rich countries of Western Asia has been one of the most significant features of intra-Asian migration since 1975. Even before the spurt of growth following the 1973 increases in oil prices, 70 per cent of the labour force in the member States of the Gulf Cooperation Council (GCC), namely, Bahrain, Kuwait, Oman, Qatar, Saudi Arabia and the United Arab Emirates, was foreign, consisting mainly of persons of Arab origin, primarily Egyptians and Palestinians and Jordanians. A small number of persons from Southern Asia had also migrated to Western Asia since the development of the oil industry in the 1930s (Seccombe and Lawless, 1986). However, the major increase in the number of migrants and the diversification of their countries of origin took place after 1973. India and Pakistan were among the first countries to respond to the ambitious development plans of the GCC countries

*Department of Community Medicine, Faculty of Medicine, Kuwait University, Safat.

which involved the admission of growing numbers of foreign workers. They were later joined by Bangladesh, the Philippines, the Republic of Korea and Thailand. By 1985, it was estimated that 63 per cent of all workers in the six GCC member States were from Southern or South-eastern Asia (table 40).

Birks, Seccombe and Sinclair (1988) estimate that in 1985 there were 1.5 million Arab workers in the GCC countries, representing 30 per cent of the total foreign labour force in the region (table 40). Other estimates suggest a number as high as 3.45 million for the early 1980s, with almost 2 million Egyptian and 680,000 Yemenites (Nagi, 1986). The major groups of Arab migrants in Western Asia during the 1980s included (*a*) Egyptians employed in Iraq and Jordan who worked mainly in unskilled and agricultural jobs, especially in Jordan; (*b*) Palestinians and Jordanians employed in relatively well-paying jobs in the trade and management sectors in Kuwait and other GCC countries; and (*c*) Yemenites employed mainly in agricultural and unskilled work in Saudi Arabia. Among the Arab countries, Jordan and Oman were both the origin and the destination of considerable numbers of migrants (Shah, 1992).

Among the GCC countries, Saudi Arabia is the largest in terms of both population and territory. Consequently, it hosted the highest proportion of foreign workers in the Council (68.4 per cent) in 1985. The dependence of Saudi Arabia on foreign labour has almost tripled since 1970 when only 27 per cent of the total labour force in the country was foreign. By 1980, the proportion of foreign workers had increased to 53 per cent and by 1985 to 79 per cent (Abella, 1994). Since there are no official figures on the number of foreign workers or of foreigners residing in Saudi Arabia, discrepancies in estimates are abundant. For example, ESCWA (1985) reported that in 1984 the foreign labour force in Saudi Arabia numbered 1,323,858 persons, whereas Birks, Seccombe and Sinclair (1988) cited a number two and a half times as large (3,522,700) for 1985. Although there is a difference of one year in the reference dates of those figures, it is unlikely that the pace of labour migration could have increased so dramatically between 1984 and 1985. In fact, Abella (1994) notes that labour migration to Western Asia peaked during 1983 and slackened until 1985. Such large discrepancies need to be addressed and resolved.

Since 1975, the preponderance in the total number of migrants of persons originating in the major Southern Asian countries has shifted in favour of citizens from Eastern and South-eastern Asian countries, as indicated by the figures for the average annual number of workers processed in the different countries of origin (table 41). During 1975-1979, more than 50 per cent of all Asian workers intending to go abroad originated in Southern Asia but their share declined to 29 per cent during 1985-1989. Data on the annual flows of Asian workers to Western Asian countries indicate that Indonesia, the Philippines, the Republic of Korea and Thailand had contributed only 2 per cent of processed migrant workers in 1975, but that their share increased rapidly during the late 1970s so as to account for more than 50 per cent of all such workers in 1980 (table 42). That position was maintained throughout the 1980s and in 1989 those four countries accounted for 51 per cent of all migrants departing to work in Western Asia. In contrast, India and Pakistan, which had been the sources of 97 per cent of all Asian workers departing to work in Western Asia in 1975, contributed only between about one quarter and one third of the flow of Asian workers during the 1980s. Bangladesh and Sri Lanka experienced major increases in the flow of migrants to Western Asia, accounting for 22 per cent in 1989.

TABLE 40. DISTRIBUTION OF FOREIGN WORKERS IN MEMBER STATES OF THE GULF COOPERATION COUNCIL, BY REGION OF ORIGIN, 1985
(*Percentage*)

Region of origin	*Bahrain*	*Kuwait*	*Oman*	*Qatar*	*Saudi Arabia*	*United Arab Emirates*	*Total*	*Number of persons*
Arab countries	7.8	46.5	6.6	23.2	32.8	16.0	30.1	1 547 500
Southern Asia	73.2	44.6	89.4	65.3	32.0	74.8	43.0	2 214 600
South-eastern Asia	11.1	5.7	1.5	5.7	27.4	4.2	20.3	1 043 900
Other regions	7.9	3.2	2.5	5.8	7.8	5.0	6.6	340 800
Total	100.0	100.0	100.0	100.0	100.0	100.0	100.0	..
Number of persons	96 900	543 900	314 100	70 700	3 522 700	598 500	..	5 146 800

Source: J. Birks, I. Seccombe, and C. Sinclair, "Labour migration in the Arab Gulf States: patterns, trends, and prospects" *International Migration* (Geneva), vol. 26, No. 4 (1988), p. 274.

TABLE 41. DISTRIBUTION BY ORIGIN OF AVERAGE ANNUAL NUMBER OF MIGRANT WORKERS REGISTERED BY THE MAIN LABOUR-EXPORTING COUNTRIES OF ASIA, 1975-1989

(Percentage)

Region or country of origin	*1975-1979*	*1980-1984*	*1985-1989*
Southern Asia	54.3	43.5	29.4
Bangladesh	4.9	5.1	7.1
India	19.1	22.7	13.2
Pakistan	26.3	12.9	7.5
Sri Lanka	4.0	2.8	1.6
Eastern and South-eastern Asia	45.7	56.5	70.6
Indonesia	1.7	2.4	6.1
Philippines	21.6	31.8	45.4
Republic of Korea	20.6	16.5	9.5
Thailand	1.8	5.8	9.6
TOTAL	100.0	100.0	100.0
Number of persons	351 300	1 039 000	1 013 200

Source: Calculated from *World Population Monitoring, 1991: With Special Emphasis on Age Structure* (United Nations publication, Sales No. E.92.XIII.2), table 86.

The Philippines needs to be singled out as the country that dominates labour migration in Asia and particularly that directed to Western Asia. During 1985-1989, the average annual number of registered migrant workers leaving the Philippines was about 460,000, that is, 45 per cent of all migrant workers originating in Asia (table 41). Among the 459,000 who migrated in 1989, 355,000 were land-based while the rest were seamen. Over two thirds of all land-based Philippine migrant workers (241,000) went to Western Asia, although other countries in Asia also attracted about one quarter of all of them (table 43).

The migration of workers of the Republic of Korea, in contrast, has shown a substantial decline, with the number of migrating workers passing from 151,000 in 1982 to about 21,000 in 1988. Both Indonesia and Thailand have made substantial progress in placing their workers in the Western Asian labour market. A majority of Indonesian migrant workers are women who work in domestic service mainly in Saudi Arabia (Appleyard, 1991a; Abella, 1995). Among the countries of Southern Asia, India and Pakistan have lost ground in terms of the number of workers departing and Bangladesh and Sri Lanka have displayed a fluctuating pattern with a substantial increase registered in 1989 (table 43).

It should be noted, however, that the data in tables 41 and 42 are based on official statistics gathered by the countries of origin on migrants securing permits to work abroad. Data for the Republic of Korea, for instance, are more likely to be complete not only because its systems of data collection are more advanced but also because workers of the Republic of Korea tend to migrate under formal company contracts which are easier to monitor. For other countries, the growing number of workers who leave clandestinely without securing the appropriate permits implies that the data cited above underestimate true worker migration levels. In Pakistan, for instance, official estimates put the stock of workers of Pakistan abroad in 1988 at 542,000, whereas Stahl and Azam (1990) provide an estimate of 1,147,000, taking account of unregistered migration. In the case of Sri Lanka, it is estimated that only 40 per cent of all migrant workers are registered by official sources (Bandara, 1991).

In terms of the destinations chosen by migrants, Filipinos, persons from the Republic of Korea and Thais showed a trend towards diversification in the late 1980s. Thus, 34 per cent of migrant workers from the Republic of Korea leaving their country in 1988, 32 per cent of those leaving from the Philippines in 1989 and 30 per cent of those leaving from Thailand in 1989 went to Asian countries outside of Western Asia. In contrast, over 94 per cent of the registered migrant workers originating in Southern Asia have consistently gone to Western Asia, the proportion being about 99 per cent for Bangladesh and almost 100 per cent for Pakistan (table 43).

Although China has potentially a very large labour supply, labour migration from China has been relatively modest. It is estimated that during the 1980s about 400,000 people worked in foreign countries, often as a result of contracts by State-owned international corporations and labour companies (Fan, 1991). Among the reasons that have limited China's involvement in the Western Asian labour markets are the lack of diplomatic relations with most of the GCC countries and China's inability to enter into joint ventures which are required by the host countries (Abella, 1991).

TABLE 42. ANNUAL NUMBER OF MIGRANT WORKERS FROM SOUTHERN AND SOUTH-EASTERN ASIA TO WESTERN ASIA, 1975-1989

Year	Total number	Pakistan and India	South-eastern and Eastern Asia	Bangladesh and Sri Lanka
			(Percentage)	
1975	93 100	97.3	2.0	0.7
1976	152 500	71.6	24.5	3.9
1977	311 000	66.9	26.4	6.7
1978	359 500	55.1	36.3	8.6
1979	424 000	45.5	42.7	11.8
1980	526 700	34.8	54.1	11.1
1981	688 400	31.8	52.0	16.2
1982	968 700	37.3	49.1	13.6
1983	986 800	34.8	53.8	11.4
1984	843 000	34.6	53.2	12.2
1985	793 000	30.6	55.1	14.3
1986	683 500	24.5	58.1	17.4
1987	740 200	25.4	57.7	16.9
1988	809 200	30.5	53.7	15.8
1989	795 900	27.2	50.7	22.1

Source: Manuel Abella, "International migration in the Middle East: patterns and implications for sending countries", paper presented at the Informal Expert Group Meeting on International Migration, Geneva, Switzerland, 16-19 July 1991, p. 20.

B. EMERGING LABOUR MARKETS IN EASTERN AND SOUTH-EASTERN ASIA

Since the 1950s, labour migration within Eastern and South-eastern Asia has been relatively small. In recent years, however, its pace has increased considerably. Eastern and South-eastern Asia have become the most economically dynamic regions in the contemporary world (Lim, 1991). A redirection of migration from Western Asia to Japan and the newly industrializing economies (NIEs)—Hong Kong, the Republic of Korea, Singapore and Taiwan Province of China—has begun. In addition, labour migration to Brunei Darussalam and Malaysia has risen. Investment by Japan and the NIEs in the relatively less developed Asian countries of the Pacific Basin (Indonesia, Malaysia and Thailand) has been a catalyst for the growth of migration within the region. As Japan and the NIEs have reached higher levels of development, they have shifted some of their production to less technologically advanced, but labour-rich countries (Appleyard, 1991a). Industrial growth has generated a labour demand in the receiving countries that could not always be met by the indigenous population.

Several migration streams have emerged. Undocumented flows are an important feature of migration within Eastern and South-eastern Asia. The major flows include: (*a*) Filipinos and others to Japan; (*b*) Malaysians to Singapore; (*c*) Indonesians to Malaysia; (*d*) Filipinos and South Asians to the Republic of Korea; (*e*) Chinese to Hong Kong; and (*f*) Filipinos and others to Taiwan Province of China. Each of these flows will be discussed below.

1. *Migration to Japan*

Ideologically, Japan still represents the archetypal closed society where immigrants are admitted only under exceptional circumstances (Zlotnik, 1994a). However, Japan's sustained economic growth combined with an ageing population, low birth rates, and the growing reluctance of indigenous workers to take up arduous and menial jobs has led to serious labour shortages. It has been estimated that the number of job vacancies in Japan exceeds the number of job seekers by 35 per cent, the construction industry alone having a shortage of 100,000 workers (Stahl, 1991). Certain policy responses aimed at redressing such imbalances have emerged. First, a law allowing the admission of persons of Japanese descent as long-term residents was adopted in June, 1990. It has resulted in the immigration of between 30,000 and 50,000 Brazilians and Peruvians of Japanese descent (Lim, 1991). Second, the admission of legal workers increased from 34,000 to 81,000 between 1982 and 1988 and the categories under which workers can be

TABLE 43. DISTRIBUTION OF MIGRANT WORKERS BY COUNTRY OF ORIGIN AND DESTINATION, 1976-1989

Country of origin/receiving region	*1976-1979*		*1980*		*1985*		*1988*		*1989*	
	Number	*Percentage*	*Number*	*Percentage*	*Number*	*Percentage*	*Number*	*Percentage*	*Number*	*Percentage*
Philippines										
Western Asia ...	141 185	70.8	132 044	83.9	253 867	79.2	267 035	69.3	241 081	67.8
Other Asia	33 287	16.7	17 708	11.2	52 835	16.5	92 648	24.1	86 196	24.3
Other regions ...	24 905	12.5	7 642	4.9	13 789	4.3	25 434	6.6	28 069	7.9
Total (land-based)	199 377	100.0	157 394	100.0	320 494	100.0	385 117	100.0	355 346	100.0
Seamen	144 411	..	57 196	..	52 290	..	85 913	..	103 280	..
Total	343 788	..	214 590	..	372 784	..	471 030	..	458 626	..
India										
Western Asia ...	..	..	..	..	160 396	98.4	165 880	97.7	120 561	95.1
Other regions ...	..	..	..	..	2 639	1.6	3 964	2.3	6 225	4.9
Total	..	..	..	..	163 035	100.0	169 844	100.0	126 786	100.0
Thailand										
Western Asia ...	27 784	91.3	20 761	96.6	61 660	88.5	91 905	77.3	87 627	69.9
Other Asia	2 421	7.9	723	3.4	7 937	11.4	21 593	18.1	31 536	25.2
Other regions ...	234	0.8	0	0.0	88	0.1	5 459	4.6	6 151	4.9
Total	30 439	100.0	21 484	100.0	69 685	100.0	118 957	100.0	125 314	100.0
Republic of Korea										
Western Asia ...	261 505	78.6	120 535	96.6	72 907	90.0	21 542	65.8	..	..
Other Asia	17 985	5.4	4 095	3.3	5 590	6.9	8 215	25.1	..	..
Other regions ...	53 340	16.0	157	0.1	2 533	3.1	2 995	9.1	..	..
Total (land-based)	332 830	100.0	124 787	100.0	81 030	100.0	32 752	100.0	..	..
Seamen	111 832	..	21 649	..	39 215	..	50 230	..	..	..
Total	444 662	..	146 436	..	120 245	..	82 982	..	..	..
Pakistan										
Western Asia ...	426 198	99.1	115 922	97.9	82 250	99.9	81 409	99.8	..	..
Other Asia	14	0.0	2	0.0	10	0.0	..	0.0	..	..
Other regions ...	3 715	0.9	2 473	2.1	73	0.1	136	0.2	..	..
Total	429 927	100.0	118 397	100.0	82 333	100.0	81 545	100.0	..	..
Bangladesh										
Western Asia ...	68 012	98.4	29 815	97.5	76 785	98.8	67 404	98.9	100 432	98.7
Other Asia	133	0.2	672	2.2	839	1.1	..	0.0	..	0.0
Other regions ...	961	1.4	86	0.3	70	0.1	717	1.1	1 286	1.3
Total	69 106	100.0	30 573	100.0	77 694	100.0	68 121	100.0	101 718	100.0
Indonesia										
Western Asia ...	7 651	73.7	11 501	71.0	48 280	85.2	53 208	83.2	..	..
Other Asia	884	8.5	1 227	7.6	5 930	10.5	6 485	10.1	..	..
Other regions ...	1 843	17.8	3 458	21.4	2 468	4.3	4 305	6.7	..	..
Total	10 378	100.0	16 186	100.0	56 678	100.0	63 998	100.0	..	..
Sri Lanka										
Western Asia ...	57 300	..	28 600	..	..	..	17 793	93.8	..	..
Other Asia	..	..	..	..	..	..	989	5.2	..	..
Other regions ...	..	..	..	..	..	..	191	1.0	..	..
Total	57 300	..	28 600	..	12 374	..	18 973	100.0	11 079	..
TOTAL	1 385 600		576 266		954 828		1 075 450		823 523	

Sources: Charles W. Stahl, "South-North migration in the Asia-Pacific region", *International Migration* (Geneva), vol. 29, No. 2 (June 1991), special issue, pp. 163-193; Reginald T. Appleyard, *International Migration: Challenge for the Nineties* (Geneva, International Organization for Migration, 1991), pp. 56-57.

admitted increased from 18 to 28 (United Nations, 1992). A large proportion of the legal migrant workers admitted annually by Japan has consisted of female entertainers whose number increased from 24,000 in 1982 to 71,000 in 1988 (United Nations, 1992).

It is estimated that 273,000 unregistered foreign nationals were illegally employed in Japan at the end of 1990 (Mori, 1991). Measures to deal with undocumented migration and illegal employment were passed as part of a new Immigration Control Act that entered into force in 1991. The distribution by citizenship of undocumented aliens in Japan cannot be ascertained with certainty, although it appears that the largest group consists of Philippine and Thai women, followed by men from the Republic of Korea illegally employed in shipyards and industrial plants, men from Pakistan and Bangladesh and Chinese men. These observations are based on recorded detentions of illegal migrants which may or may not represent the total undocumented population. Despite the increase in migration during the 1980s, foreign workers still constitute only a very small proportion of the labour force (less than 1 per cent).

2. *Migration to Singapore*

In 1990, Singapore had an estimated 120,000-180,000 foreign workers of whom 10,000 were daily commuters from Johore (Stahl, 1991; Fong, 1991). With a total labour force of 1.3 million, foreigners constitute about 9-14 per cent of it. Fong (1991) estimates that in 1990 there were 150,000 unskilled workers and 30,000 professional and skilled workers on employment passes. These numbers do not include undocumented migration.

Traditionally, Malaysians have constituted the largest number of foreign workers in Singapore. In recent years, the sources have diversified to include Filipinos, Indians, Indonesians and Thais. Malaysian workers are allowed in all sectors but are concentrated in manufacturing. The Philippines is a source of workers engaged in domestic service (35,000), while Thailand has sent about 20,000 construction workers to Singapore. While a majority of foreign workers are unskilled, government policy encourages the migration of relatively educated workers.

Singapore is also the source of several thousand skilled workers who are employed in neighbouring countries (Indonesia and Malaysia) either on short-term assignments or as employees of Singaporean-owned firms.

3. *Migration to Malaysia*

Malaysia is both the origin and the destination of labour migration flows. In addition to being the source of workers employed in Singapore, it hosts more than 1 million foreign workers (including undocumented migrants). It is estimated that one half of Sabah's 1.4 million population consists of undocumented migrants, mainly from Indonesia. The Government estimates that there are an additional 350,000 Indonesians working illegally in peninsular Malaysia (Stahl, 1991).

Undocumented Indonesians enter Malaysia at several points by ferry. They find employment in three main sectors: plantation estates, government agricultural land schemes and construction. Labour shortages in Malaysia's construction industry are partly a result of the emigration of Malaysians to work in Singapore. It is estimated that 70 per cent of the workers in Singapore's construction industry are foreigners. On the other hand, in 1987, about 50 per cent of the 360,000 construction workers in Malaysia were believed to be Indonesians (Fong, 1991).

Indonesia, Malaysia and Singapore have become progressively interlinked economically through capital and manpower transfers. A growth triangle consisting of Johor in Malaysia, the Riau islands (especially Batam) in Indonesia and Singapore was proposed in 1989 by Singapore's Deputy Prime Minister with a view to maximizing the economic growth and competitiveness of each country. Joint ventures among the three countries are expanding at a rapid pace. One of the eventual labour force implications of the development of the growth triangle and the narrowing of wage differentials between the three countries may be a reduction in migration between them. Emerging patterns will depend on several factors, including rates of economic growth, levels of migration to the traditional countries of immigration (Australia, Canada and the United States of America) and the growth of the service sector.

4. *Migration to the Republic of Korea*

During the late 1970s and early 1980s, the predominant form of migration affecting the Republic of Korea was the export of contract labour to Western Asia which peaked in 1982 at about 150,000 workers and declined to less than 10,000 in 1990. The number of migrant workers in 1990 was 55,774 of whom 42,574 were employed on foreign vessels (Park, 1991). Since about 1985, worker migration has decreased because of both the worsening economic conditions in Western Asia and the competition from cheaper labour sources in South-eastern Asia, as well as the improving economic situation in the Republic of Korea itself which is reducing incentives to migrate. Wages in the Republic of Korea have increased more than 50 per cent since 1987 (Lim, 1991).

Labour migration to the Republic of Korea has been limited. The existing policy does not allow unskilled workers to enter the country. Of those who in 1990 were admitted for employment, more than 90 per cent entered for business or capital investment. However, it is believed that in recent years a number of unskilled foreign workers have been employed illegally in the country. The number of undocumented migrants apprehended increased from 255 in 1988 to 1,918 in 1990 (Park, 1991) and it is believed that they represent only a small proportion of all undocumented migrants. Among those apprehended in 1990, about one fourth were from the Philippines, followed by Pakistan, Bangladesh and the Islamic Republic of Iran. They had worked as language teachers, maids, restaurant workers, and as unskilled workers at construction sites and garment factories.

5. *Migration to Hong Kong*

According to the 1981 Census of Hong Kong, the Crown Colony had over 2 million foreign-born residents, more than 90 per cent of whom were from China. At least since the 1950s, Hong Kong has received significant inflows of Chinese from the People's Republic. The 1970s witnessed a sharp rise in undocumented migration from China, mostly due to the fact that Chinese migrants could regularize their status in Hong Kong if they managed to reach the urban areas of the Colony. Such a rise prompted British authorities to remove that possibility in 1980. It is estimated that net Chinese migration to Hong Kong dropped from 79,000 in 1976-1981 to 25,000 during 1981-1986 (Zlotnik, 1994b). In May, 1988, 13,300 illegals were apprehended, 4,000 of whom were imprisoned (Stahl, 1991).

Measures to reduce migration to Hong Kong have been accompanied by the relocation of Hong Kong industries to China. It is estimated that in 1988, about 1 million jobs were created in the Pearl River delta through industrial linkages and subcontracting. The Hong Kong Financial Secretary reported that in 1992 about 3 million Chinese workers in south China were employed in concerns owned by Hong Kong interests.

6. *Migration to Taiwan Province of China*

With gross national product (GNP) per capita growing at 14 per cent per annum and a total fertility of 1.7 children per woman, Taiwan Province of China is faced with serious labour shortages that are increasingly being filled by undocumented workers (Stahl, 1991). Estimates of the number of such workers vary from 20,000 to 300,000. Abella (n.d.) suggests that an estimate of 40,000 is reasonable. Migrants to Taiwan Province of China originate mainly in the People's Republic of China, Malaysia, the Philippines and Thailand. Most are employed in the construction industry and in domestic service.

C. Policies of host countries

The Governments of most host countries, be they in Western, Eastern or South-eastern Asia, are generally concerned about high numbers of migrants and about undocumented migration. Hence, policies to limit legal migration and to discourage or eliminate undocumented migration are common. The latter include fines, deportation and even imprisonment. In addition, most receiving countries attempt to attract only high-quality foreign workers. Thus, the admission of unskilled workers is usually discouraged, especially in Japan and the NIEs. As already mentioned, both the NIEs and Japan have tended to relocate their labour-intensive industries in labour-rich countries. Such a strategy has not been possible in Western Asia, since foreign workers in that region have been used for the development of infrastructure, services and the oil industry, all of which cannot be moved.

Countries in Eastern and South-eastern Asia have to different extents formalized their policy to restrict legal migration and stem undocumented migration. Japan, having experienced increasing undocumented flows, decided to regulate the process through an amendment of the existing Immigration Control Act. The amendment, promulgated in December 1989, came into force on 1 June 1991. One of its main objectives is to control undocumented migration by establishing tough sanctions on employers who hire undocumented workers (up to three years of imprisonment and a maximum fine of 2 million yen). Under the original Act, only undocumented workers were liable for imprisonment of up to six months or a fine of up to 200,000 yen (Mori, 1991). The new Act increases the fine on undocumented aliens to 300,000 yen and the possible imprisonment term to a maximum of three years. As a result of the new Act, some 20,000 undocumented workers turned themselves in to the Japanese authorities (Stahl, 1991).

In 1988, the Government of Sabah in Malaysia introduced legislation to curb undocumented migration. Amnesty was granted to undocumented workers who would be issued work permits if they wanted to return to Malaysia. Some 200,000 undocumented Indonesian migrants took advantage of that opportunity (Stahl, 1991).

In the early 1980s, Singapore announced its intention to repatriate all foreign workers, a plan that it had to give up in view of its continued labour needs. It has had, however, a vigorous policy to control the skill and educational level of the foreign workers admitted legally. In order to discourage unbridled migrant inflows, the Government imposed a levy on firms hiring foreign workers. On 1 August 1990, the levy was $300 for all employees other than maids, for whom it was $230. In order to encourage the import of skilled workers, the Government is considering setting a higher levy for unskilled workers ($350) than for those with skills ($250) (Stahl, 1991). In March 1989, Singapore enacted a new law to control undocumented migration. The new law establishes three months of mandatory imprisonment and three strokes of the cane for migrants who lack the necessary permits to stay and work in the country (Fong, 1991).

In the Republic of Korea, the Government has still not reached a consensus regarding the legal admission of foreign workers. Current immigration law does not allow the admission of unskilled foreign workers. Certain ministries, such as the Ministry of Trade and Industry, agree with the business sector's claim that there is a need for foreign labour. However, the Ministry of Justice opposes this position. Until 1990, the Ministry of Labour supported the Ministry of Justice. Then, it announced that it would consider the admission of foreign coal miners, though not that of other workers. A final decision on the matter has been postponed until guarantees for the departure of foreign workers can be secured (Park, 1991).

Just like the countries of Eastern and South-eastern Asia, the labour-importing countries of Western Asia also have concerns about the relatively high numbers of foreign workers in their midst. The share of foreign workers in the labour force of most GCC countries is around 80 per cent. Planning documents and laws have codified the importance of reducing dependence on foreign labour. Strict regulations regarding undocumented migrants have always existed. In Kuwait, undocumented workers who are apprehended face imprisonment and deportation. In recent months, following the liberation of the country from Iraqi occupation, several cases of Iranian undocumented workers crossing the Gulf by boat have been reported and dealt with firmly.

As part of the policies to control "guest workers", the GCC countries encourage the use of fixed-term contracts which are renewable. A migrant worker can only change jobs if the contract is legally transferred from one employer to another. Another way of controlling the movement of workers is to require employees to deposit their passports with employers. In Kuwait, those employed in Government ministries must obtain an exit permit each time they leave the country. Families of workers are allowed to join them only under certain conditions. For example, only those persons earning about US$ 1,500 per month if employed in the public sector or US$ 2,200 if employed in the private sector may take their families to Kuwait. Children of foreigners have no right to the host country's citizenship, even if they are born in the country. In 1989, one third of the non-Kuwaiti population was born in Kuwait, largely of Palestinian parents. That sub-group had acquired virtual permanent status in Kuwait and many Palestinian workers had stayed in the country for about 18 years (Shah and Al-Qudsi, 1989). After the Iraqi invasion, a large majority were not allowed to remain.

D. Policies of countries of origin

The labour-sending countries usually share three concerns around which most policies and programmes have centred. First, they want to increase, or at least maintain, the number of their workers working abroad in a market that is highly competitive and becoming increasingly so. Second, they want to protect the rights of their workers abroad. Third, they want to regulate the process so that intermediaries can be checked and prevented from cheating potential migrants.

In order to fulfil the above concerns, Governments have developed various legal and administrative arrangements, generally establishing separate bureaux to regulate the emigration of workers and streamlining procedures for the issuing of travel permits. Most sending countries have established minimum wages and basic standards that must be met by employers abroad. In countries where emigration takes place primarily through private recruitment agents, extensive rules to regulate the work of the agents exist. Several Governments have placed restrictions on the emigration of certain subgroups specified in terms of age, sex and skills. In addition, the major countries of origin in Asia have labour attachés in Western Asian countries to promote migration opportunities and oversee the welfare of migrants. Labour attachés assist in cases where disputes between the employer and employee arise.

In a review of the policies and programmes of sending Asian countries, Shah and Arnold (1986) concluded that an inherent basic issue in regard to the policies for regulating labour migration to Western Asia was their somewhat contradictory nature. Since a major objective is to increase the number of workers abroad, Governments of countries of origin may be constrained with respect to accepting relatively unfavourable contractual terms in order to remain competitive. They may also refrain from protesting against the mistreatment of workers to avoid creating an embarrassment for the country of employment. A reflection of the increased competition associated with the shrinking Western Asian market and an unabated supply of labour is the trend towards lower wages in the host countries. In June 1986, Pakistan announced a 15 per cent cut in the minimum wage standard, while Bangladesh lowered its standard by 20 per cent (Birks, Seccombe and Sinclair, 1988). Sri Lanka, which has traditionally given free rein to private initiative, has also begun to formalize its recruitment procedures, setting up a Bureau of Foreign Employment in 1985 (Eelens and Speckmann, 1990). While wages in other sectors may have gone down since the liberation of Kuwait, the Sri Lankan Embassy has tried to secure higher wages for housemaids (US$ 136) by making a public announcement of the wage level. Prior to the invasion of Kuwait, the Embassy did not enforce a minimum wage and maids usually earned about US$ 100 per month.

Ideally, countries of origin would have more power if they could collaborate in setting and enforcing minimum wage standards and standards of treatment of workers abroad. During the 1980s, no advances were made towards such collaboration. In the host countries, problems such as non-payment or arbitrary reduction of wages, violation of contracts, increased number of working hours without adequate remuneration, subhuman living conditions, social isolation, and ill-treatment, particularly of domestic workers, are all well known and persistent. In the countries of origin, the unscrupulous and criminal treatment of potential migrants by recruitment agents continues to be a serious problem. Agents continue to overcharge and provide false information or promises of jobs that may not even exist.

E. Female migration: its nature and impact

About one half of all international migrants in the world are women (United Nations, 1995). The increase in female migration is one of the most significant features of Asian migration in the last decade. Domestic service is the single most important category of employment among female migrant workers in the region. Countries such as Indonesia, the Philippines and Sri Lanka have responded rapidly to the increased demand for female workers in Western Asia and other countries in the region. In Kuwait, 103,501 women were employed as domestic workers in 1989, constituting 5.1 per cent of the total population of the country. Most (98 per cent) were Asian. Among all employed Asian women, 85 per cent were domestic workers compared with only 5 per cent among Arab women (Kuwait, 1989). In Saudi Arabia, there were 219,000 Asian female workers in 1986 (Abella, 1995). In Hong Kong, the number of domestic helpers increased from 880 to 70,335 between 1974 and 1990 (Lim, 1991). In Kuala Lumpur, 10,000 Philippine workers were employed as domestic workers in the late 1980s (Stahl, 1991). The entertainment industry has also been an important employer of migrant women in Eastern and South-Eastern Asia. Consensus appears to exist concerning the fact that demand for both domestic workers and entertainers is unrelated to the pace of economic development and exists primarily for reasons of luxury and social status.

Abella (1995) describes five distinct flows involving the migration of Asian women: (*a*) the flow from Southern and South-eastern Asia to Western Asia involving an annual outflow of some 95,000 women migrating through legal channels and another 50,000-60,000 migrating clandestinely every year; (*b*) the flow from Indonesia, the Philippines and Thailand to the NIEs involving about 62,000 women annually; (*c*) the flows of Philippine and Thai women working mainly as entertainers in Japan where, in 1988, 71,000 Philippine women were admitted legally; (*d*) the migration of skilled and professional women to Western Asia that involves at least some 28,000 Philippine women (mostly nurses); and (*e*) the migration of Asian women to Australia, Canada, the United States and Europe, whose numbers are estimated at 14,000-18,000 migrating legally, and 35,000-50,000 migrating through irregular channels. These numbers suggest that more than 320,000 Asian women migrate as workers every year, the majority being within Asia.

In all Southern Asian countries except Sri Lanka, men far outnumber women among migrant workers. In Indonesia, the Philippines and Sri Lanka, women have constituted the majority of migrant workers during the 1980s. In Sri Lanka, the Ministry of Plan Implementation reported that 57 per cent of all migrants in 1981 were women (Athukorala, 1990). That percentage may have increased, as shown by a survey of returning migrants of low socio-economic status in Colombo and Gampala where 70 per cent of the returning migrants were women (Eelens, 1995). Women outnumbered men 3.4 to 1 among migrant workers from Indonesia and 12 to 1 among Philippine migrants going to Eastern and South-Eastern Asian in the late 1980s (Lim, 1991).

Female migration has increased not only because of growing demand in the countries of destination, but also because of the fact that it is often cheaper for a woman to migrate than for a man. Thus, agents' fees are lower for women than for men in Sri Lanka (Eelens and Speckmann, 1990). Because of cultural affinities, Muslim women have been in greater demand in countries like Saudi Arabia, thus providing added opportunities for Muslim Indonesians. Philippine women have migrated in large numbers to destinations both in Western Asia and in South-eastern Asia. Among those who left in 1987, 44 per cent were domestic workers, 17 per cent were entertainers and another 17 per cent were professionals such as doctors and nurses (Abella, 1995).

Because of the nature of the jobs open to them, female migrants, being easy subjects of exploitation, often find themselves in vulnerable positions (Eelens, 1995). Governments of countries of origin have therefore been especially concerned about the protection of female migrant workers. Some countries, such as Bangladesh, Pakistan and, more recently, India, have effectively banned the migration of female domestic workers by establishing age limits. Sri Lanka has imposed no restrictions on the migration of female domestic workers, but the Philippines, which is perhaps the major exporter of female workers, did put a ban on the migration of female domestic workers in 1982. Since then, it has selectively lifted the ban once bilateral agreements on the protection of domestic workers were reached (United Nations, 1992). However, in most countries where women constitute large proportions of the outflow of migrant workers, the restrictions established are more symbolic than real. Authorities in the Philippines and Thailand have found ways of exempting many employment categories of women from such prohibitions. The enforcement of regulations to restrict female migration depends on their compatibility with the prevailing sociocultural milieu, with labour-market realities and the political systems of countries of origin (Abella, 1995).

Governments of countries of origin have only limited possibilities of protecting female migrants abroad. In desperate situations, as when domestic workers are abused by employers, their embassies provide food and shelter and try to salvage the situation. Sometimes, embassies issue new passports to enable women to return to their country when their own passports have been retained by their employers. Embassies also intervene in negotiations between migrants and employers, and request the help of the local police if necessary. However, in most cases the employers' word is given more credence than that of the migrant women.

Increased female migration appears to be a response to changing lifestyles in the host countries. As families become more affluent, there is a social need for buying more leisure. This seems to be true in Western Asia as well as in the NIEs. In Kuwait, after liberation from the Iraqi occupation in February 1991, some of the first migrants to be recalled were domestic workers. Between liberation and July 1992, a total of 128,603 visas were issued for domestic workers, 73 per cent of those visas being for women and 41 per cent for Sri Lankan nationals (Ministry of Social Affairs and Labour, verbal communication). The demand for domestic workers has become so pervasive that even relatively poor households engage them. In 1986/87, one third of households earning less than US$ 850 per month had at least one domestic worker, whereas among the richest Kuwaiti households, with incomes of US$ 6,800 or more, 90 per cent had at least one domestic worker. In some cases, the presence of a domestic worker was needed to allow the labour force participation of the women in the household, but a majority of the domestic servants (64 per cent) were employed in households where no woman worked (Shah, Al-Qudsi and Shah, 1991; Shah and Al-Qudsi, 1990).

Another change in respect of lifestyles is the growing emphasis on entertainment in Japan. It remains to be seen whether the import of entertainers to Japan marks the beginning of a trend in the NIEs. If so and if the demand for domestic workers expands with the rising affluence of the NIEs, an increase in female migration may be expected. Countries like the Philippines and Sri Lanka, with established traditions of migration and well-developed networks to support the move, will continue to be at centre stage, while other countries, like Indonesia, may expand their markets further. The increase in female migration will be relatively unrelated to the demand for male migrants, since the latter is more closely associated with economic growth than with sociocultural factors.

In Asia, as in all other developing regions, women are usually concentrated in low-paid, low-status jobs, and that is still the pattern with respect to female international migrants. In some cases, the migration of women may be critical for the survival of the family left behind and in others, it may lead to the improvement of the family's living standard in terms of housing, education of children and access to health care. It is hazardous to make any generalizations about the positive or negative impacts of female migration since it occurs under very varied situations. However, as the United Nations Expert Group Meeting on International Migration Policies and the Status of Female Migrants concluded, in most instances migrant women have proved to be active agents of change and adaptation rather than passive victims of circumstances (United Nations, 1995). There remains a need to enforce certain minimum standards for the protection of migrant women in domestic and entertainment occupations, difficult as it is to do. More effective implementation of existing policies should be pursued, hopefully through a closer cooperation between host countries and countries of origin.

F. Consequences for the development of countries of origin

Evaluating the developmental impacts of migration for countries of origin is problematic for several reasons. Migration is only one among a host of complex economic, political and social factors that affect the development process. Research on the topic has hardly addressed the issue of migration within robust models of socioeconomic growth (Appleyard, 1989). In addition, lack of data or their poor quality severely hampers the studies possible. For example, in the view of Russell (1986), data problems relative to remittances are so great that their systematic analysis seems hardly worth the effort. An analysis of the literature on the subject led Seccombe (1985) to characterize it as "descriptive and judgemental". Lastly, the economic characteristics of labour-exporting countries may vary so widely that a migratory flow of a given volume and composition may be beneficial to one country, neutral with respect to another and detrimental to a third (Stahl, 1991).

With the above qualifications in mind, one may summarize the major arguments of the debate on the impacts of migration on the development process as follows. The two variables that are typically considered in assessing the economic impacts of migration are remittances and the short- and long-term effects of migration on the labour market of the countries of origin. A third possibility that is harder to quantify is the social and "human" costs of migration and it is not addressed here.

1. *Remittances*

The issue of remittances is at the heart of the debate on whether labour migration is good or bad. There is a wide-ranging consensus that remittances sent by workers to their families may have positive as well as negative effects (Russell, 1986). Among the positive macrolevel effects, the easing of foreign exchange constraints which contributes to the growth of the economy is crucial. In some Asian countries, remittances have become the most important source of foreign exchange earnings. In 1987, remittances represented 8.1 per cent of gross domestic product (GDP) in Pakistan, 5.6 per cent in Sri Lanka,

4.5 per cent in Bangladesh and 2.3 per cent in the Philippines. In certain Arab countries, such as (northern) Yemen and Jordan, remittances corresponded to even higher percentages of GDP, 40.9 per cent and 22.5 per cent, respectively (Abella, 1994). Whether remittances boost the growth of the economy depends ultimately on the rates of savings and investment and the productivity of investment. In certain countries (for example, Yemen) where remittances are used mainly for consumption (often of foreign goods), remittances may create only dependency; but in other countries where the economy has the necessary absorptive capacity and where structural adjustments of the labour force are possible, the effects of remittances may be positive and substantial. Jordan has been cited as a successful example of the latter type of country (Birks and Sinclair, 1980).

At the household level, even though generalizations cannot be made for each country in question, there is evidence to show that migration may reduce income inequality (Appleyard, 1989). In the case of Pakistan, Adams (1992) concluded that the impact is neutral since the migrants are widely distributed throughout the socio-economic strata and the gains from migration are therefore widespread.

Among the negative aspects of remittances is their unpredictability, the fact that substantial portions may be spent on consumption (hence increasing inflation), and that a large part may be "hidden" and thus not amenable to government regulation and control. The unpredictable nature of remittances was amply highlighted by the Gulf war and the sudden return of about 2.2 million Arab and Asian expatriate workers to their countries of origin. The numbers of workers returning to India, Pakistan and Sri Lanka were 170,000, 90,000 and 73,000, respectively (Abella, 1994). The number of Arab workers who moved as a result of the Gulf war and the political decisions following it included 800,000 Yemenites, 400,000 Palestinians/Jordanians and 500,000 Egyptians. In Egypt, it was estimated that a total of 1.5 million persons were affected, including dependants. The public expenditure necessary for education, health and job creation was estimated to be 5 billion or 6 billion dollars for which the country had to rely on foreign aid (Appleyard, 1991b).

In terms of expenditure on consumption and the ensuing inflationary impacts, a consensus exists that migrants are better savers than investors. Most surveys of household behaviour refute the popular impression that migrants waste their income in frivolous consumption. In Southern Asian countries, migrant households save about 40 per cent of the remittance income. The proportion of savings put into productive investments is relatively low, however. Most investment—40.6 per cent in Sri Lanka, 57 per cent in Thailand, 45 per cent in Bangladesh and 50 per cent in Pakistan—goes into the purchase of housing and land (Lim, 1991). Investment in housing is likely to have had a positive impact on creating employment and generating "multiplier effects" in other related industries, while at the same time raising the price of land (Abella, 1994).

A basic problem in terms of the Government's inability to regulate investment relates to the individualized nature of earnings. Choucri (1986) claims there is mounting evidence that in addition to the formal channels through which income is sent, there are informal channels and remittances flow largely through such informal channels. The hidden economy is so large that in some labour-exporting countries the value of its transactions is several times the value of export earnings. In Egypt, the remittances captured by official sources amount to 3.3 billion dollars but the actual amount may be closer to 20 billion. Similarly, in the Sudan only 10-15 per cent of remittances are made through official channels (Choucri, 1988). Among certain groups of migrants from Pakistan, informal networks, such as those based on the *hundi* (a negotiable instrument, bill of exchange or promissory note used especially in the internal finance of trade), that provide better exchange rates are also known to be common. In Pakistan, a study carried out in 1986 showed that only 57 of all remittances were sent home through official channels (Abella, 1994). An assessment of the prevalence of such informal systems in other Asian countries and the amounts remitted through them is needed for a realistic view of the impacts of remittances on development.

While the debate on the net effects of remittances on development continues, it may be stated that benefits are greatest if the following conditions are met: (*a*) existence of a diversified economic structure; (*b*) an adequate supply of labour; and (*c*) a financial system capable of handling remittances effectively (Appleyard, 1989).

2. *Labour-market impacts*

In terms of the impact of migration on the labour market of the countries of origin, it is again difficult to make generalizations both because there is a paucity of data and because the effects depend on specific labour-market conditions, underlying demographic factors, the occupational distribution of migrants and the overall structure of the economies in question (Russell, 1992). The major issues concerning the labour-market impacts relate to the effects on unemployment, labour shortages, skill depletion and deskilling of the labour force. Even though findings are not consistent for various countries, a general consensus seems to exist that emigration from Asian countries with a surplus of labour has acted as a safety valve for the increasing pressures on local labour markets. In the presence of high levels of underemployment and unemployment, emigration has had no obvious unfavourable impact on output or output growth (Amjad, 1989).

While the above conclusion is generally accepted for the large Southern Asian countries, selective shortages of skilled workers have been mentioned by several countries. In the case of some Arab countries, like Yemen, the negative developmental and labour force consequences have been particularly severe. With the migration of 800,000 Yemenites, labour shortages became so severe that children had to take over the full-time jobs of adults. In 1978, it was estimated that about 20 per cent of all jobs were vacant, while technical departments suffered vacancy rates of up to 50 per cent (Fergany, 1983). The country had to import workers to satisfy its own needs. Although remittances allowed the country to buy rather than produce what it needed, it became dependent on them.

Concerns have been expressed about the loss of human capital when the best-qualified workers emigrate and are replaced by less-qualified workers. Labour shortages in certain sectors, such as construction, have pushed wage rates up in various countries. In addition, workers who need to learn their skills through apprenticeship to a master may no longer be trained once the master migrates and no replacement can be found, as was the case in Pakistan (Ahmad, 1982). Furthermore, some deskilling may occur when workers are employed in occupations below their qualifications, as is the case with Philippine women who have completed their higher education and work as domestic servants (Lim, 1991). One third of Sri Lankan workers in Western Asia have accepted jobs that rank lower in status and require lower qualifications than those they had prior to migration (Athukorala, 1990).

Most countries, but especially the NIEs, are concerned about the emigration of their highly skilled personnel, including professionals and entrepreneurs, to developed countries. Singapore, where the Cabinet Minister described the brain drain as a time bomb, has instituted several mechanisms to stem the flow (Lim, 1991). Hong Kong, in particular, has experienced relatively high emigration of the highly skilled resulting from the anticipated return of the Crown Colony to Chinese rule in 1997. Singapore has tried to attract Hong Kong emigrants by giving them preferential treatment in respect of educational requirements and easier business terms. Singapore thus hopes to attract about 100,000 persons from Hong Kong (United Nations, 1992). Concerns about the brain drain have also been expressed by the Arab countries. It was estimated that by the early 1980s, 50 per cent of all Arab scientists and engineers with Doctor of Philosophy (Ph.D.) degrees had already emigrated (Tabbarah, 1988). Countries with high unemployment rates among the educated, however, can hardly claim that the brain drain has significant economic effects.

G. Refugee movements and impacts on receiving countries

At the beginning of the 1990s, there were about 17 million refugees in the world, 87 per cent of whom had found asylum in developing countries. Asia had more than 7 million refugees in early 1989, almost 80 per cent originating in Afghanistan (table 44). In 1989, about 60 per cent of the 5.6 million Afghan refugees who had left as a result of the Soviet invasion of their country in December 1979 were in Pakistan and the rest (except for about 4,900 in India) in the Islamic Republic of Iran. When refugees started moving into the North-West Frontier Province of Pakistan, the Government of Pakistan did not try to seal the borders on Muslim brethren in distress. Instead, the refugees were housed in camps and were well treated. When the Union of Soviet Socialist Republics withdrew its troops from Afghanistan during 1988-1989, efforts to achieve the voluntary repatriation of Afghan refugees began with the assistance of the Office of the United Nations High Commissioner for Refugees (UNHCR). However, the war in Afghanistan continued after the country gained independence and it is still going on even after the communist regime in the country was replaced by the mujahideen Government. Thus, instead of repatriation, an additional 70,000 refugees were registered by May 1989 (United Nations, 1992).

Over the years, substantial foreign assistance was made available to help refugees and to make them as self-sufficient as possible. Afghans have engaged in a variety of economic activities in Pakistan. Some have bought property and businesses in Peshawar. Others are employed as labourers and may have replaced some of the migrants to Western Asian countries. Geographically, they are no longer restricted to the campsites. Some can be found as far south as Karachi, pursuing transportation and other businesses (Shah and Arnold, 1990). About 200,000 Afghan refugees have settled in the Mianwali district of the Punjab (Arnold, 1989). UNHCR believes that a proportion ranging from one quarter to one third of all Afghan refugees will not return to Afghanistan.

In the Islamic Republic of Iran, the Government's policy has allowed most Afghan refugees to become part of the local economy by settling, taking up employment and benefiting from subsidized food rations, free education, and medical care on equitable terms with nationals (United Nations, 1992). Although the Afghan presence in Pakistan has led to some communal tensions (for example, the clashes in Karachi between the Pathans and the local community in 1987 and 1988 and a series of bombings in public places during 1987 believed to be associated with Afghan terrorists (Shah and Arnold, 1990)), the presence of Afghan refugees does not seem to be an important issue in Pakistan at present. It is generally assumed that the Afghan refugees will eventually leave. However, Afghans are beginning to adopt the local language and cultural patterns and might eventually become assimilated to the population of Pakistan, especially since they share several of the language and ethnic characteristics of the people of the North-West Frontier Province.

A large group of migrants not shown in table 44 consists of Palestinian refugees, 2.3 million of whom were recorded by the United Nations Relief and Works Agency for Palestine Refugees in the Near East (UNRWA) in 1989. About 38 per cent of them were in Jordan, 37 per cent in the occupied West Bank and Gaza Strip, and the remaining 25 per cent in Lebanon and the Syrian Arab Republic. In addition, about 400,000 Palestinians with Jordanian passports were living in Kuwait prior to the Iraqi invasion. Because of the Palestine Liberation Organization (PLO) allegiance to Iraq during the Gulf war, a majority of the Palestinians either left during the occupation or had to leave Kuwait before July 1992 since their residence permits were not renewed even though many of them had been born and raised in Kuwait. Most of these people went to Jordan, that is to say, between 300,000 and 350,000 Palestinians moved to that country as a result of the Gulf war.

Within Jordan, the Palestinian refugees are integrated into the economy and society more than in any other Arab State (Wardwell, 1990). Only one quarter of them resided in camps in the late 1980s. One of the consequences of the refugee experience has been the relatively high premium that Palestinians have placed on educating their children. In Jordan, both Jordanians and Palestinians have emphasized skill development to improve their competitiveness in terms of jobs in other Western Asian countries. Consequently, Jordanians and Palestinians are among the most highly qualified workers in the region. The Jordanian economy has benefited considerably from the presence of Palestinians who have taken up the jobs vacated by nationals emigrating to other Western Asian countries. Nevertheless, Jordan has experienced considerable labour shortages and has become a labour importer, especially of agricultural workers. In some parts of the country, 65 per cent of the agricultural labour force consists of foreigners, 95 per cent of whom are Egyptian (Wardwell, 1990).

The third group of refugees originating in Asia consists of the Indo-Chinese (Cambodians, Vietnamese and Laos). Between 1975 and 1988, nearly 1.5 million Indo-Chinese refugees were resettled in third countries, almost one half of them in the United States. However, by the late 1980s, there were still 500,000 Indo-Chinese refugees in Eastern and South-Eastern Asia, including some 97,000 Vietnamese "boat people". Indeed, the increasing outflow of refugees from Viet Nam during the 1980s prompted a series of deterrent responses from the main receiving countries in the region, including detention of new arrivals, the summary "push back" of asylum-seekers arriving over land and the "push off" of those arriving by boat. Concern about these developments led the international community to convene a second International Conference on Indo-Chinese Refugees held in Geneva in June 1989. The Conference adopted a Comprehensive Plan of Action (CPA) (United Nations, 1989) which called for the expansion of orderly departures from Viet Nam, the discouragement of clandestine departures by boat and the continued resettlement of Vietnamese outside the region (United Nations,

TABLE 44. NUMBER OF REFUGEES IN ASIA, BY COUNTRY OR AREA OF ASYLUM, AND BY ORIGIN, 1985-1989

Country or area of asylum	*Origin*	*Early 1985*	*Early 1988*	*Early 1989*
Bangladesh	Various	—	41	42
China	Indo-Chinese	279 750	280 600	248 018
Hong Kong	Viet Nam	11 896	9 532	25 749
India	Afghanistan	5 846	5 175	4 866
	Iran (Islamic Republic of)	1 215	1 440	1 656
	Sri Lanka	—	129 750	90 000
	Various	92	98	92
	TOTAL	7 153	136 463	96 614
Indonesia	Indo-Chinese	9 453	2 453	2 352
Iran (Islamic Republic of)	Afghanistan	1 800 000	2 350 000	2 300 000
	Iraq	—	—	90 000
	Various	100 000	410 000	410 000
	TOTAL	1 900 000	2 760 000	2 800 000
Japan	Indo-Chinese in transit	1 290	515	487
	Indo-Chinese resettled	—	5 590	5 800
	Various	—	103	133
	TOTAL	1 290	6 208	6 420
Lao People's Democratic Republic	Cambodia	1 200	—	—
Lebanon	Various	2 900	5 900	4 400
Macau	Indo-Chinese	727	518	440
Malaysia	Filipino	90 000	90 000	90 000
	Indo-Chinese	8 853	9 120	14 210
	Total	98 853	99 120	104 210
Nepal	Various	—	20	88
Pakistan	Afghanistan	2 500 000	3 156 000	3 255 000
	Various	—	3 000	3 000
	TOTAL	2 500 000	3 159 000	3 258 000
Papua New Guinea	Indonesia[a]	10 946	9 566	8 816
Philippines	Indo-Chinese	1 960	3 219	5 030
(Refugee Processing Centre)	Indo-Chinese	12 907	9 740	15 734
	Various	—	229	264
	TOTAL	14 867	13 188	21 028
Republic of Korea	Viet Nam	58	63	123

TABLE 44 (*continued*)

Country or area of asylum	*Origin*	*Early 1985*	*Early 1988*	*Early 1989*
Singapore	Indo-Chinese	249	311	203
	Various	—	184	273
	TOTAL	249	495	476
Syrian Arab Republic	Lebanon	33 600	—	—
Thailand	Cambodia	41 619	22 974	19 905
	Lao People's Democratic Republic	82 094	74 984	74 960
	Viet Nam	4 726	14 535	13 470
	Non-Indo-Chinese	—	247	299
	TOTAL	128 439	112 740	108 634
Turkey	Various	2 600	2 200	4 450
Viet Nam	Cambodia	21 000	25 000	25 000
Yemen[b]	Ethiopia	—	1 000	1 000
	Democratic Yemen[b]	—	70 000	70 000
	TOTAL	—	71 000	71 000
Other Asian countries	Various	—	230 000	230 000
	GRAND TOTAL	5 024 981	6 924 107	7 051 860

Source: *World Population Monitoring, 1991: With Special Emphasis on Age Structure* (United Nations publication, Sales No. E.92.XIII.2), table 89.

NOTE: According to sources at the Office of the United Nations High Commissioner for Refugees, the figures presented here were provided mainly by Governments of asylum countries based on their own records and methods of estimation.

[a]Irian Jaya.

[b]On 22 May 1990, Democratic Yemen and Yemen merged to form a single State. Since that date they have been represented as one State Member of the United Nations with the name "Yemen". For some statistical data that predate the merger, it has been necessary to refer occasionally to the former States of Yemen and Democratic Yemen.

1996). In addition, the CPA led to the adoption of individual screening procedures for Vietnamese asylum-seekers in all countries of the region. Thus, whereas prior to 15 March 1989 Vietnamese asylum-seekers had been granted refugee status on a group basis, as of that date proof of a well-founded fear of persecution on an individual basis became necessary to secure refugee status. Application of individual screening has since resulted in growing numbers of asylum-seekers who do not qualify for refugee status and who may have few options aside from returning voluntarily to their country.

H. FUTURE TRENDS

Conclusions about likely future trends in migration between Asian countries must take account of the broad demographic and socio-economic scenarios expected for the countries of both origin and destination. In the countries of origin, especially those in Southern Asia, the rates of population and labour force growth continue to be high. In both Bangladesh and Pakistan, the labour force will increase by more than 3.5 per cent annually during 1990-2000 (table 45). Even in most of the countries of South-eastern Asia, annual rates of labour force growth are well above 2 per cent. In most of Southern Asia, particularly in Pakistan, family planning has had only limited success and high rates of growth are expected to continue for the foreseeable future. The pressures of unemployment and underemployment in countries like Bangladesh are enormous: in 1987 it had an unemployment rate of 12 per cent and the prospects of creating jobs for future labour force entrants are bleak (Abella, 1994). Amid such pressures there exist information and support networks that facilitate migration and the growth of a rather efficient "recruitment

TABLE 45. PROJECTED GROWTH OF INDUSTRIAL EMPLOYMENT IN SELECTED ASIAN COUNTRIES

Country	*Employment in industry as a proportion of total employment (Percentage)*	*Growth of labour force 1990-2000 (Percentage)*	*Growth of industrial employment 1981-1990 (Percentage)*	*Number of industrial sector jobs per new entrant into the labour force*
Japan	33.6	0.4	4.4	3.7
NIEs				
Hong Kong	38.7	0.9	5.0	2.1
Korea, Republic of	34.3	1.8	11.2	2.1
Singapore	35.6	0.6	4.6	2.7
Taiwan Province of China	42.2	1.7	11.5	2.9
Southern Asia				
Bangladesh	12.0	3.6	5.0	0.2
India	15.0	2.1	5.9	0.4
Pakistan	19.0	3.7	6.8	0.4
South-eastern Asia				
Indonesia	15.0	2.4	5.2	0.3
Malaysia	23.3	2.9	6.5	0.5
Philippines	14.5	2.8	4.2	0.2
Thailand	12.3	1.8	8.7	0.6

Source: Charles W. Stahl and Reginald T. Appleyard, "International manpower flows in Asia: an overview", *Asian and Pacific Migration Journal* (Quezon City, Philippines), vol. 1, Nos. 3 and 4 (1992), p. 430.

industry". In addition, societal attitudes towards migration as a way of improving one's chances in life continue to be highly positive. Furthermore, the Governments of countries of origin encourage migration and are even willing to lower minimum wage standards in order to maintain a competitive edge. Thus, a large pool of labour geared for migration is available in the region and is not expected to diminish in the near future.

In contrast, in the countries of destination, whether Japan, the NIEs or those in Western Asia, the present and future demand for labour remains unsatisfied. Western Asia, with its strong dependence on foreign workers, will continue to require them for several years, mainly because indigenous populations are small, lack the needed skills and have a preference for white-collar administrative jobs, especially in the public sector. In Kuwait, for instance, the proportion of economically active indigenous men in the production sector has declined steadily, from about 30 to 11 per cent between 1965 and 1988. In contrast, 76 per cent of nationals of both sexes were employed in the public sector in 1975, and that proportion increased to 91 per cent of economically active men and 97 per cent of economically active women by 1989. The private sector is therefore almost completely manned by foreign workers and the natives are not learning the skills necessary to replace them. In fact, as already noted, a sector that is wholly dependent on expatriates is that involving domestic workers and drivers. Consequently, foreign female workers from Muslim countries are a group for which there is continued, and perhaps even growing, demand. There is also a growing demand for workers in maintenance and cleaning services, occupations considered unworthy of Kuwaitis. Assuming that such attitudes and patterns characterize other GCC countries, the demand for foreign workers will persist as long as Governments and citizens remain affluent. In terms of the skill composition of foreign workers, the demand for construction workers will further decline as infrastructure development is completed, while the demand for skilled workers will increase.

The Governments of labour-importing countries in Western Asia recognize that their needs for foreign manpower are bound to continue for some time, although there is a repeated call for the "indigenization" of the labour force. In some countries, such as Kuwait, the Government has also emphasized the need to import Arab (rather than Asian) workers. However, in reality the opposite has happened. The proportion of Asians in the combined labour force of the GCC countries, only 28 per cent in 1975, had increased to 63 per cent in 1985. In Saudi Arabia, the most important labour-importing country in the region, almost 60 per cent of the expatriate workers were Asian in 1985, and this figure represented a sharp increase from the corresponding 1975 figure of only 6 per cent. Some analysts believe that, as a consequence of the Gulf war and the increased polarization of the Arab world, Asian workers will be preferred over less tractable Arab ones. Furthermore, among Asians, non-political, non-vocal workers from non-Muslim countries who do not have strong feelings about Iraq's occupation might be the ones who are preferred (Addleton, 1991). Egypt, having been part of the coalition that helped liberate Kuwait, will probably gain further outlets for its workers. Thus, positions vacated by Palestinians and Jordanians in Kuwait and by Yemenites in Saudi Arabia may be filled progressively with Egyptians. A policy that would discourage the migration of families and the eventual settlement of the foreign population was recently announced by the Prime Minister of Kuwait. The preference for a temporary migrant labour force is shared by all GCC countries.

Among the Eastern and South-eastern Asian countries that have begun to import labour, Japan is the linchpin, even though the others have rapidly growing economies and their need for labour is likely to increase (Appleyard, 1991a). Over one third of the total labour force in Japan and the NIEs works in industry, and labour demand in the industrial sector is still high (table 45). Judging from the estimates put forward by Japanese firms, it is likely that the total current requirement for workers may be around 600,000, although some sources put the figure at 1 million (Appleyard, 1991b). It is estimated that by the end of this decade Japan will have about 2.7 million more jobs than workers (Stahl, 1991). Despite these expected needs, the Government of Japan does not favour the admission of unskilled workers and has reinforced policies against undocumented migration. The Japanese image of self-sufficiency is probably a factor against the import of foreign workers. Public reaction to foreigners is also likely to be an important factor in the political decisions on the subject. Yet, in Japan as in many other developed countries, certain menial jobs are not considered fit for the increasingly well-off Japanese worker. A recent survey found that about 57 per cent of the respondents interviewed gave a qualified yes to the question of whether to import unskilled workers (Mori, 1991). Although the public debate on the issue is continuing, a policy to encourage large-scale Asian migration is not likely. In the meantime, undocumented migration will probably continue to satisfy domestic needs and the relocation of some industry to labour-surplus countries will remain an important means of achieving economic gains without the accompanying social and economic costs of labour migration.

The Korean Institute of Labour reports a labour shortage equal to 5.53 per cent of total employment in the Republic of Korea. In the industrial sector, a shortage of 6.85 per cent for production workers was estimated. Shortages of 15 per cent in clothes manufacturing and 9 per cent in coal mining have also been reported (Park, 1991). The demand for foreign workers and the policies against their admission have led to a heated debate. No agreement has been reached about the best way to handle labour shortages. Trade unions oppose the admission of foreign workers since they might depress domestic wages. Business groups, on the other hand, are lobbying in favour of their admission. In the view of Park (1991), the problem of labour shortages can largely be solved by using indigenous human resources in a more efficient way and by the adjustment of society to a new environment. Employment of the 1.69 million potential workers (mostly women), reduction of discrimination against production jobs and a slowdown in the construction industry that is already occurring are some of the factors that might reduce the need for foreign workers. The selective admission of foreign workers for the coal mines might be supported as well as the admission of Chinese and Russians of Korean origin.

Taiwan Province of China has both one of the fastest growing economies and a fertility that has been declining since the 1960s. It is therefore facing acute labour shortages. In 1989, Taiwan Province of China had an estimated labour shortage of 200,000 workers and an estimated annual future shortfall of 50,000 workers (Stahl, 1991). Migration to Taiwan Province of China is expected to increase in view of the Government's decision to allow the employment of foreign workers on large-scale infrastructure development projects. Taiwan Province of China is likely to need large numbers of workers in construction, textiles, transport and services (the hotel industry, in particular). If current trends continue, most workers would come from South-eastern Asian countries.

Singapore has relied on foreign workers ever since it was founded in 1813. Its policies towards the admission of migrant workers have fluctuated with economic conditions. The continuing strength of Singapore's economy and the slow growth of its labour force suggest a continued and increasing reliance on expatriates. Recent legislation allows the admission of workers for certain sectors, such as retail trade, banking and community services. Firms in the construction sector and shipbuilding will be allowed to fulfil 50 per cent of their needs from foreign sources. Ensuring the high quality of foreign workers is a primary concern of the Government and a minimum level of education has been specified for those planning to work in the service sector (Stahl, 1991). As a way of rationing the number of workers employed in the country, a system for auctioning foreign labour permits to the highest bidder has been proposed recently (Lim, 1991).

Over the next few years, an increased globalization of the Asian labour force is likely to occur as economies become more integrated and the inequalities between Asian countries increase. A policy option that countries will continue to use is the relocation of industries and capital to labour-surplus countries. Other options, such as the increased labour force participation of women and retirees, may be used. It has been estimated that if the age-specific labour force participation rates for women in Japan and the NIEs converged to the levels currently prevalent in the United States, the number of women aged 20-64 in the labour force would increase by 10 per cent in Japan, 27 per cent in the Republic of Korea, 33 per cent in Taiwan Province of China and 44 per cent in Singapore (Bauer, 1990).

Possibilities for the migration of Asians to the developed countries outside Asia are bound to decrease in the wake of economic stagnation, the large numbers of undocumented migrants already present in those countries and mounting public discontent about high immigration levels. The vehement public reactions in Germany against migrants from Asia and Africa, and the increasingly tighter regulations against migration to European countries suggest declining trends in future. With barriers in the West increasing and the pressures for emigration in certain Asian countries rising, Asian labour will continue to seek markets in the Asian region itself. If admission policies continue to be stringent, undocumented migration will grow. The networks to support such migration already exist. Women migrating to work in domestic service or in entertainment activities will continue to be a particularly vulnerable group. There is a need for the countries of origin and destination to collaborate in maximizing the benefits and minimizing the negative consequences of migration, a phenomenon that has become a necessity for the growth and development of both types of countries. In doing so, it is important that those countries not to lose sight of the human dimensions of the process as the future unfolds.

REFERENCES

Abella, Manolo (1991). Manpower movements in the Asia Region. Paper presented at the Second Japan-ASEAN Forum on International Labour Migration in East Asia, Tokyo, Japan, 26 and 27 September.

________ (1994). International migration in the Middle East: Patterns and implications for sending countries. In *International Migration: Regional Processes and Responses,* Miroslav Maçura and David Coleman, eds. Economic Studies, No. 7. Sales No. GV.E.94.0.25. New York and Geneva: United Nations, Economic Commission for Europe.

________ (1995). The sex selectivity of migration regulations governing international migration in Southern and South-eastern Asia. *International Migration Policies and the Status of Female Migrants.* Sales No. E.95.XIII.10. New York: United Nations.

Adams, R. H. (1992). The effects of international migration and remittances on income distribution in rural Pakistan. International Food Policy Research Institute (June). Mimeographed.

Addleton, J. (1991). The impact of the Gulf War on migration and remittances in Asia and the Middle East. *International Migration* (Geneva), vol. 29, No. 4, pp. 509-526.

Ahmad, M. (1982). Emigration of scarce skills in Pakistan. International Migration for Employment. Working Paper, No. 5. Geneva: International Labour Organization.

Amjad, Rashid (1989). *To the Gulf and Back: Studies on the Economic Impact of Asian Labour Migration.* New Delhi: ILO/Asian Regional Team for Employment Promotion (ARTEP).

Appleyard, Reginald T. (1989). Migration and development: myths and reality. *International Migration Review* (Staten Island, New York), vol. 23, No. 3 (fall), pp. 486-499.

________ (1991a). *International Migration: Challenge for the Nineties.* Geneva: International Organization for Migration.

________ (1991b). Summary report. *International Migration* (Geneva), vol. 29, No. 2 (June), special issue ("South-North Migration"), pp. 333-340.

Arab Times (1992). Japanese anguish over foreigners in bath houses (5 July), p. 9.

Arnold, Fred (1989). Revised estimates and projections of international migration 1980-2000. Policy, Planning, and Research Working Papers. Population and Human Resources Department, World Bank (August). WPS 275.

Athukorala, P. (1990). International contract migration and the reintegration of return migrants: the experience of Sri Lanka. *International Migration Review* (Staten Island, New York), vol. 24, No. 2 (summer), pp. 323-346.

Bandara, U. W. (1991). Recent developments in labour out-migration and country responses by Sri Lanka delegate. Paper presented at the International Labour Organization Meeting on the Implications of Changing Patterns of Asian Labour Migration, Kuala Lumpur, Malaysia. May.

Bauer, J. (1990). Demographic changes and Asian labor markets in the 1990s. *Population and Development Review* (New York), vol. 16, No. 4 (December), pp. 615-645.

Birks, J. S., and C. A. Sinclair (1979). Migration and development: the changing perspective of the poor Arab countries. *Journal of International Affairs* (New York), vol. 33, No. 2, pp. 285-309.

__________ (1980). Arab labour markets: a broad assessment. In *Arab Manpower: The Crisis of Development*. London: Croom Helm.

Birks, J. S., I. J. Seccombe, and C. A. Sinclair (1988). Labour migration in the Arab Gulf States: patterns, trends, and prospects. *International Migration* (Geneva), vol. 26, No. 4, pp. 267-285.

Choucri, N. (1986). The hidden economy: a new view of remittances in the Arab world. *World Development* (Boston, Massachusetts), vol. 14, No. 6, pp. 697-712.

__________ (1988). Migration in the Middle East: old economics or new politics? *Journal of Arab Affairs* (Fresno, California), vol. 7, No. 1, pp. 3-4.

Eelens, Frank (1995). The migration of Sri Lankan women to Western Asia. In *International Migration Policies and the Status of Female Migrants*. Sales No. E.95.XIII.10. New York: United Nations.

__________, and J. D. Speckmann (1990). Recruitment of labour migrants for the Middle East: the Sri Lankan case. *International Migration Review* (Staten Island, New York), vol. 24, No. 2 (summer), pp. 297-322.

Fan, Z. (1991). Status quo of China's emigration of scientific and technical labour. Paper presented at the Second Japan-ASEAN Forum on International Labour Migration in East Asia, Tokyo, 26 and 27 September 1991.

Fergany, Nader (1983). The impact of emigration on national development in the Arab region: the case of the Yemen Arab Republic. *International Migration Review* (Staten Island, New York), vol. 16, No. 4 (winter), pp. 757-780.

Fong, Pang E. (1991). International labour migration and structural change in Indonesia, Malaysia and Singapore. Paper presented at the Second Japan-ASEAN Forum on International Labour Migration in East Asia, Tokyo, 26 and 27 September 1991.

Kuwait (1989). *Directory of Civil Information: Population and Labor Force*. Kuwait: Public Authority for Civil Information (July).

Lim, Lin L. (1991). International labour migration in Asia: patterns, implication and policies. Paper presented at the Informal Expert Group Meeting on International Migration, Geneva, Switzerland, 16-19 July 1991.

Mori, H. (1991). Structural change in contemporary Japanese labour market and immigrant workers. Paper presented at the Second Japan-ASEAN Forum on International Labour Migration in East Asia, Tokyo, Japan, 26 and 27 September 1991.

Nagi, H. M. (1986). Determinants of current trends in labor migration and the future outlook. In *Asian Labor Migration: Pipeline to the Middle East*, F. Arnold and N. M. Shah, eds. Boulder, Colorado: Westview Press.

Park, Y. (1991). Foreign labour in Korea: issues and policy options. Paper presented at the Second Japan-ASEAN Forum on International Labour Migration in East Asia, Tokyo, Japan, 26 and 27 September 1991.

Russell, Sharon S. (1986). Remittances from international migration: a review in perspective. *World Development* (Boston, Massachusetts), vol. 14, No. 6, pp. 677-696.

__________ (1992). International migration in Europe, Central Asia, the Middle East and North Africa: issues for the World Bank. Paper prepared for the Population and Human Resources Division, Technical Department, Europe and Central Asia, Middle East and North Africa Regions, World Bank, 3 April.

Seccombe, I. J. (1985). International labor migration in the Middle East: a review of literature and research, 1974-84. *International Migration Review* (Staten Island, New York), vol. 19, No. 2 (summer), pp. 335-352.

__________, and R. J. Lawless (1986). Foreign worker dependence in the Gulf, and the international oil companies: 1910-50. *International Migration Review* (Staten Island, New York), vol. 19, No. 2 (summer), pp. 335-352.

Shah, Nasra M. (1992). Arab labor migration: trends, characteristics and problems. Paper presented at the IOM Seminar on International Migration in Egypt and the Arab World, Cairo, Egypt, 2-4 November.

__________, and Fred Arnold (1986). Government policies and programs regulating labor migration. In *Asian Labor Migration: Pipeline to the Middle East*, F. Arnold and N. M. Shah, eds. Boulder, Colorado: Westview Press.

__________ (1990). Pakistan. In *Handbook on International Migration*, W. J. Serow and others, eds. New York: Greenwood Press.

Shah, Nasra M., and S. S. Al-Qudsi. (1989). The changing characteristics of migrant workers in Kuwait. *International Journal of Middle East Studies* (Cambridge, Massachusetts), vol. 21, No. 1, pp. 31-55.

__________ (1990). Female work roles in a traditional oil economy: Kuwait. In *Female Labor Force Participation: Research in Human Capital and Development*, vol. 6, R. Frank, I. Serageldin and R. Sorkin, eds. Greenwich, Connecticut: JAI Press.

Shah, Nasra M., S. S. Al-Qudsi and M. A. Shah. (1991). Asian women workers in Kuwait. *International Migration Review* (Staten Island, New York), vol. 25, No. 3 (fall), pp. 464-486.

Stahl, Charles W. (1991). South-North migration in the Asia-Pacific region. *International Migration* (Geneva), vol. 29, No. 2 (June), special issue ("South-North Migration"), pp. 163-193.

__________, and F. Azam (1990). Counting Pakistanis in the Middle East: problems and policy implications. *Asian and Pacific Population Forum* (Honolulu, Hawaii), vol. 4, No. 2 (summer), pp. 1-32.

__________, and Reginald T. Appleyard (1992). International manpower flows in Asia: an overview. *Asian and Pacific Migration Journal* (Quezon City, Philippines), vol. 1, Nos. 3 and 4, pp. 417-476.

Tabbarah, Riad (1988). Human resources development and its population dimension in the Arab world. *Population Bulletin of ESCWA* (Baghdad, Iraq), No. 32 (June), pp. 3-29.

United Nations (1989). Report of the Secretary-General on the Office of the United Nations High Commissioner for Refugees: International Conference on Indo-Chinese Refugees. A/44/523, annex, sect. II.

__________ (1992). *World Population Monitoring, 1991: With Special Emphasis on Age Structure*. Population Studies, No. 126. Sales No. E.92.XIII.2.

__________ (1995). Report of the meeting (chapter I). In *International Migration Policies and the Status of Female Migrants*. Sales No. E.95.XIII.10.

__________ (1996). *World Population Monitoring, 1993: With a Special Report on Refugees*. Sales No. E.95.XIII.8.

__________, Economic and Social Commission for Western Asia (ESCWA) (1985). Demographic and related socio-economic data sheets. Baghdad: ESCWA.

Visaria, Pravin, and Leela Visaria (1990). India. In *Handbook on International Migration*, W. J. Serow and others, eds. New York: Greenwood Press.

Wardwell, J. M. (1990). Jordan. In *Handbook on International Migration*, W. J. Serow and others, eds. New York: Greenwood Press.

Zlotnik, Hania (1994a). Comparing migration to Japan, the European Community and the United States. *Population and Environment: A Journal of Interdisciplinary Studies* (New York), vol. 15, No. 3 (January), pp. 173-186.

__________ (1994b). Migration to and from developing regions: a review of past trends. In *Alternative Paths of Future World Population Growth. What Can We Assume Today?*, Wolfgang Lutz, ed. London: Earthscan Publications, Ltd.

XIII. MIGRATION BETWEEN DEVELOPING COUNTRIES IN SUB-SAHARAN AFRICA AND LATIN AMERICA

*Sharon Stanton Russell**

Although they are continents apart, and both are vast and heterogeneous, sub-Saharan Africa and Latin America share a number of characteristics that invite comparisons in terms of their experiences in relation to intra-regional migration. In both, such population movements are sizeable, involving undocumented as well as documented migrants who have moved for a wide variety of reasons. In both cases, economic and demographic differentials among countries have stimulated migration for employment, in turn establishing patterns that have shaped the directions of subsequent refugee flows. Both regions have experienced severe economic crises during the 1980s as well as political upheavals that have given rise to important population flows and blurred the distinction between economic and political migrants. Albeit in different ways, both are distinctive among world regions in their expressed views and policies towards international migration. Lastly, in both sub-Saharan Africa and Latin America, a dearth of timely quantitative data impedes the systematic analysis of international migration patterns and trends.

The present paper examines the evidence regarding migration between the developing countries of sub-Saharan Africa and Latin America and explores a number of major factors bearing upon both past migration within each region and likely prospects for future flows, including economic differentials, political instability, the effects of governmental policies towards migration in general as well as those affecting the movement of highly skilled personnel, and the role of regional trading blocs.

A. Patterns and trends in intraregional migration

1. *Sub-Saharan Africa*

Statistical information on international migration in sub-Saharan Africa is limited and often of poor quality. Flow data on international migration between countries in the region are virtually non-existent. However, using the database of the Population Division of the United Nations Secretariat on the international migrant stock in developing countries and taking into account refugee data as reported by the Office of the United Nations High Commissioner for Refugees (UNHCR), estimates of the total migrant stock in sub-Saharan Africa are available for 1965, 1975 and 1985 (Skoog, 1994). According to those estimates, the number of international migrants in the region rose from 7.1 million in 1965 to 10.3 million in 1975 and reached 11.3 million in 1985, when they represented 2.7 per cent of the total population of sub-Saharan Africa. The basic data used to derive these estimates refer to the foreign-born or, in some cases, the foreign population enumerated by censuses carried out in the countries of the region. Because censuses aim at the universal coverage of the population of a country, there is no reason to believe that they cover only legal migrants. Indeed, since much of the migration between sub-Saharan African countries has traditionally occurred outside official control, it is likely that most of the African migrants enumerated by censuses are undocumented. There is the possibility, of course, that censuses may underestimate the migrant stock, either because they selectively fail to canvas certain types of migrants or because those enumerated fail to report their migrant status. In particular, censuses are unlikely to capture temporary migrants or those who engage in circular migration. They are also unlikely to reflect properly the stock of refugees. Therefore, the latter has been added as an independent component of the migration stock. Indeed, refugee migration is one of the fastest growing types of migration in sub-Saharan Africa. According to UNHCR statistics, the number of refugees in the region increased from 2.6 million in early 1980 to 5.2 million by early 1993 (United Nations, 1996).

Refugee movements have been particularly acute in Eastern Africa, where they dominate other types of migration. Since 1990, Mozambique has been Africa's major source of refugees, accounting for a total of nearly 1.2 million in early 1991, and surpassing Ethiopia, Somalia and the Sudan as a source of refugees in the region. By early 1993, Mozambique still led the group,

*Research Scholar, Center for International Studies, Massachusetts Institute of Technology, Cambridge, Massachusetts.

being the source of 1.3 million refugees who had mostly found asylum in neighbouring Malawi. It was followed by Ethiopia, which was the source of 781,000 refugees, and Somalia, whose expatriate refugee population reached 716,000 in early 1993 because of continued conflict in that country (UNHCR, 1993). Indeed, the crisis in Somalia has been so acute that the country ceased being a haven for refugees in 1992. Before that, it had been hosting some 700,000 Ethiopians. Ethiopia itself increased its intake of refugees during the 1990s, hosting over 700,000 in early 1991, a number that declined to 430,000 by 1993. The Sudan, which has also been a source and a recipient of refugees, had a refugee stock of over 700,000 by early 1993 (United Nations, 1996). Other important generators of refugees in the region have been Burundi and, particularly, Rwanda. In early 1993, there were an estimated 430,000 Rwandese refugees in sub-Saharan Africa.

There have been and continue to be other, smaller-scale but important migration flows in Eastern Africa (Russell, Jacobsen and Stanley, 1990). Since before independence, workers had moved from Burundi, Rwanda and Zaire to Kenya, Uganda and elsewhere in East Africa, as contract labour replaced old East-West slave trading. Recently, the main flows between these countries have involved refugees, both seeking asylum and being repatriated. Similarly, earlier labour flows from Mozambique to South Africa, Zambia and Zimbabwe have been followed by refugee movements from Mozambique to the same destinations, although the numbers involved have been small in comparison with the number of refugees going to Malawi. The United Republic of Tanzania has also received both workers and refugees from other sub-Saharan African countries and exported labour to other countries.

In Middle Africa, historical patterns of international migration have been linked to religious factors, tribal expansion, the slave trade, migrations of nomadic groups, and movement across "artificial" boundaries set by the colonial powers which divided people belonging to the same tribal group. More recently, migration has been directed predominantly to poles of economic growth, such as the subregion's three oil-producing countries, namely, Cameroon, the Congo and Gabon (Russell, Jacobsen and Stanley, 1990). However, the country with the largest migrant stock in Middle Africa is Zaire, where 1.4 million foreigners were enumerated in 1984, 594,000 of whom were African (table 46). Before the political upheavals of 1991, the mineral deposits of Zaire and infusions of investment capital had created jobs that attracted many skilled and unskilled migrant workers. The second major destination has been Cameroon, where nearly 220,000 foreign-born persons were enumerated in 1976. Those born in Africa originated in the Central African Republic, Chad and Nigeria and tended to work in palm plantations. The Congo, with about 100,000 foreign-born migrants enumerated in 1984, had the highest proportion of migrants in Middle Africa (over 5 per cent); and they had originated in a number of countries, including Angola, Benin, Cameroon, the Central African Republic, Mali, Senegal and Zaire.

While migration in Middle Africa is often characterized as being largely male and temporary, an analysis of the sex ratio of migrant groups suggests that the composition of migration can vary considerably by origin and country of destination and over time. Migrants from Zaire in Cameroon had the highest sex ratio of any foreign-born group enumerated in the 1976 census (2.6 men per woman). Yet, Zairian-born migrants enumerated in the Congo in 1984 had a sex ratio of 0.9 men per woman (Cameroon, 1976; Congo, 1984.)

The figure for the refugee population in Middle Africa has hovered around 500,000 since 1988. By far the major source country has been Angola, although hopes for the establishment of democratic order by late 1992 (now in jeopardy) reduced the number of Angolan refugees to 340,200 in 1991, down from 436,000 the previous year (United States Committee for Refugees, 1992).

Western Africa is the subregion having the highest concentration of international migrants. Indeed, Western Africa has long been considered a region where people move as freely as goods. Whereas pre-colonial migrations were often group movements caused by trade, internecine warfare, slave raids, famine, drought, and the spread of different religions, colonial labour recruitment policies and the development of plantation agriculture brought a shift to the largely spontaneous movement of individuals. Levels of both seasonal and undocumented migration are reportedly higher in Western Africa than elsewhere in sub-Saharan Africa, and population movements in the subregion are notably volatile (Makinwa-Adebusoye, 1992).

Ghana was the favoured destination of Western African migrants until the late 1960s when an extended period of negative economic growth started. Over the past two decades, Côte d'Ivoire has supplanted Ghana as the major pole of attraction for migrants from Burkina Faso, Ghana, Guinea, Mali, the Niger and elsewhere.

TABLE 46. MIGRANTS IN SELECTED SUB-SAHARAN AFRICAN COUNTRIES

Region or country of enumeration	*Reference date*	*Migrants* African[a]	*Migrants* Other	*Migrants* Total	*Non-migrants*	*Not stated*	*Total population*	*Migrants as percentage of total population*
Middle Africa								
Angola	1983	7 892	7 338	15 230	8 184 770	..	8 200 000[b]	0.2
Cameroon	1976	185 558	14 630	218 069	6 914 889	..	7 132 958	3.1
Central African Republic	1975	41 362	3 221	44 583	1 699 451	36 995	1 781 029	2.5
Congo	1984	45 703	5 464	96 639	1 260 055	552 306	1 909 000	5.1
Zaire[c]	1984	594 400	791 900	1 386 300	29 345 100	..	30 731 400	4.5
Eastern Africa								
Burundi	1979	79 902	2 820	82 851	3 945 569	129	4 028 549	2.1
Kenya	1979	..	..	157 371	15 169 560	130	15 327 061	1.0
Madagascar[c]	1975	1 078	52 237	53 315	7 549 710	765	7 603 790	0.7
Malawi	1977	281 806	6 938	288 744	5 257 554	1 162	5 547 460	5.2
Mozambique[c]	1980	..	..	39 142	11 634 858	..	11 674 000	0.3
Rwanda[c]	1978	36 789	5 122	41 911	4 788 569	1 047	4 831 527	0.9
Sudan	1973	..	..	227 906	12 015 614	1 870 480	14 114 000	1.6
Uganda[c]	1969	486 300	56 114	542 414	8 998 319	2 812	9 543 545	5.7
United Republic of Tanzania[c]	1978	..	..	415 684	17 096 927	..	17 512 611	2.4
Zambia	1980	184 742	46 612	231 354	5 430 646	..	5 662 000	4.1
Southern Africa								
Botswana[c]	1981	8 471	6 886	15 619	925 381	..	941 000[b]	1.7
South Africa	1985	1 404 975	462 084	1 862 192	..	..	23 386 000[b]	8.0
Swaziland	1976	21 946	4 212	26 460	468 074	466	495 000	5.3
Western Africa								
Benin[c]	1979	..	..	41 284	3 286 937	2 779	3 331 000	1.2
Burkina Faso	1975	107 517	13 275	110 681	5 517 411	10 000	5 638 092	2.0
Côte d'Ivoire[c]	1975	1 437 319	37 124	1 474 469	5 203 580	31 951	6 710 000	22.0
Gambia	1973	53 300	1 254	54 554	437 636	1 309	493 499	11.1
Ghana[c]	1970	547 149	14 983	562 132	7 997 181	..	8 559 313	6.6
Guinea-Bissau	1979	12 043	888	12 931	755 069	..	768 000	1.7
Liberia	1974	47 654	11 804	59 458	1 443 910	..	1 503 368	4.0
Mali	1976	72 365	4 549	146 089	6 248 829	..	6 394 918	2.3
Mauritania[c]	1977	23 007	5 161	28 168	1 310 832	..	1 339 000	2.1
Senegal[c]	1976	93 072	25 710	118 782	4 879 103	..	4 997 885	2.4
Sierra Leone[c]	1974	67 164	8 826	79 414	2 655 745	724	2 735 883	2.9
Togo	1970	..	..	143 620	1 807 380	..	1 951 000	7.4
Other Africa[d]		20 581	42 264	35 272	1 825 339	4 673	1 860 611	1.9

[a]Migrants of African origin including those born in Northern Africa and those classified in the category "Other Africa".

[b]World Bank estimate for mid-1973 obtained from "Population growth and policies", 1986, mimeographed. Data on foreign-born for Angola refer only to persons enumerated in the capital, Luanda.

[c]Country where migrants are identified in terms of citizenship. In all other countries, migrants are defined as foreign-born persons.

[d]Including the Comoros, Mauritius, Réunion, Sao Tome and Principe, and Seychelles.

The 1988 census of Côte d'Ivoire enumerated over 3 million foreigners in the country, who constituted over 26 per cent of the population, by far the highest concentration in sub-Saharan Africa. In relative terms, the Gambia is the second most important migrant-receiving country in the region, with its foreign-born population accounting for about 12 per cent of the total in 1985. The migrant population in the Gambia includes significant numbers of persons born in Guinea, Guinea-Bissau, Mali and Senegal.

Nigeria, being the largest country of Western Africa in terms of population size, has relatively few international migrants. However, international migration flows to the country increased substantially during the late 1970s and early 1980s as a result of the oil boom and the entry into effect of the protocol on freedom of movement signed in 1980 by member States of the Economic Community of West African States (ECOWAS). By 1982, there were an estimated 2 million-2.5 million foreigners in Nigeria, the majority from Benin, Chad, Ghana, Togo and the Niger. However, economic and political adversities changed the situation dramatically. In 1983 and again in 1985 foreigners were expelled by government order and an estimated 1.5 million left (Afolayan, 1988). By the late 1980s, Nigeria was better known as an exporter of professionally and technically trained personnel than as a magnet for migrants from abroad (Russell, 1993).

Refugee flows have increased dramatically in Western Africa in recent years. As of 1988, Guinea-Bissau was the main source of refugees in Western Africa and accounted for only 5,000. The subregion hosted a total of 21,000 refugees, most originating in Middle Africa. However, only two years later, Western African nations were providing asylum to over 800,000 people, originating primarily in other Western African countries, especially Liberia, which accounted for nearly 730,000 refugees, and Mauritania, which accounted for a further 60,000 (United States Committee for Refugees, 1991).

In Southern Africa, international migration has been dominated by temporary flows of workers to the Republic of South Africa, principally from Botswana, Lesotho and Swaziland but also from Mozambique, Malawi and Zimbabwe. In addition to the migration of mineworkers, organized through labour recruiters, there was until 1963 considerable clandestine migration, which included accompanying women and children. In 1963, South Africa began imposing strict immigration controls, which curtailed both undocumented and family migration, and prohibited any migration originating in Botswana, Lesotho and Swaziland except for the purpose of working in the mines or in agriculture.

Since the mid-1970s, South Africa has sought to reduce its dependence on foreign labour and to establish a permanent, more skilled labour force from domestic sources. The results are most evident in mining: although total employment in the mines has grown, and documented migrants are increasingly concentrated in mining, the proportion of foreign workers decreased from 78 per cent in 1974 to about 40 per cent in 1984-1986 (United Nations, 1990; Migrant Labour: Quest for Reform, 1987). As a result, the supply of labour in the sending countries of Southern Africa now exceeds demand (De Vletter, 1985). In Botswana alone, the number of mine labour recruits dropped from 40,390 in 1976 to 19,648 in 1986, and the proportion of first-time workers among all recruits dropped from 25 per cent in 1976 to 1.6 per cent in 1985 (Taylor, 1990). Yet, with the changing political climate in South Africa, there are indications that African citizens from as far away as Ghana, Nigeria and Zaire are entering the country in increasing numbers (British Broadcasting Company, 1992; Foundation for Contemporary Research, 1992).

In 1990, less than 1 per cent of all sub-Saharan African refugees originated in Southern Africa. South Africa remains the major country of asylum, hosting an estimated 201,000 refugees, nearly all from Mozambique. However, South Africa is also the subregion's largest source of refugees, some 40,000 of whom are dispersed in the various countries of Southern Africa. UNHCR expects to assist in the repatriation of 30,000 of those refugees under an official repatriation programme begun in 1990.

2. *Latin America*

As in Africa, the best information on international migration to Latin America is obtained from censuses. According to estimates derived from that source, in 1985 Latin America and the Caribbean had a stock of approximately 6.5 million international migrants which represented 1.6 per cent of the total population of the region (Skoog, 1994). Consequently, in relative terms, Latin America had considerably less international migrants than sub-Saharan Africa. Of those 6.5 million international migrants, 4.6 million were in South America.

In comparison with sub-Saharan Africa, where most international migrants originate within the region, only about 41 per cent of the international migrants present in Latin America around 1980 originated within the region: the majority were settlers of European origin (United Nations, 1990). In addition, whereas most migration originating in countries of sub-Saharan Africa is directed to other countries in the region, most of the migration originating in Latin America and the Caribbean is directed towards Northern America, particularly to the United States of America. In terms of the stock of emigrants abroad, Mexico is the main country of origin in Latin America. Around 1980, some 2.2 million persons born in Mexico were enumerated by the censuses of other countries in the Americas, the vast majority (99 per cent) in the United States. The second major country of origin, Colombia, recorded about 673,000 persons abroad in the early 1980s. If undocumented migration is included, the actual number of Colombians abroad at the time is estimated to have been of the order of 860,000 (United Nations, 1990).

Three countries, namely Argentina, Brazil and Venezuela, have been the main Latin American destinations of international migrants. Argentina, with nearly 2 million foreign-born inhabitants constituting 6.8 per cent of total population in 1980, has long been the main destination for migrants from neighbouring Bolivia, Paraguay, Uruguay and, to a lesser extent, Chile (Balán, 1985). However, the relative composition of migration flows originating in neighbouring countries has changed over time for both economic and political reasons.

As of 1914, the largest group of migrants from neighbouring countries to Argentina was made up of Uruguayans, including Europeans whose secondary migration to Argentina had been prompted by technological changes in rural production. Later, the expansion of sugar production in north-western Argentina attracted growing numbers of Bolivian seasonal workers, while Paraguayans migrated into north-eastern Argentina to work in the harvesting of tea and cotton. Following the Chaco War of 1932-1935, migration from both countries rose, as the recruitment of labour became more systematic and even coercive. Emigration from Paraguay grew following the 1947 civil war in that country and Paraguayans became the largest migrant group from neighbouring countries into Argentina.

After the Second World War, migrants from neighbouring countries began to settle in the urban areas of Argentina. The prolonged recession that Uruguay experienced as of the mid-1950s led many Uruguayans to emigrate (Balán, 1985). Those settling in Argentina were mostly of urban origin, educated, of middle-class background and in the economically active ages. They found jobs as salaried workers, as employees in the private sector and even as labourers.

Emigration from Uruguay peaked between 1973 and 1975, and had mostly ceased by 1983 (Balán, 1985). The prolonged recession that began in Argentina in 1974 contributed to that decline: during 1974-1983 more than 300,000 industrial jobs were lost and increasing numbers of Argentines emigrated because of both the worsening labour market and growing political repression. It is estimated that some 650,000 Argentines emigrated between 1974 and the reinstatement of a constitutional regime in December 1983 (Balán, 1985).

Conditions in Argentina also affected migration from Bolivia and Paraguay, which involved primarily workers with low-level skills, predominantly male in the case of Bolivia, but involving a more balanced distribution by sex in the case of Paraguay. Starting in 1974, Argentina introduced explicit policies to repatriate citizens from those countries. At the same time, the Paraguayan economy expanded and internal migration provided a viable alternative to international relocation. Consequently, the emigration of Paraguayans to Argentina declined. In contrast, the outflow of Bolivians to Argentina grew towards the end of the 1970s (Balán, 1985; Dandler and Medeiros, 1988).

Emigration from Chile to Argentina has been less significant than that from other neighbouring countries, although it broadened and increased with the overthrow of the Allende Government in 1973. Traditionally, Argentina had been the major destination of Chilean professionals and technicians and, especially after Chile's agrarian crisis in the 1950s and 1960s, of workers with little education.

Among the Latin American countries, Brazil ranks second to Argentina in the number of foreign-born, which amounted to 1.2 million in 1980 and constituted 1 per cent of the total population. Although the majority of Brazil's foreign-born originate in European countries, the country has been an important destination for migrants from Argentina, Bolivia, Chile, Paraguay, Peru, Uruguay and Venezuela (United Nations, 1990). Brazil is also a country of emigration, with a long history of migration directed towards Argentina, Paraguay and Uruguay, and relatively large numbers of Brazilian farmers have settled in Paraguay in recent years (Balán, 1985).

The third major destination country of the region, Venezuela, had over 1 million foreign-born persons in 1981, who constituted 7.2 per cent of the total population. Colombia is by far the main source of migrants to Venezuela, and the latter is the destination of about three quarters of all Colombian emigrants. In addition, Venezuela attracts sizeable numbers of migrants from Chile, Costa Rica, the Dominican Republic, Ecuador, Peru and Uruguay (United Nations, 1990). Because migration from Colombia to Venezuela has involved many undocumented migrants, estimates of the true number of Colombians in its neighbouring country have varied widely. Around 1979, it was claimed that the number of Colombians in Venezuela could be as high as

1.5 million (Bleier, 1988). Yet, estimates derived from information allowing the indirect estimation of lifetime emigration from Colombia indicated that there were around 600,000 Colombians in Venezuela in 1980, a number that compares well with the 500,000 or so enumerated by the 1981 census of Venezuela and indicates that the likely levels of undocumented migration of Colombians in that country were not as high as many speculated (Zlotnik, 1989).

During the 1980s, Central America became an important theatre for international migration, as conflict in the region led to sizeable flows of refugees and displaced persons. According to estimates based on census data and on reports on the number of refugees in the region as of 1985, the Central American countries and Mexico had approximately 1 million migrants at that time (Skoog, 1994). However, those numbers do not reflect well the number of displaced persons in the region, many of whom were outside of their countries but were not recognized as refugees. As of early 1990, 1.2 million persons were considered to be either refugees or displaced persons in the region, though only 128,000 were receiving UNHCR assistance (United Nations, 1996). Many had left the region to seek asylum in Northern America where, if unsuccessful, they remained as undocumented migrants. It has been estimated that by 1982 about 10 per cent of the population of Central America (excluding Mexico) were living outside their countries of origin and that before 1978 only a small percentage had migrated (Stanley, 1991).

Until 1978, most international migration within Central America involved the movement of workers. Indeed, the region had experienced nearly a century of gradually increasing labour migration, which intensified with new developments in agriculture in the 1960s. Historically, Salvadorans were recruited to work in the banana plantations and the mines of Honduras, and emigration grew as the availability of land in El Salvador for small tenant farmers decreased. By 1969, an estimated 300,000 Salvadorians were in Honduras, most as undocumented migrants (Stanley, 1991). Their presence was an important element in the 1969 war between the two countries, which led to the repatriation of most Salvadorans and the closure of the border between El Salvador and Honduras. Labour flows were then directed to south-western Guatemala, where up to 300,000 undocumented Salvadorans worked on the coffee, cotton and sugar plantations. Nicaraguans had also been involved in intraregional labour migration, mostly by migrating to Costa Rica and Mexico, while Guatemalans had a long history of engaging in seasonal migration to southern Mexico.

Between 1980 and 1984, an estimated 350,000 Central Americans fled their homelands as civil war and political violence resulted in what is probably the most important forced population movement in the history of Central America (Torres-Rivas, 1985). The main flows of refugees involved Nicaraguans moving to Costa Rica and Honduras; Salvadorans to Honduras; and Guatemalans to Belize and Mexico (Aguilar Zinser, 1991). While in many cases these refugee flows followed earlier migration paths, their political nature altered their composition. Thus, earlier Central American migrants to Mexico had been predominantly young men of urban origin, with levels of schooling higher than average for the region. After 1978, migration flows from Central America involved increasing proportions of women, children and the elderly, still of urban origin, but with lower educational levels and having been engaged in service rather than professional occupations (O'Dogherty, 1989).

Although violence has continued in Guatemala, and fragile peace has come to Nicaragua and El Salvador only within the past two years, the voluntary repatriation of Salvadorans began as early as 1987 and has increased and broadened under the process initiated at the International Conference on Central American Refugees (CIREFCA) held in May 1989. Between January 1990 and March 1992, some 87,000 persons repatriated voluntarily, mainly to El Salvador and Nicaragua, and to a lesser extent, Guatemala (United Nations, 1992a). Still, Central American Governments estimated that, as of 1989, 893,000 undocumented persons remained externally displaced within the region (Conferencia Internacional sobre Refugiados Centroamericanos, 1989).

In the Caribbean, international migration is mainly directed to Northern America yet an estimated 854,000 international migrants were present in the region in 1985 (Skoog, 1994). Because of the region's small population, migrants accounted for a relatively high proportion of it, nearly 3 per cent (United Nations, 1989; Skoog, 1994). Intraregional flows are not generally well documented. Significant ones have taken place from the Dominican Republic to Venezuela and Puerto Rico, where the majority of Dominican migrants are women (United Nations, 1990; Baerga and Thompson, 1990). The long-standing labour migration of Haitians to the Dominican Republic expanded during the 1980s to include those fleeing political violence at home. As of 1991, when the total population of Haiti was approximately 6 million, it was estimated that over 1 million Haitians resided in the Dominican Republic, mostly as undocumented migrants working on sugar plantations (United Nations, 1992b).

Migration is common among the island nations of the English-speaking Caribbean. Certain northern and eastern Caribbean States have begun to experience a "migration transition", that is to say, they have been transformed from net labour exporters to net importers, as a consequence of developments in the tourist and offshore banking industries. The Bahamas, the Cayman Islands and the United States Virgin Islands have all been affected, and other islands are beginning to experience a similar transition (McElroy and De Albuquerque, 1988).

B. Has migration between developing countries increased?

During the 1980s, the number of refugees in both Africa and Latin America grew. In Africa, the number rose from 2.7 million in early 1980 to 4.4 million in early 1990 and had reached 5.4 million by early 1993 (United Nations, 1996). In Latin America, the numbers reported early in the 1980s were not truly comparable with those reported towards the end of the decade but an increasing trend probably persisted until the peace process began to bear results. In terms of the numbers reported, the stock of 300,000 refugees present in the region early in 1982 had risen to about 900,000 by early 1993 (United Nations, 1996).

Trends in other types of migration are more difficult to document, not only because distinctions between "economic" and "political" migrants are blurred, but also because there is a paucity of data allowing an assessment of trends. In sub-Saharan Africa, only a handful of countries have stock data from at least two censuses. In the Congo, for instance, those data indicate that the number of foreign-born persons increased from 54,000 to 97,000 between 1974 and 1984 (Skoog, 1994). In Côte d'Ivoire, the number of foreigners increased from 1.5 million in 1975 to over 3 million in 1988. However, since those data reflect the citizenship of persons and not their place of birth, there is no guarantee that they are all migrants. Indeed, among the 3 million foreigners enumerated in 1988 in Côte d'Ivoire, only 1.7 million were foreign-born. Because of such problems, it is not possible to affirm categorically that the true number of migrants in Africa has been increasing. Nevertheless, estimates of the stock of foreign-born or foreign persons in African countries suggest that between 1975 and 1985 their number increased by about 950,000, representing a 9 per cent rise over the decade (Skoog, 1994).

Data allowing an assessment of trends in the migrant stock are more readily available for Latin America. As presented in table 47, intercensal data for 13 Latin American countries, albeit incomplete, show increases in both the number and the proportion of migrants from within the region. Between 1970 and 1980 alone, the number of migrants originating in countries of the region rose by approximately 800,000 and, as a proportion of the total migrant population in the region, migrants originating in Latin American countries increased from about one quarter to over two fifths.

Case studies support the view that migration between developing countries takes many forms and that, increasingly, it involves short-term movements of various types that add volatility to the phenomenon. In addition, the fact that most developing countries lack explicit policies favouring the admission of international migrants, particularly of unskilled migrant workers from neighbouring countries, implies that much of the international migration occurring between developing countries is poorly controlled and remains undocumented. As the case of Nigeria indicates, intraregional international migration is quite responsive to economic disparities, so that it can rise quickly if opportunities exist. However, Governments are far from impotent in controlling migration and their actions can have major impacts. In the case of Nigeria, although the exact numbers of migrants will never be known, it is nevertheless clear that sizeable numbers entered the country attracted by the favourable economic circumstances associated with the oil boom of the late 1970s and that between 1 million and 2 million left when the Government decided to expel them in the first half of the 1980s. Nigeria was thus transformed almost between one day and the next from a country of net immigration to one of net emigration. Similar transformations have been all too common in the developing world during the past 40 years.

Both sub-Saharan Africa and Latin America are regions that have been negatively affected by the economic developments of the past 15 years. In the case of Latin America, Murillo Castaño (1992) argues that the economic crisis has eroded the economic differences that characterized countries linked by international migration flows and has led to a relative reduction of the magnitude of international migration. Others read the evidence

TABLE 47. NUMBER OF MIGRANTS ORIGINATING IN LATIN AMERICA AND THEIR PERCENTAGES WITH RESPECT TO THE TOTAL MIGRANT POPULATION AND THE TOTAL POPULATION OF SELECTED LATIN AMERICAN COUNTRIES, 1960-1980

Country of enumeration	Sex	*1960 round of censuses*			*1970 round of censuses*			*1980 round of censuses*		
			As percentage of			*As percentage of*			*As percentage of*	
		Number of migrants	*Total migrant population*	*Total population*	*Number of migrants*	*Total migrant population*	*Total population*	*Number of migrants*	*Total migrant population*	*Total population*
Argentina	Male	252 274	17.8	2.5	..	..	..	383 180	40.4	2.8
	Female	214 986	18.1	2.1	..	..	..	378 809	39.6	2.7
	Total	467 260	17.9	2.3	533 850	24.3	2.3	761 989	40.0	2.7
Brazil[a]	Male	31 233	4.6	0.1	31 993	5.5	0.1	48 044	10.0	0.1
	Female	29 613	5.2	0.1	30 878	6.1	0.1	44 398	10.3	0.1
	Total	60 846	4.9	0.1	62 871	5.8	0.1	92 442	10.1	0.1
Chile	Male	14 527	24.7	0.4	14 861	30.9	0.3	34 554	85.0	..
	Female	13 419	29.2	0.4	15 607	36.9	0.3	38 752	97.3	..
	Total	27 946	26.7	0.4	30 468	33.7	0.3	73 306	91.1	0.6
Costa Rica	Male	..	..	..	8 587	72.3	1.1	36 961	81.4	3.1
	Female	..	..	..	7 631	73.4	0.9	36,489	83.8	3.0
	Total	..	..	..	16 218	72.8	1.0	73 450	82.6	3.0
Ecuador	Male	..	..	..	..	..	..	25 009	65.7	..
	Female	..	..	..	..	..	..	26 139	70.0	..
	Total	..	..	..	57 195	..	0.9	51 148	67.8	0.6
El Salvador[a]	Male	3 585	50.6	0.3	7 679	66.5	0.4	..	..	..
	Female	5 231	60.4	0.4	9 165	70.9	0.5	..	..	..
	Total	8 816	56.0	0.4	16 844	68.8	0.5	..	..	..
Guatemala[b]	Male	..	..	..	12 274	71.3	0.5	12 801	72.6	0.4
	Female	..	..	..	16 065	79.5	0.6	18 113	80.2	0.6
	Total	..	..	..	28 339	75.7	0.5	30 914	76.9	0.5
Mexico	Male	7 400	6.2	0.0	10 304	10.6	0.0	17 607	13.1	0.0
	Female	7 626	7.4	0.0	10 414	11.1	0.0	18 460	13.7	0.0
	Total	15 026	6.7	0.0	20 718	10.8	0.0	36 067	13.4	0.0
Nicaragua[c]	Male	6 925	..	0.9	11 460	..	1.2	..	..	..
	Female	6 151	..	0.8	10 558	..	1.1	..	..	..
	Total	13 076	..	0.9	22 018	..	1.2	..	..	..
Paraguay	Male	..	..	..	32 691	78.4	2.8	78 403	88.7	5.2
	Female	..	..	..	31 880	80.9	2.7	71 295	89.4	4.7
	Total	..	..	..	64 571	79.6	2.7	149 698	89.1	4.9
Peru	Male	9 372	25.3	0.2	..	..	..	20 448	45.9	0.2
	Female	12 230	41.3	0.2	..	..	..	23 760	56.8	0.3
	Total	21 602	32.4	0.2	26 105	38.9	0.2	44 208	51.2	0.3
Uruguay	Male	..	..	..	14 289	22.1	..	13 231	27.5	0.9
	Female	..	..	..	20 670	30.8	..	18 770	34.2	1.2
	Total	..	..	..	34 959	26.5	..	32 001	31.1	1.1

TABLE 47 (*continued*)

Country of enumeration	Sex	1960 round of censuses: Number of migrants	1960: As percentage of Total migrant population	1960: As percentage of Total population	1970 round of censuses: Number of migrants	1970: As percentage of Total migrant population	1970: As percentage of Total population	1980 round of censuses: Number of migrants	1980: As percentage of Total migrant population	1980: As percentage of Total population
Venezuela[d]	Male	58 982	20.9	1.5	104 079	32.6	1.9	278 176	51.2	3.8
	Female	52 064	29.0	1.4	115 171	41.5	2.1	299 275	60.5	4.1
	Total	111 046	24.1	1.5	219 250	36.8	2.0	577 451	55.7	4.0
Total	Male	384 298	14.4	0.4	..	..	..	948 414	39.1	1.0
	Female	341 320	15.6	0.4	..	..	..	974 260	41.7	1.0
	Total	725 618	15.0	0.4	1 133 406	29.9	0.6	1 922 674	40.4	0.7

[a]Data classified by nationality.
[b]Data for the 1973 census classified by nationality.
[c]Data for Latin America referring to "other countries" since a detailed tabulation by place of birth is not available.
[d]Data for the 1961 census classified by nationality.

differently. For example, the expected massive return of Colombian migrants from Venezuela as the latter experienced economic recession did not occur (Bleier, 1988). Yet the total foreign-born population present in Venezuela failed to increase between 1981 and 1990, indicating that net migration was considerably lower during the 1980s than in previous decades.

Unfortunately, the results of the 1990 round of censuses are not yet available for most of the countries of sub-Saharan Africa and Latin America. Consequently, it is still not possible to determine more conclusively whether or not a general increase in the migrant stock took place in those developing regions during the 1980s. However, trends in the migrant stock are poor indicators of migration flows whose volume may have increased in both directions while net gains over a decade remain small. An increase in temporary migration can therefore be compatible with low growth in the migrant stock. On the basis of the information currently available, it is not possible to establish whether international mobility has increased more than net gains would indicate. There is reason to believe that such mobility has indeed risen. The following section will examine the reasons that have been advanced to justify belief in an increase in international migration between developing countries in all its various forms.

C. MAJOR FACTORS AFFECTING INTRAREGIONAL MOBILITY

Since many factors influence international migration between developing countries, only those most likely to have shaped recent movements and to be relevant for an assessment of future developments are considered here, including economic differentials; political instability; government policies towards migration; human resource policies affecting mobility of the highly skilled; and regional trade agreements. Additional factors include rates of population and labour force growth, the expansion of transportation and communications facilities, and the role of migrant networks.

1. *Economic differentials*

Both Africa and Latin America experienced deep economic crises during the 1980s. In sub-Saharan Africa as a whole, per capita income and food production fell since 1980, while ecological degradation accelerated. The region lost a substantial part of its share in world export markets and its debt burden grew faster than that of other developing regions. In a number of countries, previous gains made in human resource development have eroded and open urban unemployment has increased. Although the crisis has been general throughout Africa, its effects have been unevenly distributed. Between 1961 and 1987, the annual growth rate of per capita gross domestic product (GDP) was below 2 per cent in 22 countries (in some, it was negative) and higher than 2 per cent in 16. In most countries, economic growth has not kept pace with population growth, which averaged 3.2 per cent per annum during the 1980s. Thus, disparities between countries have widened and per capita incomes now vary by a factor of 20 (World Bank, 1989).

In some respects, the economic crisis in Latin America has been even more profound because the decline was so sudden and from a higher base. While rates of real GDP growth exceeded 4 per cent in the 1970s, they dropped precipitously in the early 1980s, were negative in 1982 and 1983, and, being below 2 per cent per annum for the decade as a whole, were lower than the region's rate of population growth which stood at 2.1 per cent per year (World Bank, 1991; PAHO, 1985). The region's debt service, equivalent to over 30 per cent of its total exports of goods and services in 1989, was the highest of any world region (World Bank, 1991). As in Africa, the effects of the crisis were unevenly distributed. Annual rates of GDP growth were below 2 per cent in 14 countries and higher than 2 per cent in 7 (World Bank, 1991).

To the extent that international migration is a response to both low overall rates of economic growth and differentials between countries, one would expect that intra-regional mobility would have increased during the volatile 1980s. The prospects for the 1990s and beyond are uncertain, but recent World Bank projections provide some insights. Annual average per capita GDP growth for 1990-2000 is projected to be 0.5 per cent in sub-Saharan Africa and 2.0 per cent in Latin America (World Bank, 1991), suggesting that the economic push for increased migration is not likely to diminish, especially in sub-Saharan Africa. In both regions, the growth of the labour force will add to the pressure to migrate. Thus, between 1990 and 2010, the size of the economically active population of Latin America is projected to increase by more than 55 per cent and that of Africa by more than 75 per cent (Russell and Teitelbaum, 1992). In both cases, labour force growth will be unevenly distributed among countries and subregions.

Economic differentials between countries are also unlikely to diminish substantially in the near future. Terms of trade for primary producers are expected to be considerably worse on average, while conditions for oil-exporters and countries that have pursued wide-ranging economic and political reforms are more likely to improve (World Bank, 1991). Africa and Latin America contain countries in both categories. Lastly, as suggested by the United States Commission for the Study of International Migration Cooperative Economic Development (1990), accelerated development itself may induce increased migration over the short term, that is, at least over a decade or two.

2. *Political instability*

As the overview of migration patterns presented above indicates, political instability has long been a factor leading to population movements within Africa and Latin America. During the 1980s, political crises in those regions led to a major increase in the number of refugees and externally displaced persons. This development should perhaps come as no surprise: among other things, rapid population growth in both Africa and Latin America has meant that considerably more people are likely to be affected by political instability or conflict when it arises.

Although future prospects for politically generated migration within Africa and Latin America are difficult to predict, factors conditioning such movements are more readily identified. First, political crises are often inextricably linked to economic crises and, to the extent that countries experience continuing or worsening economic conditions, both political instability and international migration are likely to increase. Second, hopes for greater regional stability resulting from the end of the cold war have been dampened as underlying regional and national conflicts have grown more salient. Third, even where peace or a more stable and democratic form of governance has been established, international mobility may not decline. The repatriation of refugees itself entails increased mobility. Furthermore, people returning to places ravaged by conflict may find that emigration is once more necessary to secure their livelihoods. In some cases, the establishment of a constitutional Government may induce migrant inflows not only because of the return of exiles but also because other migrants' possibly being attracted by the return to stability. The new migratory streams converging to South Africa are but one case in point. Alternatively, nascent efforts to establish peaceful governance may falter, generating new refugee outflows.

Lastly, despite promising signs that rates of population growth are declining in Latin America and in parts of Africa, both regions are still expected to experience significant increases in total population size. The population of Africa alone is projected to grow from 642 million in 1990 to over 1.1 billion by the year 2010, experiencing an addition of 500 million people (United Nations, 1992b). While the effects of population growth on migration depend upon a host of other factors, undeniably the pool of potential migrants and refugees is growing.

Some recent developments may serve to modify the volatility and volume of future politically induced population flows. One of the most promising is the emergence of regional and multilateral efforts to manage political crises better and earlier. The 1980s and early 1990s have witnessed a number of new approaches. In Latin America, these include the formation of the Contadora Group and the leadership role played by UNHCR and the United Nations Development Programme (UNDP) in the International Conference on Central American Refugees (CIREFCA) process to mediate peace in Central America and provide new long-range development-oriented strategies to assist the region's uprooted peoples. In Africa, the application of bilateral and international pressure was crucial to reducing the prospects of full-scale political conflict in Kenya. UNHCR has introduced the concepts of "preventive diplomacy" and "humanitarian intervention" into international discussions about ways to address prospective refugee outflows, and recently made a concerted effort to apply the lessons learned from the CIREFCA process to other regions.

The net contribution of these and other developments to the future of politically induced migration remains to be seen. On balance, the likelihood and volume of such movements are probably greater in Africa than in Latin America, where economic conditions, progress towards stable governance, and the success of regional approaches to conflict management all appear to be more promising.

3. *Government policies towards migration*

The prospects for future migration between the countries of Africa and Latin America will depend on how Governments perceive such movements and the extent to which they establish and enforce policies dealing explicitly with international migration flows. On balance, countries in both sub-Saharan Africa and Latin America are less likely than those in other world regions to take actions aimed at impeding migration or to adopt policies aimed at reducing the inflow of migrants. However, there are marked contrasts between the two regions in terms of perceptions regarding international migration (United Nations, 1992b). Fewer than 2 per cent of African countries view immigration as too low, in contrast with over 12 per cent of Latin American countries; the latter figure reflects the long immigration tradition or the low fertility levels experienced by some Latin American countries (for example, Argentina, Uruguay and Venezuela). Correspondingly, over 15 per cent of Latin American countries, in contrast with fewer than 2 per cent of African countries, have policies to increase immigration. Similar though relatively small proportions of African and Latin American countries view immigration as too high (21 per cent and 18 per cent, respectively).

A sizeable proportion of Latin American countries (42 per cent) perceive emigration as too high, in contrast with only 15 per cent of African countries. In keeping with this perception, nearly 41 per cent of Latin American countries have policies to reduce migration outflows. Interestingly, the proportion of African countries reporting policies to reduce emigration is higher than the proportion perceiving it as a problem, and Africa is the only region other than Asia where countries report having policies to increase emigration.

If perceptions and policies are any indication of future movements, reduced migration may be expected to countries seeking to lower immigration, which include Burundi, Côte d'Ivoire, Djibouti, Gabon, Ghana, Guinea-Bissau, Sierra Leone and the Sudan in Africa; and Antigua and Barbuda, the Bahamas, the Dominican Republic, Costa Rica, Honduras and Venezuela in Latin America (United Nations, 1992b). There are considerably fewer countries seeking to increase immigration and all but one (Equatorial Guinea) are in Latin America (Argentina, Bolivia, Guyana, Paraguay and Uruguay).

Additional migratory pressures may be expected from those sub-Saharan African countries seeking to increase emigration (Rwanda and Cape Verde), as well as from the 18 other African countries that are intervening to maintain existing levels of emigration. While no Latin American country has explicit policies to increase emigration, Brazil views existing levels as too low, and Venezuela as well as several Caribbean countries are intervening to maintain existing levels (United Nations, 1992b).

In a context where migration has been relatively unconstrained, the effectiveness of government policies can be assessed at best on the basis of anecdotal evidence. Certain interventions have had clear-cut effects, especially expulsions by government order, as was the case in Honduras in 1969, Ghana in the same year, Uganda during the early 1970s and Nigeria in 1983 and 1985. Policies promoting the immigration of overseas persons were effective in the cases of economically attractive countries, such as Argentina, Brazil and Uruguay earlier in the century, and Venezuela more recently. However, Bolivia and Paraguay have not had much success in attracting immigrants. No Latin American country has actively promoted the immigration of citizens from neighbouring countries. However, with the wane of overseas immigration, Latin American migrants have taken the place of Europeans and have tended to be more responsive to economic and political conditions in the potential country of destination than to the country's migration policies. Much of the intraregional mobility in both Africa and Latin America has been undocumented, and such migration has been particularly difficult to manage. Some countries, such as Argentina, Belize and Venezuela, have implemented the granting of amnesties for undocumented aliens as a means of gaining some degree of control over the migration of these aliens (Balán, 1992; Palacio, 1988; Zlotnik, 1989). Others, including Nigeria and Zambia, have tightened frontier controls to forestall the entry of undocumented migrants (Brennan, 1984).

Many countries that experienced substantial immigration in the 1980s have only recently instituted controls, and their success remains to be assessed. Some, such as Gabon, have introduced sophisticated computerized systems to control migration (Azonga, 1986). Such measures are not always effective, especially if controls are exercised selectively. Thus, when in the 1960s Venezuela favoured the immigration of Europeans, such immigrants were subject to control. However, virtually no control was exerted over the numerous Colombian migrants who crossed the border without proper documentation to fill jobs in agriculture (Bleier, 1988).

There is reason to believe that current and future efforts by developed countries in Europe and Northern America to stem migration from developing countries may be more effective than in the past, when restrictions on legal entry for employment resulted in migrants' seeking to enter by other means, especially the requesting of asylum. Border controls have become tighter, asylum procedures are being streamlined, and there is a greater willingness to turn away potential migrants. If such measures are successful in discouraging continued migration to the developed countries, migration between developing countries may well increase.

4. *Human resource policies and the mobility of the highly skilled*

Migration policies are not the only governmental policies affecting population mobility. In addition to economic and political conditions, development strategies and policies pertaining to education, training and employment are particularly important in shaping the conditions for migration, especially in regard to the movement of highly skilled personnel.

For both Africa and Latin America, there is only fragmentary evidence as to the number of highly skilled personnel who have migrated to developed countries. As of 1987, the United Nations Conference on Trade and Development (UNCTAD) estimated that about 70,000 people or 30 per cent of the high-level manpower stock in Africa were residing legally in member States of the European Union (United Nations, Economic Commission for Africa, 1988). Other estimates suggest that the level of emigration of skilled personnel has been higher. Thus, Adepoju (1991) reports that over 110,000 skilled Nigerians had taken jobs abroad between 1987 and 1989. That number seems to be on the high side, considering that the net migration gain recorded by the major receiving countries in the developed world included only 156,000 Africans during the same period (Zlotnik, 1994). Although those receiving countries do not exhaust the possible destinations of Nigerian migrants, the comparison serves to set the estimates proposed in perspective. Inconsistencies in this area abound. Clearly, if the UNCTAD estimates of the high-level manpower stock in Africa are correct, Adepoju's estimates would imply that over half of that stock is Nigerian. Differences in the definition of who qualifies as "high-level" or "skilled" are likely to be at the root of these discrepancies.

Censuses provide more reliable information on the stock of highly skilled migrants. In the United States, the 1990 census indicates that 819,000 persons born in Latin America were working in managerial and professional specialty occupations at the time of enumeration, 232,000 of whom had entered the United States since 1980 (United States Department of Commerce, 1993). The equivalent figures among persons born in Africa were 48,000 and 31,000, respectively. The total stock of foreign-born persons in managerial and professional occupations was 2,364,000 of whom 746,000 had arrived since 1980. That is, Latin American immigrants constituted about 30 per cent of the highly skilled employed immigrants who had entered the United States since 1980, whereas African immigrants accounted for only 4 per cent of that group.

Some studies have used flow statistics to ascertain the level of highly skilled migration to the United States. However, those statistics must be interpreted with caution since the category of skilled immigrants does not exhaust those having advanced skills who may be admitted in other categories (as spouses of United States citizens, for instance). In 1990, the number of immigrants admitted as professionals or highly skilled workers included 764 from Africa and 1,071 from Latin America (United States Department of Justice, 1991). These numbers are considerably lower than those of highly skilled personnel admitted on a temporary basis who usually work under a variety of temporary visa categories whose statistical accounting is far from straightforward. Thus, the nearly 155,000 admissions of temporary workers and trainees from Latin America made during 1987-1990 included the arrival of both highly skilled personnel and unskilled workers to fill specific vacancies on a temporary basis (Zlotnik, 1996). Because multiple entries during a year are counted as different admissions, the number of admissions cannot be equated with the number of persons involved and is therefore a poor indicator of the true level of short-term skilled migration to the United States.

In both Africa and Latin America, it is thought that the intraregional migration of the highly skilled has been increasing since the late 1970s. In the case of Africa, the numbers involved have not been estimated but three reasons are put forward to justify the postulated trend: opportunities for migration to developed countries have declined; economic differentiation among African countries has increased; and the number of persons with higher educational attainment has expanded faster than the economies of many countries, leading to disparities between the supply and the demand for skilled manpower and, consequently, to the emigration of those unable to find work at home (Gould, 1985). Since skilled personnel usually have access to the means needed to emigrate, they usually are the first to leave, especially in times of economic and political crisis.

In Latin America, census data allowing the measurement of trends in the stock of highly skilled migrants are generally not available, but a diversification of the migratory flows of the highly skilled to new destinations within the region has been observed since the 1970s. Of all the highly skilled emigrants originating in the region, 35 per cent are estimated to be in other Latin American countries (Fernández Lamarra, 1992). According to census data, nearly 74,000 highly skilled emigrants were abroad in 11 Latin American countries in the early 1980s. Nearly three quarters of them were in Argentina, Brazil and Venezuela, and the major source countries were Argentina, Colombia, Chile, Paraguay, Peru and Uruguay, which together accounted for three quarters of all skilled emigrants (Centro Latinoamericano de Demografía (CELADE), 1991).

Various reasons have been offered for the rising emigration of highly skilled personnel in Latin America. As in Africa, educational policies have contributed to the increase of the supply available. In Argentina, for instance, educational policies produced many professionals trained to be competitive in the international arena. Because wage levels at home were low relative to world market prices for such skills, a considerable number emigrated (Balán, 1985). Other factors that led to emigration included (*a*) the multiple linkages with developed countries in education and training programmes; (*b*) the limited local opportunities for postgraduate training; (*c*) the orientation of research to topics relevant for developed countries; (*d*) the integration of international research teams around such topics; (*e*) the lack of equipment or academic incentives; and, lastly, (*f*) the political persecution to which many professionals were subject under military regimes. The presence of transnational corporations has also been a major contributing factor leading to the increased international mobility of professional, technical and managerial personnel within the region (Centro Latinoamericano de Demografía (CELADE), 1991; Fernández Lamarra, 1992).

The extent to which the emigration of the highly skilled has impeded or slowed development in sub-Saharan Africa and Latin America has been a subject of considerable debate, and attitudes towards such movement have shifted over the past several decades, as the supply, characteristics and destination of skilled migrants have changed. Migration directed to the developed countries of Europe and Northern America has generally been characterized as a brain drain, the term "brain drain" implying the exploitation of the less affluent countries by the more affluent ones. The migration of skilled personnel to other developing countries, which has gained momentum since the late 1970s, is described in more neutral terms as "reverse transfer of technology" or "cooperative exchange of skills between developing countries".

Emigration of the highly skilled does not necessarily have negative effects upon development in source countries, especially if there is an ample supply of such personnel. However, development may be impeded if the emigration is substantial and critical sectors of the economy are affected. As regards Africa, shortages of skilled manpower have been reported in Ethiopia as a result of insufficient output from the higher-education establishments and of the outflow of qualified personnel since the revolution. Uganda lost more than half its high-level manpower during the regime of President Amin. Recent emigration from Nigeria has affected medicine, the universities, and the airline industry. Similarly, Zambia has lost substantial numbers of university lecturers and public sector physicians, mainly to other countries of Southern Africa (Russell, 1993). It should be noted that emigration of less skilled workers can also hinder development in the country of origin, particularly if the workers leaving have training in areas experiencing labour shortages.

In Latin America, concern over the loss of qualified personnel has been reflected in efforts to promote their return. Thus, in 1975, the Programme for the Return of Qualified Latin American Nationals was established under the auspices of the International Organization for Migration (IOM). Following the restoration of democratic institutions in Argentina, Chile and Uruguay, all three countries have established national commissions to promote the return of qualified human resources. While these efforts have been more successful in effecting the return of those who left for political reasons than of those who left for economic reasons, the proportion of highly skilled emigrants who have returned is relatively small and others have continued to emigrate (Fernández Lamarra, 1992).

Increasingly, Latin American countries are perceiving emigration of the highly skilled in the context of both regional development and the need to integrate the various sectoral policies affecting human resources. Proposals put forward by the Economic Commission for Latin America and the Caribbean (ECLAC) in 1990 and in April 1992, known by the title "Productive Transformation with Equity", are aimed at formulating integrated policies for the training, development and use of human resources. It is recognized that these strategies should constitute the framework for future policies and actions in the area of migration, especially as they pertain to the highly skilled (Centro Latinoamericano de Demografía (CELADE), 1991; Fernández Lamarra, 1992).

5. *Role of regional trading groups*

Regional trade agreements have the potential to influence international migration profoundly, especially if they include provisions for the free movement of people. Although voluntary international population movements in Africa are largely spontaneous, some occur within the framework of regional agreements. At least 17 regional trade organizations have emerged since the early 1960s, prompting Ricca (1989) to assert that no other continent has seen the emergence of so many groups in so short a time as Africa. Those with provisions for the free movement of people include the Communauté économique de l'Afrique de l'Ouest (CEAO); the Economic Community of the Great Lakes Countries (Communauté économique des pays des Grands Lacs (CEPGL)); the Economic Community of West African States (ECOWAS); and the Union douanière et économique de l'Afrique centrale (UDEAC) (Central African Customs and Economic Union).

The regional group that has most influenced international migration in Africa is probably ECOWAS, established by treaty in 1975.[1] In May 1979, the member States of ECOWAS signed the Protocol on the Free Movement of Persons, Right of Residence and Establishment, to be phased in progressively over 15 years. During the first phase (1980-1985), citizens of member States were granted the right to be admitted by other member States provided they had a valid travel document and a health certificate and were intending to stay for less than 90 days. Visas were no longer required for stays of that length but official permission was required to remain beyond that period. The second phase (1986-1990) involved the abolition of residence permits for citizens of member States residing in other member States and more liberal rights in respect of their taking up employment were established. The third and current phase (1991-1995) extends to citizens of member States the right to establish commercial or industrial enterprises in other member States.

After signature of the Protocol, which coincided with a dramatic increase in oil exports from Nigeria, there was an unprecedented volume of migration directed to that country. According to official statistics, inflows increased from 76,000 in 1979 to over 203,000 in 1980 and remained above 150,000 during each of the following two years. However, many migrants entered without being controlled, through unofficial entry points and bush paths. Without valid travel documents, they stayed beyond 90 days, and took up employment and commercial activities without authorization (Whannou, 1986; Afolayan, 1988). It is thought that most migrants did not understand the restrictions established by the Protocol and assumed it gave them the right to unlimited stay and access to employment. As noted earlier, by the time the expulsions began in 1983, over 2 million international migrants were estimated to be present in Nigeria.

It is difficult to disentangle the effects of the Protocol itself from those of economic factors. According to Afolayan (1988), export revenues accounted for nearly 93 per cent of the increase in migration to Nigeria over the period 1970-1982. However, there were also problems with the Protocol itself. The first was ambiguity over what would happen to citizens of ECOWAS member States who stayed beyond 90 days without permission. Second, ECOWAS did not achieve its objectives as rapidly as planned; despite the aim of economic integration, wide disparities among member States remained, with Nigeria exporting more to its trading partners than it imported. Nigeria has implemented the second phase of the Protocol, but it has been ensuring better control of migrants, providing public education as to the rights and obligations of those moving, and enforcing tightened border security. In 1989, ECOWAS introduced a common travel document to facilitate movement between member States (United Nations, 1992b, para. 762).

Latin America also has its share of regional bodies to promote trade and economic integration, which include the Andean Group; the Rio Group or Group of Eight; the Southern Cone Common Market (Mercado Común del Sur (MERCOSUR)); and the Group of Three or G3. Of these, only the Andean Group or Andean Pact, begun in 1969, has addressed international migration between member States, which in the case of the Pact currently include Bolivia, Colombia, Ecuador, Peru and Venezuela. Two agreements that deal with the issue have been formulated: the Convenio Simón Rodríguez (CSR), signed in 1973; and the Estatuto Andino de Migraciones (EAM), drafted in 1977. The CSR seeks to define and harmonize labour and social policies to facilitate population mobility, but it has not been implemented. The EAM intends to revitalize the principles and tenets of the CSR and to create offices dealing with labour migration whose mandate would include producing a labour statute for the region, facilitating labour movements between member States, and enhancing control over such movements. The provisions of the EAM have only been partially implemented (Murillo Castaño, 1992).

In 1991, the migration authorities of the member States of the Andean Pact met to prepare recommendations for action on a number of specific proposals, including the elimination of visas for citizens of member States planning to stay in a member State other than their own for up to 90 days; the establishment of a regional system for migration information and control; measures to facilitate the mobility of highly skilled personnel; the use of personal identification documents for travel between member States; the establishment of a common set of visas for third-country nationals and common passports for citizens of Andean Pact countries; and measures to facilitate the residence in member States other than their own of citizens of Andean Pact countries (Peru, Ministerio del Interior, Dirección de Migraciones y Naturalización, 1992).

As these examples illustrate, it is difficult to assess the effects of regional trade accords on international migration between member States. In both sub-Saharan Africa and Latin America, economic factors may well have induced migration even in the absence of international instruments facilitating international mobility. It is even more difficult to project the future impact of such instruments. On balance, they are likely to foster an increase of intraregional migration, especially when existing regional agreements establish some measure of freedom of movement in areas where migration is already well established. Furthermore, even in the absence of such provisions, international mobility is likely to rise as trade increases. Thus, since the introduction in January 1992 of a new customs union between Colombia and Venezuela, trade between the two countries has increased by 30 per cent and at Venezuela's busiest border crossing point, the number of vehicles crossing daily has been estimated at some 30,000, with the result that the three customs agents on duty can do little more than count cars (*The New York Times*, 1992). Indeed, trade in goods is likely to be accompanied by trade in services which cannot flourish if people are not allowed to move from one country to another on a temporary basis (Feketekuty, 1988). Some indication of the importance of international migration for trade in services is reflected in the fact that, as of 1989, the level of official remittances for the world as a whole was equivalent to nearly 9 per cent of world trade in services (Russell and Teitelbaum, 1992).

Economic theory argues that, in the long run, trade and investment can reduce economic disparities between trading partners and trade can become a substitute for migration by inducing the growth of export industries and related employment in the countries of origin of migrants. However, such a process, if it occurs at all, takes time to bear fruit. Frequently, as in the case of Nigeria in ECOWAS, one member State dominates the trading bloc and becomes a pole of attraction for workers, whose migration occurs before the forces of economic equalization have time to act. In addition, there are powerful economic inducements for labour mobility, especially with regard to remittances, whose global value grew substantially in the 1980s and was estimated to exceed US$ 71 billion in 1991, representing revenues second only to those obtained from trade in oil and far exceeding the value of official development assistance, which amounted to US$ 51 billion in 1988 (Russell, 1992; Russell and Teitelbaum, 1992). In selected countries, official remittances, which by no means cover all those actually received by a country, are equivalent to a substantial proportion of merchandise exports and imports. Thus, in Burkina Faso, their value amounts to nearly 69 per cent of net revenue obtained from trade in goods; in Mali, the equivalent figure is 34 per cent; and in Colombia, it is nearly 8 per cent. As a proportion of merchandise imports, remittances amount to nearly 29 per cent in Burkina Faso, over 26 per cent in Mali and 10 per cent in Colombia (Russell and Teitelbaum, 1992). Consequently, wage differentials and exchange rate advantages will have to narrow considerably before the inducements leading to international migration are counteracted.

D. Conclusion

In conclusion, the evidence available suggests that intraregional migration in sub-Saharan Africa and Latin America is sizeable and that, despite the economic difficulties experienced since the late 1970s, it is likely to have increased. Economic differentials between the countries of both regions are likely to persist and to induce continued and possibly growing migratory flows, especially from those countries whose rates of population and labour force growth remain high in relative terms. In addition, migration caused by civil strife and conflict is not likely to abate in the near future, despite the efforts being made to promote peace and democracy in both sub-Saharan Africa and Latin America. Although international and regional action to manage (if not stem) politically induced outflows is being stressed, the concrete measures it entails usually involve keeping displaced persons and refugees near the zone of conflict, thus fuelling flows to countries in the immediate vicinity. Migration between neighbouring countries is also likely to result from the efforts being made to achieve greater economic integration among groups of States. Both the success of trading blocs in fostering exchanges between member States and the traditionally low degree of Government intervention in controlling migration are likely to result in continued flows of persons seeking to improve their life chances.

Note

[1]The member States of ECOWAS are Benin, Burkina Faso, Cape Verde, Côte d'Ivoire, the Gambia, Ghana, Guinea, Guinea-Bissau, Liberia, Mali, Mauritania, the Niger, Nigeria, Senegal, Sierra Leone and Togo. CEAO is an international member.

References

Adepoju, Aderanti (1991). South-North migration: the African experience. *International Migration* (Geneva), vol. 29, No. 2 (June), pp. 205-222.

Afolayan, A. A. (1988). Immigration and expulsion of ECOWAS aliens in Nigeria. *International Migration Review* (Staten Island, New York), vol. 22, No. 1 (spring), pp. 4-27.

Aguilar Zinser, Adolfo (1991). CIREFCA. *The Promises and Reality of the International Conference on Central American Refugees: An Independent Report.* Washington, D.C.: Georgetown University, Center for Immigration Policy and Refugee Assistance, Hemispheric Migration Project.

Azonga, Tikum Mbah (1986). Towards 'Gabonisation'. *West Africa* (United Kingdom), No. 3586 (May), pp. 1100-1101.

Baerga, María del Carmen, and Lanny Thompson (1990). Migration in a small semiperiphery: the movement of Puerto Ricans and Dominicans. *International Migration Review* (Staten Island, New York), vol. 24, No. 4 (winter), pp. 656-683.

Balán, Jorge (1985). International migration in the Southern Cone. Washington, D.C.: Georgetown University, Center for Immigration Policy and Refugee Assistance (October). Mimeographed.

_________ (1992). The role of migration policies and social networks in the development of a migration system in the Southern Cone. In *International Migration Systems: A Global Approach*, Mary M. Kritz, Lin L. Lim, and Hania Zlotnik, eds. Oxford: Clarendon Press.

Bleier, Elisabeth Ungar (1988). Impact of the Venezuelan recession on return migration to Colombia: the case of the principal urban sending areas. In *When Borders Don't Divide: Labor Migration and Refugee Movements in the Americas*, Patricia R. Pessar, ed. New York: Center for Migration Studies.

Brennan, Ellen M. (1984). Irregular migration: policy responses in Africa and Asia. *International Migration Review* (Staten Island, New York), vol. 18, No. 3 (fall), pp. 409-425.

British Broadcasting Company (BBC) (1992). News report, 29 October.

Cameroon (1976). *Recensement général de la population et de l'habitat d'avril 1976.* Yaoundé: Ministère de l'économie et du plan, Direction de la statistique et de la comptabilité nationale, Bureau central du recensement.

Centro Latinoamericano de Demografía (CELADE) (1991). *La Movilidad de Profesionales y Técnicos Latinoamericanos y del Caribe.* Santiago, Chile.

Congo (1978). *Recensement général de la population du Congo 1974*, tome IV, *Tableaux statistiques détaillés.* Brazzaville: Ministère délegué auprès du Premier ministre chargé du plan, Centre national de la statistique et des études économiques, Direction des statistiques démographiques et sociales.

_________ (1984). *Recensement général de la population du Congo 1984*, tome III. Brazzaville: Ministre Délegué auprès du Premier Ministre Chargé du Plan, Centre National de la Statistique et des Etudes Economiques, Direction des Statistiques Démographiques et Sociales.

Conferencia Internacional sobre Refugiados Centroamericanos (1989). Map and table. First International Meeting, CIREFCA, Guatemala City, 29-31 May 1989.

Dandler, Jorge, and Carlos Medeiros (1988). Temporary migration from Cochabamba, Bolivia to Argentina: patterns and impact in sending areas. In *When Borders Don't Divide: Labor Migration and Refugee Movements in the Americas*, Patricia R. Pessar, ed. New York: Center for Migration Studies.

De Vletter, Fion (1985). Recent trends and prospects of Black migration to South Africa. Working paper MIG WP.20. Geneva: International Labour Office, International Migration for Employment (January).

Feketekuty, Geza (1988). *International Trade in Services: An Overview and Blueprint for Negotiations.* Cambridge, Massachusetts: Ballinger Publishing Company.

Fernández Lamarra, Norberto (1992). Human resources, development, and migration of professionals in Latin America. *International Migration* (Geneva), vol. 30, Nos. 3/4, pp. 313-334.

Foundation for Contemporary Research (1992). African transurban programme: funding proposal. Belleville, South Africa: Foundation for Contemporary Research.

Gould, W. T. S. (1985). International migration of skilled labour within Africa: a bibliographical review. *International Migration* (Geneva), vol. 23, No. 1 (March), pp. 5-27.

Makinwa-Adebusoye, Paulina (1992). The West African migration system. In *International Migration Systems: A Global Approach*, Mary M. Kritz, Lin L. Lim, and Hania Zlotnik, eds. Oxford: Clarendon Press.

McElroy, Jerome L., and K. De Albuquerque (1988). Migration transition in small Northern and Eastern Caribbean States. *International Migration Review* (Staten Island, New York), vol. 22, No. 3 (fall), pp. 30-58.

Migrant Labour: Quest for Reform (1987). *Financial Mail* (South Africa), No. 103, 13 March, pp. 31-32.

Murillo Castaño, Gabriel (1992). Comparisons among national policy-making patterns towards migration in Latin American sending States: a North-South interdependence approach. Paper presented at the Center for Latin American and Caribbean Studies, New York University, 10 April. Mimeographed.

The New York Times (1992). Venezuela joins the drug route. Thursday, 5 November, p. A.4.

O'Dogherty, Laura (1989). *Central Americans in Mexico City: Uprooted and Silenced.* Washington, D.C.: Georgetown University, Center for Immigration Policy and Refugee Assistance, Hemispheric Migration Project.

Palacio, Joseph O. (1988). Illegal aliens in Belize: findings from the 1984 Amnesty. In *When Borders Don't Divide: Labor Migration and Refugee Movements in the Americas*, Patricia R. Pessar, ed. New York: Center for Migration Studies.

Pan American Health Organization (PAHO) (1985). *The Economic Crisis and its Impact on Health and Health Care in Latin America and the Caribbean.* Washington, D. C.

Peru, Ministerio del Interior, Dirección de Migraciones y Naturalización (1992). La migración en el Perú: características, problemas y propuestas, 1989-1991. Paper presented at the Tenth International Organization for Migration Seminar on Migration, Special Topic: Migration and Development, Geneva, 15-17 September 1992. Document No. 13.

Ricca, Sergio (1989). *International Migration in Africa: Legal and Administrative Aspects.* Geneva: International Labour Office.

Russell, Sharon Stanton (1992). Migrant remittances and development. *International Migration* (Geneva), vol. 30, Nos. 3/4, pp. 268-288.

__________ (1993). International migration. In *Demographic Change in Sub-Saharan Africa*, Karen A. Foote, Kenneth H. Hill and Linda G. Martin, eds. Washington, D.C.: National Academy Press.

__________, K. Jacobsen and W. D. Stanley (1990). *International Migration and Development in Sub-Saharan Africa*, vol. I, *Overview.* World Bank Discussion Papers, Africa Technical Department Series, No. 101. Washington, D.C.: World Bank.

Russell, Sharon Stanton, and Michael S. Teitelbaum (1992). *International Migration and International Trade.* World Bank Discussion Papers, No. 160. Washington, D.C.: World Bank.

Skoog, Christian (1994). The quality and use of census data on international migration. Paper presented at the XIII World Congress of Sociology, Bielefeld, Germany, 18-23 July.

Stanley, William D. (1991). State responses to Central American migration: the role of ideology, domestic politics, and foreign relations. Paper presented at the Conference on the Impact of International Migration on the Security and Internal Stability of States, Center for International Studies, Massachusetts Institute of Technology, Cambridge, Massachusetts, 5-6 December. Mimeographed.

Taylor, John (1990). The reorganization of mine labor recruitment in Southern Africa: evidence from Botswana. *International Migration Review* (Staten Island, New York), vol. 24, No. 2 (summer), pp. 250-272.

Torres-Rivas, Edelberto (1985). Report on the condition of Central American refugees and migrants. Washington, D. C.: Georgetown University, Center for Immigration Policy and Refugee Assistance (July). Mimeographed.

United Nations (1989). *World Migrant Populations: The Foreign Born.* Sales No. E.89.XIII.7A. Wall chart.

__________ (1990). *World Population Monitoring, 1989: Special Report: The Population Situation in the Least Developed Countries.* Population Studies, No. 113. Sales No. E.89.XIII.12.

__________ (1992a). Update on the CIREFCA Process, United Nations General Assembly, Executive Committee of the High Commissioner's Programme, Forty-third session, 21 September. A/AC.96/INF.176.

__________ (1992b). *World Population Monitoring, 1991: With Special Emphasis on Age Structure.* Population Studies, No. 126. Sales No. E.92.XIII.2.

__________ (1996). *World Population Monitoring, 1993: With a Special Report on Refugees.* Sales No. E.95.XIII.8.

__________, Economic Commission for Africa (1988). An enabling environment to retain Africa's high-level manpower. Paper presented at the International Conference on the Human Dimension of Africa's Economic Recovery and Development, Khartoum, Sudan, 5-8 March.

United Nations High Commissioner for Refugees (UNHCR) (1993). *The State of the World's Refugees. The Challenge of Protection.* London and New York: Penguin Books.

United States Commission for the Study of International Migration and Cooperative Economic Development (1990). *Unauthorized Migration: An Economic Development Response.* Washington, D.C.: U. S. Government Printing Office.

United States Committee for Refugees (1991). *World Refugee Survey, 1991.* Washington, D.C.

__________ (1992). *World Refugee Survey, 1992.* Washington, D.C.

United States Department of Commerce, Bureau of the Census (1993). *1990 Census of Population. The Foreign-Born Population in the United States.* Washington, D.C.: U. S. Government Printing Office.

United States Department of Justice (1991). *1990 Statistical Yearbook of the Immigration and Naturalization Service.* Washington, D.C.: U. S. Government Printing Office.

Whannou, Georges Abiodun (1986). Free movement of people. *West Africa* (United Kingdom), 30 June, pp. 1367-1368.

World Bank (1989). *Sub-Saharan Africa, from Crisis to Sustainable Growth: A Long-Term Perspective Study.* Washington, D.C.: World Bank.

__________ (1991). *World Development Report, 1991: The Challenge of Development.* New York: Oxford University Press.

Zlotnik, Hania (1989). Estimación de la emigración a partir de datos sobre la residencia de hijos sobrevivientes: el caso de Colombia. *Memorias de la Tercera Reunión Nacional sobre la Investigación Demográfica en México*, vol. I. Mexico City: Universidad Nacional Autónoma de México and Sociedad Mexicana de Demografía.

__________ (1994). International migration: causes and effects. In *Beyond the Numbers: A Reader on Population, Consumption, and the Environment*, Laurie Ann Mazur, ed. Washington, D.C.: Island Press.

__________ (1996). Policies and migration trends in the North American system. In *International Migration, Refugee Flows and Human Rights in North America: The Impact of Free Trade and Restructuring*, Alan B. Simmons, ed. Staten Island, New York: Center for Migration Studies.

Part Five

SOCIAL, ECONOMIC AND POLITICAL ASPECTS OF INTERNATIONAL MIGRATION

XIV. THE PROCESS OF INTEGRATION OF MIGRANT COMMUNITIES

*Stephen Castles**

Most developed countries have experienced large-scale immigration from less developed areas since 1945. Whatever the original intentions of the Governments and migrants concerned, many migrants have become long-term or permanent settlers. The integration of migrants into the societies of the receiving countries has therefore become an important issue of public policy, with significant effects on the labour market, on housing, education and social welfare, as well as on political institutions and those related to national identity. The present paper discusses concepts of integration and ethnic group formation, examining the experiences and policies of both traditional countries of immigration and the Western European countries that have recruited foreign labour. Using cross-national comparisons, it shows both the common factors and the differences in the process of integration in the various countries. Lastly, the relative success of the various approaches is discussed. Much of the analysis presented here is based on Castles and Miller (1993).

A. THE CONCEPT OF MIGRANT INTEGRATION

"Integration" is widely used to refer, as a general term, to the process by which immigrants become incorporated into the receiving society. That usage will be followed in this paper, although it might be more correct to speak of various forms of migrant incorporation, of which integration is just one possible variant. An assessment of integration cannot restrict itself to government policies, but must look at a wide range of social processes, such as incorporation into social, economic and political structures; the degree and nature of migrant participation in the institutions of the society; and the emergence of various forms of inequality. It is important to examine what social conditions facilitate or hinder full incorporation of migrants into the social fabric (Breton and others, 1990, p. 3). The outcome of the process may not be the absorption of immigrants but rather the formation of ethnic groups (Rex and Mason, 1986). In principle, it is possible to differentiate among four possible approaches to migrant incorporation, each of which implies different outcomes. These are commonly referred to as assimilation, integration, exclusion and multiculturalism.

1. *Assimilation*

Assimilation refers to the incorporation of migrants into society through a one-sided process of adaptation, in which migrants are expected to give up their distinctive linguistic, cultural or social characteristics and become indistinguishable from the majority population. Assimilation is generally seen as a process of individual adaptation to prevailing values, norms and behavioural forms. A theory of migrant assimilation was first provided by the Chicago School of Sociology in the 1920s. It is summed up in Park's concept of the "race relations cycle", in which groups pass through the stages of contact, conflict, accommodation and assimilation (Park, 1950). Successful completion of this process can lead to a restoration of the overall cohesiveness of society. In the common-sense version, immigrants must speak the majority language and learn to behave like the majority if they want to be accepted. Assimilation presupposes the willingness and ability of immigrants to discard their distinctive traits and the willingness of the dominant group to accept new members. It also assumes a high degree of cultural homogeneity and consensus on values and norms among the population of the receiving society.

Assimilation was the prevailing approach to immigrant incorporation in the United States in the early part of this century, at a time of massive immigration and urbanization. It also represented the conventional wisdom in several countries that experienced mass immigration after 1945, including Australia, Canada and the United Kingdom of Great Britain and Northern Ireland. Assimilation implies that the role of the State is to create conditions favourable to individual adaptation, but not to

*Centre for Multicultural Studies, University of Wollongong, Australia.

the maintenance of distinct ethnic groups. This aim usually implies a policy of "benign neglect", in which Government does not provide special services for immigrants (except, perhaps, for initial settlement services or language training). Insistence on the use of the dominant language and attendance in normal schools by migrant children is seen as a way of transferring the majority culture and values. The aim of assimilation is inherent in current human capital approaches to immigration, which argue that the State should simply leave matters to market mechanisms, which will select the most suitable immigrants and encourage successful adaptation (Borjas, 1990).

2. *Integration*

Integration refers to a process of mutual accommodation involving immigrants and the majority population. The concept implies that immigrant groups will cease over time to be distinctive in culture and behaviour, but entails the view that the adaptation needed is a two-way process in which minority and majority groups learn from each other and take on aspects of each other's cultures. Integrationist approaches developed in the 1960s in response to several factors. First, it became evident that recent migrants were not simply becoming assimilated as individuals, but were tending to form social, cultural and political associations, and to maintain the use of their mother tongue. Second, it became clear that migrants were becoming concentrated within particular occupations and residential areas, so that ethnic background and class were linked. Third, as immigrants formed groups and associations, they became increasingly critical of the denial of the granting of legitimacy to their cultures and languages.

As a result, official policies shifted. The integrationist approach became a tenet of official policy in Australia (Collins, 1991) and Canada (Fleras and Elliott, 1992). It also became the guiding principle for educational policy in the United Kingdom. In the United States of America, the shift was rather towards changing the role of the State, so as to remove barriers to individual participation through equal opportunity and affirmative action legislation. The legitimacy of ethnic organizations and the inevitability of some degree of cultural and linguistic maintenance in the first and second generations were accepted. Ethnic group solidarity came to be seen as an instrument of successful settlement and adaptation, rather than as a sign of maladaptation to the receiving society.

3. *Exclusion*

Exclusion refers to a situation in which immigrants are incorporated into certain areas of society (above all the labour market) but denied access to others (such as welfare systems, citizenship and political participation). Exclusion may take place through legal mechanisms (such as refusal of naturalization and sharp distinctions between the rights of citizens and non-citizens) or through informal practices (racism and discrimination). Through those means, immigrants become rightless ethnic minorities, which are part of society but excluded from the State and the nation. Since such ethnic minorities are usually socio-economically disadvantaged, this situation implies a strong and continuing link between class and ethnic background. Former "guest worker" recruiting countries in Western Europe have tended to adopt exclusionary models.

4. *Multiculturalism*

Multiculturalism refers to the development of immigrant populations into ethnic communities that remain distinguishable from the majority population with regard to language, culture, social behaviour and autonomous associations over a long period (usually several generations). It differs from exclusion in that immigrants are granted more or less equal rights in most spheres of society, without being expected to give up their diversity, although usually with an expectation of conformity to certain key values. Multiculturalism implies the willingness of the majority group to accept and even welcome cultural difference, and to change social behaviour and institutional structures accordingly. General policies of multiculturalism exist in Australia, Canada and Sweden, while in the Netherlands and the United Kingdom, multicultural policies exist in specific sectors, such as education or welfare.

5. *Ethnic group formation*

Although exclusion and multiculturalism appear to be opposites, they are in fact facets of a single process: ethnic group formation. A number of structural factors in developed countries cause migrants to assume specific (and often inferior) social positions. At the same time, group solidarity in the face of a new (and sometimes hostile) environment gives a subjective dimension to group formation. Whether the ethnic group takes the form of an excluded minority or an accepted community depends largely on the attitudes of the majority population and the actions of the State. Ethnic group position is a product of both "other-definition" and "self-definition". Through other-definition, the dominant group ascribes undesirable characteristics and assigns inferior social positions to minority groups. Self-definition refers to the consciousness among group members of belonging together on the basis of shared cultural and social characteristics. The relative strengths of other- and self-definition can vary. Some minorities are constituted mainly on the basis of their members' cultural and historical consciousness (or ethnicity). Others are constructed mainly through processes of exclusion (or racist discrimination) set in motion by the majority.

Ethnicity may be understood as referring to the sense of belonging to a group, based on ideas of common origins, history, culture, experience and values (Fishman, 1985; Smith, 1986). Ethnicity is sometimes viewed as being primordial, that is, pre-social—almost instinctual—something one is born into (Geertz, 1963). However, some scholars speak of situational ethnicity, whereby members of a group decide to "invoke" ethnicity as a criterion for self-identification. The markers chosen for ethnic boundaries are variable, generally emphasizing cultural characteristics, such as language, history, customs and religion, but sometimes including phenotype or physical characteristics (popularly referred to as race) (Wallman, 1986). Similarly, studies of ethnic revival by sociologists Glazer and Moynihan (1975) and Bell (1975) emphasize the instrumental role of ethnic identification, which is used to strengthen the solidarity of the group in order that it may compete more effectively for market advantages or for the allocation of resources by the State.

Whether ethnicity is primordial, situational or instrumental need not concern us here: the point is that ethnicity leads to a person's identification with a specific group. However, its visible markers, such as phenotype, language, culture, customs, religion or behaviour, may also be used as criteria by other groups for the purpose of exclusion. Becoming an ethnic minority is not an automatic result of immigration, but rather the consequence of specific mechanisms of marginalization, which affect different groups in different ways. Exclusionary practices against ethnic minorities are known as "racism", which may be defined as the process whereby social groups categorize other groups as different or inferior on the basis of phenotypic or cultural markers. This process involves the use of economic, social or political power, and generally has the purpose of legitimating exploitation or exclusion of the group so categorized.

Racism consists in making predictions about people's character, abilities or behaviour on the basis of socially constructed markers of difference, and acting upon those predictions. The process whereby the power of the dominant group is sustained by structures such as laws or administrative practices that exclude or discriminate against the dominated group, is known as institutional or structural racism. Racist attitudes and discriminatory behaviour on the part of members of the dominant group are referred to as informal racism. In many cases, supposed biological differences are not the only markers for exclusion; culture, religion or language are also used.

The historical explanation for racism lies in traditions, ideologies and cultural practices that have developed through ethnic conflicts associated with nation-building, as well as through colonial expansion (Miles, 1989). The reasons for recent increases in racism in developed countries must be sought in fundamental economic and social changes. Since the early 1970s, world economic restructuring and increasing international cultural interchange have been experienced by some social groups as a direct threat to their livelihood, lifestyle and national identity. Since these changes have coincided with the implantation of new ethnic minorities, the tendency has been to perceive the newcomers as the cause of the threatening changes.

Ethnic group formation is obviously of great importance. If migrant settlement generally leads to the formation of ethnic groups, then notions of fairly rapid assimilation or integration are mistaken. Immigration countries have to face up to the long-term prospect of pluralist societies. The crucial question then is whether structural factors and policies lead to excluded ethnic minorities or to accepted ethnic communities. Ethnic minorities are likely to lead a disadvantaged and conflictual life on the margins of societies that seek to maintain the myth of cultural homogeneity. Ethnic communities, on the other hand, can be integral parts of societies that are consciously multicultural. This dichotomy refers to ideal types, while most real societies fall somewhere in-between, as will be shown below.

It is also vital to realize that migrants do not enter consensual societies with homogeneous cultures and universally accepted values. Rather they are incorporated into social structures marked by differentiation on the lines of class, gender, position in the life cycle and so forth. Nor are the migrant or ethnic groups themselves homogeneous: they are structured according to similar criteria. Thus, the study of integration means looking at the way complex social groups interact with each other, form new social networks and modify their cultures and forms of action.

B. Policy and experience in traditional countries of immigration

Since 1945, the countries with the largest permanent-settler inflows have been the United States, Canada and Australia, in order of importance. Other traditional immigration countries, such as Argentina, Brazil and New Zealand, have received considerably smaller migrant inflows and will not be considered here. The United States, Canada and Australia have certain common features. They are white-settler societies in which mainly British colonists destroyed or dispossessed aboriginal populations. Mass immigration has been part of the strategy for nation-building and industrialization. All three have regarded most entrants as permanent settlers and have allowed family reunion. Immigrants have readily been granted citizenship, while their children have been born as citizens. All three countries preferred British immigrants, but admitted other Europeans too. Asians and other non-Europeans were excluded until the 1960s, but now form a large proportion of new entrants.

1. *United States of America*

United States society is a complex ethnic mosaic constructed by five centuries of immigration (Feagin, 1989). Until 1965, most migrants came from Europe. The 1965 Immigration Act removed national origin quotas and thus opened the door for the diversification of immigration in terms of origin. New settlers have come mainly from Latin America and Asia. The main Latin American source country is Mexico, with many migrants entering as undocumented agricultural workers. The main Asian countries of origin are China, India, the Philippines, the Republic of Korea and Viet Nam. Current policy establishes high immigrant intakes and a larger share of economic migrants. The 1990 census enumerated a total population of 249 million, of whom 80 per cent were white, 12 per cent black, 1 per cent American Indian, 3 per cent Asian or Pacific islander and 4 per cent of other races. Hispanics (who can be of any race) made up 9 per cent of the total population (United States Bureau of the Census, 1991). In 1980, there were 14 million foreign-born persons in the United States, amounting to 6 per cent of the total population (Briggs, 1984). By 1990, that number had grown to nearly 20 million.

Before 1945, European migrants and African-Americans joined the labour force as unskilled industrial workers, thus experiencing employment disadvantages and residential segregation. Over the years, many "white ethnics" have been able to achieve upward mobility, but African-Americans have become increasingly segregated. Distinctions between blacks and whites in terms of income, unemployment rates, social conditions and education are still extreme. Members of some recent immigrant groups, especially those from Asia, have high educational and occupational levels, while those from Latin America lack education and skills (Portes and Rumbaut, 1990).

The integration of migrants into the economy and society has been largely left to market forces. The egalitarian character of United States society has been seen as providing the best possible chances for immigrant groups to achieve the "American dream". Nevertheless, the Government has also played a role by making it easy to obtain United States citizenship and by using compulsory education as a way of transmitting the English language and American values. Legislation following the civil rights movement of the 1950s and 1960s led to measures to prevent discrimination and secure equal opportunities for ethnic minorities. However, commitment to equal opportunity and anti-poverty measures declined during the Reagan-Bush era, leading to increased community tension and racism.

2. *Canada*

The 1986 census enumerated 3.9 million foreign-born persons, of whom 2.4 million were from Europe, 623,000 from other countries in the Americas and

693,000 from Asia. In recent years, Asians have been the largest group of immigrants, followed by migrants originating in Latin America and the Caribbean. The five-year immigration plan announced in 1990 established annual admission levels of 250,000 persons for the period 1992-1995 (Canada, Employment and Immigration Canada, 1991). According to 1986 data on the "ethnocultural origins" of Canada's 25 million people, 34 per cent were of British origin, 24 per cent of French origin, 5 per cent of British and French origin combined and 38 per cent of other origin. In 1986, about 6 per cent of the total population was regarded as belonging to "visible minorities" (in other words, were non-Europeans), of whom 2 per cent were Native Peoples, 1 per cent South Asians, and 1 per cent blacks.

After 1945 a policy of assimilation or "Anglo-conformity" was pursued towards non-British settlers, especially towards persons from Southern Europe. In the 1960s, there was a move towards integrationist approaches for both immigrants and aboriginal peoples. In 1971, multiculturalism was proclaimed as Canada's official policy and a Minister of State for Multiculturalism was appointed. In 1988, Canada became the first country in the world to pass a Multiculturalism Act, which proclaimed multiculturalism as a central feature of Canadian society and laid down principles for cultural and language preservation, reducing discrimination and enhancing intercultural awareness and understanding.

Multiculturalism was seen as a central strategy for all government policies that might affect immigrants. This involved recognition of the need for institutional change in areas such as law enforcement, legal services, health services and education. Anti-racist and equal opportunity measures were integral parts of the policy of multiculturalism, as were measures to improve the situation of immigrant and "visible minority" women (Fleras and Elliott, 1992). The Employment Equity Act of 1986 required all federally regulated employers to assess and report on the composition of their workforces in order to correct the disadvantages faced by women, visible minorities, Native Peoples and the disabled. Despite such policies, community relations appeared to be deteriorating in the 1980s. Among the visible signs of conflict were discrimination against Native Peoples and racially motivated assaults against blacks and Asians. The unwillingness of the authorities to respond to racist attacks has been a major cause of politicization and resistance among visible minorities. Police treatment of minorities has become a major issue (Stasiulis, 1988).

3. *Australia*

In 1947, the Australian Government started a large-scale immigration programme to increase the population and stimulate industrial development. The Government wanted British migrants, but not enough came. In the late 1940s, most migrants came from Eastern and North-western Europe, and in the 1950s and 1960s, Southern Europeans predominated, while in the 1970s and 1980s, the end of the White Australia Policy opened the way for migrants from Asia, the Middle East and Latin America (Collins, 1991). Australia moved from being an almost monocultural country to being one of considerable diversity. By 1991, 23 per cent of the population were immigrants, and another 20 per cent were Australian-born persons with one or both parents immigrants. Two per cent of the population were Aborigines or Torres Strait Islanders. In 1991, there were 3.9 million foreign-born persons, of whom 2.4 million came from Europe, including 1.2 million from the United Kingdom, 898,000 from Asia, 288,000 from New Zealand, 186,000 from Africa and 158,000 from the Americas (Organisation for Economic Cooperation and Development, 1992, table 26).

Australia wanted permanent settlers: family reunion and naturalization were encouraged. The initial five-year waiting period for naturalization was reduced to three and then to two years. Children born to legal immigrants in Australia are automatically citizens. The Australian model for managing diversity has had two main stages. At first there was a policy of assimilation, based on the doctrine that immigrants could be absorbed culturally and socially so as to become indistinguishable from Anglo-Australians. Migrants were treated as "New Australians", who were to live and work with Anglo-Australians and rapidly become citizens. There was no special educational provision for migrant children. Cultural pluralism and the formation of "ethnic ghettos" were to be avoided at all costs; but by the 1960s it became clear that this policy of assimilation was not working owing to processes of labour-market segmentation, residential segregation and community formation. Moreover, political parties were beginning to discover the political potential of the "ethnic vote".

The result was a shift to multiculturalism in the 1970s, based on the idea that ethnic communities, which maintained their own language and culture, were legitimate and that their existence was compatible with Australian citizenship, as long as respect for the basic values and democratic institutions of Australia was safeguarded. Multiculturalism also means recognition of the need for special laws, institutions and social policies to overcome the barriers to full participation in society (Castles and others, 1990; Castles, 1992). Multicultural policies are implemented by an array of special agencies. The Office of Multicultural Affairs in the Department of the Prime Minister is responsible for promoting multiculturalism at the federal level and there are similar bodies at the state level. Health, education and other departments have multicultural units to ensure that services are appropriate to the needs of the various ethnic groups. The Government provides services for migrants, such as reception centres, help in finding work, language courses, educational support for children, and translation and interpretation services. Ethnic organizations participate in the planning of services through consultative bodies.

C. Policy and experience in the former labour-importing countries of Europe

Between the late 1950s and the early 1970s, several Western-bloc countries in Europe experienced large migration inflows. Two types of migration predominated: the movement of foreign workers from the European periphery and that of "colonial workers" to the former colonizing country. Both began as labour migration and eventually led to family reunification and permanent settlement, but the legal arrangements for entry and integration have been very different among the countries concerned. As table 48 shows, foreign populations started to grow again in the late 1980s, through renewed admission of workers, continued family reunification and the growing entries of asylum-seekers. Foreign residents make up a significant percentage of the population in many Western-bloc European countries and, because in several of the host coutries citizenship is granted on the basis of *jus sanguinis* (law of consanguinity), the growth of the foreign population is due both to continued net migration gains and to natural increase.

1. *United Kingdom*

In 1990, there were 1.9 million foreign citizens in the United Kingdom (3.3 per cent of the total population). The largest single group comprised 638,000 Irish citizens, followed by 155,000 Indians, 102,000 United States citizens and 75,000 Italians (Organisation for Economic Cooperation and Development, 1992). The ethnic minority population which consists mostly of British citizens of Afro-Caribbean and Asian origin, totals 2.7 million (4.7 per cent of the population). Most have been born in the United Kingdom and are therefore second- or third-generation descendants of immigrants who entered the country beginning in the 1950s. About 1.4 million have their origins in the Indian subcontinent (Bangladesh, India and Pakistan) and 455,000 in the Caribbean. Consequently, the overall population of overseas origin may be estimated at 4.5 million (about 8 per cent of the total population).

Most Commonwealth immigrants entered as British subjects and enjoyed all rights once admitted. This situation was ended by the 1971 Immigration Act and the 1981 British Nationality Act, which put new Commonwealth immigrants on a par with foreigners. Irish settlers enjoy virtually all rights, including the right to vote. It is relatively easy for foreigners to obtain citizenship after five years of legal residence in the United Kingdom. Since 1965, a series of Race Relations Acts have been passed, outlawing discrimination in public places, in employment and in housing. A Commission for Racial

TABLE 48. FOREIGN POPULATION RESIDENT IN SELECTED EUROPEAN COUNTRIES
(*Thousands*)

Country of residence	*1980*	*1985*	*1990*	*Percentage of total population, 1990*
Austria	283	272	413	5.3
Belgium	..	845	905	9.1
Denmark	102	117	161	3.1
France	3 714[a]	..	3 608	6.4
Germany	4 453	4 379	5 242	8.2
Italy	299	423	781	1.4
Luxembourg	94	98	..	27.5[b]
Netherlands	521	553	692	4.6
Norway	83	102	143	3.4
Sweden	422	389	484	5.6
Switzerland	893	940	1 100	16.3
United Kingdom	..	1 731	1 875	3.3

Source: OECD, *SOPEMI 1991: Trends in International Migration* (Paris, 1992), p. 131.

NOTES: These figures refer to the foreign population and therefore exclude naturalized immigrants (particularly important in France, Sweden and the United Kingdom) and immigrants from colonies or former colonies who have the citizenship of the receiving country (particularly important for France, the Netherlands and the United Kingdom).

[a]Referring to 1982.

[b]Referring to 1989.

Equality (CRE) has been set up to enforce anti-discrimination laws and promote good community relations (Banton, 1985).

Labour-market segmentation developed in the 1950s and 1960s, with Asians and Afro-Caribbeans concentrated in the least desirable jobs. Today, black workers still have low average socio-economic status and high unemployment. Ethnic minorities are heavily concentrated in the most run-down areas of the inner cities. In the 1960s, policies towards black immigrants oscillated between calls for cultural assimilation and the demand for repatriation. Schooling was seen as an important instrument for the assimilation of the second generation. In the 1970s, a number of integration policies were adopted, including special measures to overcome the disadvantages of migrant children at school. The emphasis was put on adaptation as a long-term group process and social policies were changed to provide special funding for local government areas with high migrant populations. By the 1980s, multicultural education became official policy in many areas, providing appropriate recognition to ethnic group cultures in teaching and curriculum-building. Social policy initiatives were designed to draw ethnic organizations into youth and community work. However, some black organizations argued that such measures were designed to co-opt the leadership and stifle protest. They called instead for anti-racist policies.

Organized racist groups, such as the National Front, grew rapidly during the 1970s and 1980s. Their electoral success was limited, but they recruited members of violent youth subcultures, such as skinheads. Racist violence became a major problem. A 1981 Home Office survey found that the rate of attacks on Asians was 50 times that of attacks on white people, and the rate of attacks on blacks was 36 times that of attacks on white people (Home Office, 1989). The discontent of black youth exploded in riots in inner-city areas in 1980-1981 and again in 1985-1986 (Beynon, 1986). Growing conflicts in inner-city areas led to government measures to combat youth unemployment, make education more accessible to minorities, improve living conditions in urban areas and change police practices. Task forces were set up for disadvantaged areas, spending under the Urban Programme was substantially increased, and a large scale Youth Training Scheme was introduced (Layton-Henry and Rich, 1986; Solomos, 1989).

2. *France*

After 1945, the French Government recruited migrant workers in Southern Europe. There were also inflows from the colonies or former colonies in Northern Africa and, to a lesser extent, from Western Africa and the Caribbean. Labour recruitment was stopped in 1974, but entries for family reunification and other forms of migrant inflows continued. The 3.6 million foreign residents present in 1990 constituted 6.4 per cent of France's total population. The main groups were Portuguese (646,000), Algerians (620,000), Moroccans (585,000), Italians (254,000), Spaniards (216,000), Tunisians (208,000) and Turks (202,000) (Organisation for Economic Cooperation and Development, 1992). In addition, there were over 1 million immigrants who had become French citizens, and up to 500,000 French citizens of African, Caribbean and Pacific island origin from Overseas Departments and Territories.

Citizens of member States of the European Union (EU) enjoy all basic rights, except the right to vote. Immigrants from Poland, Yugoslavia and other non-member European countries lack the privileges of EU citizens, and many have an irregular legal situation. People of non-European birth or parentage, whether French citizens or not, constitute the ethnic minorities. These include Algerians, Tunisians and Moroccans, young Franco-Algerians, black Africans, Turks and settlers from the Overseas Departments and Territories. Foreign residents have had relatively easy access to citizenship through naturalization and children born in France to foreign residents have had the option of becoming French citizens at age 18. The 1972 Law against Racism prohibits racial incitement and discrimination in public places or in employment, but there is no special body to enforce the law and the number of prosecutions is small (Costa-Lascoux, 1991). There is little support for multicultural policies in France. The prevailing view is that immigrants are best integrated by becoming French citizens, a move that presupposes a certain degree of cultural assimilation. Special policies for ethnic groups and recognition of their leaders are seen as potential obstacles to integration (Weil, 1991).

The *bidonvilles* (shanty towns) that developed around French cities in the 1960s have disappeared, but there is still residential concentration of foreigners or ethnic groups in inner-city areas and in the *grands ensembles* (public housing areas on the periphery of cities). The work situation of ethnic minorities is marked by low status, insecure jobs and high unemployment rates (Verbunt, 1985). In the late 1980s, growing racism and serious social problems led to a series of special programmes to improve housing and education and to combat unemployment among youth.

The position of ethnic minorities in French society has become highly politicized. Immigrants have taken an active role in major strikes, and demanded civil, political and cultural rights. The second generation (known as *beurs*) and Muslim organizations are emerging political forces (de Wenden, 1987). Youth discontent with unemployment and police practices led to riots in Lyons, Paris and other cities in the 1980s. Immigration is a central issue in party and electoral politics. The *Front national* (on the extreme right) mobilizes on issues of immigration and cultural difference. By the early 1990s, it had become a major political force (Lapeyronnie and others, 1990).

3. *Germany*

The foreign resident population of the Federal Republic of Germany grew from 4.5 million in 1980 to 5.2 million in 1990 (figures refer to Western Germany only). The main groups in 1990 were Turks (1.7 million), Yugoslavs (652,000), Italians (548,000), Greeks (315,000) and Poles (241,000) (Organisation for Economic Cooperation Development, 1992). The Federal Republic of Germany has experienced several migratory movements, including large inflows of people of German ethnic origin from Eastern Europe after the Second World War and again since 1985. Those migrants have the right to German citizenship and have generally received special treatment. The discussion here focuses only on foreign migrants and their descendants who do not have the right to citizenship. Admission of foreign workers began after 1955, when the Government signed labour recruitment agreements at various times with Greece, Italy, Morocco, Portugal, Spain, Tunisia, Turkey and Yugoslavia. The migrants were regarded as "guest workers" who would work in Germany for a few years and then go home. They were not expected to settle and family reunion was discouraged (Castles and Kosack, 1973).

Migrants became manual workers in construction or in factories. Most were men, but there was a substantial female minority. Employers in the textile and clothing, electrical assembly and food processing industries often preferred female labour. In the early 1970s, Turkish workers became the largest single migrant group. The German Government stopped labour recruitment in 1973, hoping that surplus workers would leave. Although many did leave, others stayed and family reunification increased. By the late 1970s it was clear that permanent settlement was taking place. However, the federal Government continued to assert that Germany was "not a country of immigration". Since the mid-1980s, there has been a growth of temporary labour migration (mostly short-term) from Poland. The fact that these workers now play a major role in construction, in domestic work and other informal sector activities in Germany indicates a resumption of historical patterns going back to the nineteenth century.

During the late 1980s, the number of persons filing applications for asylum in Western Germany increased markedly, reaching 100,000 in 1986, 193,000 in 1990 and 256,000 in 1991 (Organisation for Economic Cooperation and Development, 1992). Although a significant proportion of those filing asylum applications originated in Africa and Asia, since 1989 the major increase has been associated with East-West flows of Europeans. Until recently, the German constitution guaranteed that anyone reaching German territory could file an application for asylum and those filing asylum claims were permitted to stay pending an official decision on their application, a process that could take several years. In 1990, 96 per cent of the asylum applications considered were rejected. However, legal and practical difficulties precluded the deportation of most rejected applicants and many have stayed. Most Eastern European asylum-seekers belong to ethnic minorities, which include Jews and Gypsies (Organisation for Economic Cooperation and Development, 1991). In 1992, there was an upsurge of asylum-seekers from war zones in the former Yugoslavia. Many of the asylum-seekers from Romania and the former Yugoslavia are Gypsies, an ethnic group that became the main target of racist attacks in Germany during 1992. Much of this violence was organized by groups of the extreme right and was concentrated in the territory of the former German Democratic Republic.

The myth of temporary residence still shapes the legal status of foreigners in Germany, except in the case of citizens of other member States of the European Union who enjoy social and economic parity with German citizens. Other foreigners are excluded from a whole range of rights and services, and are denied political participation and representation. In addition, they face considerable barriers to obtaining citizenship. By the mid-1980s, over 3 million foreigners in the Federal Republic of Germany had been present in the country for at least 10 years and therefore qualified for naturalization in terms of length of residence. However, only about 14,000 persons naturalized per year (Funcke, 1991). In addition, children born in Germany to foreign parents do not have the right to German citizenship, although the Foreigners Law of 1990 has made it somewhat easier for them to obtain it. Even second-generation migrants can, under certain circumstances (conviction for criminal offences or long-term unemployment), be deported.

In Germany, the continuous public debate on "foreigners policy" is characterized by three basic positions. The right, encompassing both the extreme right and much of the ruling Christian Democratic Union (CDU), calls for a continuation of exclusionary policies, including restrictions on family reunification, the limitation of foreigners' rights and the widespread use of deportation powers. In contrast, most members of the Social Democratic Party (SPD) and some parts of the CDU recognize that settlement is irreversible, and call for assimilation or integration policies. The third position holds that assimilation is no longer possible in view of the emergence of ethnic communities, and that the adoption of multicultural models is the only solution. This view is held by the Green Party, some parts of the SPD, the churches and the trade unions (Castles, 1985; Leggewie, 1990). The German model is still essentially exclusionary, but there is a growing realization that a democratic society cannot permanently deny citizenship and participation to a large section of the population. On the other hand, the growth of the racist extreme right hampers moves towards improving the position of migrants.

4. *Sweden*

Until 1945 Sweden was a fairly homogeneous country, with only a small Lapp minority (about 10,000 people today). After 1945, labour migration was encouraged. Foreign worker recruitment was stopped in 1972, but migration for family reunification and refugee resettlement continued. In 1990, there were 484,000 foreign residents who constituted 5.6 per cent of the total population. Most foreigners originated in other Scandinavian countries, especially in Finland whose citizens accounted for 185,000 of the foreigners in Sweden. The largest non-Scandinavian groups were Yugoslavs (41,000), Iranians (39,000), Turks (26,000) and Chileans (20,000). Many immigrants have become Swedish citizens. In 1986, the total population of immigrant background was estimated to number 920,014 persons or about 11 per cent of the total population. Among them, 250,138 persons had been born in Sweden (Lithman, 1987).

Migrant workers are overrepresented in manufacturing, and in service occupations using lower-level skills. They are underrepresented in agriculture, health care and social work, administrative and clerical work, and commerce. Migrants have unemployment rates about twice as high as those of Swedes. Most migrants have settled in the cities and people of the same nationality tend to cluster in certain neighbourhoods, thus fostering linguistic and cultural maintenance.

Sweden is the only European country with a declared policy of multiculturalism, first introduced in 1975. The policy emphasizes the need to ensure that migrants have the same living standards as Swedes and that they have the right to choose between retaining their own cultural identities and adopting the Swedish norm. Sweden has a comprehensive set of social policies designed to integrate migrants into society. Since 1975, foreign residents have had the right to vote and stand for election at the local and regional levels. The waiting period for naturalization is two years for Scandinavians and five years for everybody else, while children born to foreign resident parents can obtain Swedish citizenship upon application. In 1986, an Act against Ethnic Discrimination came into force. Migrants have the right to 400 hours of Swedish instruction with financial assistance. Children of migrants can receive pre-school and school instruction in their own language, within the normal curriculum. Other measures include translation and interpretation services, information services, grants to migrant organizations, and special consultative bodies (Hammar, 1985).

However, multiculturalism has been under attack in recent years. The increase in the number of asylum-seekers in the late 1980s led to strains in housing and other services (Ålund and Schierup, 1991). The *Sverigepartiet* (SP or Sweden Party), on the extreme right, started anti-immigrant campaigns in 1986. In the late 1980s there was an increase in racist violence, including arson and bomb attacks on refugee centres. In

December 1989, the Government decided that asylum should henceforth be granted only to applicants meeting the criteria set by the United Nations Convention relating to the Status of Refugees (United Nations, 1957) (Larsson, 1991).

5. *Other European countries*

As the cases discussed above illustrate, there is considerable variability among European countries and the rest are equally heterogeneous regarding both their migration experience and their policies towards migrants. Belgium and Switzerland have policies rather similar to those of Germany. Migrants are not regarded as permanent settlers, lack rights and find it difficult to become citizens. However, permanent settlement is unavoidable, as indicated by the fact that most foreign residents in Switzerland hold long-term "establishment permits". In Belgium, as in France and Germany, Muslim groups tend to be the most isolated from mainstream society and there have been a series of youth-related disturbances in major cities in recent years.

The Netherlands' approach to migration has many similarities with the Swedish model. The Minorities Policy, introduced in 1983, accepted the need for specific social policies to integrate minorities and recognized that it was necessary to deal not just with individuals but with ethnic groups. "Integration with preservation of cultural identity" became the slogan. Anti-discrimination legislation and local voting rights for foreign residents were introduced. However, the increasing socio-economic disadvantages of migrants has led to a re-examination of policies. The 1991 Action Programme on Minorities Policy pays little attention to the former goal of minority group emancipation and participation, emphasizing instead measures to reduce economic and social deprivation, to prevent discrimination and to improve the legal situation of minorities.

D. Comparative dimensions

1. *Migration policies and integration*

Migration policies may have considerable consequences for the integration process. Among the various receiving countries, three groups can be distinguished. The traditional countries of immigration, namely, Australia, Canada and the United States, have encouraged permanent migration and treated most immigrants as future citizens, permitting family reunification and granting them secure residence status. Sweden, despite its very different historical background, has followed similar policies. The second group includes France, the Netherlands and the United Kingdom where migrants from former colonies have received preferential treatment and have often been citizens at the time of entry. Their permanent migration has generally been accepted and family reunification has been permitted. Those countries have also tended to grant permanent residence to migrants from European countries and to allow their family reunification and naturalization. The third group of countries consists of those that have tried to cling to rigid "guest worker" models, especially Germany and Switzerland. They have allowed family reunification on a restricted basis, have been reluctant to grant secure residence status, and have highly restrictive naturalization laws.

However, these distinctions are neither absolute nor static. There has been a convergence of policies in European countries: the former colonial countries have become more restrictive, while the former guest-worker countries have become less so. This evolution has gone hand in hand with a new type of differentiation: the member States of the European Union agreed to grant a privileged status to intra-community migrants in 1968, and further improvements in that status became a reality in 1993 with the creation of the single European market; at the same time, entry and residence have become far more difficult for citizens from non-member States, especially those from developing countries. Moreover, the current emotional debate on the "refugee influx" from both developing countries and former Eastern-bloc countries has had a considerable impact. Increased racism, restrictions of the rights of foreigners, demands for their repatriation and the emergence of a "fortress Europe" mentality cannot but worsen the social and political position of existing minorities.

Migration policies affect other policy issues, including labour-market rights, security of residence and naturalization. Moreover, ideologies of temporary migration create expectations within the receiving population. If a temporary sojourn turns into settlement, and the Governments concerned refuse to acknowledge this, then migrants are usually blamed for the resulting problems. Migration policies also help shape the consciousness of the migrants themselves. In countries where migrants are granted secure residence status and civil rights, they are able to develop long-term perspectives for themselves and their families. Where the myth of the short-term sojourn is maintained, the position of migrants becomes essentially contradictory: return to the country of origin may be difficult or impossible, but permanence in the host country is doubtful. Migrants in such a position cannot plan for a future as part of the wider society. The result is isolation, separatism and emphasis on their differences with respect to the rest of society.

2. *Labour-market position*

Trends towards labour-market segmentation by ethnicity and gender were intrinsic in the type of labour migration practised until the mid-1970s: institutional discrimination, established through rules against job changes, refusal to recognize overseas qualifications or exclusion from public employment, was a major cause of disadvantage. Informal discrimination, embodied in the unwillingness of employers to hire or promote migrant workers, also played a part. The situation has changed since the 1970s: new migrants are more diverse in terms of education and occupational status. There is a trend towards polarization: highly skilled persons are encouraged to enter and are seen as an important factor in technology transfer. Unskilled migrants are unwelcome as workers, but sometimes enter through family reunification or as refugees. Their contribution to occupations involving low-level skills, casual work, the informal sector and small businesses is economically important, but officially unrecognized.

When people migrate from poor to rich countries, without having sufficient information about the country of destination or being unable to rely on social networks, and lacking proficiency in the local language, they are likely to join the labour market at its lowest rungs. Such entry levels would not be too important if migrants had a fair chance of experiencing upward mobility as time went on. However, that chance depends on whether the State encourages the continuation of labour-market segmentation through its own practices or whether it takes measures to give migrants equal opportunities. Some countries have active policies to improve the labour-market position of migrants through language courses and vocational training, as well as through equal opportunity and anti-discrimination legislation. These countries include Australia, Canada, Sweden and, to a lesser extent, France, the Netherlands and the United Kingdom. The United States has also enacted equal opportunity, affirmative action and anti-discrimination legislation, but little in the way of language, education and training measures. This strategy is consistent with the laissez-faire approach to social policy that characterized the Reagan-Bush era.

The former guest-worker countries, Germany and Switzerland, do offer some education and training for foreign workers and foreign youth, but at the same time they restrict the labour-market rights of migrants. During the period of labour recruitment, work permits often bind foreign workers to specific jobs. Most workers in those two countries now have the right to job mobility, but in many cases employers must still give priority to citizens. That is to say, an employer cannot hire a foreign worker for a job if a citizen is available or, in the case of Germany, if a citizen of a member State of the European Union is available. Foreigners have full equality in the labour market only if they have acquired long-term residence rights, a status that is becoming increasingly common.

3. *Migrant women*

Migrant women's settlement experience is and remains distinct from that of men. In the early stages of labour migration, there is a tendency to concentrate on the situation of male workers. However, in most receiving countries, a growing proportion of migrant workers are women, many of whom have been admitted under family reunification. Migrant women usually find themselves in the lowest rungs of labour markets segmented according to ethnicity and gender. They frequently work as unpaid family members in small businesses, thus providing the competitive edge that makes the survival of such enterprises possible. In some sectors, notably the garment industry, complex patterns of division of labour on ethnic and gender lines have developed. Male ethnic entrepreneurs use female labour from their own groups, but are themselves dependent on large retail corporations (Waldinger and others, 1990; Phizacklea, 1990). As Phizacklea (1983) points out, it is particularly easy to label female migrant workers as inferior, just because their primary roles in patriarchal societies are defined as being those of wife and mother, making them dependent on a male breadwinner. They can therefore be paid lower wages and controlled more easily than men.

For many women (especially those from Muslim societies), migration to a developed country may mean a radical break with previous forms of social behaviour. Their important position as producers in rural households and the support they get from other women within their community may disappear, leading to social isolation and dependence on their husbands or male relatives. Male roles, by contrast, being based on work and social interaction outside the home, are easier to maintain or re-establish. Cultural factors as well as child-care needs make it much harder for women to participate in language or other educational courses. Access to health, maternal and other social services is also an issue of special concern to migrant women.

Government agencies have to provide special services suited to meet the needs of migrant women of varying backgrounds. Such services exist to some extent in most receiving countries, although provision tends to be more systematic and comprehensive where migrant women have been able to secure a consultative role through their own associations. Countries with multicultural policies have, on the whole, been more willing to recognize migrant women's needs, although even in those countries there is still much to be done. Countries that have been reluctant to accept family reunification and permanent settlement have also been slow to recognize the needs of migrant women.

Migrant women play a particularly important part in the process of community formation. The significance of women's family and educational roles in reproducing and maintaining their languages and cultures has been emphasized in many studies (Vasta, 1990). Family formation and the emergence of social networks in the host country help to improve the social and economic situation of migrants. These processes also provide the basis for community organization and a measure of protection against racism. The formation of social networks is an important part of the integration process (Boyd, 1989) and women play a special and decisive role in their creation and maintenance.

4. *Residential segregation and community formation*

Residential segregation is to be found in all the receiving countries. It is extreme in the United States, where in certain areas there is almost complete separation between blacks and whites, and sometimes between either of those groups and Asians or Hispanics too. In other countries, migrant groups are highly concentrated in certain city neighbourhoods, though they rarely form the majority of the population. Causes of residential segregation include low income and lack of local networks, discrimination and the desire of migrants to group together for cultural maintenance and for protection against racism. Institutional practices often encourage residential segregation. Many migrant workers were initially housed by employers or public authorities, a practice that encouraged clustering, for when workers left their initial accommodation they tended to seek housing in the vicinity. Public housing allocation policies may also cause concentration.

Residential segregation is a contradictory phenomenon. Migrants may be socially disadvantaged by concentration in areas with poor housing and social amenities but they frequently want to be together in order to enjoy mutual support, rebuild family and neighbourhood networks, and maintain their languages and cultures. Ethnic neighbourhoods allow the establishment of small businesses and agencies that cater to migrants' needs, as well as the formation of associations of all kinds. Residential segregation is thus both a precondition for and a result of community formation.

Some members of the majority population perceive residential segregation as a deliberate and threatening attempt to form "ethnic enclaves" or "ghettos". Racism is often a self-fulfilling prophecy: it justifies itself by portraying migrants as alien groups, which will "take over" the neighbourhood. By forcing migrants to live together for protection, racism creates the ghettos it fears. One official reaction has been dispersal policies, designed to reduce ethnic concentrations. However, in the absence of the economic opportunities and political structures needed to overcome the powerful forces of marginalization, dispersal policies are difficult to implement.

5. *Social policy*

As migrants moved into the inner cities and industrial towns, social conflicts with lower-income groups of the majority population developed. Migrants have been blamed for rising housing costs, declining housing quality and deteriorating social amenities. In response, a whole set of social policies has been formulated in most receiving countries. In the early stages, social and community-work measures were often initiated by non-governmental organizations, including church, trade union and other voluntary agencies. As time went on, migrant associations played an increasingly important role and, as the realities of migrant settlement gained recognition, there was a growing involvement of government agencies, which often collaborated with voluntary organizations. In cases where the Government has been slow to act, the role of non-governmental organizations has remained important, and the character of their involvement has become increasingly political.

In France, for instance, since the late 1960s urban renewal policies have included measures to make public housing more accessible to migrants. Yet, the concept of the *seuil de tolérance* (threshold of tolerance), according to which the migrant presence should be limited to a maximum of 10 or 15 per cent of residents in a housing estate or 25 per cent of students in a class, has effectively been used as an exclusion tool (Verbunt, 1985), being based on the premise that the concentration of migrants itself is a problem and that dispersal is a precondition for assimilation. By the 1980s, the situation of ethnic minorities in the inner-city areas and the large public housing estates around the cities had become the central issue of social policy because such areas were characterized by persistent unemployment, social problems and conflict between ethnic minorities and disadvantaged sections of the French population. The social policies of the time focused mostly on urban youth (Lapeyronnie and others, 1990) but, as Weil (1991) concludes, they largely failed mostly because, although they were meant to achieve the integration of migrants into French society, they in fact linked all the problems of towns and neighbourhoods to the presence of migrants. Consequently, instead of promoting integration, social policy encouraged the concentration of minorities, thus slowing integration, encouraging the formation of ethnic communities, and strengthening religious and cultural affiliation.

Social policy towards migrants is clearly a complex issue that requires careful analysis in each country. Such policy has often reinforced the trend towards segregation. Thus, the "dual strategy" pursued in German education has led to special classes for foreign children, causing both social isolation and poor educational performance. Housing allocation policies in the United Kingdom are intended to be non-discriminatory, yet they have sometimes led to the emergence of "black" and "white" housing estates. On the other hand, in Sweden, the children of migrants have the right to instruction in their mother tongue. The official view is that such instruction does not lead to separation, but rather encourages "active bilingualism", which makes it easier for the children of migrants to succeed in school and work (Lithman, 1987).

Again, it is possible to suggest a rough classification. Australia, Canada, the Netherlands and Sweden have pursued active social policies in favour of migrants, linked to broader models of multiculturalism (or minorities policy, in the case of the Netherlands). The basic assumption has been that special social policies do not lead to segregation or separatism but, on the contrary, constitute the precondition for successful integration. In Australia, there has been a policy of "mainstreaming" migrant services, that is, providing those services through normal government agencies, rather than through special bodies dealing exclusively with migrants.

A second group of countries are those that, on principle, reject special policies for migrants. United States authorities oppose special measures directed towards migrants because such measures are seen as unnecessary government intervention. Nevertheless, equal opportunity, anti-discrimination and affirmative action measures deriving from civil rights laws have benefited immigrants. The French Government has rejected special policies on the principle that migrants should become citizens and that any special treatment would hinder or slow the process of naturalization. Yet, there have been a number of special social policies. The United Kingdom has also developed a range of social policies in response to the urban crisis and youth riots, despite the ideological rejection of such measures by the Conservative leadership.

The third group of countries consists of the former guest-worker recruiters. In the early period of recruitment, the German Government delegated the provision of special social services to charitable organizations linked to the churches and the labour movement. Although foreign workers were granted the same rights as citizens in regard to work-related health and pension benefits, they were excluded from some welfare rights. Application for social security payments on the grounds of long-term unemployment or disability could lead to deportation. In Switzerland, there are few social measures directed specifically at migrants. Provision of support in emergency situations is left largely to voluntary efforts (Hoffmann-Nowotny, 1985).

6. *Racism and resistance*

Racist harassment and attacks have become major problems for ethnic minorities in all the main receiving countries. In the United Kingdom, racist violence organized by groups like the National Front became a problem in the 1970s. In London, 2,179 racial incidents were reported to the police in 1987, including 270 cases of serious assault, 397 of minor assault, 483 of criminal damage, 47 of arson and 725 of abusive behaviour (Home Office, 1989). In France, the *Front national* (extreme right) has been able to mobilize resentment caused by unemployment and urban decline, and to crystallize it around the issues of migration and cultural differences. German reunification was followed by outbursts of racist violence during 1991-1992. The United States has a long history of white violence against African-Americans, and the Ku Klux Klan is still a powerful force. Asians, Arabs and other minorities are also frequent targets (Anti-Defamation League, 1988). Even countries that pride themselves on their tolerance, like Canada, the Netherlands and Sweden, report a growing incidence of racist attacks. The European Parliament's Committee of Inquiry into Fascism and Racism in Europe found that migrant communities are constantly subject to displays of distrust and hostility, to discrimination and, in many cases, to race-related violence that may lead to murder (European Parliament, 1985).

Racism is directed particularly against people of non-European background, such as Asian migrants in Australia, Canada and the United States; Afro-Caribbean and Asian minorities in the United Kingdom; Northern Africans and Turks in most Western European countries; and asylum-seekers of non-European background just about everywhere. Phenotypic difference ("race") seems to be the main criterion for hostility towards migrants. A survey carried out on behalf of the Commission of the European Communities in all member States in 1989 found strong feelings of distance from and hostility towards non-Europeans, particularly Arabs, Africans and Asians. Overall, one European in three believed that there were too many people of another nationality or race in his or her country, with such feelings being most marked in Belgium and Germany (Commission of the European Communities, 1989).

Islam provides a particularly important marker of difference. The largest non-European groups of Islamic background are in France (Northern Africans) and Germany (Turks). There are also large Muslim groups in Belgium and the Netherlands (Turks and Moroccans) and in the United Kingdom (Bangladeshis and Pakistanis). European fear of Islam and hostility to Muslims have a tradition going back to the crusades. In recent years, anxiety about fundamentalism and loss of modernity and secularity has played a major role, but it could be argued that such fears are based on racist ideologies rather than on social realities. The strengthening of Muslim affiliations is often a protective reaction of groups discriminated against, so that fundamentalism is something of a self-fulfilling prophecy. Moreover, stereotypes of Islam are usually undifferentiated and inaccurate, and ignore the great diversity of religious values and behavioural forms in the different countries of origin.

Racist campaigns, harassment and violence can be a major constraint to integration. One form of migrant resistance to racism is residential clustering and ethnic community formation. Such communities develop their own social, political and religious institutions, which can increase isolation from the mainstream of society. Reactions of members of ethnic minorities to racism vary widely. There is often a gulf between the experiences of the migrant generation and those of their children, who have grown up and gone to school in the new country. Ethnic minority youth become aware of the contradiction between the prevailing ideologies of equal opportunity and the reality of discrimination and racism in their daily lives. Such perceptions can lead to the emergence of counter-cultures and political radicalization.

Ethnic minority youth are in turn perceived by those in power as embodying a "social time bomb" or a threat to public order, which has to be contained through institutions of social control, such as schools, welfare agencies and the police. The youth disturbances that took place in many European cities in the 1980s epitomized such conflicts. The United States has also experienced violent protests by minorities. The Los Angeles riots of May 1992 were provoked by police brutality towards a black motorist, which went unpunished by the courts. The aggression of rioters was directed not only against whites, but also against Koreans and Cubans, who have taken on middleman roles in areas where large white-owned companies fear to engage in trade.

7. *Citizenship*

The analysis so far has shown important differences in models of migrant integration, not only between "traditional immigration countries" and "labour importers" but also within each group. The variations are linked to fundamental differences in concepts of the nation State and citizenship. Some countries make it very difficult for immigrants to become citizens, and others grant citizenship but only at the price of cultural assimilation, whereas the members of a third group make it possible for immigrants to become citizens while maintaining distinct cultural identities. These practices correspond closely with other policies: countries unwilling to grant citizenship often deny the reality of settlement, restrict migrant rights and follow exclusionary social policies. Countries with assimilationist policies regarding citizenship accept permanent settlement, but tend to reject special social and cultural policies for immigrants. Countries that grant citizenship and accept cultural pluralism generally also see the need for special social policies for immigrants.

The extent of naturalization as the means of acquiring citizenship varies considerably among countries. Based on the data in table 49, the ranking (from highest to lowest) of receiving countries according to the number of naturalizations per thousand foreign residents would be as follows: Australia, Sweden, the United Kingdom, the Netherlands, France, Switzerland, Germany and Belgium. It has not been possible to calculate naturalization rates for Canada and the United States, owing to lack of data on the foreign resident population. However, the large absolute numbers of naturalizations in those two countries make it reasonable to assume that their naturalization rates fall somewhere between those of Australia and Sweden.

Even in countries where the goal of naturalization is easily achieved, some immigrants are unwilling to give up their original citizenship, partly because of a refusal to cut symbolic links with the place of birth, but also because of practical reasons connected with military service or land ownership rules in the country of origin. The ideal solution in such cases would be the granting of dual citizenship. Although many Governments reject dual citizenship because of fears of "divided loyalties", it is becoming increasingly common, especially for children of mixed marriages. Another solution is to create special residence categories, which confer some of the rights of citizenship (as reflected; for example, by "establishment permits" in Switzerland and "residence entitlements" in Germany). Granting local voting rights to foreign residents is also a means of recognizing "quasi-citizenship". However, such approaches are unsatisfactory because they create new divisions among full citizens, quasi-citizens and foreigners.

Another important issue is the status of the second generation. In Germany and Switzerland, where the principle of *jus sanguinis* dictates who is a citizen, children of foreign parents who have been born and have

TABLE 49. NUMBER OF NATURALIZATIONS AND NATURALIZATION RATE IN SELECTED COUNTRIES

Country	*Reference date*	*Foreign resident population (thousands)*	*Number of naturalizations*	*Naturalizations per thousand foreign resident population*
Australia	1988	1 427	81 218	56.9
Belgium	1989	881	1 878	2.1
Canada	1989	..	87 476	..
France	1989	3 752	49 330	13.1
Germany	1988	4 489	16 660	3.7
Netherlands	1990	642	12 700	19.8
Sweden	1989	456	17 552	38.5
Switzerland	1989	1 040	10 342	9.9
United Kingdom	1988	2 550	64 600	25.3
United States	1989	..	233 777	..

Sources: OECD, *OECD Continuous Reporting System on Migration Report (SOPEMI), 1990* (Paris, 1991); Philip J. Muus, *Migration, Minorities and Policy in the Netherlands: Recent Trends and Developments. Report for SOPEMI* (Amsterdam, University of Amsterdam, Department of Human Geography, 1991); and Australian Bureau of Statistics, *Census 86. Australia in Profile: A Summary of Major Findings* (Canberra, Australia, 1988).

NOTES: For Australia, the foreign resident population in 1988 was estimated by assuming that it grew at the same rate as the foreign-born population from 1986 to 1988. The 1986 base figure was obtained from the Australian census conducted that year. For France, the foreign resident population used refers to 1985, but the number of naturalizations refer to 1989. For the United Kingdom, the foreign resident population used refers to 1989. Consequently, the possible comparisons have merely indicative value, since definitions vary from country to country.

grown up in the receiving country are denied not only the security of residence but also a clear national identity. Because those children usually have the citizenship of their parents, they are formally citizens of a country they may never have seen and they may even be deported there under certain circumstances. In contrast, in countries such as Australia, Canada, the United States and the United Kingdom, where nationality laws are based on the principle of *jus soli* (literally, "right of the soil"; rule of law by which a child's citizenship is determined by his or her place of birth), citizenship is granted to those born in the country's territory and, consequently, although the children of migrants may have different cultural identities, they have a secure legal basis on which to make decisions about their life perspectives. Intermediate forms with respect to the granting of citizenship, such as the provision of a choice of citizenship at age 18 (as in France and Sweden), are reasonably satisfactory. The allowing of dual citizenship would provide a better solution in not forcing individuals to make difficult and irrevocable decisions about their national identity.

8. *Linguistic and cultural rights*

Maintenance of language and culture is seen as a need and a right by most migrant groups. Many ethnic associations are concerned with language and culture: they teach the mother tongue to the second generation, organize festivals and carry out rituals. Language and culture not only serve as a means of communication, but take on a symbolic meaning that is central to ethnic group cohesion. In most cases, the native language is maintained by the first two or three generations, after which there is a rapid decline in its usage. The significance of cultural symbols and rituals may last considerably longer.

Migrant languages and cultures become symbols of otherness and markers for discrimination. Many people regard cultural differences as a threat to a supposed cultural homogeneity and to national identity. Such a position is rationalized by the assertion that the official language of the receiving country is essential for economic success and that the cultures of migrants are inadequate in a modern secular society. The alternative view is that migrant communities need their own languages and cultures to develop identity and self-esteem. Cultural maintenance helps create a secure basis that contributes to group integration into the wider society, while bilingualism brings benefits in terms of learning and intellectual development.

Policies and attitudes on cultural and linguistic maintenance vary considerably. Some receiving countries have a history of multilingualism. Canada's policy of bilingualism is based on two "official languages", English and French. Multicultural policies have led to limited recognition of and support for migrant languages, but they have hardly penetrated into mainstream contexts, such as broadcasting. Switzerland has a multilingual policy for its founding languages, but does not provide an official recognition for the languages of migrants. Australia and Sweden, both accepting the principle of linguistic and cultural maintenance, provide interpretation and translation services and classes in the native language of migrants, as well as support for ethnic cultural organizations. Australia has a "national policy on languages", concerned with both community languages and languages of economic significance. Multicultural radio and television programmes are funded by the Government. Australia, Canada and Sweden all have multicultural education policies.

In the United States, monolingualism is being eroded by the growth of the Hispanic community, a development that has prompted a backlash in the form of a "United States English movement" which calls for a constitutional amendment to declare English the official language of the United States. Despite official rejection of linguistic pluralism in the United States, it has proved essential to establish a range of multilingual services. Monolingualism is also the basic principle in France, Germany, the Netherlands and the United Kingdom. Nevertheless, all these countries have been forced to introduce language services to meet the needs of migrants regarding the use of the judicial system, the bureaucracy and health services. The multilingual character of inner-city schools has led to special measures for migrant children and to a gradual shift towards multicultural educational policies.

E. Conclusion

The cross-national comparison of migrant integration can be summed up by saying that in all receiving countries there are major groups that do not become integrated into mainstream society sufficiently rapidly. Ethnic group formation takes place everywhere, but under conditions that vary considerably from country to country. Consequently, there are different outcomes: in some countries, ethnic groups become marginalized and excluded minorities, whereas in others, they take the form of ethnic communities that are accepted as part of a pluralist society.

Exclusion is most severe in the former guest-worker countries, such as Germany and Switzerland. Multicultural models are to be found in countries with explicit policies of permanent settlement and pluralism, above all, Australia, Canada and Sweden. The United States, as a country of permanent settlement without multicultural policies, comes close to conforming to the pluralist model, but without espousing its explicit political goals. The Netherlands also comes close to conforming to the multicultural model, though in a weaker form. Between the extremes of exclusion and multiculturalism lie countries, like France and the United Kingdom, that recognize the reality of permanent settlement, but are unwilling to accept pluralism as a long-term goal. However, there are important differences between France and the United Kingdom, particularly with regard to the role of the State in managing cultural difference. France continues to emphasize political integration through the acquisition of citizenship, while some British policies tacitly acknowledge the need for special social and cultural policies for minorities.

One must stress, however, that the reality in each country is considerably more complex and contradictory than this brief account suggests. Nevertheless, some general conclusions emerge. The first is that policies of temporary migrant labour recruitment are almost certain to lead to permanent settlement and to the formation of ethnic groups. No State that bases its policies on the rule of law and human rights can prevent this development. Countries that have recently begun to make use of migrant labour, both in Southern Europe and in Eastern and South-eastern Asia would do well to reflect upon this conclusion.

The second is that the character of ethnic groups will, to a large extent, be determined by what the State does in the early stages of migration. Policies that try to deny the reality of immigration by tacitly tolerating large-scale illegal movements lead to social marginalization, the formation of minority groups and racism. Policies that try to maintain a permanent division between foreigners and citizens by denying rights to the former and preventing them from acquiring citizenship, also lead to a divided society. The best chance for successful settlement and peaceful interaction between ethnic groups lies in policies that accept permanent settlement and family reunification, and that make it easy for immigrants to become citizens.

A third conclusion is that ethnic groups established as a result of migration need their own associations and social networks, as well as their own languages and cultures. Policies that deny legitimacy to such associations lead to the phenomena of isolation and separatism. Cultural rights should therefore be recognized as part of citizenship, along with the more customary civil, political and social rights.

A fourth conclusion is that successful integration requires active policies emanating from the State, including settlement services that provide initial accommodation, help in finding work and language training. Social policies are also needed to provide for the special needs of migrants who stay over long periods, especially regarding education, health and social services, the provision of interpretation and translation facilities, and care for elderly migrants. Migrant women are often particularly disadvantaged and need special services and amenities. The guiding principle of social policy should be that its long-run aim is not to provide separate systems of service provision, but rather to change mainstream systems so as to make them better able to respond to the needs of a culturally diverse population. Lastly, the State needs to introduce legislation for the purpose of removing any barriers that prevent the full participation of migrants in society. Such measures would include anti-discrimination and equal opportunity legislation, as well as laws to combat racist violence and vilification. Special agencies are needed to monitor the implementation of such measures.

These conclusions suggest that policies of multiculturalism constitute the best path towards migrant integration. Although European scholars and policy makers have often asserted that the experience of the "traditional countries of immigration" has little relevance for European countries, this claim appears increasingly dubious. European labour-importing countries have become countries of permanent settlement, whatever their original intentions. They therefore need to consider the models expressly developed to integrate permanent settlers, rather than cling to myths of temporariness. The case of Sweden shows, moreover, that a multicultural model can be implemented in a European country with monocultural traditions.

The same applies to countries that have recently become migrant receivers, such as Greece, Italy, Japan, Spain and some of the newly industrializing economies of South-eastern Asia. Some Governments have tended to take a laissez-faire approach towards migrants, tolerating the entries of undocumented workers because employers need the labour. Others have set up guest-worker systems for temporary labour recruitment. Since there has been no official expectation of settlement, social policies and legal frameworks to support integration have largely been absent. The result has been growing marginalization of migrants, to the point that, in some countries, shanty towns are visible around big cities. Those changes are exploited by the propaganda machine of the extreme right and lead to hostility towards migrants. The parallels with earlier developments in countries like France and Germany are obvious.

However, it would be wrong to conclude that multicultural models offer easy solutions for Western European countries or for new receivers of migrants. The growth of racist violence and the success of the mobilization of the extreme right around issues of immigration and cultural difference make it very hard for Governments to move towards improvements in migrants' rights or to accept the perspective of cultural pluralism. The central problem is the fear elicited by assumptions about the potential for growing migration from poor countries with growing populations to rich countries with stagnant populations. It is clear that the successful integration of existing migrants depends on the development of international strategies to narrow the gap between North and South or between East and West, thus reducing the perceived threat of unwanted migration.

REFERENCES

Anti-Defamation League (1988). *Hate Groups in America*. New York: Anti-Defamation League of B'nai B'rith.

Ålund, A., and C. U. Schierup (1991). *Paradoxes of Multiculturalism*. Aldershot: United Kingdom Avebury.

Australian Bureau of Statistics (1988). *Census 86. Australia in Profile: A Summary of Major Findings*. Canberra.

Banton, M. (1985). *Promoting Racial Harmony*. Cambridge, United Kingdom: Cambridge University Press.

Bell, D. (1975). Ethnicity and social change. In *Ethnicity: Theory and Experience*, N. Glazer and D. P. Moynihan, eds. Cambridge, Massachusetts: Harvard University Press.

Beynon, J. (1986). Spiral of decline: race and policing. In *Race, Government and Politics in Britain*, Z. Layton-Henry and P. B. Rich, eds. London: Macmillan.

Borjas, George J. (1990). *Friends or Strangers: The Impact of Immigration on the US Economy*. New York: Basic Books.

Boyd, Monica (1989). Family and personal networks in migration. *International Migration Review* (Staten Island, New York), vol. 23, No. 3 (special silver anniversary issue), pp. 638-670.

Breton, R., and others (1990). *Ethnic Identity and Equality*. Toronto: University of Toronto Press.

Briggs, V. M., Jr. (1984). *Immigration Policy and the American Labor Force*. Baltimore, Maryland and London: Johns Hopkins University Press.

Canada, Employment and Immigration Canada (1991). *Annual Report to Parliament: Immigration Plan for 1991-1995, Year Two*. Ottawa.

Castles, Stephen (1985). The guests who stayed: the debate on "foreigners policy" in the German Federal Republic. *International Migration Review* (Staten Island, New York), vol. 19, No. 3, pp. 517-534.

________ (1992). Australian multiculturalism: social policy and identity in a changing society. In *Nations of Immigrants: Australia, the United States and International Migration*, G. P. Freeman and J. Jupp, eds. Melbourne: Oxford University Press.

________, and G. Kosack (1973 and 1985). *Immigrant Workers and Class Structure in Western Europe*. London: Oxford University Press.

Castles, Stephen, and Mark J. Miller (1993). *The Age of Migration: International Population Movements in the late 20th Century*. London: Macmillan.

Castles, Stephen, and others (1990). *Mistaken Identity. Multiculturalism and the Demise of Nationalism in Australia* (second edition). Sydney: Pluto Press.

Collins, J. (1991). *Migrant Hands in a Distant Land: Australia's Post-War Immigration* (second edition). Sydney: Pluto Press.

Commission of the European Communities (1989). *Eurobarometer: Public Opinion in the European Community, Special: Racism and Xenophobia*. Brussels.

Costa-Lascoux, J. (1991). *De l'immigré au citoyen*. Paris: La Documentation française.

de Wenden, C. (1987). *Citoyenneté, nationalité et immigration*. Paris: Arcantère Editions.

European Parliament (1985). *Committee of Inquiry into the Rise of Fascism and Racism in Europe: Report on the Findings of the Inquiry*. Strasbourg.

Feagin, J. R. (1989). *Racial and Ethnic Relations*. Englewood Cliffs, New Jersey: Prentice-Hall.

Fishman, J. A. (1985). *The Rise and Fall of the Ethnic Revival: Perspectives on Language and Ethnicity*. Berlin, New York and Amsterdam: Mouton Publishers.

Fleras, A., and J. L. Elliott (1992). *Multiculturalism in Canada: The Challenge of Diversity*. Scarborough, Ontario: Nelson.

Funcke, L. (1991). *Bericht der Beauftragten der Bundesregierung für die Integration der ausländischen Arbeitnehmer und ihrer familienangehörigen*. Bonn: German Government.

Geertz, C. (1963). *Old Societies and New States. The Quest for Modernity in Asia and Africa*. Glencoe, Illinois: Free Press.

Glazer, N., and D. P. Moynihan, eds. (1975). *Ethnicity: Theory and Experience*. Cambridge, Massachusetts: Harvard University Press.

Hammar, Tomas (1985). Sweden. In *European Immigration Policy: a Comparative Study*, Tomas Hammar, ed. Cambridge, United Kingdom: Cambridge University Press.

Hoffmann-Nowotny, Hans-Joachim (1985). Switzerland. In *European Immigration Policy: a Comparative Study*, Tomas Hammar, ed. Cambridge, United Kingdom: Cambridge University Press.

Home Office (1989). *The Response to Racial Attacks and Harassment: Guidance for the Statutory Authorities. Report of the Inter-Departmental Racial Attacks Group.* London: Home Office.

Lapeyronnie, D., and others (1990). *L'Intégration des Minorités Immigrées, Étude Comparative: France et Grande-Bretagne.* Paris: Agence pour le développement des relations interculturelles.

Larsson, S. (1991). Swedish racism: the democratic way. In *Race and Class*, vol. 32, No. 3, pp. 102-111.

Layton-Henry, Z., and P. B. Rich, eds. (1986). *Race, Government and Politics in Britain.* London: Macmillan.

Leggewie, C. (1990). *Multi Kulti: Spielregeln für die Vielvölkerrepublik.* Berlin: Rotbuch.

Lithman, E. L. (1987). *Immigration and Immigrant Policy in Sweden.* Stockholm: Swedish Institute.

Miles, R. (1989). *Racism.* London: Routledge.

Morokvasic, M. (1984). Birds of passage are also women. *International Migration Review* (Staten Island, New York), vol. 18, No. 4, pp. 886-907.

Muus, Philip J. (1991). *Migration, Minorities and Policy in the Netherlands: Recent Trends and Developments. Report for SOPEMI.* Amsterdam: University of Amsterdam, Department of Human Geography.

Organisation for Economic Cooperation and Development (1991). *OECD Continuous Reporting System on Migration Report (SOPEMI) 1990.* Paris.

__________ (1992). *SOPEMI 1991: Trends in International Migration.* Paris.

Park, R. E. (1950). *Race and Culture.* London: Collier Macmillan.

Phizacklea, A. (1990). *Unpacking the Fashion Industry: Gender, Racism and Class in Production.* London: Routledge.

__________, ed. (1983). *One Way Ticket? Migration and Female Labour.* London: Routledge and Kegan Paul.

Portes, Alejandro, and R. G. Rumbaut (1990). *Immigrant America: A Portrait.* Berkeley, California, Los Angeles, California, and London: University of California Press.

Rex, J., and D. Mason, eds. (1986). *Theories of Race and Ethnic Relations.* Cambridge, United Kingdom: Cambridge University Press.

Smith, A. D. (1986). *The Ethnic Origins of Nations.* Oxford: Blackwell.

Solomos, J. (1989). *Race and Racism in Contemporary Britain.* London: Macmillan.

Stasiulis, D. K. (1988). The symbolic mosaic reaffirmed: multiculturalism policy. In *How Ottawa Spends 1988/89*, K. A. Graham, ed. Ottawa: Carleton University Press.

United Nations (1957). *Treaty Series*, vol. 189, No. 2545.

United States Bureau of the Census (1991). Census Bureau releases 1990 Census counts on specific racial groups. *Commerce News.* Washington, D. C.: United States Department of Commerce, 12 June, pp. 1-2.

Vasta, E. (1990). Gender, class and ethnic relations: the domestic and work experiences of Italian migrant women in Australia. *Migration* (Berlin), No. 7, pp. 115-137.

Verbunt, G. (1985). France. In *European Immigration Policy: A Comparative Study*, Tomas Hammar, ed. Cambridge, United Kingdom: Cambridge University Press, pp. 127-164.

Waldinger, R., and others (1990). *Ethnic Entrepreneurs: Immigrant Business in Industrial Societies.* Newbury Park, London, and New Delhi: Sage.

Wallman, S. (1986). Ethnicity and boundary processes. In *Theories of Race and Ethnic Relations*, J. Rex and D. Mason, eds. Cambridge, United Kingdom: Cambridge University Press.

Weil, P. (1991). *La France et ses étrangers.* Paris: Calmann-Lévy.

XV. GROWING ECONOMIC INTERDEPENDENCE AND ITS IMPLICATIONS FOR INTERNATIONAL MIGRATION

*Lin Lean Lim**

Countries are economically dependent on each other because they possess different factor endowments and production capabilities. Interdependence is rooted in the structural imbalances within and between developed and developing countries. It makes economic sense, based on the principle of comparative advantage, to engage in exchange, either through specialization in production or through the movement of the factors of production to areas of higher returns. These exchanges and flows enhance the economic interdependence of countries. Conventional neoclassical theories stress that the liberalization of trade and capital flows leads to factor market adjustments and obviates the need for international labour flows. However, over the past four decades, increased volumes of international trade and investment have not proved to be substitutes for the movement of labour. In a world increasingly characterized by a tendency towards both globalization and regionalization, where growing interdependence coexists with the involuntary delinking of poor countries from the more advanced ones, migration pressures have been increasing and the volume of migration has expanded.

The growing economic interdependence among nation States fosters and is in turn fostered by international migration. Where there have been barriers to economic integration or where the involuntary delinking of poor countries from their more developed counterparts has occurred, the unauthorized movement of people across national borders has been an increasingly significant form of adjustment. Migration, like other exchanges and flows between countries, represents both an opportunity and a source of vulnerability for interdependent nation States.

Two main explanations are offered for the fundamental paradox that the economic integration of countries, which is supposed to transmit the benefits of economic growth from one country to another and thus facilitate structural adjustment, increases rather than reduces migration pressures. First, the modalities of socio-economic development associated with rising interdependence are essentially disruptive and dislocating. Development has almost always resulted in considerable migration, first within countries and particularly from rural to urban areas, and in some cases between countries. In addition, differences in the timing and relative pace of the underlying economic restructuring and structural adjustments experienced by different countries inevitably create migration pressures. The disparities between developed and developing countries remain so large that even extraordinary development gains by countries of origin may not eliminate migration pressures, especially because there are other systematic linkages in operation.

The second explanation is that international trade and foreign investment create objective and subjective bridges between trading and investing partners that activate the flows of labour, whether legal or illegal, between particular countries. Thus, the economic penetration of a developing country by a developed country through industrial relocation, for instance, may not only lead to internal imbalances in the developing country concerned, but may also have the unintended effect of exposing its citizens to international production and consumption standards that might raise expectations and lead to the consideration of migration. Close economic ties also imply well-developed transportation and communication systems that, by reducing the costs of information and travel, help focus the attention of potential migrants on specific destinations. Where there are prior historical, cultural or political ties between the countries concerned, the potential for migration is easily activated, especially if the Governments of developing countries, aided by a host of profit-oriented intermediaries, actively promote and organize the export of labour. Once activated, migration flows tend to become self-sustaining, at least over a period, because of the operation of social networks.

*Regional Office for Asia and the Pacific, International Labour Organization, Bangkok, Thailand.

The implications of economic interdependence on international migration are examined within a dynamic systems framework according to which there is an unequal, though integrated, system of relations between countries, and the systematic processes linking countries of origin and destination of migrants are not only economic but also political, social and cultural in nature. International flows of people are only imperfectly explained by differences in economic growth rates and job opportunities. Nor have migration flows arisen spontaneously out of poverty. Economic success by itself has not led to zero emigration, especially in the short term, and labour shortages in countries of destination have not been the main pull factor attracting migrants. While elemental market forces give rise to migration pressures, actual movements have been initiated and sustained by a number of dynamic processes, including trade, foreign direct investment, foreign aid and flows of technology that build linkages between countries.

The forms of interdependence have been changing over time. Increasingly, it is technology, rather than quantitative factor endowments, that determines the competitive advantage of countries, and remittances from migrants in conjunction with earnings from tourism, rather than earnings from the export of agricultural or manufactured goods, now account for significant shares of the trade of many developing countries. Rich countries prefer to cope with their demographic imbalances by using a rotating stock of temporary workers from selected countries rather than by opening their doors to permanent immigrants. Geopolitical realignments and new arrangements have carved up the world into regional blocs that combine an outward-orientation with an inward-orientation that includes, in some cases, the free movement of people. Demographic differentials between developing and developed countries are more marked and the nations of the world are now moving along a three-speed road in terms of economic growth, with some groups moving in opposite directions (Emmerij, 1993). The Organisation for Economic Cooperation and Development (OECD) countries plus the newly industrializing economies of Eastern and South-eastern Asia constitute the "world economy" and are moving forward at a robust pace. Low-income countries, especially in Africa, have been sliding backwards. In between, many Latin American countries plus China and India are advancing slowly and still face the danger of sliding back. The economies in transition are also facing difficult economic circumstances and it is not yet clear when their economies will enter a period of sustained growth. The implications of these differences for international migration have to be examined in a context where the tendencies towards globalization and regionalization coexist and where the growing interdependence of nation States has not prevented some countries from being involuntary delinked from the world economy.

A. The world economy during 1945-1970 and the movement of labour

The international trading and financial systems constructed after the Second World War were based on the assumption that if the movements of the factors of production, especially labour, were restricted, a partial substitute for approaching some equalization in income levels among countries could be found in free trade. Since it was understood, though not explicitly stated, that labour migration would not be free, the main emphasis was on the liberalization of trade and capital flows. After the General Agreement on Tariffs and Trade (GATT) came into force in 1948, trade barriers were progressively removed in the 1950s and 1960s, especially with regard to manufactured goods, thereby promoting a rapid industrial specialization among countries. Considerable opening of national markets was achieved through successive rounds of multilateral trade negotiations.

The period between 1950 and the first oil-price shock in 1973 represents the "golden age" of economic growth. During that period, real gross domestic product (GDP) in the industrialized countries of the Western bloc increased by an average of nearly 5 per cent per year or almost twice as rapidly as it has since 1973. Strong economic growth was bolstered by a number of factors that improved incentives and capabilities in the advanced economies and initiated international economic integration. As Soltwedel (1993) notes, a critical factor in the exceptionally strong economic growth experienced during the post-war period was an institutional framework that put emphasis on freeing market incentives and strengthening capabilities. Many barriers were swept away in the process of trade liberalization and liberalization was also extended to movement of capital and foreign direct investment.

The economies of the world became increasingly integrated as global trade grew faster than both production and income until the 1970s. Just as global trade expanded, foreign direct investment flows grew at approximately twice the rate of growth of the world economy (Le Bideau, 1991, p. 1). Capital moved increasingly to countries yielding the highest returns; technological advances spread among a rising number of economies; and the international migration of labour, at all levels of skills, grew rapidly. During 1950-1970, pull factors dominated push factors in explaining international migration, a state of affairs that has generally not been repeated since then (Donges, 1983, p. 13).

Virtually throughout the advanced industrialized countries, internal and international migration made labour not only abundant but also mobile, facilitating the growth of new firms and industries in the countries of destination and easing employment problems in the countries of origin. It is estimated that between 1950 and 1973, the market economies of Europe registered a net migration gain of nearly 10 million people as compared with a net loss of 4 million during 1914-1949 (Soltwedel, 1993). Western Germany was one of the major importers of foreign labour before 1973; the share of foreign workers in the total labour force increased from 1.3 per cent in 1960 to 11.7 per cent in 1973 (Donges, 1983). The United States of America, which played a crucial role in the development of a world economic system, admitted large numbers of immigrants during the 1960s after nearly five decades of low-to-moderate inflows. In the 1960s and early 1970s, the United States passed legislation aimed at opening its own and other countries' economies to the flow of capital, goods, services and information. The expansion of the United States economic and military activities in Asia and the Caribbean also helped create conditions that mobilized potential migrants. In the case of Mexico, the migration process had its origins in geopolitical and economic intervention by the United States which first restructured the neighbouring nation and then proceeded to organize dependable labour outflows from it. Labour movements across the border were a well-established routine before they became redefined as migration and then as undocumented migration (Portes, 1991).

For developing countries, incorporation into the world economy and the promotion of economic growth were an essentially disruptive process. Economic growth was inevitably accompanied by the dislocation of stable social structures, the transformation of rural or agrarian societies, and a growing outward-oriented, modern industrial sector with limited linkages with the rest of the domestic economy. Dualistic structures emerged, with the modern sector becoming increasingly capital-intensive and having high labour productivity, and the traditional, largely rural, sector struggling to cope with labour surpluses and low productivity. The modern sector thus became linked to the economies of the industrialized nations rather than to the national hinterlands. Growth in the modern sector did not "trickle down" to the other sectors of the domestic economy, at least not in the short term. At the same time, people's aspirations for higher incomes and consumption rose with their growing exposure to modern products, and their capacity to move increased as incomes rose and educational opportunities expanded. Internal migration escalated, particularly that from rural to urban areas, and internal migrants often became international migrants at a later stage (Massey, 1988; Sassen, 1988).

These dynamics help to explain why, although respectable rates of economic growth were sustained over a relatively long period, many developing countries experienced growing emigration. "The transformations intrinsic to the development process are at first destabilizing. They initially promote rather than impede migration" (Commission for the Study of International Migration and Cooperative Economic Development, 1990, p. 34). A major potential for migration is structurally built into the development process, and the greater the degree of economic integration between a given developing country and potential countries of destination, the greater the potential for international migration (Massey, 1991). For instance, the Dominican Republic's rapid economic growth in the mid-1960s coincided with the departure of large numbers of persons to the United States. There was also a steady increase of emigration from Mexico during the years of high economic growth (Weintraub and Stolp, 1986).

In addition, the worldwide exchanges of goods and services, along with the flows of capital and manpower, not only led to unprecedented levels of prosperity in the industrialized world, but also permitted other countries to advance their economic development, as was the case with Japan and the newly industrializing economies of Asia (Hong Kong, the Republic of Korea, Singapore and Taiwan Province of China). Countries in the latter group are competing increasingly with the industrialized countries of Europe and Northern America.

1. *Inward-oriented versus outward-oriented development: the implications for migration*

Trends during the golden age suggest that the mode of development, in particular the manner and extent of integration into the world economy, is more likely than the rate of economic growth to influence emigration from developing countries. In the search for development strategies that would reduce migration, Portes (1982, pp. 84-87) proposed an inward-oriented strategy that would include the production of basic manufactured goods, limited reliance on imported models and patents, the development of domestic technology, and the regulation of externally induced consumer pressures. The core of his argument was that export-led development would promote migration because it exposed the citizens of developing countries to the fashions, consumption standards and expectations of advanced countries; it contained an inbuilt bias against increases in the returns of labour, and it made countries dependent on insecure external markets. However, Portes himself presented little supporting empirical evidence for his arguments, and the experience of developing countries does not seem to support his conclusions.

At least until the 1970s, the countries of Latin American and the Caribbean pursued policies that limited their integration into the world economy. They emphasized industrialization for import substitution and protectionism against foreign competition by imposing both high tariff and non-tariff barriers to trade. Because the domestic markets of most countries were too small to support an efficient industrial base, regional integration schemes developed to widen the scope of import substitution. The Central American Common Market, the Latin American Free Trade Association, and the Caribbean Community and Common Market, among others, stressed the need for protected regional markets. Not only did countries that were export pessimists benefit less from the transmission of economic growth from other countries, but they also became important sources of migrants. Whatever success their development policies may have had in stimulating industrialization and overall economic growth, it is now generally accepted that they failed to improve income distribution or to create adequate job opportunities for their growing populations, thus reinforcing the migration tendencies that already existed. Donges (1983, pp. 13-14) concludes that import substitution strategies not only failed to achieve the highest economic growth possible but also dampened considerably the capacity of the economy, and in particular that of the newly established manufacturing industry, to absorb local labour. Under such conditions, unemployment became chronic and the low chances of finding a job in their country induced workers to migrate abroad. The direction of the resulting migration flows depended on the location of economic growth poles and the ease with which migrants could gain admission to the potential countries of destination.

Some Asian countries, especially the newly industrializing economies, were more successful at employment creation. Although their development policies favoured import substitution, their industrialization programmes emphasized also export promotion. The newly industrializing economies have become major international competitors and seem not to have experienced migration pressures to the same degree as certain Latin American countries. In fact, Hong Kong and Singapore have long been importers of foreign labour, but the Republic of Korea and Taiwan Province of China have experienced moderate emigration levels, directed mostly to the United States. In addition, the most recent newly industrializing economies of Malaysia and Thailand have both been exporters of labour. A major recent development is that Japan, Malaysia, and Taiwan Province of China have also begun to receive migrant workers, often in an unauthorized fashion. The increasing migration flows within the region have coincided with increasing flows of capital. Hong Kong, Japan, and Taiwan Province of China, among others, are important exporters of capital to other countries both within and outside Eastern and South-eastern Asia.

The cases of the newly industrializing economies indicate that the relation between the mode of development and international migration is complex. As noted above, migration from those countries was important at the early stages of development and only after economic growth was sustained for some time have migration inflows become more important than migration outflows. It is likely that, in the case of the newly industrializing economies, their very economic success may have provided the impetus for migration outflows by improving the skills of the population and increasing incomes, thus providing both the human capital that was in demand in potential countries of destination and the capital that was needed to finance the costs of migration. Thus, Singapore, despite its impressive economic performance, has been losing through emigration almost as many citizens per 100,000 population as the Philippines, whose annual gross national product (GNP) per capita is less than 7 per cent that of Singapore. For Taiwan Province of China, too, the three generations exposed to sustained economic growth include a broad class of people with enough money to do what they want and what many of the newly rich want is to move abroad (*The Nation*, "Emigration becomes big business in Taiwan", 8 October 1990).

It is important to underscore the fact that the Asian newly industrializing economies had the advantage of joining the world economy during the 1960s and 1970s, when economic growth was strong in the industrialized world. The United States played a major role by absorbing the manufactured products exported by those countries, many of which were produced by the overseas subsidiaries of American multinational corporations. The exporting success of the Asian newly industrializing economies and Japan even reversed the trade balance with the United States and the latter currently has a large and growing trade deficit which it has been trying to reduce through protectionist measures. Countries that adopted the outward-looking mode of development more recently have been facing greater constraints in respect of penetrating the world market. The lacklustre economic growth that has characterized the industrialized countries of the West since the 1970s and their growing protectionism have fostered competition among developing countries to gain export markets and attract foreign direct investment. Outward-looking development strategies, therefore, do not provide the magic solution to the problems associated with the pursuit of development, including migration.

B. "Oil shocks" of the 1970s and migration to Western Asia

During the 1970s, as the comparative advantage of industrialized countries eroded, they began restricting the import of labour-intensive products from developing countries. The GATT rounds on trade liberalization, including the Tokyo Round of multilateral trade negotiations (1973-1979), maintained high barriers on the industrial products of greatest export potential for developing countries, especially textiles and clothing. The Multi-Fibre Arrangement which entered into force in 1974 set global limits to the annual growth of textile imports and forced developing countries to restrain their exports accordingly. Under the Generalized System of Preferences (GSP), which has been in operation since the early 1970s, the tariff concessions that industrialized countries unilaterally granted to developing countries tended to treat labour-intensive imports in a less preferential manner.

An alternative response by developed countries to their loss of comparative advantage could have been to allow the inflow of foreign workers in order to increase their labour supply, contribute to reducing wages and thereby slow down any further adverse shifts in their comparative advantage. However, the choice between import protection and labour migration was never seriously considered. The social and political situation of the 1970s led the Governments of most industrialized countries to institute restrictions on the import of both goods and labour.

The pace of economic integration slowed down, especially as a result of the oil shocks of 1973 and 1979. The effects of those shocks showed that global interdependence could foster the transmission of both positive and adverse economic trends among countries. The economic growth of industrialized Western-bloc countries declined sharply after the first oil shock and culminated in negative growth and mounting unemployment in 1975. In order to help industry adjust to the changed energy market, many Governments intervened to a greater extent than before in both domestic and international product and factor markets. The industrialized countries returned to sectoralism and bilateralism in international trade, thereby increasing protectionism and inviting retaliation by other countries. The growth of world trade decelerated and developing countries faced deteriorating terms of trade. A number of the industrialized countries imposed major restrictions on the admission of foreign workers and established programmes to foster their return.

However, a new phase of labour migration was initiated with the importation of foreign workers by the oil-producing countries of Western Asia, that embarked on major development programmes financed by mounting oil revenues. Members of the Gulf Cooperation Council and the Islamic Republic of Iran relied on project-tied migration to engage large numbers of foreigners to build roads, hospitals, harbours and so on. Foreign enterprises contracted to carry out those projects were charged with securing the needed labour abroad and ensuring its departure once the project was completed. The same underlying factor that prompted developed countries to impose restrictions on labour inflows led the oil-exporting countries to import labour. These developments were of major importance for the countries supplying the necessary labour. From the mid-1970s to the early 1980s, the single most important factor affecting domestic employment and the balance-of-payments situation in a number of Asian countries was the outflow of contract workers to Western Asia and the inflow of remittances which helped to pay the rising oil bill (Amjad, 1989).

Until 1975, other Arab countries had been the main source of foreign labour for the countries of the Gulf Cooperation Council. Between 1975 and 1979, there was a sharp increase in the presence of workers from Southern Asian countries, mainly India and Pakistan. Pakistan became heavily dependent on labour migration and remittances and was thus subject to a number of vicissitudes. During the late 1970s to the early 1980s, as much as one third of Pakistan's labour force increase found employment in Western Asia and remittances represented some 70 per cent of total exports of goods and non-factor services (Amjad, 1989). Starting in 1979, the oil-producing countries began recruiting workers from other Asian countries, including the Philippines, the Republic of Korea and Thailand. More recently, Indonesia and Sri Lanka joined the labour suppliers.

It is worth noting that there was only a limited history of active economic linkages between the oil-exporting countries of Western Asia and the labour-exporting countries of the rest of Asia before the mid-1970s. Yet, the impact of the oil price hikes of 1973 on the balance-of-payments position of many of those countries was such that they found it necessary to export labour to help pay for essential imports. In addition, the oil-exporting countries deliberately recruited workers from cheap and plentiful sources. Even countries that did not depend on imported oil, such as Indonesia, took advantage of the possibility of exporting excess labour to Western Asia. These developments show that a sudden increase in the demand for labour can activate migration between countries that had not been integrated previously and that the labour flows themselves forged interlinkages. Political and cultural factors also played a role. Thus, the shift from Arab to other Asian sources of labour was partly due to the desire of oil-producing countries to limit the presence of foreign Arab citizens in their territories. In addition, the reliance on Indonesian female workers was prompted by the desire to ensure that domestic workers followed the Muslim faith.

C. The new international division of labour

As Tapinos (1993) has noted, the industrialized countries of Europe closed their borders to worker migration for reasons related to changes in the national and international economic environment, including the entry into the labour market of the baby-boom cohorts, increased female economic activity rates, slower economic growth, and industrial restructuring. A problem of replacement and relocation then arose. There was talk of a "new international division of labour" involving the reduction of the number of foreign workers by providing incentives for their return to the countries of origin, the relocation of certain productive activities to developing countries, a return to specialization more in line with natural factor endowments, the development of trade, and increased investment in developing countries.

During the 1970s, the growing tendency towards industrial relocation was based on two factors: the technical possibility of splitting the manufacturing process into separate segments and specific production functions that could be located in different countries; and the comparative economic advantage of doing so associated with the low wages prevalent in certain countries or with the access to local markets made possible by setting up assembly plants in countries with high tariff barriers or import quotas. The most common form of relocation has been that of assembly operations which, being labour-intensive, were moved to developing countries. Such relocation has involved primarily textiles and clothing, electronics and the automotive industry. Components manufactured in the industrialized countries are sent to low-wage countries to be assembled into the finished product and re-imported by the country of origin or exported to another industrialized country.

The relocation operations were to form the basis of a new international division of labour that would replace the old one based on trading manufactures for primary products (Frobel, Heinrichs and Kreye, 1980). Relocation operations were thus closely linked to trade between developed and developing countries and were an important influence on the dynamics of industrial development and employment in the developing countries endowed with an abundant labour supply. The insertion of those countries into the international subcontracting system or "global sourcing system" of labour-intensive industries was seen as the most effective way of ensuring development and job creation and, therefore, by logical extension, of reducing international migration pressures by taking capital and technology to labour rather than the other way round.

This process worked relatively well in the first tier of newly industrializing economies in Asia. The export-processing zones in Mexico also experienced the dynamics of development and job creation. However, their implications for international migration must be assessed taking account of the fact that most of the employment created was for women aged 18-24 who had previously not been in the labour force and who would have been less likely to migrate in the absence of those opportunities (Bailey and Parisotto, 1993). The main participants in the new industrial division of labour have been young women entering the labour force for the first time and originating mainly in rural areas. They have been preferred over male workers because they can be paid lower wages, are generally docile, and willing to accept the rigid discipline and tedious monotony of assembly line production, and are efficient and malleable (Lim, 1988). The entry of young rural women into the labour force has been both socially and economically disruptive. Men, being faced with increased competition from women, began considering overseas labour markets. Thus, in countries like Malaysia or Thailand, out-migration from rural areas in the 1970s became increasingly shaped by employment opportunities for women and overseas employment for men.

In the cases of small countries or countries with limited human resources, the establishment of export-processing zones has itself stimulated international migration. Thus, Guatemalans work in the export-processing zones in Belize, Sri Lankans in Mauritius, and Chinese in Australia (Bailey and Parisotto, 1993). In addition, the types of industries or industrial processes relocated abroad are fairly mobile themselves and can move to new locations if wages or quota restrictions increase. Consequently, the employment created by the new industrial division of labour has been unstable and the mobility of industry may itself lead to the international migration of skilled personnel.

The traditional countries of immigration have promoted the relocation of industry in countries of origin to promote their development and employ people at home. The United States, for instance, adopted the Caribbean Basin Initiative and promoted the establishment of *maquiladoras* (in-bond assembly plants) in Mexico. However, the effects of industrial relocation on employment creation have generally been small. According to International Labour Organization (ILO) estimates, out of 65 million workers employed in multinational enterprises worldwide, only 22 million (of which 7 million worked in developing countries) were employed in countries different from those in which the parent company was located. Among all workers employed abroad by multinational enterprises, approximately 18 per cent worked in Latin America, 13 per cent in Asia and a marginal 2 per cent in Africa. In addition, employment by multinational enterprises was heavily concentrated in a few developing countries: 18 countries absorbed about 86 per cent of all foreign direct investment going to the developing world. These data confirm that the contribution of the operations of multinational enterprises to employment creation in developing countries has been limited and that such employment is unlikely to effectively counter all the factors responsible for migration outflows (Bailey and Parisotto, 1993).

Not only has the employment impact of industrial relocation been limited for the developing countries, but the short-term impact is more likely to have stimulated migration. Thus, although the United States-Mexican Border Industrialization Programme provides relatively well-paid jobs by Mexican standards for about half a million Mexican workers in 2,000 mostly United States-owned factories, the workers are mostly women each of whom is employed for only about five years (Martin, 1991). More than a decade of the *maquiladoras* (in-bond assembly plants) programme has failed to promote a major transfer of technology or to create lasting linkages with the rest of the economy, and the programme has therefore not been successful in stemming migration to the United States (Bailey and Parisotto, 1993). In addition, the fact that young women are the ones most likely to get jobs in border areas often acts as an incentive for other family members, in particular men, to seek work on the United States side of the border (Portes, 1991).

Since the beginning of the 1980s, further progress in redefining a "new industrial division of labour" in favour of potential migrant-sending countries has been hampered because, as Mouhoud (1993) argues, the diffusion of technological change in developed countries, based on information technologies, has eroded the comparative advantage associated with the availability of cheap labour; changes in demand conditions (increased versatility and differentiation) are causing firms to locate their production units close to markets in order to reduce delivery times for finished products; and non-tariff protectionist measures against imports from low-wage countries have been proliferating. This neo-mercantilism in North-South relations has resulted from both the desire to protect technological property and opposition to the relocation of industrial operations on the part of Governments, trade unions and even firms.

A strategy on the part of developed countries to relocate industry in developing countries as a means of slowing migration flows or a strategy on the part of developing countries to emulate the development model of the Asian newly industrializing economies now seems to have little chance of success (Mouhoud, 1993). The disappearance of the expression "new international division of labour" is also indicative of the failed objectives of the exercise (Tapinos, 1993). Such failure, however, has not meant that capital movements through foreign direct investment are no longer important. Since the mid-1980s, unprecedented levels of international investment have been the major force behind both the ever-increasing ties between countries and their growing financial and economic integration, but capital flows have been increasingly separated from considerations about employment creation and the promotion of development in countries of emigration. As Martin (1991) notes, multinational corporations invest abroad to earn profits and not to deter migration. With so many countries vying for investment funds, countries of origin and international organizations are not in a position to require multinational corporations to invest in a manner that minimizes the potential for migration (Martin, 1991).

Böhning (1991) concludes similarly that, because investment decisions are determined by short-to-medium-term estimates of likely returns, investment flows tend to be channelled to countries other than those from which migrants leave because the latter typically lack attractive economic prospects or political stability. Governments can rarely use this option, since often they themselves do not invest abroad for economic reasons and public enterprises in developed countries are free to decide for themselves where to invest. Given that most foreign direct investment decisions are in the hands of the private sector, Government incentives to direct funds to certain countries can have only a marginal effect at best (Böhning, 1991).

D. New forms of interdependence under structural adjustment and the implications for international migration

During the early 1980s, the world underwent a major recession, with developing countries experiencing falling exports, increasingly serious balance-of-payments deficits and debt servicing problems, aggravated by rising real interest rates and the unwillingness of commercial banks to resume lending on any scale. The industrialized countries of the Western bloc underwent a shorter recession than developing countries and after 1983 were able to achieve sustained economic growth, albeit at a slow pace and, for the most part, without a return to full employment. Asia was less affected by the crisis than other developing regions, either because of the size and limited integration in the world economy of the countries concerned (as in the case of China and India) or because of the diversification of their manufactured exports (as in the case of the Association of South-East Asian Nations (ASEAN) countries). The regions most severely hit by the crisis were Africa and Latin America. For Africa, the 1980s represented a further stage in a steady decline that had started before, but for Latin America it reversed a long period of rapid growth that had been sustained largely by foreign capital (International Labour Organization, 1989).

In the changed international environment, "structural adjustment" based on supply-side economics became the order of the day. Assiduously propagated among the industrialized countries by the Organisation for Economic Cooperation and Development (OECD) and among developing countries by the international financial institutions, the essence of the approach was that the market should determine resource allocation and economic outcomes, that macroeconomic policy should be geared primarily to monetary stability, and that Government should concentrate on the preservation of a legal framework in which business could be done (Standing, 1991). The systematic review of microeconomic policies by most countries focused on increasing competition in product markets; strengthening the responsiveness of factor markets; deregulating labour markets and creating flexible job structures; and increasing the efficiency and effectiveness of the public sector. These reforms, in particular the first two, have contributed to increasing the interdependence among countries because exposure to international trade is the single most effective means of enhancing competition and prompting firms to respond to the opportunities arising from technological advances and broader changes in economic circumstances. Trade liberalization, especially for manufactured goods, has been advocated as being an important element of structural adjustment to increase efficiency in the productive base of developing countries.

Adjusting efficiently to the incentives and signals coming from the product market requires a capacity to move labour and capital towards new opportunities and to ensure that they are used effectively wherever they are employed. The transnational integration of markets has made the conditions and level of employment in individual nation States increasingly dependent on a country's competitiveness. As with product and financial markets, the global labour market today is experiencing a combination of globalization and regionalization, and the coexistence of an outward-orientation with an inward-orientation, and of free movement with barriers (Emmerij, 1993). In domestic labour markets, the emphasis on deregulation and flexible job structures has both increased the instability of jobs and thus contributed to migration pressures and created niches in the informal sector that undocumented foreign workers are ready to fill. Increasing the efficiency of the public sector has involved a reduction of State activities and the privatization of public enterprises. Because the State has been the leading employer in many countries, some of those losing their jobs or those entering the labour market with intermediate educational levels have had to look beyond national borders for jobs that meet their aspirations.

The implications of structural adjustment have to be considered in relation to growing demographic imbalances between the developed and developing countries. The varying pace at which the demographic transition has occurred in different regions and its effects on the population age structure and on labour force participation rates have led to unprecedented differences in the growth rate of the labour force between regions, but particularly between developed and developing countries. Thus, the developing countries will account for almost the entire increase in the world's labour force over the next 25 years. While the average annual growth of the labour force in developed countries has been 0.4 per cent since 1985, the equivalent rate has been 2.1 per cent for all developing countries excluding China. It has been estimated that during the 1990s, for every job created in the developed world, 12 will have to be created in developing countries. In the 1960s, the equivalent ratio was 1 to 4 (Golini, Righi and Bonifazi, 1993). That is, developing countries are confronted with the challenge of providing employment for an unprecedented number of new workers every year. The labour force in Southern Asia, for instance, is expected to increase by 153 million persons during the 1990s.

In contrast, some developed countries expect a reduction in their labour forces. Japan, for instance, is likely to experience negative labour force growth by the turn of the century. The newly industrializing economies of Hong Kong and Singapore are expected to experience a reduction of the local labour force by the year 2010. In many industrialized countries, the labour force participation rate of women is already so high that they can no longer be regarded as constituting a labour reserve that may be called upon to increase the labour supply. Given the rapidly ageing populations of those countries, the challenges of structural adjustment will have to be met as the sizes of cohorts of younger persons shrink and the members of large cohorts of older, less adaptable and mobile workers move into retirement.

1. *Impact of structural adjustment on trade and its implications for international migration*

During the 1980s, the volume of world trade rose faster than that of world output. There was, however, a concomitant weakening of the open multilateral trading system through both specific actions of Governments and more general policies. Thus, although tariffs were reduced, developed countries have resorted increasingly to non-tariff barriers. As Soltwedel (1993, p. 16) notes, the fact that measures forcing trade into bilaterally regulated channels are widespread, chronic and increasingly taken for granted is a worrisome development. Many Governments view managed trade as a necessary strategy, despite their repeated commitments to the multilateral trading system. Even as developing countries are being urged to diversify their production base and to shift the orientation of their economies towards exports, developed countries have been succumbing to protectionist measures. Furthermore, according to Martin (1991, p. 7), an industrialized world subdivided into European, Northern American and East Asian trading blocs may make it harder for developing countries to realize export-led economic growth during the 1990s and may lead to increased international migration within blocs as barriers around other blocs grow. The European Union provides the clearest example of such developments.

Emmerij (1993) underscores the fact that unemployment is particularly high precisely in those regions that are most threatened by involuntary delinking from the world economy. According to him, such a situation will increase migration pressures, prompting both the unemployed and the highly skilled to look for employment opportunities abroad, mostly in the industrialized countries.

Trade appears to have been a substitute for migration among the industrialized countries, particularly those that have regional trading arrangements. Thus, within the European Union, where the free movement of workers has been linked to free trade, less than 2 per cent of the labour force consists of citizens of member States working in other member States (Martin, 1991). However, trade does not appear to have been a substitute for migration between developed and developing countries. There is no certainty that migration and the remittances it generates will necessarily lead to development. Countries that became major sources of migrants during the 1960s, such as Algeria, Mexico, Morocco, Tunisia and Turkey, are still far from being developed and remain countries of emigration. Migration pressures have been building up as developing countries find it difficult to solve their employment problems on their own. They must enhance their ability not only to export but also to attract foreign direct investment and, in some cases, foreign aid. Although prestigious advisory bodies, such as the United States Commission for the Study of International Migration and Cooperative Economic Development (1990), have advocated trade expansion, especially that in labour-intensive goods, as the single most important long-term remedy with respect to reducing migration pressures, as noted above, trade barriers against developing-country products have been rising and migration to the industrialized countries has been a way to overcome trade disadvantages.

The changing composition of trade has also affected migration flows. The increasing trade in services, especially banking, insurance and information and media services, has implied greater international mobility of the providers of such services. Thus, the newly industrializing economies of Asia, in particular Hong Kong and Singapore which have limited physical and human resources, have been building themselves up as financial service centres, thus attracting management consultancy firms, banks, insurance companies, data-processing firms and media organizations that have been recruiting professionals from developed countries (*Business Times*, 1 April 1991).

2. *Recent capital flows and the implications for international migration*

During 1975-1985, foreign direct investment expanded less rapidly than during the post-war era, but since about 1985, it has been growing rapidly again. This change is associated with a number of major developments in the world economy that also have important implications for international migration. First, a growing number of countries are sources of foreign direct investment and there has also been an increase in the number of recipients of such capital flows. The roles of countries as receivers or sources of foreign direct investment have changed considerably since 1970. The United States, for instance, has become a major recipient of foreign direct investment. During 1985-1989, it accounted for about 20 per cent of all outward flows, but received over half of the total OECD foreign investment flows. It has also been the single largest destination of foreign investment from France, Germany, the Netherlands and the United Kingdom (Le Bideau, 1991). Investors have sought to benefit from the strong growth of the large United States market and from United States technology, rather than from a cheap labour supply. To the extent that capital inflows result in employment creation in the receiving country, the United States labour market has benefited therefrom.

During the past decade or so, Japan has been the single largest source of foreign direct investment. In 1989, Japan's US$ 44 billion foreign direct investment accounted for about one third of the total foreign investment from OECD countries (Le Bideau, 1991). Consequently, the pattern of Japanese foreign investment and its implications for international migration are especially important. Compared with the practice in other countries, Japanese foreign direct investment originates in both large and small companies. Furthermore, only 38 per cent of overseas employees of Japanese multinationals are in industrialized countries. The rest are in developing countries, mainly ASEAN member States. Therefore, Japanese foreign direct investment has had a greater impact on job creation in developing countries than in other developed countries. There have been important changes in Japanese foreign direct investment over the years. In the 1960s, such investment was largely concentrated in the Republic of Korea and Taiwan Province of China, but as labour costs in those countries rose over the 1980s, Japanese industries relocated to Indonesia, Malaysia, the Philippines, Singapore and Thailand. More recently, as the labour markets in Malaysia and Singapore tightened, Japanese foreign direct investment has shifted to Southern Asian countries, especially Sri Lanka.

The globalization strategies pursued by Japan have also resulted in an increase of migration inflows. Sassen (n.d., p. 16) argues that, just as in the case of the United States, as Japan becomes an ever more important actor in the world economy and a key investor in Eastern and South-eastern Asia, it creates a transnational space for the circulation of its goods, capital and culture that sets up conditions conducive to the circulation of people and to the formation of an international labour market. Since 1985, the inflows of migrants to Japan from countries such as Bangladesh, Malaysia, Pakistan, the Philippines, the Republic of Korea, Sri Lanka and Thailand have been growing. Japan's economic presence in the countries of origin has been instrumental in exposing potential migrants to Japanese mores, culture and economic success, encouraging them to seek jobs in the new land of opportunity. Demographic trends in Japan itself have also favoured migration. Having the world's highest life expectancy and fastest ageing population, Japan expects to experience a reduction of its labour force early in the twenty-first century. The tightness of the labour market has already prompted Japan to admit foreign workers. Their number grew from the 22,000 admitted in 1976 to almost 95,000 in 1990, among whom over 90 per cent were Asian (Shigemi, 1991). However, most labour migration to Japan has not been authorized. The Ministry of Labour estimates that the number of undocumented foreign workers in the country was 210,000 in November 1991, about twice as many as the year before. Nevertheless, Japan continues to espouse a policy stance consistent with its image as an archetypal closed society and refuses to admit legally the large numbers of foreign workers its economy needs (Zlotnik, 1994).

Foreign direct investment also influences international migration through the dynamics of the internal labour markets of multinational corporations, whose employees work in different countries. Thus, Japanese multinationals post Japanese managers, professionals, technicians and other highly qualified workers abroad to oversee the relocated production or transfer technology. In Thailand, for instance, there are an estimated 30,000 Japanese, and 5,000 more work in Malaysia where there are nearly 250 Japanese manufacturing operations (International Labour Organization, 1992). In addition, Japanese, multinationals move "trainees" from their overseas subsidiaries to work in the parent company. In 1990, some 38,000 foreigners were admitted to Japan as trainees in Japanese companies. They constituted the largest single category of entries after tourists. In 1992, Japan established the Japan International Training Cooperation Organization (JITCO) to arrange for the annual admission of 100,000 workers from developing countries who would receive a year's training in selected Japanese industries, including construction and the automobile industry.

Taiwan Province of China has also emerged as an important foreign investor in the world economy. Taiwanese enterprises have become the largest source of foreign direct investment in the Philippines and are more important than United States companies in Thailand (Abella, 1990). As a result, the number of Taiwanese has been growing in several ASEAN member States. Similarly, Hong Kong and Singapore have increasingly resorted to outward processing in neighbouring countries, while concentrating higher-value-added and more-capital-intensive production in their territories. Since China began admitting foreign direct investment in 1978, Hong Kong manufacturers, traders and hoteliers have been moving their operations across the border to the "special economic zones" in order to take advantage of the abundant supply of cheap labour. During the 1980s, the steady reduction of labour-intensive manufacturing in Hong Kong was accompanied by an increase in Southern China where nearly 2 million workers are employed in enterprises owned by Hong Kong entrepreneurs. The Government of Singapore has also been offering attractive tax incentives and investment protection for Singaporean companies that shift labour-intensive operations abroad, mainly to Southern Peninsular Malaysia and to the Indonesian island of Batam.

3. *Casualization of labour markets and international migration*

The employment situation has not improved markedly during the past 20 years. Developing countries have to cope with overt unemployment, with the fact that many of those who work receive such low wages that they remain in poverty, and with the lack of attractive jobs to satisfy the aspirations or expectations of a better-educated labour force (Emmerij, 1993). The retrenchment of the public sector in many countries has intensified the unemployment problem. In Africa and Latin America, unemployment rose to unprecedented levels with the economic recession of the 1980s, reaching between 20 and 40 per cent.

In Latin America, an increase of unemployment rates into the double digits and a substantial loss of purchasing power took place at a time when the region was more integrated than ever into the world economy and its people were more aware than ever of the gap between their standards of living and those in developed countries. Such developments could not but increase migration pressures given the growing inability of people to meet their consumption expectations domestically (Portes, 1991).

One of the most striking responses to the unemployment problem in developing countries has been the entry of women into the international market for contract labour. As men lost their jobs at home or were no longer in high demand as foreign workers because of the downturn in the construction industry in the oil-rich countries of Western Asia, women opted to work overseas, mainly as domestic workers, nurses, entertainers or in other occupations in the service sector. The participation of women in international labour migration has been seen as a family survival strategy that both maximizes family earnings and minimizes the risks associated with dependence on the local economy (Katz and Stark, 1986; Massey, 1988). The number of women participating in international labour migration has been growing and the countries of destination are no longer restricted to those in Western Asia (United Nations Secretariat, 1995). Thus, in the 1980s, Philippine women accounted for less than one third of all Philippine migrants in Western Asia, but they outnumbered Philippine men by 12 to 1 in the flows directed to certain Asian destinations, such as Hong Kong and Malaysia. In 1983, women accounted for 42 per cent of all Indonesian contract workers leaving the country, but by 1988 they constituted 77 per cent of the total outflow. Women of Sri Lanka have also constituted a majority of migrant workers originating in that country. Breaking with tradition, they have left their families behind to take overseas jobs as domestic workers.

Despite the growing unemployment in many developed countries, foreign workers have been able to find jobs in those countries because of recent changes in their labour markets. Deregulation and increased emphasis on cost-cutting measures have prompted employers to make use of more flexible work arrangements, involving part-time work, subcontracting and the growth of the informal sector. These new methods of operation explain the existence of a demand for foreign labour in certain sectors even when unemployment is rising (Tapinos, 1993).

As Sassen (n.d.) has noted, the casualization of the labour force is an important process facilitating the incorporation of undocumented migrants. Casualization implies less restrictions on employers and tends to lower the direct and indirect costs of labour. In the United States, for instance, the increase in low-wage jobs has resulted in part from the same international economic processes that channelled investment and manufacturing jobs to low-wage countries. As industrial production moved abroad to low-wage areas in the developing world, the traditional United States manufacturing organization based on high wages has eroded and been replaced, at least in some industries, by a downgraded manufacturing sector characterized by semi-skilled or unskilled production jobs and extensive subcontracting. In addition, the rapid growth of the service sector has created large numbers of low-wage jobs. Both processes facilitate the absorption of undocumented migrants (Sassen, n.d., p. 17).

Similarly, in the major Japanese cities, such as Tokyo and Osaka, both high-income and low-income jobs have been increasing, the latter as a result of the casualization of the labour market that has created conditions for the absorption of undocumented migrant workers. The Japanese themselves avoid working in dirty, dangerous and difficult jobs that pay low wages. Smaller, labour-intensive industries, such as construction and services, face labour shortages and, unlike the large Japanese production enterprises, do not have the option of relocating operations overseas. They have become the main employers of foreign workers. Thus, most of the undocumented male migrants apprehended in Japan since 1987 were working in construction and factory jobs, while a growing number of undocumented female migrants were working for small or medium-sized factories. However, those apprehended may not be representative of all undocumented migrants. There are indications that varied forms of subcontracting, including growing numbers of home-based workers, and the increase of part-time or temporary jobs in the service sector have significantly widened the employment opportunities for undocumented migrants.

E. GLOBALIZATION OR REGIONALIZATION? IMPLICATIONS FOR MIGRATION

The creation of regional trading blocs operating within an integrated world economy has implications for international migration since, like production and trade, it is a global phenomenon. Recent decades have witnessed a diversification of migration flows, especially in terms of countries of origin and destination. Yet, the trend towards the globalization of migration flows is not incompatible with their regionalization. Because neighbouring countries or countries within the same region are more likely to have historical, cultural and political ties with one another than with countries further away, most international migration occurs within regions. The regional character of international migration can be further reinforced by the creation of economic groupings, such as the European Union, where the free movement of citizens of member States became a reality on 1 January 1993. On the same date, under agreements concluded between the European Union and the seven countries of the European Free Trade Association (EFTA), which are subject to national ratification, the largest free-trade area in the world was created and the gradual introduction of freedom of movement of citizens of member States is expected.

The trend towards regionalization has also been strengthened by the passage of the North American Free Trade Agreement (NAFTA) between Canada, Mexico and the United States; and by recent initiatives to expand the Association of South-East Asian Nations (ASEAN) to become the more encompassing Asia-Pacific Economic Cooperation (APEC). In Africa, the largest and most influential economic grouping is the Economic Community of West African States (ECOWAS), established in 1975. Other economic groupings in the developing world include the South Asian Association for Regional Cooperation (SAARC), the Central American Common Market, the Latin American Free Trade Association and the Caribbean Community and Common Market. The creation of regional economic blocs has been accompanied by a combination of free trade within and protectionism towards the rest of the world. Their impact on international migration has to be assessed by considering the effects of preferential trade and production arrangements among member States and any policies regarding the movement of citizens of member States. Furthermore, it is important to consider whether regional economic integration has been successful in stimulating the economic development of member States.

Given its importance and the major effort made recently to increase its economic and political integration, the experience of the European Union has attracted considerable attention. It is noteworthy that the Treaty of Rome, creating the European common market, included provisions allowing the free movement of workers within the then European Economic Community. They set in motion a process that is expected to culminate in the complete removal of all controls at the internal frontiers of the European Union. However, the process leading to the free movement of all European Union citizens within the Union's territory has been halting at best and has generally not led to significant increases in intraregional migration. Indeed, Union citizens do not constitute the majority of international migrants within the European Union. It is estimated that around 1990, the 12 member States of the European Union were hosting nearly 13 million persons who were not citizens of the States wherein they resided, and only 5.1 million of whom were citizens of other member States. As Hovy and Zlotnik (1994) point out, over the past 30 years, freedom of movement within the European Union has been extended selectively, so that when it became a political reality the economic factors leading to intraregional migration had mostly disappeared. Tapinos (1993) argues that by removing tariff barriers and quotas on the movement of goods before freedom of movement was allowed, standards of living tended to be equalized within Union territory thus reducing the incentive to migrate. The convergence of demographic trends among member States, in particular the sharp declines in fertility experienced by Southern European countries also reinforced the trend towards lower intraregional migration.

However, further economic integration is likely to increase short-term migration, especially that of skilled personnel. Growing investment in member States, whether from within or outside the European Union, stimulated by trade liberalization and the potential for economic growth, will contribute to accelerating that type of population mobility. Intraregional investment has been increasing, doubling in 1986 when the Single European Act was passed and Portugal and Spain became members of the then European Community (Le Bideau, 1991). However, increases in trade and investment within the European Union have reduced those flowing to developing countries. Emmerij (1993) notes that the move towards further economic integration set in motion by the Single European Act has been one of the crucial factors underlying the involuntary delinking of between 15 and 30 per cent of the world's population from the world economy. For some of the citizens of low-income countries that are adversely affected by the protectionism of the European Union, migration to the developed world is a means of overcoming their disadvantages.

Within the European Union, the spatial mobility of unskilled and semi-skilled workers has not increased because the demand for them is low in most member States and demographic factors have been reducing their supply within the Union. There is, however, an ample supply of labour outside the Union. The Southern European countries, in particular, have been attracting undocumented migrant workers. In addition, the collapse of communism has been accompanied by growing migration pressures in the former Eastern-bloc countries. Already, citizens from those countries account for an important proportion of persons seeking asylum in Western-bloc countries. These developments have led to increasing restrictions on the admission of migrants by Western-bloc countries. However, as Emmerij (1993) notes, there is no reason to believe that the importance of market forces in shaping migration flows will diminish. It is likely that the conjunction of restrictive admission policies, the operation of the social networks created by past migration, the easy access to European markets because of low transportation costs, and the persistent demand for foreign labour in certain sectors will fuel undocumented migration.

In Africa, the Economic Community of West African States (ECOWAS) has probably been the most successful grouping in terms of achieving economic integration, and still its achievements are at most modest. Since its establishment as a customs union, ECOWAS has adopted as an essential principle the free movement of labour, goods, services and capital. Based on that principle, the member States of ECOWAS signed a Protocol on the Free Movement of Persons, Right of Residence and Establishment that establishes "community citizenship" for citizens of member States and spells out the details for the elimination of border-crossing requirements (Makinwa-Adebusoye, 1992). The Protocol was to be implemented in three stages: the first phase granted the right of entry to citizens of member States; the second granted the right of residence to community citizens; and the third, which began to be implemented in 1991, granted community citizens the right to establish business ventures in member States other than their own. There is, however, little evidence about the effect that the Protocol has had on international migration within the ECOWAS region.

Within Asia, various political, security and trade arrangements have been made to strengthen and formalize the subregional groupings of ASEAN, APEC and SAARC, and as countries align themselves, there has been increasing integration and interdependence within the blocs and distinction from, if not competition with, other blocs. For instance, APEC can be considered the response of countries in the Asian Pacific Rim to the forging of the European Union. To the extent that the trend towards regionalization has strengthened the integration of the Asian Pacific Rim countries, the weaker SAARC countries have found it harder to achieve export-led economic growth or to break into the regional labour market, and are thus experiencing stronger migration pressures. To date, however, the Asian subregional groupings, unlike the European Union, have not favoured the free movement of labour nor do they offer preferential treatment to workers of other member States. This reluctance is partly responsible for the increase in clandestine migration within Asia. However, it is likely that future subregional integration measures will extend from the liberalization of trade and capital flows to the movement of labour, or at least lead to policies that make explicit allowance for labour movements. Already, some bilateral agreements have been concluded. Thus, that between Indonesia and Malaysia covers the proper documentation of Indonesian workers who have entered Malaysia illegally; and that between Hong Kong and the Philippines establishes a model contract for domestic workers.

It is worth noting that the most successful regional economic blocs are those among countries that have been doing relatively well. Yet, it is precisely low-income countries that should be pursuing regional economic integration as a means of promoting their own economic development (Emmerij, 1993). However, to date, the attempts by low-income countries to create or join such blocs have largely been unsuccessful.

The various trends towards regionalization have to be seen within the broader context of global interdependence, if for no other reason that some people in the poorer regions are likely to migrate, even through unauthorized channels, to regions where their lot can be improved. There are also other trends towards globalization countervailing or accompanying the regionalization processes. First, technological innovations are now spreading so rapidly that countries making technological breakthroughs are no longer able to maintain their competitive edge for long before other countries gain access to the same technology. Perhaps more importantly, technological developments, such as those in the use of microelectronics, numerically controlled machines, robotics and other information processing technologies, have tended to reduce the significance of labour-cost differences between developed and developing countries, thus explaining, at least in part, the return of certain industries to industrialized countries where they have been able to regain their competitiveness through technological advances. Technology has also made more services tradable and trade in services, particularly in banking, insurance, consultancies and computerization, has been expanding worldwide. The move to liberalize trade in services in the Uruguay Round of multilateral trade negotiations of the GATT has important implications for migration. Since the service to be provided is generally engendered by the person providing it, relaxation of the rules of entry for the person providing the service will have to accompany the liberalization of trade in services among countries. Another factor affecting migration is the fact that the operations of multinational corporations and business companies, especially those dealing in trade and the provision of services, have transcended not only national but regional boundaries to create worldwide networks. Multinational corporations have been described as constituting a "stateless" and "extra-national" phenomenon managed by businessmen who are "citizens of the world" (Knorr, 1975). In addition, there has been a globalization of information and the impact of the media in linking even the most distant parts of the globe. The worldwide tourist industry is also a factor linking countries and constituting an information channel about conditions in other countries. Many undocumented migrants enter host countries as tourists.

F. Economic interdependence, national independence and the international political economy

Since relations among nation States are not shaped by economic forces alone, changing non-market interactions significantly affect migration. Growing economic interdependence should be analysed in relation to the dynamics and implications of the international political economy. Sovereign States play an intrusive role by attempting to influence the effects of internationalization processes to their own advantage but without adversely affecting their political relations with other countries. In this context, the regulations of migrant-receiving countries are crucial; but in recent times a growing number of Governments, aided by a whole host of profit-oriented intermediaries, have played active and innovative roles in expanding global labour markets and thus promoting migration.

For instance, the Korean Overseas Development Office was set up in 1965 to recruit workers for overseas markets. Targets for the export of manpower were assigned to embassies of the Republic of Korea in potential receiving countries and the promotional activities of embassy officials were officially rewarded. The Government of the Republic of Korea also backed the efforts of enterprises of the Republic of Korea to set up collective, project-tied contract migration. Similarly, the Indonesian Office of Overseas Employment has been charged with encouraging, controlling and coordinating the recruitment and deployment of workers abroad so as to meet the targets for overseas employment and remittances specifically built into the Indonesian Five-Year Development Plans. The Philippines Overseas Employment Administration, which has been operating for over 18 years, carries out pre-departure orientation seminars catering to entertainers, domestic workers, nurses and mail-order brides, and those intending to resettle permanently in host countries.

Some developing countries are highly dependent on overseas employment to occupy their burgeoning labour force and on remittances to boost their foreign exchange earnings. Remittances from migrant workers constitute the major export-income earner for Bangladesh, Pakistan and the Philippines. For Bangladesh, remittances represented some 60 per cent of merchandise exports in 1988. Even for Thailand, remittances from overseas workers constitute one of the top three export revenue earners and relatively closed economies such as China, Mongolia and Viet Nam have also relied increasingly on labour exports (Lim, 1991). The dependence of Asian countries on international labour migration was starkly demonstrated by the events surrounding the Gulf war: the forcible return of migrants from the Persian Gulf sent shock waves through their economies. Recognition of their common vulnerability has prompted them to discuss measures to avoid detrimental competition among themselves and to cooperate in enforcing common minimum standards, combating illegal intermediaries, improving systems of recruitment and deployment, preparing and selecting workers to be sent abroad, and instituting social protection measures for the migrant workers (International Labout Organization, 1991).

Receiving countries, on their part, are not eager to increase their dependence on foreign workers. Unlike other global flows, the spatial mobility of people has not only economic but also humanitarian, political and cultural implications which often bring domestic and international interests into conflict. Therefore, despite growing interdependence, receiving countries still zealously exercise the right to control their labour market by deciding who may enter their territory and under what conditions. However, while attempting to regulate and restrict the inflow of unskilled workers, Governments generally wish to maintain relatively open borders and easy access to the flows considered to be in the national interest, including those of persons entering for business, investment, tourism, technical assistance or education.

As global or regional economic integration proceeds, nation States experience growing limitations on their ability to control markets as they see fit. Economic interdependence means that economic activity and employment in one country can grow at the expense of employment in other countries and that economic considerations must be tied to foreign policy interest. For instance, the stance of Taiwan Province of China on illegal migration has been very much dictated by its concern about maintaining good relations with the resource-rich ASEAN countries which represent an important source of trade and also are the major destination of Taiwanese foreign direct investment. Migration clearly cannot be treated as a process autonomous from other international processes. With national labour markets intricately linked to the global and regional markets for goods and capital, countries are aware that their actions in controlling migration can affect other markets that they may wish to expand.

In conclusion, the greater the integration of nation States into the international political economy, the more they have to relinquish their sovereignty and embrace a collective decision-making process. Adjustment problems being global rather than national, national actions are by themselves inadequate. Coordinated international action is required to confront the challenges and responsibilities that structural imbalances and underdevelopment present to nation States. At the global or transnational level, the first priority should be to halt the involuntary marginalization of poor countries from the world economy and to integrate them into the main thrust of economic development. Admittedly, the search is still on for "migration-reducing development strategies", and it is very likely that development, supported by appropriate trade, investment and aid policies, will be destabilizing for some decades to come; but in the long run, the achievement of balanced economic development that creates productive employment opportunities in the countries where potential migrants live is the main, if elusive, goal.

References

Abella, M. (1990). Structural change and labour migration within the Asian region. Paper presented at the Expert Group Meeting on Cross-national Labour Migration in the Asian Region: Implications for Local and Regional Development, Nagoya, Japan, November 1990.

Amjad, R., ed. (1989). *To the Gulf and Back: Studies on the Economic Impact of Asian Labour Migration*. New Delhi: International Labour Organization, Asian Regional Team for Employment Promotion (ARTEP).

Bailey, P. J., and A. Parisotto (1993). Multinational enterprises: what role can they play in employment generation in developing countries? In *The Changing Course of International Migration*. Paris: Organisation for Economic Cooperation and Development.

Böhning, W. R. (1991). *Integration and Immigration Pressures in Western Europe*. Geneva: International Labour Organization.

Charlton, S. E. (1984). *Women in Third World Development*. Boulder, Colorado: Westview Press.

Commission for the Study of International Migration and Cooperative Economic Development (1990). *Unauthorized Migration: An Economic Development Response*. Washington, D. C.

Donges, J. B. (1983). Labour, capital and technology in an interacting world. Paper prepared for the Conference on Population Interactions between Poor and Rich Countries organized by Harvard University and the Draeger Foundation, Cambridge, Massachusetts, 6-8 October 1992.

Emmerij, L. J. (1993). The international situation, economic development and employment. In *The Changing Course of International Migration*. Paris: Organisation for Economic Cooperation and Development.

Frobel, F., J. Heinrichs and O. Kreye (1980). *The New International Division of Labour*. Cambridge, United Kingdom: Cambridge University Press.

Golini, A., A. Righi and C. Bonifazi (1993). Population vitality and decline: the North-South contrast. In *The Changing Course of International Migration*. Paris: Organisation for Economic Cooperation and Development.

Hiemenz, U., and K. W. Schatz (1979). *Trade in Place of Migration: An Employment-Oriented Study with Special Reference to the Federal Republic of Germany, Spain and Turkey*. Geneva: International Labour Organization.

Hovy, B., and H. Zlotnik (1994). Europe without internal frontiers and international migration. *Population Bulletin of the United Nations* (New York), No. 36. Sales No. E.94.XIII.12.

International Labour Organization (1989). *Recovery and Employment: Report of the Director-General, International Labour Conference, 76th Session, 1989*. Geneva.

_________ (1991). ILO Meeting on the Implications of Changing Patterns of Asian Labour Migration, Kuala Lumpur, 14-15 May 1991. Summary of conclusions. Bangkok: International Labout Migration, Asian Regional Programme on International Labour Migration.

_________ (1992). *World Labour Report, 1992, No. 5.* Geneva: ILO.

Katz, E., and O. Stark (1986). Labour migration and risk aversion in less developed countries. *Journal of Labour Economics* (Chicago), vol. 4, pp. 131-149.

Knorr, K. (1975). *The Power of Nations: The Political Economy of International Relations*. New York: Basic Books.

Le Bideau, J. L. (1991). The economic climate for foreign direct investment. Paper presented at the Organisation for Economic Cooperation and Development (OECD) International Conference on Migration, Rome, 13-15 March.

Lim, L. L. (1988). *Economic Dynamism and Structural Transformation in the Asian Pacific Rim Countries: Contributions of the Second Sex.* Nihon University Population Research Institute (NUPRI) Research Paper Series, No. 45. Tokyo: Nihon University Population Research Institute.

_________ (1989). Processes shaping international migration flows. In *International Population Conference,* vol. 2. New Delhi and Liège, Belgium: International Union for the Scientific Study of Population (IUSSP).

_________ (1991). International labour migration in Asia: patterns, implications and policies. Paper presented at the Economic Commission for Europe and United Nations Population Fund Informal Expert Group Meeting on International Migration. Geneva, 16-19 July.

Makinwa-Adebusoye, P. (1992). The West African migration system. In *International Migration Systems: A Global Approach,* M. M. Kritz, L. L. Lim and H. Zlotnik, eds. Oxford: Clarendon Press, pp. 63-79.

Martin, P. L. (1991). Trade, aid and migration: a conference report. Mimeographed.

Massey, Douglas (1988). Economic development and international migration in comparative perspective. *Population and Development Review* (New York), vol. 14, No. 3, pp. 383-413.

_________ (1991). Economic development and international migration in comparative perspective. In *Determinants of Emigration from Mexico, Central America and the Caribbean,* Sergio Díaz-Briquets and Sidney Weintraub, eds. Boulder, Colorado: Westview Press.

Mouhoud, E. M. (1993). Enterprise relocation, North-South economic relations, and the dynamics of employment. In *The Changing Course of International Migration.* Paris: Organisation for Economic Cooperation and Development.

Pastor, R. A. (1985). Introduction: the policy challenge. In *Migration and Development in the Caribbean: The Unexplored Connection,* R. A. Pastor, ed. Boulder, Colorado: Westview Press.

Portes, A. (1982). International labour migration and national development. In *U.S. Immigration and Refugee Policy: Global and Domestic Issues,* M. Kritz, ed. Lexington, Massachusetts: Lexington Books.

_________ (1991). Unauthorized immigration and immigration reform: present trends and prospects. In *Determinants of Emigration from Mexico, Central America and the Caribbean,* Sergio Díaz-Briquets and Sidney Weintraub, eds. Boulder, Colorado: Westview Press.

Sassen, S. (1988). *The Mobility of Labour and Capital: A Study in International Investment and Labour Flows.* Cambridge, United Kingdom: Cambridge University Press.

_________ (n.d.). Immigration in Japan and the US: the weight of economic internationalization. Mimeographed.

Shigemi, K. (1991). Recent changes in Japanese Immigration Control Act and recent developments in the immigration of foreign workers to Japan. Paper presented at the International Labour Organization Meeting on the Implications of Changing Patterns of Asian Labour Migration, Kuala Lumpur, May 1991.

Soltwedel, R. (1993). Structural adjustment, economic growth and employment. In *The Changing Course of International Migration.* Paris: Organisation for Economic Cooperation and Development.

Standing, G. (1991). Structural adjustment and labour market policies: towards social adjustment? In *Towards Social Adjustment Labour Market Issues in Structural Adjustment,* G. Standing and V. Tokman, eds. Geneva: International Labour Organization.

Tapinos, Georges (1993). Can international co-operation be an alternative to the emigraiton of workers? In *The Changing Course of International Migration.* Paris: OECD.

Turnham, D., and D. Erocal (1991). The supply of labour, employment structures and unemployment in developing countries. Paper prepared for the OECD International Conference on Migration, Rome, 13-15 March.

United Nations Secretariat (1995). Measuring the extent of female international migration. In *International Migration Policies and the Status of Female Migrants.* Sales No. E.95.XIII.10. New York: United Nations.

Weintraub, S., and C. Stolp (1986). The implications of growing economic interdependence for future trends in international migration. Paper presented at the Conference of National Experts on the Future of Migration, Paris, 13-15 May 1986.

Zlotnik, Hania (1994). Comparing migration to Japan, the European Union and the United States. *Population and Environment,* vol. 15, No. 3 (January), pp. 173-188.

XVI. SAFEGUARDING THE INSTITUTION OF ASYLUM

*Astri Suhrke**

Asylum, as it exists today in law and practice, originated in ancient customs to protect individuals and groups from particular forms of hardship. The specific forms taken by asylum typically reflected the conflict between the needs of the claimant and the interests of the receiving State. In moral-philosophical terms, this conflict was formulated in terms of the rights of individuals versus thc rights of communities. In political terms, the conflict and its outcome reflected the power structures within the historical context in which refugee movements unfolded. Current issues of asylum must likewise be understood in relation to fundamental issues of law and moral philosophy as well as prevailing structures of international relations. The present paper reviews the institutionalization of asylum after the Second World War and identifies the main mechanisms for international protection available today. These mechanisms are assessed with respect to their adequacy in providing asylum and the extent to which they represent a reasonable framework for mediating the enduring conflicts between the needs of individuals and the interests of States.

For some time now, the institution of asylum has been under pressure in many of the developed countries of the West. In developing countries, the main problem is to secure sustenance for large numbers of refugees who, with various degrees of formality, have obtained immediate asylum. This difference has been reinforced by the fundamental change in international politics stemming from the end of the cold war. Contemporary asylum and refugee conditions thus bear the imprint of the emergence of a new historical period: that of the post-cold war world.

*Department of Social Sciences and Development, Christen Michelsen Institute, Bergen, Norway.

A. Legal framework and its implications for refugee flows after the Second World War

1. *Right to asylum*

The essence of asylum is to enjoy protection outside the reach of the persecuting agent, customarily taken to be the State of origin. This has traditionally meant either territorial asylum, that is, protection on the territory of another State, or diplomatic asylum in the Latin American tradition, that is, within the diplomatic mission of another State which, according to the principle of extraterritoriality, constitutes a piece of sovereign territory of that State (Grahl-Madsen, 1966; Goodwin-Gill, 1983). If the right to seek asylum is abridged by, for instance, preventing access to places where applications for asylum can be lodged, the potential refugee cannot enter and thus cannot escape. The ability to seek asylum therefore appears as the most basic right, preceding and remaining distinct from the outcome of a process that determines eligibility according to a given definition of "refugee" (Plender, 1989).[1]

The 1951 United Nations Convention relating to the Status of Refugees (United Nations, 1957)[2]—which remains the principal piece of international law on refugee matters—does not establish an individual's right to seek and enjoy asylum. That wording was included instead in the Universal Declaration of Human Rights (General Assembly resolution 217 A (III)), article 14 (1) which, as a declaration, has less weight than a legally binding convention. In positive international law, therefore, the right to asylum appears as a State's right to give and not an individual's right to seek.

Recognizing this lacuna, international lawyers and organizations such as the Carnegie Endowment for International Peace drafted conventions that specifically dealt with asylum as it affected the rights and duties of both States and individuals. In 1977, a United Nations Conference on Territorial Asylum assembled in Geneva but it did not adopt a final document. Since then, the matter has been "allowed to rest" (Encyclopedia of Public International Law (EPIL), 1985, p. 45; Grahl-Madsen, 1980). Several nations have statutory provisions on the right to asylum in municipal laws; some also have constitutional provisions—most notably article 16 of the German Constitution which, before being amended in 1993, guaranteed asylum for the politically persecuted—but international law remains inconclusive.

It remains inconclusive rather than entirely closed because the 1951 Convention has a provision relating to asylum in its article 33. That provision prohibits refoulement, that is, the return of refugees to an area where their lives or freedom would be endangered. Since an individual who is denied access to seeking asylum might in fact risk persecution, the failure to consider his application may have the same result as refoulement. Article 33 is, however, only an "embryonic provision" regarding an individual's right to seek asylum (EPIL, 1985, p. 48). This situation gives rise to important legal battles, which is precisely what is occurring as asylum becomes increasingly restricted in Europe and Northern America. Asylum advocates contend that, in order to prevent a possible refoulement, a State is obliged to permit individuals to apply for asylum so that they have the possibility of making their case in the first place. Advocates of reluctant States argue that provisions relating to the form of exclusion cannot constitute a basis for admission, in so far as a person must be admitted in order to be excluded (see, for instance, Haitian Refugee Center v. James Baker III, Secretary of State).

2. *Definition of "refugee"*

The other legal battleground over protection concerns the definition of "refugee". Two types of asylum mechanism had been delineated by European law and practice by the end of the Second World War. One entailed protection of individual exilees, mostly embodied in positive national law through prohibitions on extradition. The other entailed protection of masses of people fleeing from war, as first developed in international practice in connection with the League of Nations' aid to 800,000 Russian refugees after the 1917 revolution. As regards these two possible approaches, the 1951 Convention centred on individual exile, although it defined persecution in terms of general sociological categories rather than with respect to particular nationalities, as had been the case with the League.

The Convention categories were a reflection of the classic refugee movements in European history as well as those resulting from the evolving cold war. The political activist was captured by the term "persecution owing to political opinion". Members of targeted minority groups were characterized, based on ascriptive criteria, as belonging to groups defined in terms of race, religion, nationality or social group. When the original time and space limitations on the "refugee" definition established by the Convention were removed in 1967, only one type of refugee was not formally covered, namely, the victim of generalized violence such as that stemming from war or civil conflict.

While in some respects liberal compared with the League's instruments, the United Nations Convention had clearly been designed to circumscribe the obligations of States and limit the number of persons who, under a universal definition of "refugee", could press their claims. The definitional emphasis on persecution served this purpose by postulating a targeted and discriminatory treatment of individuals, that is, by singling out the presumably exceptional case. Moreover, the Convention was customarily taken as requiring individual screening of eligibility and thus preventing mass inflows.

During the 1980s, the definitional question centred on the distinction between economic and political refugees. Facing mass inflows from developing countries, developed countries claimed that the persons involved were mere "economic refugees" and hence excludable. The label—like the dichotomy it reflected—was controversial. More correctly, the distinction between economic and political reasons for migration should be seen as covering a wide spectrum, with pure types existing only at the extremes and corresponding to the so-called migrant workers and political dissidents. In situations likely to produce mass outflows, economic deprivation typically exists alongside political oppression and civil conflict. Such coexistence is not a new phenomenon,[3] yet it is likely to be increasingly common in the contemporary world. In regions where demographic and environmental pressures make life more difficult, political upheavals may cause major outflows. In such situations, the claim that there is one basic cause of flight, be it economic, political or environmental, is untenable.

Some minimal definitional requirements can nevertheless be identified (Shacknove, 1985; Zolberg, Suhrke and Aguayo, 1989). Receiving states must apply consistent standards and consider the need for protection as well as the cause of flight. The definitional framework, in turn, must permit a distinction between individual plight (such as persecution) and mass outflows caused by a combination of economic and political deprivation accompanied by widespread violence or, in the extreme, war. That was indeed the view of many critics who in the 1980s argued that the 1951 Convention was too narrow to protect masses of people who had been forced to leave their countries and were living in the developing world. In the 1990s, the Convention has also been considered inadequate to deal with the refugee crisis confronting Europe in the Balkans.

3. *State rights versus human rights*

A more fundamental challenge to the Convention emerged as the cold war waned. Like refugee law generally, the Convention is basically State-centric in that it was framed in terms of the rights and duties of Statesrather than of individuals (Aleinikoff, 1991).[4] The political controversies that had shaped the drafting of the Convention and the establishment of the post-war refugee regime were State-centric as well. Questions of definition, asylum and repatriation appeared as points of conflict between socialist and democratic States and constituted yet another dividing line between the Eastern- and Western-blocs as the cold war deepened (Holborn, 1975). The former Union of Soviet Socialist Republics (USSR) fought consistently the development of a refugee regime that it feared would be used against it. Western-bloc countries supported the emerging regime partly in that spirit. Only with the demise of the cold war did the underlying conflict between individual and State rights rise to the surface in matters of asylum and refugee status. A new debate emerged, anchored to human rights law and political concepts of "right of membership" on one side and to conventional rights of States on the other (Adelman, 1988; Gibney, 1989; Warner and Hathaway, 1992).

The arguments followed the classic philosophical divide between communitarians and universalists. According to the communitarians, communities have rights because individuals do not have a social existence nor, perhaps, moral character apart from the community (Walzer, 1985; Sandel, 1984; Miller, 1988). Nation-state communities have the capacity for rational and democratic self-rule and possess a system of distributive justice for its members. Consequently, they have the right to protect members vis-à-vis outsiders. In legal terms, the argument parallels the notion of absorptive capacity as a limitation on asylum, an old concept in international law that can be traced back to Emerich de Vattel's work in the mid-1700s.

The universalist argument is, in brief, that rights inhere in individuals and not in States. States are seen as "agents", merely exercising rights on behalf of the individuals in their care, much like a lifeguard on the beach is commissioned, on behalf of all bathers, to save the drowning (Goodin, 1988). Since the right of the State is derivative from the right of the individual, the plight of the needy must take precedence over the (derived) right of the State. Given an international system that is explicitly ordered on the premise of mutually excluding sovereignties, all individuals must "belong", in the sense of being "assigned", to a State. Those who fall or are pushed out—the de facto or *de jure* stateless—have therefore the right to be assigned to other States (van Gusteren, 1988).

The solution to this classic conflict is collective (sometimes called assigned) responsibility (Shue, 1988; Goodin, 1988). Its equivalent for refugees is an international refugee regime with common norms for admission, support and burden-sharing. A particular right (in this case, the right of the individual to asylum) need not be matched by a particular duty (the duty of the neighbouring State, for instance, to provide asylum). The important consideration is that the totals add up in the sense that all asylum-seekers have reasonable access to protection somewhere and that all States bear some responsibility in relation to their resources and absorptive capacity. As Hathaway (1991) argues, a solution requires that the rights of both sides be recognized. For many of the developed countries, this was the essence of the asylum challenge as it emerged in the late 1980s. Faced with new and large populations on the move, the Western-bloc countries, whose coalition was soon to be recognized as the winner of the cold war, had the opportunity to forge new mechanisms of international cooperation so as to adjust the balance of rights in matters of asylum.

4. *Post-war flows*

For several decades, the 1951 Convention had served Europe's refugees well. It was the legal backbone of a liberal asylum regime that was buttressed by financial resources and resettlement quotas offered by the United States of America, Canada and a recovering Western Europe. In retrospect, it is evident that the refugee regime rested on the ideological foundation of the cold war. Giving asylum to people fleeing the Eastern bloc was a moral imperative and a political instrument in the struggle against communism. That the refugees were Europeans and could be readily absorbed by Western European countries experiencing the post-war economic boom was also important.

Strict exit controls in the Eastern-bloc countries conferred an "exceptional" character on the refugees who managed to leave. Yet, the numbers absorbed during the height of the cold war were truly massive, a point worth stressing in view of current restrictions. About 200,000 Hungarians fled when the 1956 uprising was crushed; 80,000 citizens of Czechoslovakia escaped after the suppressive acts of 1968; 30,000 Jews were effectively expelled from Gomulka's Poland, and 3.5 million Germans moved to the former Federal Republic of Germany (FRG) before the Berlin Wall closed up the loophole in 1961. In addition, a steady stream of assorted nationalities, estimated at 10,000 annually during the early 1960s, found its way into Western-bloc countries (Marrus, 1985). They were all welcomed. Ethnic Germans were accepted as citizens under the law of the Federal Republic of Germany. No attempt was made to determine individual eligibility during mass inflows following political upheavals. Growing doubts about whether all citizens from Eastern-bloc countries were political refugees in the Convention's sense, or primarily economic refugees, merely produced administrative adjustments to accommodate them under other categories.[5] Former European colonial powers also absorbed expatriates and sometimes their supporters who were under pressure to leave owing to wars of independence in the ex-colonies.

Outside Europe, the picture was mixed. A large and intractable refugee problem existed in Western Asia, although the main issue for the Palestinians was to recover their State rather than to obtain asylum. In Asia, some flows were accepted for political and humanitarian reasons independently of legal conventions;[6] others consisted of minorities with a homeland to which they could return, like those, for instance, moving after the 1947 partition of India. Here, the refugee issue concerned not the granting of asylum but rather the provision of material assistance.

However, the largely forgotten fate of refugees from the Chinese revolution does point to a classic case of restricted asylum. Already in 1950, they were packed so tightly into the hill-sides of Hong Kong that British authorities declared most new arrivals to be illegal migrants, subject to immediate deportation. The simple but effective policy of definitional exclusion was intermittently applied in subsequent years (Bonavia, 1983). Other States have used the same tactic, perhaps most notably the United States in dealing with Salvadorans and Haitians.

The case of the Chinese asylum-seekers reveals the enduring frailty of the asylum system in its current form. Having fled a combination of political repression and poverty, exacerbated by famine, during the Great Leap Forward in the late 1950s, the Chinese had but a weak claim to external patronage on political grounds (the United Kingdom of Great Britain and Northern Ireland had recognized and sought to improve relations with the Government of China), they had no other ethnic homeland (at least not large enough to accommodate them), they were poor, and they were very numerous. The case for granting them asylum had therefore to rest primarily on humanitarian grounds. In the event, those grounds proved insufficient.

B. Expansion of asylum as a result of crisis

In Africa and Latin America, regional mechanisms combined with innovative practices by the Office of the United Nations High Commissioner for Refugees (UNHCR) worked reasonably adequately for several years with respect to meeting the asylum challenge. When a crisis arose, the result was an expansion of the formal institutions of protection. Observers have tended to mythologize this response, citing "traditional African hospitality" and the "traditional inter-American system of asylum" as explanatory factors for the process of institutional expansion. Other conditions, however, seem more relevant.

1. *Africa*

Africa's first modern refugee crisis arose in the 1960s when wars of national liberation displaced thousands of people. The result was a strengthening of the formal support system for refugees. UNHCR steadily expanded its mandate to protect and aid displaced persons under a "good offices doctrine", which made it unnecessary to determine whether or not its beneficiaries were refugees in the Convention-related sense. The adoption of a legal framework to assist mass outflows was promoted by the many newly independent States Members of the United Nations. In addition, the Organization of African Unity (OAU) adopted a liberal convention governing asylum and the status of refugees in the region.

The 1969 OAU Convention Governing the Specific Aspects of Refugee Problems in Africa (United Nations, 1976) reflected the political realities of the continent at the time, just as the 1951 United Nations Convention had reflected post-war European concerns. The OAU Convention was formulated after a decade of increasingly violent struggles for liberation in Angola, Mozambique, Rhodesia and Southern Africa. For Africa's independent States, those wars had been just and, consequently, the need to extend maximum support to the victimized populations was not questioned (Arboleda, 1991; Oloka-Onyango, 1991). Thus, article 2 of the OAU Convention stresses the right of asylum and article 5 prohibits rejection at the border. Neither article has a counterpart in the United Nations Convention. Not only is the right to seek asylum more firmly grounded, but the grounds for granting asylum are expanded, including not only those set by the 1951 Convention but, in addition, flight from generalized violence due to external aggression, occupation or foreign domination.

The broad OAU definition of "refugee" was eminently suitable for a war situation that, having produced mass outflows of people, necessitated the collective determination of eligibility. Moreover, this was the only realistic determination procedure at that time, given the "absence of decision-making infrastructure" in receiving States (Arboleda, 1991, p. 195). Two decades later, the African asylum regime is still to be characterized by its liberal, formal-legal dimension. The law, as embodied in the OAU Convention, is probably as generous as it can be expected to be under a State-centric refugee system, drawing continuous strength from its original legitimizing ideology based on anti-colonialism and anti-racism. Furthermore, easy asylum conditions have often prevailed in practice. Cross-border ethnic ties facilitated hospitality and the much-discussed weakness of African States made it difficult to impose rigorous border controls. Under those conditions, de facto asylum could be claimed by default.

Since the early 1960s, Africa has probably hosted the largest share of the world's refugees on a continuous basis. Although there are other highly developed migration systems in the continent, they are fairly distinct from those involving refugee flows (Russell, Jacobsen and Stanley, 1991). Consequently, claims for assistance typically arise when there are mass displacements of poor people who are forced to flee from evident disasters. Since their need for relief is unquestionable, there is little room for suggestions that the asylum system is being abused by individuals making false claims (African Group in Geneva, 1990). Arguably, the patent need of the people involved also constitutes the moral pillar of the system: an elementary sense of justice—as traced by the sociologist Moore (1978) through the ages and across continents—upholds the strong rhetorical commitment to asylum in contemporary Africa and, to the extent that the sense of justice is observed, underpins that commitment's translation into practice.

Because African States are typically both the origin of refugees and their destination, symmetry in vulnerability has strengthened the institution of asylum. If mass displacements are likely to occur in all populations and States at one time or another, a strong institution of asylum is a common good. However, practice has changed over time. As victims of anti-colonial struggles have been followed by victims of abuse, civil war or ethnic violence in independent African States, both rejection at the border and mass expulsions have occurred (Zolberg, Suhrke and Aguayo, 1989; Oloka-Onyango, 1991). Once inside a country other than their own, refugees have been subjected to abuse that renders asylum meaningless[7] and has made formal asylum little better than rejection at the border. Furthermore, in several cases, asylum has been compromised by the actions of neighbouring States which have attacked the refugees theselves, the host country or both.

Critics of the African asylum system argue that the traditional and much-touted generosity of African States towards refugees cannot by itself guarantee their security (Oloka-Onyango, 1991, p. 460). Consequently, recent efforts have been made to improve the legal framework of asylum by strengthening the non-refoulement provisions of the OAU Convention (for instance, through the 1981 Charter on Human and Peoples' Rights (Banjul Charter) (adopted 27 June 1981; see OAU document CAB/LEG/67/3, rev. 5, 21 I.L.M. 58 (1982)). (Forty-two countries had acceded to the Charter as at early 1994.) It is increasingly recognized that the weakness of the African asylum system stems mostly from the very heavy refugee burden imposed on poor countries. During the 1980s, the number of refugees in Africa fluctuated between 3 million and 4 million, rising to 5 million at the end of the decade (United States Department of State, 1990) and the numbers are not expected to decline significantly as long as there are no major changes in Africa's economic, demographic or political context. Therefore, attention is focused on ways to secure external assistance, especially by establishing a special fund as was suggested by the Second International Conference on Assistance to Refugees in Africa (ICARA II), held in Geneva in 1984. In addition, the importance of formulating appropriate aid strategies to meet the needs of the refugee population in Africa is being increasingly recognized.

2. *Latin America*

In Latin America, the traditional inter-American system of asylum was indeed highly developed in law and practice. Yet, the conditions for obtaining asylum were narrowly circumscribed by regional instruments requiring evidence of political persecution, instead of being based on the presumably more generous standard of "well-founded fear" enshrined (article 1.A.2) in the United Nations Convention relating to the Status of Refugees (Yundt, 1989; Arboleda, 1991). Asylum was defined in terms of the right of States to grant protection without incurring hostilities from other States and not in terms of the individual's right to protection, and there were no prohibitions regarding rejection at the border. Fearing additional obligations, several Latin American countries hesitated to join formally the United Nations refugee regime. By the end of the 1980s, six Latin American countries had not ratified the 1951 United Nations Convention, and four had ratified the Convention retaining its original geographical limitations.

The traditional Latin American asylum system had worked reasonably well because the refugee caseload was small, the *asilados* (those receving asylum) were mostly financially secure members of the elite, and Governments recognized that people in power were all politically vulnerable and might one day need asylum themselves. Yet, relatively large flows were also accommodated. Mexico, often cited as a traditionally liberal country of asylum, admitted about 40,000 refugees from the Spanish civil war and, together with other Latin American countries, was a generous host to thousands of victims from repressive regimes in South America during the 1970s. Yet, a decisive change of policy took place when the exodus of asylum-seekers from Central America began towards the end of that decade. The magnitude of the problem increased as revolutions and civil war engulfed several Central America countries during the 1980s (Gros Espiel, 1981; Organization of American States, Inter-American Commission on Human Rights, 1983; Aguayo, 1985). Mexico alone became host to an estimated 120,000 Salvadorans, about 50,000 Guatemalans and some Nicaraguans. Only the Guatemalans were recognized as refugees.

In contrast with previous groups of refugees, those originating in Central America were mostly peasants and ordinary workers, and the manner of their treatment soon became as deeply politicized as the conflicts that had forced them to flee. The United States, which for political reasons had admitted over half a million Cubans fleeing the Castro regime between 1959 and 1979, refused to accept more than a select few of the people fleeing Central America's civil wars. Consequently, the immediate burden fell on local institutions of asylum which were unable to cope with the large numbers of asylum-seekers involved. As a result, repeated deportations were common and insecurity reigned even for those who managed to leave their country.

To prevent the full collapse of the local asylum system, UNHCR worked with liberal social forces to adjust legal norms in the region. Those initiatives, which began with a 1981 colloquium held in Mexico City with the support of the Inter-American Commission on Human Rights, culminated in the 1984 Cartagena Declaration on Refugees. Promoted by experts and representatives from 10 States in the Central American region, the Declaration widened the definition of refugee to ensure that all victims of generalized violence could claim protection. The Cartagena criteria to determine "refugee" status include flight from generalized violence, foreign aggression, international conflicts, massive violations of human rights, and circumstances seriously disturbing public order, thus going beyond comparable criteria established in the African context. However, the Cartagena Declaration lacks treaty status, and official willingness to abide by its liberal asylum norms has not always been forthcoming. Nevertheless, the process leading up to the Cartagena Declaration is of special interest in view of the restrictive European response to the potential flows of asylum-seekers caused by civil war in the former Yugoslavia that took place just 10 years later.

With regard to the asylum-seekers generated by conflict in Central America, the United States formally adopted a restrictive policy, although it eventually absorbed many of those who had entered the country illegally. For countries in the immediate vicinity, the dilemma was sharper: they had to respond somehow to the presence of tens of thousands of war victims who lacked any alternatives for protection. The final argument of humanitarianism was in terms of protection versus chaos and death. Confronted with such choices, neighbouring States opted for protection, though they disagreed sharply on the subject of Central America's wars and feared that mass inflows of asylum-seekers would adversely affect both their labour markets and their political stability.

The demise of the cold war prompted the settlement of regional conflicts such as those raging in Central America and the repatriation of refugees started. With the exception of Haiti, no new forced outflows are in the offing in the Latin American and Caribbean region. For the first time in more than two decades, refugees are no longer a major problem in the region and the formal Latin American asylum system has been strengthened to meet the eventuality of both individual applications and mass displacements.

C. Restriction of asylum as a result of crisis

By the late 1970s, the international refugee regime had experienced periodic crises, but they had generally served to widen asylum, as evidenced by the existence of the 1967 Protocol relating to the Status of Refugees, the OAU Convention, UNHCR's "good offices" doctrine, and the Cartagena Declaration. During the 1980s, however, crises had the opposite effect. Numerous restrictions on asylum were introduced and an entirely new vocabulary, with terms like "interdiction", "humane deterrence", "carrier sanctions" and "economic refugees", was created to justify them. The restrictions applied mainly to spontaneous asylum flows from developing to developed countries. European States established an elaborate international network to reduce the number of asylum-seekers. In Northern America, United States authorities sought to keep out the large number of refugees originating in Central America and the Caribbean. Canadian authorities followed the same trend by introducing in 1992 a number of amendments to the immigration law that would effectively restrict the possibility of seeking asylum.

Large refugee flows by themselves do not necessarily trigger restrictions. Five million Afghans, for instance, found ready asylum in the neighbouring Islamic Republic of Iran and in Pakistan during the 1980s. The inflow of asylum-seekers to the Western-bloc countries of Europe has been small, in numerical terms, by developing country standards. Although the number of asylum-seekers increased markedly towards the late 1980s and early 1990s, especially in Germany which received nearly 440,000 applications for asylum in 1992, restrictive responses have been evident since the mid-1980s. Similarly, Haitians seeking asylum in the United States caused a reaction quite out of proportion to their numbers: the number of Haitian "boat people" amounted to only 12,400 persons in 1980, 15,000 in early 1992, and virtually zero in the intervening years. By comparison, the legal ceiling on immigrants to the United States was 534,000, and was further increased to more than 700,000 annually by the 1990 Immigration Act (Papademetriou, 1994).

These examples illustrate a general point: although the number of persons seeking asylum may be small, they may nevertheless elicit a strong response because presumptive refugees have claims on receiving States that migrants do not. Under international law, sovereign States must at least give potential refugees a hearing and, under certain conditions, admit them. In democratic societies, pressures to do so can be significant. In principle, admission is based solely on humanitarian considerations, regardless of social or economic interests. Yet, although refugees may have skills and other endowments, they are in a fundamental sense selected by forces external to and independent of the receiving society. Consequently, asylum-seekers who arrive spontaneously in groups or successive waves pose particularly troublesome challenges to State control.

In some cases, however, the response to asylum-seekers—like that of Western-bloc countries to refugee flows from Eastern-bloc countries during the cold war—has been generous. Yet, especially in that case the humanitarian impulse benefited from political considerations regarding the hostile alliance of the States from which refugees had originated. By comparison, the inflows that led to progressive restrictions on asylum in Western Europe and Northern America during the 1980s had little political value. They were large compared with previous movements, they consisted of asylum-seekers arriving spontaneously rather than of prescreened refugees resettled in an organized manner, and they were preceded by a general crisis in the international refugee regime triggered by the sudden outflows of millions of refugees from Afghanistan, Indo-China and the Horn of Africa during the late 1970s and early 1980s. In addition, many of those seeking asylum in Western-bloc countries originated in developing countries and were suspected of not being or officially acknowledged as not being "genuine" refugees. Such developments conjured up visions of millions more to come. One might even argue that developed countries introduced asylum restrictions as a defence against the populousness and turbulence of developing countries, if the issue were not more complex.

Indeed, the earliest and sharpest restrictions on asylum during the 1980s were not those imposed in Europe, but rather those established in South-eastern Asia, where asylum opportunities for the Indo-Chinese were sharply curtailed. In the late 1980s, the collapse of socialism in the former USSR and the end of exit controls in the former Eastern-bloc countries made Western-bloc countries adjust their refugee policies to the realities of the post-cold war era. Consequently, restrictions similar to those applied to non-Europeans were imposed on asylum-seekers from Central and Eastern Europe and the former USSR. Clearly, the asylum issue was not simply a facet of the "North-South" conflict. An underlying concern was that asylum, in being used as a vehicle for migration, was being misused. That concern was a major factor behind the restrictions imposed in both Asia and Europe (Suhrke, 1983; Meissner, 1992a).

1. *Northern America*

Since the adoption of the 1980 Refugee Act, the United States refugee policy has been marked by a sharp dualism: the legal framework is relatively liberal, but practice has been highly politicized and restrictive towards unwanted groups (Loescher, 1986; Zucker and Zucker, 1987). The ideological bias in policy implementation has been well documented. While the 1980 Act incorporated the universal definition of "refugee" established by the 1951 Convention, about 95 per cent of the refugees admitted to the United States during 1980-1988 originated in communist countries. In asylum cases, the differential treatment of Cubans, who were generally admitted, and Haitians, who were generally rejected, bore no obvious relation to their respective degree of "persecution" and the same applied to the differential treatment of citizens from other countries (Helsinki Watch, 1989). Whereas patterns of admission or rejection in terms of socio-economic and racial factors may be detected as having been weak,[8] clear-cut ones have been noticeable with respect to political considerations: asylum applicants from States whose Governments were supported by the United States were declared to be economic migrants rather than refugees, even though they had a prima facie claim to protection. Examples include Salvadorans fleeing from civil war during the 1980s and Guatemalans escaping repression in the early 1980s. Persons fleeing from communist or leftist regimes, for instance, Nicaraguans under Sandinista rule, were generally granted refugee status or were, like the Cubans, paroled under special provisions.

Since the Second World War, most of the refugees admitted by the United States had initiated asylum applications in American missions abroad or had entered the country under organized resettlement schemes. Asylum-seekers arriving spontaneously originated mostly south of the Rio Grande, especially in Central America and the Caribbean. When conflict erupted there, refugee flows were largely beyond United States control. Despite strenuous efforts, the United States could not insulate itself against the consequences of social conflict in its vicinity, mainly because of the porousness of its southern land border. Only boat people from island States could be interdicted with some ease.

Thus, geography and politics often worked in opposite directions during the 1980s. United States policy consistently rejected the claims of Salvadorans for asylum arguing that most were "economic refugees". Yet perhaps half a million succeeded in entering and staying in the country, usually as undocumented migrants. For Central American countries, the escape of local refugees to the United States was a positive factor easing their own refugee burdens. However, the system imposed great hardship and uncertainty on the individual refugee and generated significant tension between the United States and its Latin American neighbours, especially Mexico.

Being faced with the phenomena of civil war and mass population outflows, the United States could have responded by setting up a formal system of temporary protection and collective determination of eligibility. Such a response was suggested by experts at the time (Hartman, 1988; Gallagher, Martin and Weiss-Fagen, 1989), especially in view of the fact that the United States had already resorted to a unilateral practice of that kind. Indeed, Extended Voluntary Departure (EVD) status could be granted on a temporary and collective basis to individuals from eligible countries. Eligibility criteria centred on generalized violence, rather than on the more discriminatory and narrow criterion of "persecution". Customarily given to persons already in the United States when hostilities erupted, EVD status had been granted to Ethiopians, Nicaraguans during 1979-1980, Chinese after the events of 1989, and Bosnians as of 1992.

Owing to concerted political pressure, an equivalent status was eventually granted to Salvadorans. Yet, the position of the United States and, more generally, the intense politicization of the Central American refugee question made it impossible to develop a common, regional system of temporary protection. Optimally, such a system would streamline criteria and the conditions for protection, and establish rules for burden-sharing in the region. Divided over the question of whom to give protection to, and faced with United States opposition to any form of common regime, the smaller and poorer States of the region did not go beyond the declaration-related innovations of Cartagena and the United States did not support even those.

Until 1990, the United States used administrative discretion to grant certain groups permission to stay temporarily. To reduce the potential for partisan application, critics obtained a legislative basis for the granting of Temporary Protected Status (TPS) in an amendment to the Immigration Act of 1990 (Helton, 1990). By legislative action, TPS could be collectively bestowed on citizens of particular States. Such provisions helped to solve the immediate problems of Salvadorans, who were granted TPS, but considerable administrative discretion-related and associated problems still remain.

A major issue was the eventual return of those granted temporary permission to stay once conflict in their countries of origin subsided.[9] Given the poor economic conditions prevalent in those countries, repatriation was expected to be problematic and might have to be encouraged by aid for reconstruction and development. Failing that, permanent absorption in the United States seemed to be the inevitable outcome of geography and history, the consequence of being a rich country with poor neighbours. Yet, the prospects of eventual repatriation constituted a necessary element in the political debate geared to secure temporary protection for the affected population in the first place. By 1992, the outcome of the first major test case was uncertain: the United States had agreed to postpone repatriation of Salvadorans pending implementation of the December 1991 Peace Agreement and the further economic reconstruction of El Salvador.

The controversial case of the Haitian boat people also raised fundamental questions of access to asylum and eventual repatriation. Because Haitians had fled conditions of deprivation and systemic poverty and a ruthless regime, asylum claims under prevailing legal norms and democratic conditions invited long and costly court battles. Deciding to cut the process short, the United States Government instructed the Coast Guard in 1981 to interdict Haitians at sea and return them to Haiti after cursory hearings on the high seas. As critics noted, it was a clear case of restricting access to seeking asylum, similar to efforts under way in Europe. The criteria for eligibility were narrowly interpreted: of more than 12,000 Haitians interdicted between 1981 and 1987, only two were granted political asylum (United States Committee for Refugees, 1987).

The Haitian case generated intense legal and political battles. By some standards, it was a case of discrimination, especially in the context of the admission of Cubans. By other standards, the complexity of the case defied easy answers, raising questions about both access to asylum and the misuse of the institution, and about the consistency of the asylum adjudication process versus the expected rise in admissions should the Cuban standard be applied. As conditions in Haiti worsened, the situation became clearer: the 1991 coup led to widespread violence, acute deprivation and new population outflows. The optimal response suggested itself: temporary protection pending the outcome of diplomatic efforts to restore some degree of political normalcy or the political status quo ante (Meissner, 1992b). That is to say, the situation called for procedures similar to those adopted in response to other mass outflows caused by violent social upheavals. The United States Government, however, did not initially adopt such approach. Rather, it moved from the use of regular and fairly stringent asylum procedures to interdiction and deportation.

Given the major role played by the United States in the operation of the international refugee regime, its interdiction of Haitian boats created fears that such actions would legitimize similar measures elsewhere, especially in South-eastern Asia. However, massive denials of access to asylum by South-eastern Asian countries had started two years earlier and asylum policy in the region had been shaped by considerations other than United States practice (Sutter, 1990). Furthermore, the United States' partisan policy regarding asylum had undermined any claim it had to moral leadership in asylum and refugee matters. As for Europe, restrictive measures were being adopted for unrelated reasons.

2. *Europe*[10]

After declining somewhat during the early 1980s, the number of asylum applications filed in European countries increased rapidly, passing from 70,000 in 1983 to nearly 700,000 in 1992 (table 50). A refugee system that had traditionally responded to crises (wars, repression, pogroms, revolutions) and was based on the notion that States were under a special obligation to protect refugees because, unlike migrants, they constituted exceptional cases, was being confronted with an ongoing phenomenon.

From the outset, the flows of asylum-seekers seemed problematic. Being no longer a region that favoured the admission of migrant workers, Western-bloc Europe did not welcome the new arrivals on economic grounds. The non-European origin of many of the asylum-seekers was readily exploited by political fringe movements, especially as unemployment rose. There was also the legitimate concern that the asylum system was being used as a means of access to the labour market, especially because, since labour migration was stopped in 1973-1974, asylum had represented one of the few legal channels for admission to most Western-bloc European countries. Numerous asylum-seekers came from countries that had been major sources of migrant workers for Western Europe, such as Turkey and Yugoslavia, and there was evidence that migration networks operated in the case of some African countries, such as Ghana and Zaire (table 51).[11]

Yet, many asylum-seekers were nationals of countries experiencing civil strife, war or violent minority-majority conflicts as in the cases of the Islamic Republic of Iran, Lebanon and Sri Lanka. Asylum-seekers represented a very heterogeneous group, a fact reflected in the outcome of screening.[12] During 1987-1988, 40 per cent of the

TABLE 50. INDICATIVE NUMBERS OF ASYLUM APPLICANTS IN SELECTED DEVELOPED REGIONS OR COUNTRIES, 1983-1992
(*Thousands*)

Region or country where asylum application was filed	*1983*	*1984*	*1985*	*1986*	*1987*	*1988*	*1989*	*1990*	*1991*	*1992*	*Total*
Europe	70.1	102.5	168.0	201.7	183.1	234.0	346.1	441.4	537.2	693.9	2 974.8
European Community	51.2	72.7	134.2	164.3	130.9	170.8	230.6	325.7	421.8	558.5	2 259.0
Nordic countries	3.2	12.3	15.4	17.3	26.8	26.3	35.1	36.2	34.2	92.2	298.3
Eastern-bloc countries	1.9	2.8	2.0	2.8	3.1	4.3	34.1	20.8	12.2	8.7	92.3
Other Europe	13.8	14.7	16.4	17.3	22.3	32.6	46.3	58.7	69.0	34.5	325.2
Northern America	31.1	31.4	25.0	41.9	52.1	100.7	123.5	110.2	86.9	141.7	744.3
Australia	0.0	0.0	0.0	0.0	0.0	0.0	0.5	3.6	16.0	4.1	24.2
Japan	0.8	0.6	0.5	0.4	0.2	0.3	0.8	0.2	0.4	0.1	4.3
TOTAL	102.0	134.5	193.5	244.0	235.4	335.0	470.9	555.4	640.5	839.8	3 747.6

Source: United Nations High Commissioner for Refugees, *The State of the World's Refugees: The Challenge of Protection* (London, Penguin, 1993), annex I.5.

TABLE 51. COUNTRIES OF ORIGIN RANKED ACCORDING TO THEIR CONTRIBUTION TO APPROXIMATELY THREE FOURTHS OF ALL ASYLUM APPLICATIONS SUBMITTED IN EUROPE DURING 1983-1989

(Percentage)

1983[a]		*1984*		*1985*		*1986*	
Sri Lanka	8.85	Sri Lanka	13.41	Sri Lanka	17.66	Iran	18.96
Iran, Islamic Rep. of	8.04	Iran, Islamic Rep. of	11.60	Iran, Islamic Rep. of	12.71	Poland	10.28
Turkey	7.99	Poland	10.96	Turkey	9.91	Turkey	9.81
Poland	7.83	Turkey	10.71	Poland	8.50	Lebanon	6.13
Czechoslovakia	6.64	Various[b]	5.29	Lebanon	5.23	Ghana	5.28
Viet Nam	5.94	Czechoslovakia	5.26	Ghana	5.13	Sri Lanka	5.24
Cambodia	5.04	Ghana	5.16	Various[b]	4.06	Various[b]	4.91
Ghana	4.80	Romania	3.53	Other Africa	4.00	Other Africa	4.37
Zaire	4.68	Lebanon	3.41	India	3.37	India	4.01
Romania	4.47	Ethiopia	3.39	Czechoslovakia	2.88	Romania	3.49
Other Africa	3.79	Other Africa	3.33	Pakistan	2.86	Pakistan	2.51
Hungary	3.03					Czechoslovakia	2.36
India	2.95						
Chile	2.72						
TOTAL	76.77		76.03		76.31		77.34
Number of countries	14		11		11		12

1987		*1988*		*1989*		*1983-1989*	
Turkey	14.53	Poland	18.96	Turkey	16.87	Turkey	13.28
Poland	11.60	Turkey	15.13	Romania	13.61	Poland	11.58
Iran, Islamic Rep. of	11.23	Yugoslavia	10.29	Poland	9.78	Iran, Islamic Rep. of	10.12
USSR (former)	5.59	Iran, Islamic Rep. of	8.36	Yugoslavia	7.53	Sri Lanka	7.38
Other Africa	5.29	Romania	4.57	Sri Lanka	5.75	Romania	6.35
Sri Lanka	4.40	Other Africa	4.03	Iran, Islamic Rep. of	4.33	Yugoslavia	4.95
Yugoslavia	4.24	Sri Lanka	3.52	Lebanon	4.28	Other Africa	4.24
Hungary	3.84	Lebanon	2.90	Other Africa	4.22	Lebanon	3.87
Romania	3.70	Zaire	2.89	Zaire	3.10	Ghana	3.62
Chile	3.67	Chile	2.77	Somalia	2.75	Zaire	2.83
Ghana	3.34	Hungary	2.55	Bulgaria	2.31	Czechoslovakia	2.83
Zaire	3.22			Ghana	2.31	India	2.54
Czechoslovakia	2.82					Pakistan	2.45
TOTAL	77.48		75.97		76.83		76.04
Number of countries	13		11		12		13

Source: Bela Hovy, "Asylum migration in Europe: patterns, determinants and the role of East-West movements", paper presented at the International Conference on Mass Migration in Europe: Implications in East and West (Vienna, 5-7 March 1992), based on UNHCR data.

[a]Figures including Indo-Chinese who arrived in 1983 under French resettlement quotas.

[b]Referring to a category of asylum applicants with no further indication of nationality.

claims considered were decided in favour of the applicant, a substantially higher proportion than that registered by the United States during most of the 1980s (Widgren, 1990b). The heterogeneous nature of the applications filed required an elaborate screening mechanism. Despite efforts to shorten and streamline the process, a complicated case could take two or three years to be decided before all avenues for consideration and appeal were exhausted. As public debate over asylum issues became increasingly acrimonious, so did the legal and political battles over asylum, and costs mounted.[13]

For a large and prosperous Europe, the financial costs involved, just as the number of asylum-seekers, were far from overwhelming. However, behind the figures lay the fear of uncontrolled inflows passing through the asylum gate (Rogers, 1991; Joly, 1992). With the disintegration of the Soviet Union and the relaxation of exit controls in Eastern-bloc countries, such concern was heightened towards the end of the 1980s. The number of asylum applications from citizens of Eastern-bloc countries rose from less than 60,000 in 1987 to over 150,000 in 1990 (table 52). The largest contingents of asylum-seekers from Eastern-bloc countries originated in Poland, Romania, Yugoslavia, the former Czechoslovakia and Hungary, in order of importance.

Consequently, as the 1990s decade began, the implications of the question of who was a refugee had been radically altered. Given that economic deprivation and politically oriented violence coexisted in large parts of the world, a broad interpretation of that question could open the door to massive inflows. Even a reasonably strict interpretation of the 1951 Convention would permit millions to press a valid claim for protection in Europe, including victims of oppressive or totalitarian regimes who, as a category, had been admitted from Eastern Europe and the USSR during the cold war, and members of vulnerable minorities, who constituted the classical "target" refugees groups in European history.

Ethnic minorities in conflict represented a particularly problematic category because they were numerous and also had a potential claim to refugee status under the 1951 Convention. In an earlier period of imperial disintegration and State formation in Europe, a concentration principle had been at work as ethnic diasporas moved into their respective "homelands" (*Annals of the American Academy of Political and Social Sciences*, 1939). In the 1990s, such developments were unlikely. Even minorities with a homeland might consider that economic and political opportunities outweighed ethnic affinity. For instance, although Turkey generally followed the German practice of admitting ethnic kin (Aarbakke, 1990; Organisation for Economic Cooperation and Development, 1992), not all Bulgarian Turks chose to establish themselves in Turkey. Some went to Germany or Sweden. Similarly, the ethnic Albanians who, starting in the late 1980s, left the province of Kosovo in ex-Yugoslavia went mostly to Western-bloc countries and not to Albania. Numerous minorities in the successor States of the former USSR might respond similarly under pressure. Such prospects clearly chilled Western-bloc Europe (Widgren, 1990a).

Already by the mid-1980s, the growing backlogs in asylum applications and the movement towards European unity had put the issue in the political agenda. The 1986 Single European Act, which stipulated that free movement of capital and labour within the European Community should be a reality by early 1993, precipitated a restrictive streamlining of the asylum process that led to the Schengen Agreement (on the gradual abolition of checks at the common borders of Belgium, France, the

TABLE 52. ANNUAL ASYLUM APPLICATIONS FROM EUROPEAN COUNTRIES, 1983-1990
(*Thousands*)

Country of origin	*1983*	*1984*	*1985*	*1986*	*1987*	*1988*	*1989*	*1990*	*Total*
Poland	5.6	11.1	14.2	20.7	21.1	44.3	33.8	15.0	165.8
Romania	3.2	3.6	4.5	7.0	6.7	10.7	47.0	78.1	160.8
Yugoslavia	0.9	1.0	2.1	2.7	7.7	24.1	26.0	31.7	96.2
Czechoslovakia	4.7	5.3	4.8	4.8	5.1	4.6	7.4	1.7	38.5
Hungary	2.2	2.3	3.1	4.1	7.0	6.0	3.4	0.9	28.9
Bulgaria	0.3	0.4	0.4	0.5	0.4	0.6	8.0	12.5	23.1
USSR (former)	0.1	0.1	0.1	0.1	10.2	0.5	1.3	5.1	17.5
Albania	0.5	0.4	0.5	0.3	0.2	0.2	0.2	5.9	8.2
Other	1.3	0.1	0.1	0.4	0.1	0.1	0.4	0.3	2.8
TOTAL	18.8	24.4	29.8	40.6	58.5	91.1	127.6	151.2	541.9

Sources: Bela Hovy, "Asylum migration in Europe: patterns, determinants and the role of East-West movements", in *The New Geography of European Migration*, Russell King, ed. (London and New York, Belhaven Press, 1993).

Federal Republic of Germany, Luxembourg and the Netherlands) and the Dublin Convention (Convention Determining the State Responsible for Examining Applications for Asylum Lodged in One of the Member States of the European Communities), popularly known as the outer walls of "fortress Europe". Those documents defined the core of a policy with the following elements: access to seeking asylum was restricted, the criteria for eligibility were narrowed, the determination process was shortened, the deportation of rejected claimants made more efficient, and the rules among the signatory States were harmonized. Refugee status under the 1951 Convention became increasingly difficult to obtain and eligibility under the wider criteria of humanitarian grounds was also tightened. By 1991, asylum was granted to only one quarter of those whose applications for asylum had been considered, 8 per cent receiving Convention status and 17 per cent being granted asylum on other grounds (Widgren, 1992).

Yet, the strategy that seeks to uphold a limited and formal criterion that extracts one element, namely, political persecution, from an intricate set of interrelated causes is inherently vulnerable. The democratic and legal processes taking place in Western-bloc Europe opened endless political battles and polarized the public debate, sometimes leading to violence (Kaye, 1992). Moreover, the asylum process appeared to be not only costly but essentially futile. Although only one quarter of the asylum applications considered resulted in the granting of asylum, an estimated 80 per cent of all asylum-seekers remained, as State authorities yielded to technical and political obstacles that prevented the forced repatriation of persons whose asylum claims had been rejected.[14]

The only remaining avenue of control, if this remained the objective, was to limit access to seeking asylum in the first place. Already by the mid-1980s, certain States had begun practising what in South-eastern Asia had been called humane deterrence by providing to asylum-seekers inhospitable quarters, reduced social benefits and limited work permits while the asylum application was being considered, and by establishing visa requirements and sanctions on carriers that brought in passengers without appropriate documents (Schneider, 1989; Kaye, 1992). The aim was to discourage "economic refugees" from clogging the asylum process. The problem of "manifestly unfounded or abusive applications" was also recognized in the UNHCR's Executive Committee (UNHCR, 1983), and gave rise to the controversial term "abusive asylum-seeker".

The Schengen Agreement and the Dublin Convention were formulated in the context of a process designed to enhance control over the external borders of the European Community in order to prevent drug smuggling, terrorism and undocumented migration. Since few asylum-seekers are in a position to obtain proper documentation before flight, most appear as undocumented migrants until they can lodge an asylum application. Consequently, measures to control undocumented migration affect access to asylum and persons seeking asylum are often equated with illegal migrants and are increasingly perceived as problems for the State to contend with rather than as individuals in need of protection (Joly, 1992). The Schengen Agreement and the Dublin Convention seek to strengthen control of the external borders of the Community by harmonizing visa policies and carrier sanctions and, perhaps of greater significance, ensuring that asylum-seekers lodge only one application within the European Community (Loescher, 1989; Plender, 1989).[15] To prevent asylum-seekers from moving or being compelled to move from one European country to another in search of asylum, the Dublin Convention grants each asylum-seeker only one chance to apply for asylum by placing the exclusive responsibility for the processing of the application and the provision of support on the country where the applicant first appears.[16]

With the end of the cold war, certain Eastern-bloc countries were incorporated into the evolving European asylum regime. The intention was to control East-West migration in all its aspects by giving more attention to "root causes", and exchanging information about travel routes, border controls, visa requirements, carrier sanctions, and mechanisms for the repatriation of illegal migrants. Thus, the Berlin and Vienna processes, which consist of a series of conferences and continuous consultations set up after ministerial meetings held in 1991, incorporated Eastern-bloc countries and developed measures to deal with citizens from those countries as well as with transiting travellers. In 1990-1991, the former Czechoslovakia, Hungary and Poland tightened their asylum policies by adhering strictly to the 1951 Convention's concept of persecution and were thus able to bring down the number of asylum applications both in Eastern Europe and beyond (Organisation for Economic Cooperation and Development, 1992).

In an extraordinary exercise of international consultations, the European States brought Northern America and Australia into the emerging European regime. Informal but increasingly institutionalized deliberations about asylum and refugee issues have proceeded within a separate forum called Informal Consultations (IC), with its secretariat in Geneva (Informal Consultations (IC), 1991).

Altogether, the restrictions in place delineate an international division of labour designed to make it more difficult for presumptive refugees to reach Europe's "civil space". Restrictions apply both to persons from the former Eastern-bloc countries and to non-Europeans. At present, European asylum procedures vary greatly. Until members of the European Community are truly unified, as stipulated by the 1992 Maastricht Treaty (Treaty on European Union), existing harmonization policies will favour the least-liberal level. The collective-action mechanisms established by the Schengen Agreement and the Dublin Convention penalize those States with liberal asylum systems, since they will attract more claims. The recent change in Germany's Constitution whereby the guarantee of access to asylum has been abolished is partly a consequence of these mechanisms. Moreover, restricting asylum-seekers from presenting more than one application for asylum reduces their "degrees of freedom".

The full impact of the Schengen Agreement and the Dublin Convention, which was not yet ratified at the time of writing by all member States of the European Community, remains to be seen. So far, even severe national restrictions have had limited effect. Measures introduced by most Western-bloc countries during the second half of the 1980s, including carrier sanctions, visa requirements, shortened processing and, in Germany, a 1987 act that drastically reduced social benefits for asylum-seekers, made no dent in the increasing trend with respect to the number of applications (Danish Refugee Council, 1988 and 1991). Authorities have admitted as much by seeking additional measures. One widely discussed model was the agreement concluded in 1991 between the signatories of the Schengen Agreement and Poland to return to the latter country persons with unfounded asylum claims who had entered the European Community through Poland (Widgren, 1992). Several Governments have also adopted the concept of a "safe country" to denote a State where persecution does not exist and whose citizens are therefore not eligible for asylum. Similar "safe country" standards are being applied to return asylum-seekers to countries through which they have transited and where they may lodge an asylum application.

It is a measure of the pressure to which the institution of asylum is subject that these deeply illiberal measures have been widely accepted as necessary to save the institution itself. As Hailbronner (1991, p. 4) has argued, the current asylum system is extremely inefficient, wastes resources, and has doubtful value for the protection of those whose need is greatest. Hailbronner goes on to commend the use of "safe country" profiles to weed out bogus claims like those, for instance, submitted by citizens of Eastern-bloc countries in 1990-1991 when the collapse of oppressive regimes made political persecution less likely, but exit easier.[17]

Probably the main danger for the institution of asylum is that the possibility of *seeking* asylum is being severely restricted. The point was made in the Council of Europe, in the European Parliament and by UNHCR (Council of Europe, Parliamentary Assembly, 1992). Carrier sanctions were criticized and it was feared that the harmonization of asylum policies would jeopardize individual rights and due process. Neither the Schengen Agreement nor the Dublin Convention made reference to the European Convention for the Protection of Human Rights and Fundamental Freedoms (United Nations, 1955), which implicitly recognizes the right to seek asylum (Einarsen, 1990). As human rights law has increasingly become a source for defending refugees against restrictive States, failure to cite the pre-eminent European instrument in this regard is significant. In the balance between the rights of States and those of the individual, the weight of the Schengen Agreement and the Dublin Convention lies unambiguously on the side of the State. As a report by the Council of Europe (1992) concluded, the member States of the European Community are more concerned about reducing, sharing and distributing the burden of refugees than by protecting the basic rights of asylum-seekers and de facto refugees. The Community's aim of achieving a common asylum policy is leading to a "sort of competition" between the member States to introduce the most restrictive measures towards asylum-seekers and refugees in order to become the least attractive destination. Such restrictions reinforce the view that a "fortress Europe" is being built. The report was drafted by Sir John Hunt, Conservative, of the United Kingdom.

A small but vocal group of non-governmental organizations, scholars and human rights activists organized on an all-European basis and led the attack against Europe's new asylum policy (Nobel, 1987; European Consultation on Refugees and Exiles (ECRE), 1989; Joly and Cohen, 1989; Studie-en Informatiecentrum Mensenrechten

(SIM) (Dutch Institute for Human Rights), 1989; Amnesty International, 1991; Kjærum and Horst, 1991). Taking human rights as their starting point, the group argued that the fundamental rights of asylum-seekers were endangered. Recognizing that the harmonization of asylum policies was both necessary and desirable, they nevertheless stressed that the process of harmonization should be directed at protecting asylum-seekers, not at serving the interests of States. The European Consultation on Refugees and Exiles (ECRE), the European non-governmental organization for asylum-seekers, called in 1989 for a parallel Schengen Agreement to protect, assist and resettle asylum-seekers (European Consultation on Refugees and Exiles (ECRE), 1989). The group demanded a greater role in the asylum determination process, and revived the call for a convention on asylum and for a broader definition of "refugee" including criteria similar to those contained in the definitions put forth by the OAU Convention and the Cartagena Declaration (Studie-en Informatiecentrum Mensenrechten (SIM) (Dutch Institute for Human Rights), 1989). If the result was an increase in Europe's asylum burden, so be it: from the perspective of global burden-sharing and given the existence of some 19 million refugees worldwide, Europe should rightly accept more.[18]

The debate revealed the underlying agony in European attitudes towards refugees. The contradiction between State's rights and human rights was particularly sharp because asylum had come to mean permanent residence. Although national laws often allowed only a temporary stay, return to the country of origin was rare. Moreover, the control and selection that was possible when subscribing to UNHCR quotas with respect to resettling refugees could not be exercised when asylum-seekers presented themselves at the border. Both points were illustrated, if not resolved, by the refugee crisis brought about by the disintegration of Yugoslavia.

3. *The crisis in the former Yugoslavia*

In refugee terms, Yugoslavia became Europe's Central America. The civil war in the early 1990s displaced people on the southern rim of Europe on a scale that the region had not experienced since the Second World War. As the European responses crystallized, a new orthodoxy formed: refugees should remain and be assisted as close to the centre of conflict as possible, preferably within their own country. Failing that, external asylum should be explicitly temporary (European Consultation on Refugees and Exiles (ECRE), 1992; UNHCR, 1992; United States Committee for Refugees, 1992).

The severe limitation on asylum that this view implied was underscored by the imposition of visa requirements on nationals from the war-torn area. By mid-July 1992, just before a conference held in Geneva confirmed that response as a European policy, Austria, Germany and Switzerland had effectively closed their borders to new arrivals from Bosnia and Herzegovina, requiring visa and evidence of sufficient income before allowing entry. Since such measures effectively prevented would-be asylum-seekers from reaching other European countries, the latter did not impose visa requirements. The outcome foreshadowed the emergence of a united Europe free from the victims of war as specified by article 100C(2) of the Maastricht Treaty which states that, given an emergency outside Treaty territory, member States can impose visa requirements in order to prevent mass population inflows.

While generally seeking to keep out Yugoslav refugees, Western European countries generously financed relief assistance for displaced persons within the affected area, mainly in Croatia and Slovenia, and provided relief to besieged Sarajevo and other cities. In addition, some countries, in highly controlled and symbolic humanitarian gestures, admitted small numbers of particularly vulnerable persons.

The result was soon evident. By mid-1992, out of the 2 million refugees from the former Yugoslavia, 1.7 million remained in Bosnia and Herzegovina or in Croatia and Slovenia. By far the largest group of those who had made it to Western-bloc Europe were Albanians from the autonomous republic of Kosovo, many of whom had arrived before the civil war erupted (table 53).

From the outset, UNHCR encouraged the provision of asylum by calling for temporary protection until the war ended. The experience from Croatia was encouraging; relative peace in early 1992 enabled two thirds of the 1.2 million Croatian refugees in neighbouring countries to return. Recognizing that return was likely and given the severity of the humanitarian crisis, some European Governments softened their stance. Nationals of ex-Yugoslavia who managed to enter were rarely returned. Most were not processed for formal asylum but rather were granted various forms of temporary permission to stay, sometimes on the basis of new administrative regulations, as in Italy and the Netherlands, or as a result

TABLE 53. REFUGEES AND UNITED NATIONS BENEFICIARIES FROM THE FORMER YUGOSLAVIA, 1992-1993
(*Thousands*)

Area or country of asylum	*Origin* Croatia	Bosnia	*Total*
A. *Within the former Yugoslavia (spring 1993)*			
Croatia	255	271 to 420	526 to 675[a]
United Nations Protected Areas, Croatia			87[a]
Serbia	164	294	458
Bosnia			2 280[b]
Montenegro			61
Slovenia			33
Macedonia	3	29	32
TOTAL			3 477 to 3 626
B. *Outside the former Yugoslavia (December 1992)*			
	Total		
Germany	220		
Sweden	74		
Austria	73		
Switzerland	70		
Hungary	40		
Turkey	20		
Italy	17		
Netherlands	7		
Denmark	7		
United Kingdom	4.9		
France	4.2		
Norway	3.7		
Finland	1.8		
TOTAL	674		

Source: United States Committee for Refugees, *Croatia's Crucible* (Washington, D.C., October 1992); and United States Committee for Refugees, *World Refugee Survey, 1993* (Washington, D.C., 1993); and United Nations High Commissioner for Refugees, *Information Notes on Former Yugoslavia*, July 1993.

[a]Estimated.

[b]Referring to the estimated total number of persons receiving United Nations assistance or protection in Bosnia ("United Nations beneficiaries"), thus including the population of Sarajevo.

of a decision neither to process nor to deport the persons concerned, as in Belgium and Norway. States that already hosted a migrant population of Yugoslav origin, readily issued visitor visas to family members and relatives.

The alternative to this patchwork of protection would have been an organized system of temporary protection granted on a group basis throughout Western Europe with appropriate mechanisms for burden-sharing and repatriation. The idea was familiar to legal experts (Grahl-Madsen, 1980; Melander, 1987; Einarsen, 1992) and had surfaced during the Central American refugee crisis. A formal proposal that contained all the elements of a collective, temporary regime was presented by the Swedish Government to UNHCR in April 1992. The idea was discussed at a high-level conference convened in Geneva in July 1992 by UNHCR to deal with the Yugoslav crisis. It was endorsed by various support groups and was considered with interest by Germany, Sweden and Switzerland. However, opposition from France and the United Kingdom blocked further progress. Given a general reluctance to accept refugees, individual States could effectively veto a collective regime by refusing to admit their share of refugees.

Affluent States, it seems, react similarly to the presence of refugees from war in their immediate vicinity, albeit for different reasons. The United States Government wished to deny the existence of refugees from El Salvador because they had fled from a client State. In Western-bloc Europe, political considerations should have been favourable to refugees from ex-Yugoslavia. Yet, European unhappiness with the Serbian onslaught was insufficient to prompt Governments to grant asylum to their victims. The overriding European concern was fear that mass inflows would be a burden and that the attraction of the Community's affluent civil space would make the presence of refugees permanent.

UNHCR did not aggressively advocate a more liberal European asylum regime. Having always been dependent on voluntary contributions to finance its activities, UNHCR has developed an organizational code of cautious diplomacy which prevailed in this occasion. The Balkan emergency necessitated recurring appeals for relief and supplies to prevent mass starvation in the embattled areas. It would have required a strong leadership to press developed countries for additional asylum concessions, especially since any significant offloading of displaced persons from ex-Yugoslavia would have involved resettling thousands of persons. The question to be addressed then is: Did it matter? Was not the "Yugoslav solution" to the refugee crisis optimal on political as well as ethical grounds?

D. Internalization of asylum: a new trend

The policy towards asylum is integrally linked to what follows, namely settlement or repatriation. The concept of asylum as a prelude to settlement has been firmly anchored in the practice and consciousness of Western-bloc countries. Repatriation had not been a viable alternative for émigrés in the wake of the Russian revolution, for Jews fleeing Nazism, or for Europeans escaping communism. The same principle was applied to persons from developing countries, for instance, Cubans and Indo-Chinese who left communist regimes en masse.

In the early 1990s, Western-bloc countries were ready for a change. Communism had collapsed and compassion for refugees was strained, as were the capitalist economies. Fearing cumulative asylum demands from a populous and turbulent world, the Western dominant powers suggested a new agenda. More attention was to be given to the prevention of flows by dealing with their root causes. Repatriation must be encouraged—1992 was to be the Year of Repatriation, as the United Nations High Commissioner for Refugees declared in 1991—and displaced persons should be kept and assisted as close to home as possible so as to encourage their return.

Such new trends in the global refugee regime were evident during the Yugoslav crisis and, before that, in the 1991 Gulf war. When the Iraqi Kurds rebelled during the closing phase of the war and Turkey closed its border with Iraq, the coalition led by the United States established a security zone to protect Kurdish refugees inside Iraq. Widely applauded as a model of humanitarian intervention, the immediate effect of denying asylum to the Kurds outside Iraq was nevertheless widespread death and suffering as thousands of Kurdish refugees were stranded in an inhospitable border area. Nor did the strategy address the critical political question as to who in the long run would protect the security zone. Intervention is likely to beget further intervention, setting in motion a process that, in the absence of even a rudimentary international convention regulating humanitarian interventions, is problematic (Henkin and others, 1989; Childers, 1992). On a smaller scale, UNHCR also sought to internalize aid and protection in Sri Lanka by establishing Open Relief Centres for displaced persons within the country (Rodgers, 1992).

Despite the obvious advantages of asylum based on the proximity principle, serious concerns remain. The most immediate question is whether asylum close to the centre of conflict will actually assure protection. There are also other problems. Neighbouring countries may become overloaded, and this might result in hardship for both refugees and the local population, hardship that could lead to new conflicts and, eventually, to the denial of asylum. All those developments were evident by late 1992 in Croatia. For States further away, the proximity principle may act as a sedative, relaxing their efforts to provide assistance and—even more effectively—their efforts, if any, to address the root causes of population outflows (International Labour Organization/United Nations High Commissioner for Refugees, 1992). In any event, because admitting refugees from the region has been the norm, not to do so is to send a message of indifference which may actually worsen the plight of the people affected, especially when these belong to identifiable communities targeted for attack, as in "ethnic cleansing" (Suhrke, 1992).

E. Future prospects and policy implications

The present discussion shows that pressure on contemporary asylum institutions differs significantly across both regions and countries, mainly because asylum is sought and granted with reference to conflicts that are essentially political in nature. The structure of political conflict, in turn, is highly variable and distinct from the more uniform processes of demographic growth, uneven economic development, and environmental degradation. Although such factors typically increase social tensions and may lead to emigration, they do not produce large flows of asylum-seekers unless political upheaval also occurs. Such conflict, moreover, does not necessarily stem from underdevelopment or overpopulation. Thus, areas of Central America and South-eastern Asia that until recently produced large refugee outflows have entered a period of relative peace and political stabilization, although the underlying economic and demographic conditions of the countries therein have not changed much. Other areas, including parts of Western Asia and much of Africa, will no doubt remain refugee-producing regions because of the political problems they still face. In Central and Eastern Europe and the former USSR, political disintegration has led to extreme uncertainty. The region is experiencing one major and several small civil wars and the potential for mass movements remains.

Mass outflows typically lead to pressures on the asylum institutions of neighbouring countries and contemporary communications make significant flows to destinations further away possible as well. All indications suggest that pressures on Western Europe will remain high and probably increase in the foreseeable future. In Western-bloc countries, asylum poses particular problems because they have long been the destination and not the origin of refugees. Such asymmetry weakens the institution of asylum, because granting protection becomes a one-sided obligation of the rich, just like aid transfers. Unlike foreign aid, however, providing asylum involves accepting new members into the community, often on a permanent basis. During much of the cold war, such admission was ideologically justified and reinforced by fresh memories of the refugee crisis in Europe during the 1930s and 1940s. This epoch is now closing, as the debate over article 16 of the German Constitution indicates. In the post-cold war era, the rationale for granting asylum in developed countries rests on the uncertain foundation of humanitarianism and a general interest of States in upholding international order. Moreover, the inherent suspicion that seeking asylum in affluent countries is a way of gaining access to their riches also weakens the institution.

In policy terms, three areas of challenge emerge. First, the institution of asylum itself must be made more efficient, with respect both to screening mechanisms and to the capacity to provide different types of protection, including one that is truly temporary. Second, receiving States must recognize that the refugees themselves are productive assets that can be mobilized to reduce the costs of asylum. In the African context, the possible contributions of refugees have been much discussed to justify the granting of asylum by poor States. In developed countries, this suggests closer integration of refugee and immigration policy, an issue that is being discussed in Canada. In an emergency situation, migration intakes may be reduced to accommodate more refugees and vice versa. While refugees may not meet all the criteria for admission as migrants or immigrants, the eventual substitution of refugees for migrants in terms of skills is probably possible.

However, even assuming progress in these respects, the demand for asylum will probably grow. A basic policy clarification is therefore required. As a long-time adviser to UNHCR notes, some of the current pressure on the institution of asylum stems from the notion that asylum can be used to attain programmatic rights and redress general wrongs such as religious discrimination, systemic exploitation and denial of political participation (Jaeger, 1992). The institution of asylum, however, has developed historically and legally to provide special benefits in situations of unusual hardship, involving persecution or outright conflict. In practice, the line distinguishing discrimination from persecution will vary, but at a time in which the institution of asylum itself is under considerable pressure, it is important not to lose sight of its essence.

Generally, programmatic rights must be dealt with through long-term policies that address the sources of conflict. Strategies to address root causes include development policies, the promotion of human rights and peacemaking activities, all of which are considered part of a "comprehensive refugee policy" in current United Nations discourse (International Labour Organization/United Nations High Commissioner for Refugees, 1992; Suhrke, 1992). However, as an organization, UNHCR can play only a limited role in combating the root causes. Its main contribution may be in focusing attention on the interconnections among policy, violence and refugee flows.

UNHCR can be more active in the intermediate area of prevention. Termed "preventive protection" during the Yugoslav crisis, the strategy consists of establishing UNHCR presence in or near conflict areas so as to provide relief and protection and, simultaneously, deter further violence (United Nations High Commissioner for Refugees, 1992). This strategy is closely associated with the current trend towards internalizing asylum, which has as its main objective the prevention of mass outflows. As discussed above, the implications of such a strategy are not devoid of problems.

F. The evolving institution of asylum and its implications for the refugee regime

Two types of asylum must be distinguished: individual applications for protection under the 1951 Convention or equivalent instruments, and mass displacement caused by war or other social disasters that require immediate temporary protection.

With regard to individual asylum, access to seeking asylum is at the core of the institution in its contemporary legal-philosophical form. If access is provided, it is easier to effect the second and necessary part of the operation, that is, the return of persons with unfounded claims. The problem of manifestly unfounded claims which may clog the asylum adjudication process has no easy solutions and was a main stumbling block in the attempt to adopt a convention on asylum in the late 1970s.

International cooperation can greatly reduce the costs of the asylum process, both to asylum-seekers and to the receiving society. A major problem in Europe and also partly in the Americas is that separate national jurisdictions permit asylum-seekers to file claims in several countries ("asylum shopping") and enable States to pursue beggar-thy-neighbour policies by returning refugees to the last "safe country" transited. Since liberal States will attract a disproportionate number of claims, the result is a competitive process among States to exclude persons seeking asylum.

The alternative is an international regime that grants asylum through a common process and unified criteria, which allocates the beneficiaries of asylum among member States according to need, capabilities and preferences. As Adelman (1992) has argued, such a regime would ensure shared responsibility in its authentic sense and would provide reasonable security and freedom to those granted asylum. It would be an advanced version of the present UNHCR system of quota refugees, whereby States accept the resettling of a number of refugees from first-asylum countries.

The harmonization of asylum policies in Europe since the mid-1980s represents a step in that direction. As yet, however, the European regime is a halfway house that promises reduced costs to the States but less freedom to the asylum-seekers. The competitive process to exclude persons seeking asylum still prevails. The same applies to relations between Northern America and Europe, and between Canada and the United States. There is an urgent need, therefore, to achieve greater cooperation within regional and international asylum regimes.

Historically, situations of generalized violence including civil and international war have caused large population displacements and led to the permanent migration of sizeable numbers of people. Migrants were readily admitted when there was a strong political rationale for doing so or when refugees met immigration criteria. Without such conditions, the response to large refugee flows must be temporary protection.

At present, temporary protection is discussed as the optimal solution for mass flows from conflicts such as civil war (Yugoslavia), religious-ethnic pogroms (Myanmar) and military coups (Haiti). In all cases, the conflict emerges as a temporary deviation from, if not peace, at least some degree of normalcy. There is a presumption that the status quo ante can be restored, possibly with the aid of international pressure and diplomacy. If neighbouring States host large refugee populations awaiting return, they clearly have an incentive to find an end to the conflict. The primary need is thus to provide mass protection to all victims and vulnerable groups during an interim period.

Some elements permitting a temporary protection regime to operate already exist. Certain States give various forms of temporary protection. Mass protection without individual determination of eligibility has been part of the UNHCR mandate in developing countries since the 1960s. Rudimentary regimes already exist in the form of international, typically regional, coordination to deal with refugee-producing crises. It remains to formalize ad hoc measures into predictable, collective action.

The logic is the same as that applying to individual asylum. In a crisis, refugees will flee towards the most accessible safe countries. Because the costs of denying protection can be high in terms of death and suffering, neighbouring countries find it difficult to deny asylum consistently. Yet, a system of burden-sharing would ease the task. Ad hoc arrangements in individual crisis situations do not add up to a regime; a regime in its very essence establishes norms for joint management of a common but unevenly distributed problem. The presumed result is a reasonably efficient and equitable process that enhances international order (Krasner, 1983). In the long run, it is clearly in the interest of most States to have a reliable framework for burden-sharing, since no one knows at whose border refugees will next appear.

Formalized refugee regimes, reflecting the pattern of refugee flows, will probably mean greater regionalization in response. Such regionalization will accentuate the obligation of more affluent regions and States to transfer funds to poor and hard-pressed areas. As the massive crisis in Somalia in 1992 demonstrates, acceptance in principle may not be sufficient. To avoid fragmentation, it is also advisable to retain UNHCR as the primary agent of refugee protection. Previous attempts to develop regional protection agencies have failed, notably under OAS in 1967 (Yundt, 1989). UNHCR itself suffers, however, from a structural weakness that was revealed by the Yugoslav crisis. Without a secure financial base, the organization remains hostage to the political interests of the principal donors. As a result, the only organization in the United Nations system that has a vested interest in protecting refugees is often prevented from advocating such protection forcefully. Assessed contributions from member States would constitute a secure financial base but only to the extent that members actually paid their dues.

Growing pressure on the institution of asylum in many parts of the world has generated a search for adjustment through innovative asylum mechanisms, streamlining of procedures, new aid strategies and collective action. Some of these issues are relatively new and have not been thoroughly evaluated as yet. There is a need to assess their impact so as to provide a better foundation for policy formulation. Attention should focus on the origin and nature of regional refugee regimes, systems of temporary protection, conditions for return, possibilities for the integration of refugee policy with general immigration policy, and the implications of strategies of "preventive protection". The challenges posed by the post cold-war era must be met with innovative solutions that take advantage of the growing opportunities for cooperation.

NOTES

[1]A similar logic is used in the philosophy of human rights to distinguish between basic and secondary rights (Shue, 1980).

[2]As amended by the 1967 Protocol (United Nations, 1967). Until then, the Convention had only applied to persons who were refugees as a result of events in Europe prior to 1951.

[3]As we have argued elsewhere, the exodus produced by the Irish potato famine in the 1840s was fundamentally related to the ownership of land, in itself the result of political oppression. Would this make the Irish economic or political refugees? (Zolberg, Suhrke and Aguayo, 1989, p. 32). Pronouncing on a similar case in 1980, an American judge concluded that the Haitians were political refugees because their economic plight was a result of systematic political oppression (Haitian Refugee Center versus Civiletti).

[4]The preamble does, however, refer to the Universal Declaration of Human Rights and affirms the fundamental rights of individuals.

[5]The Nordic countries led the way with the formal introduction of a "B" status for de facto or quasi-refugees, granting them permission to stay on humanitarian grounds. It was first suggested by an innovative Danish official who found that a group of Polish asylum-seekers were not "persecuted", yet realized it was politically impossible to send them back to Poland (Melander, 1979).

[6]For instance, India—not a signatory to the 1951 Convention—readily accepted thousands of Tibetan refugees in 1959.

[7]In the southern Sudan, for instance, Harrell-Bond (1986 and 1992) found numerous cases of confiscation of property, discriminatory treatment by courts, false and prolonged imprisonment, and beating, serious wounding, torture, assault, forced labour, kidnapping, rape and murder being perpetrated against refugees.

[8]Haitian boat people were both poor and black, while Cubans in their first flows were of European origin and middle-class. However, the flotilla of some 120,000 Cubans who sailed to Florida in 1980 included many belonging to the working class and primarily of non-European origin. The *Marielitos*, as they were called, were nevertheless admitted.

[9]I am indebted to Susan Forbes Martin of the Refugee Policy Group and Doris Meissner of The Carnegie Endowment, both of Washington, D.C., for discussion on this point.

[10]This section was written with the assistance of Finn Kvaale, Faculty of Law, University of Bergen.

[11]Of course, refugees like other types of migrants, are likely to foster the migration of close relatives or friends, thus promoting chain migration, as Blaschke (1989) has observed with respect to Turkish Kurds .

[12]Categories of acceptance included Convention status (persons who could establish a "well-founded fear of persecution") and other persons in need of protection. For the latter group, some States issued temporary residence permits (for instance, the exceptional leave to remain issued in the United Kingdom). Nordic countries granted residence permits on humanitarian grounds. The terms of stay were almost equivalent to those of Convention status.

[13]According to an intergovernmental study that tried to put a dollar figure on direct costs (administrative, legal and social aid, and so on), the bill for 13 Western-bloc European countries increased from around 500 million dollars in 1983 to around 7 billion dollars in 1991 (Inter-governmental Consultations on Asylum, Refugee and Migration Policies in Europe, North America and Australia, 1992). By comparison, the official development assistance (ODA) of the former Federal Republic of Germany was almost 5 billion dollars in 1989.

[14]An intergovernmental study found that only 25,000 out of 110,000 rejected applicants in 1990 had actually left the country voluntarily or "with the direct assistance of law enforcement authorities" (Inter-governmental Consultations on Asylum, Refugee and Migration Policies in Europe, North America and Australia, 1992, p. 11). The rest had "disappeared", that is to say, they probably remained as undocumented migrants or moved as asylum-seekers to other countries.

[15]A forerunner of the Dublin Convention, the Schengen Agreement was developed during the second half of the 1980s and signed by Belgium, France, Germany, Luxembourg and the Netherlands in 1990, with Italy, Portugal and Spain joining the following year and Greece being an observer. The Dublin Convention was signed by 11 of the 12 European Community member States in 1990 (Denmark did not sign) and is open for association to other European countries.

[16]This issue had for many years prevented the Council of Europe from developing a regional convention on asylum. The practice of sending asylum-seekers from one airport to another in Europe as one country tried to offload them to another proved to be the stumbling block. The Northern European countries argued that the country where the claimant first appeared should have the primary responsibility for processing an asylum application, and the Southern European countries—which were the principal points of entry for asylum-seekers from Western Asia and Africa—countered that responsibility should fall on the State that was the intended destination of the claimant (Kjærum and Horst, 1991).

[17]That the value of the practice of sending asylum-seekers back to countries deemed safe was, however, even more doubtful constituted a concern echoed in the Canadian debate (Adelman, 1992).

[18]Arguing that Western Europe's recognized refugee population in 1988 was only 0.17 per cent of Europe's population, Joly and Cohen (1989) denounced the "semi-hysteria" displayed by European Governments and the media, and noted that affluent Europe's share of the world's refugee population was indeed minute.

Aarbakke, Vemund (1990). Eksodus fra Bulgaria sommeren 1989. *Nordisk Østforum* (Oslo), vol. 1, pp. 3-20.

Adelman, Howard (1988). Refuge or asylum? A philosophical perspective. *Journal of Refugee Studies* (Oxford), vol. 1, No. 1, pp. 7-19.

_________ (1992). The safe third country in Canadian legislation. York University, Center for Refugee Studies, Toronto. Unpublished paper.

African Group in Geneva (1990). Refugees and migration flows in Africa. Paper presented by the African Group in Geneva to the United Nations High Commissioner for Refugees Working Group on Protection and Solutions. Geneva.

Aguayo, Sergio (1985). *El Exodo Centroamericano: Consecuencias de un Conflicto.* Mexico, D. F.: Consejo de Fomento Educativo.

Aleinikoff, Alexander T. (1991). The Refugee Convention at forty: reflections on the IJRL Colloquium. *International Journal of Refugee Law* (Oxford), vol. 3, No. 3 (July), pp. 617-626.

Amnesty International (1991). *Europe. Human Rights and the Need for a Fair Asylum Policy.* London: Amnesty International EUR 01\03\91. November.

Annals (1939) *of The American Academy of Political and Social Sciences* (Newbury Park, California), vol. 203 (special issue on minorities and refugees).

Arboleda, Eduardo (1991). Refugee definition in Africa and Latin America: the lessons of pragmatism. *International Journal of Refugee Law* (Oxford), vol. 3, No. 2 (April), pp. 185-207.

Blaschke, Jochen (1989). Refugees and Turkish migrants in West Berlin. In *Reluctant Hosts: Europe and Its Refugees*, Danièle Joly and Robin Cohen, eds. Aldershot, United Kingdom: Avebury.

Bonavia, David (1983). *Hong Kong 1997.* Hong Kong: South China Morning Post.

Childers, Erskine (1992). Gulf crisis lessons for the United Nations. *Bulletin of Peace Proposals* (Oslo), vol. 23, No. 2 (June), pp. 129-138.

Council of Europe, Parliamentary Assembly (1992). *Report on Europe of 1992 and Refugee Policies* (12 April). ADOC 6413. Strasbourg: Council of Europe.

Cuénod, Jacques (1989). Refugees: development or relief? In *Refugees and International Relations*, Gil Loescher and Laila Monahan, eds. Oxford: Oxford University Press.

Danish Refugee Council (1988). *The Role of Airline Companies in the Asylum Procedure*. Copenhagen.

_________ (1991). *The Effects of Carrier Sanctions on the Asylum System*. Copenhagen.

Einarsen, Terje (1990). The European Convention on Human Rights and the notion of an implied right to de facto asylum. *International Journal of Refugee Law* (Oxford), vol. 2, No. 3, pp. 361-389.

_________ (1992). Mass movements of refugees: in search of new international mechanisms. In *The Living Law of Nations*, P. Macalister-Smith and Gudmundur Alfredsson, eds. Strasbourg: Engel.

Encyclopedia of Public International Law (EPIL) (1985). Vol. 8. Amsterdam: North Holland Publishing Company.

European Consultation on Refugees and Exiles (1989). A refugee policy for Europe. In *Reluctant Hosts: Europe and Its Refugees*, Danièle Joly and Robin Cohen, eds. Aldershot, United Kingdom: Avebury.

_________ (1992). Synthesis of responses regarding the situation in former Yugoslavia. Memorandum (July). London.

Gallagher, Dennis, Susan Forbes Martin and Patricia Weiss-Fagen (1989). Temporary safe haven: the need for North American-European responses. In *Refugees and International Relations*, Gil Loescher and Laila Monahan, eds. Oxford: Oxford University Press.

Gibney, Mark, ed. (1989). *Open Borders? Closed Societies?* New York: Greenwood Press.

Goodin, Robert E. (1988). "What is so special about our fellow countrymen?". *Ethics* (Chicago), vol. 98, No. 4 (July), pp. 663-686.

Goodwin-Gill, Guy S. (1983). *The Refugee in International Law.* Oxford: Clarendon Press.

Grahl-Madsen, Atle (1966). The European tradition of asylum and the development of refugee law. *Journal of Peace Research* (Oslo), vol. 3, pp. 278-289.

_________ (1980). *Territorial Asylum.* Stockholm: Almqvist and Wiksell.

Gros Espiel, Hector (1981). El derecho internacional Americano sobre asilo territorial y extradición en sus relaciones con la Convención de 1951 y el Protocolo de 1967 sobre el Estatuto de los Refugiados. Mexico, D.F. (May). Unpublished paper.

Hailbronner, Kay (1991). The concept of "safe country" and expedient asylum procedures. Report of the Ad Hoc Committee of Experts on the Legal Aspects of Territorial Asylum, Refugees and Stateless Persons (CAHAR). *Cahar*, vol. 91, No. 2. Strasbourg: Council of Europe.

Haitian Refugee Center v. James Baker, III, Secretary of State, The. *Brief Amicus Curiae.* United States Court of Appeals, No. 91-6060 (December 1991).

Harrell-Bond, Barbara E. (1986). *Imposing Aid.* Oxford: Oxford University Press.

_________ (1992). Refugees and the reformulation of international aid policies. Background paper for United Kingdom-Japan 2000 Group. Oxford.

Hartigan, Kevin (1992). Refugee policies in Mexico and Honduras. *International Organization* (Boston), vol. 46, No. 3 (summer), pp. 709-730.

Hartman, Joan Fitzpatrick (1988). The principle and practice of temporary refuge: a customary norm protecting civilians fleeing armed conflict. In *The New Asylum Seekers: Refugee Law in the 1980s*, David A. Martin, ed. Dordrecht: Martinus Nijhoff.

Hathaway, James C. (1991). Reconceiving refugee law as human rights protection. *Journal of Refugee Studies* (Oxford), vol. 4, No. 2, pp. 113-131.

Helsinki Watch (1989). *Detained, Denied, Deported. Asylum Seekers in the United States.* New York: UTGIVER, pp. 51-52.

Helton, Arthur C. (1990). Final asylum rules: finally. *Interpreter Releases* (Washington, D.C.), vol. 67, No. 27 (23 July), pp. 789-790.

Henkin, Louis, and others (1989). *Right v. Might. International Law and the Use of Force.* New York: Council on Foreign Relations.

Holborn, Louise W. (1975). *Refugees. A Problem of Our Time.* Metuchen, New Jersey: Scarecrow Press.

Hovy, Bela (1993). Asylum migration in Europe: patterns, determinants and the role of East-West movements. In *The New Geography of European Migrations*, Russell King, ed. London and New York: Bellhaven Press.

Inter-governmental Consultations on Asylum, Refugee and Migration Policies in Europe, North America and Australia (1991). Towards international strategies: existing initiatives in a comparative perspective. Background paper prepared for the Third Meeting of the Working Group on Long-Term Perspectives and Politics, Dresden, 7-9 April 1991.

_________ (1992). Towards international recognition of the need for consistent removal policies with respect to rejected asylum-seekers. Working paper. Geneva: Inter-governmental Consultations on Asylum, Refugee and Migration Policies in Europe, North America and Australia.

International Labour Organization/United Nations High Commissioner for Refugees (1992). *Informal Summary Record.* Joint Meeting on International Aid as a Means to Reduce the Need for Emigration, Geneva, 6-8 May.

Jaeger, Gilbert (1992). Refugees or migrants? The recent approach to refugee flows as a particular aspect of migration. Brussels. Unpublished manuscript.

Joly, Danièle (1992). *Refugees. Asylum in Europe?* London: Minority Rights Group.

_________, and Robin Cohen, eds. (1989). *Reluctant Hosts: Europe and Its Refugees.* Aldershot, United Kingdom: Avebury.

Kaye, Ronald (1992). British refugee policy and 1992: the breakdown of a policy community. *Journal of Refugee Studies* (Oxford), vol. 5, No. 1, pp. 47-67.

Kjærum, Morten, and Christian Horst (1991). *Flugten til Europa.* Copenhagen: Danish Center for Human Rights.

Krasner, Stephen D., ed. (1983). *International Regimes.* Ithaca, New York: Cornell University Press.

Loescher, Gil (1986). *Calculated Kindness.* New York: The Free Press.

__________ (1989). The European community and refugees. *International Affairs* (London), vol. 65, No. 4 (autumn), pp. 617-636.

Marrus, Michael (1985). *The Unwanted.* New York: Oxford University Press.

Meissner, Doris (1992a). Managing migrations. *Foreign Policy* (Washington, D.C.), vol. 86 (spring), pp. 66-85.

__________ (1992b). Yes, they're poor - and persecuted too. *Los Angeles Times* (31 July).

Melander, Göran (1979). *Flyktning i Norden.* Stockholm: Norstedt och Söners Forlag.

__________ (1987). *The Two Refugee Definitions.* Report No. 4. Lund, Sweden: University of Lund, Raoul Wallenberg Institute.

Miller, David (1988). The ethical significance of nationality. *Ethics* (Chicago), vol. 98, No. 4 (July), pp. 647-662.

Moore, Barrington (1978). *Injustice: The Social Bases of Obedience and Revolt.* London: Macmillan.

Nobel, Peter (1987). *Protection of Refugees in Europe as Seen in 1987.* Report No. 4. Lund, Sweden: University of Lund, Raoul Wallenberg Institute.

Oloka-Onyango, Joe (1991). Human rights, the OAU Convention and the refugee crisis in Africa: 40 years after Geneva. *International Journal of Refugee Law* (Oxford), vol. 3, No. 3 (July), pp. 453-460.

Organisation for Economic Cooperation and Development (1992). *SOPEMI 1991: Trends in International Migration.* Paris.

Organization of American States, Inter-American Commission on Human Rights (1983). *Annual Report* (Washington, D.C). Ser/L/V/11.61 Doc 32 (September).

Papademetriou, Demetrios G. (1994). International migration in North America: Issues, policies and implications. In *International Migration: Regional Processes and Responses,* Miroslav Maçura and David Coleman, eds. Economic Studies, No. 7. Sales No. GV.E.94.0.25. New York and Geneva: United Nations, Economic Commission for Europe.

Plender, Richard (1989). *The Right of Asylum.* The Hague: Hague Academy of International Law, Centre for Studies and Research in International Law and International Relations.

Rodgers, Malcolm (1992). Refugees and international aid. Sri Lanka: a case study. Paper presented at the Joint ILO/UNHCR Meeting on International Aid as a Means to Reduce the Need for Emigration, Geneva, 6-8 May.

Rogers, Rosemarie (1991). Responses to immigration: the Western European host countries and their immigrants. Paper prepared for the Conference on the Impact of International Migration on the Security and Stability of States, Massachusetts Institute of Technology, Boston, 5 and 6 December.

Russell, Sharon Stanton, Karen Jacobsen and William Deane Stanley (1991). *International Migration and Development in Sub-Saharan Africa.* World Bank Discussion Papers, Nos. 101 and 102. Washington, D.C.: World Bank.

Sandel, Michael J. (1984). Justice and the Good. In *Liberalism and Its Critics,* Sandel, ed. Oxford: Blackwell.

Schneider, Robin (1989). Asylum and xenophobia in West Germany. In *Reluctant Hosts: Europe and Its Refugees,* Danièle Joly and Robin Cohen, eds. Aldershot, United Kingdom: Avebury.

Shacknove, Andrew E. (1985). Who is a refugee? *Ethics* (Chicago), vol. 95, pp. 274-284.

Shue, Henry (1980). *Basic Rights: Subsistence, Affluence and U.S. Foreign Policy.* Princeton, New Jersey: Princeton University Press.

__________ (1988). Mediating duties. *Ethics* (Chicago), vol. 98, No. 4 (July), pp. 687-704.

Studie-en Informatiecentrum Mensenrechten (SIM) (Dutch Institute for Human Rights) (1989). *Refugees in the World: The European Community's Response.* Considerations, Conclusions and Recommendations of the International Conference. SIM special, No. 10. The Hague: Netherlands Institute of Human Rights and Dutch Refugee Council.

Suhrke, Astri (1983). Indochinese refugees: the law and politics of first asylum. *Annals of the American Academy of Political and Social Science* (Philadelphia), vol. 467, pp. 102-115.

__________ (1992). Towards a comprehensive refugee policy: social conflict and refugees in the post-Cold War World. Paper presented at the Joint ILO/UNHCR Meeting on International Aid as a Means to Reduce the Need for Emigration, Geneva, 6-8 May.

Sutter, Valerie O'Connor (1990). *The Indochinese Refugee Dilemma.* Baton Rouge: Louisiana State University Press.

Tucker, Robert W., and David C. Hendricksom (1992). *Imperial Temptation.* New York: Council on Foreign Relations.

United Nations (1955). *Treaty Series,* vol. 213, No. 2889.

__________ (1957). *Treaty Series,* vol. 189, No. 2545.

__________ (1967). *Treaty Series,* vol. 606, No. 8791.

__________ (1976). *Treaty Series,* vol. 1001, No. 14691.

United Nations High Commissioner for Refugees (1983). *Conclusions on the International Protection of Refugees.* Adopted by the Executive Committee of the Programme of the United Nations High Commissioner for Refugees at its thirty-fourth session. Geneva.

__________ (1992). *A Comprehensive Response to the Humanitarian Crisis in the Former Yugoslavia.* Geneva: July.

__________ (1993). *The State of the World's Refugees: The Challenge of Protection.* London: Penguin Books.

United States Committee for Refugees (1987). *World Refugee Survey.* Washington, D.C.: United States Committee for Refugees.

__________ (1992). *Yugoslavia Torn Asunder.* Washington D.C.: United States Committee for Refugees.

United States Department of State (1990). *World Refugee Report.* Washington, D.C.: United States Government Printing Office.

van Gusteren, Herman R. (1988). Admission to citizenship. *Ethics* (Chicago), vol. 98, No. 4 (July), pp. 731-741.

Walzer, Michael (1985). *Spheres of Justice: A Defence of Pluralism and Equality.* Oxford: Blackwell.

Warner, Daniel, and James Hathaway (1992). Refugee law and human rights: Warner and Hathaway in debate. *Journal of Refugee Studies* (Oxford), vol. 5, No. 2, pp. 162-168.

Widgren, Jonas (1990a). International migration and regional stability. *International Affairs* (London), vol. 66, No. 4, pp. 749-766.

__________ (1990b). The asylum crisis in the OECD-region. Paper prepared for the Fletcher School of Law and Diplomacy, Boston.

__________ (1992). The need to improve co-ordination of European asylum and migration policies. Paper prepared for the Conference on Comparative Law of Asylum and Immigration in Europe, Trier, Germany, 12 and 13 March.

Wollny, Hans (1991). Asylum policy in Mexico: a survey. *Journal of Refugee Studies* (Oxford), vol. 4, No. 3, pp. 219-236.

Yundt, Keith W. (1989). The organization of American states and legal protection to political refugees in Central America. *International Migration Review* (Staten Island, New York), vol. 23, No. 2, pp. 201-218.

Zolberg, Aristide R., Astri Suhrke and Sergio Aguayo (1989). *Escape from Violence.* New York: Oxford University Press.

Zucker, Norman L., and Naomi F. Zucker (1987). *The Guarded Gate: The Reality of American Refugee Policy.* San Diego, California: Harcourt Brace Jovanovich.

XVII. CHANGING SOLUTIONS TO REFUGEE MOVEMENTS

*John R. Rogge**

Over four decades have elapsed since the Office of the United Nations High Commissioner for Refugees (UNHCR) was established to supersede the International Refugee Organization (IRO). At the time of its formation in 1951, UNHCR was conceived as a temporary institution whose mandate was to solve Europe's refugee problem after the Second World War (Vernant, 1953; Proudfoot, 1956; Holborn, 1975). Its principal task was to provide protection and find durable solutions for the plight of refugees and displaced persons of diverse European ancestry.[1] This was achieved primarily through resettlement in other European countries, in Northern America or in Australia. UNHCR was therefore concerned primarily with migration between developed countries. Indeed, separate agencies were set up by the United Nations in Western Asia and the Korean peninsula to cope with concurrent problems of refugees and displaced persons in those regions. One of UNHCR's most difficult and protracted tasks during those early years was to find durable solutions for vulnerable groups such as the aged, the physically or mentally handicapped, female-headed households, and Europe's many orphans. It had been expected that the Office would work itself out of a job by the mid-1950s and be disbanded.[2]

In 1956, the Hungarian crisis confronted UNHCR with a new refugee wave of major proportions and, shortly thereafter, the Office had to address for the first time two refugee outflows from developing countries: that of Algerians moving to Tunisia and Morocco; and the incipient exodus of southern Sudanese. Since then, UNHCR's ever growing role has primarily focused on protecting and finding durable solutions for refugees produced by developing countries, both in the developed world through resettlement programmes and within the developing world through local settlement or repatriation. In contrast, the outflow of refugees and self-exiled persons from Eastern Europe during 1960-1989 placed only modest demands upon the Office, since third-country resettlement for persons fleeing Europe's communist countries was almost always "guaranteed". Only with the dismantling of the former Union of Soviet Socialist Republics (USSR) and the tragic current events in the former Yugoslavia have refugee flows between developed countries again become a primary concern to UNHCR and one that is increasingly diverting attention and resources from the continuing and enormous needs in the developing world.

The issues and needs related to the current population of refugees and other involuntary migrants in developing countries are the focus of the present paper. Attention is centred on the primary causes of population displacement and how those causes are changing. The durable solutions available and the not-so-durable solutions that are increasingly adopted by default, if not by intent, are also discussed, as well as the changing attitudes towards refugees and displaced persons in countries of origin, countries of first asylum, traditional resettlement countries and the international community in general.

A. Magnitude of the refugee problem

The importance of any crisis or emergency is invariably determined by the number of people affected, as are, in most cases, the speed and level of the response. However, in almost all refugee and displaced-person emergencies, priority is seldom given to the systematic assessment of the number of persons affected. Consequently, crude estimates are the rule and, if made by local agencies in desperate need for assistance, overestimation is common. The media are also quick to propose unverified numbers, particularly when they want to dramatize the magnitude of an emergency, and all too often such numbers become accepted as fact. The net effect is that the numbers used are frequently unreliable, sometimes exaggerated and consequently misleading. Thus, many of the numbers cited in tables 54 to 57 are questionable or may vary considerably depending on the sources used.

*Disaster Research Unit, University of Manitoba, Winnipeg, Canada.

No country better illustrates the problem of reliability of refugee-related estimates than Somalia, where the number of refugees and displaced persons has always been a highly controversial issue. Indeed, in the early 1980s, the difference between Government and UNHCR refugee estimates provoked a crisis that culminated in the expulsion of the UNHCR representative from Somalia. A special commission was later formed to "negotiate" an estimate acceptable to both parties, resulting in a drop in the number of refugees from the 1.25 million claimed by the Government to 700,000, and even the latter figure was believed to be an overestimate. A similar problem exists currently with regard to the internally displaced Somalis in southern and central Somalia, whose numbers are put as high as 1.5 million by Somali "authorities". My own assessment is that the number of internally displaced persons in the region peaked in the fall of 1992 at between 550,000-650,000 (Rogge, 1992). Clearly, estimates matter in any type of crisis, since the number of people affected determines the level of need. However, given the increasing scarcity of donor support for refugees and displaced persons, such estimates must be realistic and must be appraised critically. Organizations offering assistance, whether they are international, non-governmental or national, can and must do more to ensure that more reliable data on refugees and displaced persons are generated.

Table 54 indicates the magnitude of the refugee population in the two regions hosting the largest numbers of refugees, Africa and Asia. It presents a comparison of the number of refugees according to UNHCR sources and to the United States Committee for Refugees (USCR) for late 1991. UNHCR sources indicate that the data presented are provided mainly by Governments according to their own records and methods of estimation. USCR presents no explanation on how the data were derived. In the case of Africa, the overall difference in the figures between the two sources amounts to slightly over 66,000 persons and although the correspondence on a country-by-country basis between the two sources is far from perfect, major differences arose mostly because USCR sources regarded as refugees groups that had not been reported as such by Governments. Thus, countries, like Benin, Liberia, Mauritania, Namibia and South Africa, that reported almost no refugees to UNHCR, appear as hosting sizeable numbers of refugees according to USCR statistics. When refugee populations are reported by both sources, discrepancies on their total count tend to be small, at least in relative terms, reflecting the uncertainty surrounding refugee statistics in general. An exception is the number of Rwandan refugees in Burundi which is about three times higher according to UNHCR sources than according to USCR.

In the case of Asia, the number of refugees listed under the column labelled "UNHCR" includes Palestinian refugees under the mandate of the United Nations Relief and Works Agency for Palestine Refugees in the Near East (UNRWA). Therefore the total amounts to 11.1 million in late 1991. In comparison, USCR sources put the equivalent figure at 10.5 million. The main cause of this discrepancy is the figure provided for the number of Iraqi refugees in the Islamic Republic of Iran, which amounts to 1.2 million according to UNHCR and to only 150,000 according to USCR. In addition, although there are figures for refugees reported by Kuwait according to UNHCR sources, corresponding figures are not included in the USCR. Other cases in which major discrepancies are evident usually involve the inclusion in USCR statistics of populations that are not reported as refugees by Governments. Cases in point include people from Bangladesh, Bhutan, Myanmar and Tibet in India, and people from Cambodia and Myanmar in Thailand.

Table 55 shows the countries of origin of the largest refugee populations in the world as of late 1991. It does not reflect, however, the 1992 refugee outflows from the former Yugoslavia, which by September of that year were estimated to involve 1.6 million refugees and 938,000 displaced persons (United States Committee for Refugees, 1992a), nor does it indicate the Somali exodus to Kenya, whose refugee population quadrupled during 1992 to reach 400,000. Another important refugee outflow occurring early in 1992 involved over 300,000 Rohingya refugees from Myanmar who entered Bangladesh. As table 55 shows, with the exception of Yugoslavia, the main countries of origin of refugees are located in Africa or Asia. In Latin America, the number of refugees is considerably lower. By the end of 1991, Mexico was the main country of asylum in that region, hosting 49,000 refugees, and Costa Rica followed suit with 24,000.

For some refugee populations, the period of exile has been very long. This has been the case for the Angolans, the people of Burundi, the Cambodians, the Eritreans, the Palestinians, the Rwandans, the Sahrawis and the Tibetans, whose totals reflect a sizeable second and in some cases third generation of refugees. Moreover, whereas repatriation is under way or imminent for several of those

TABLE 54. NUMBER OF REFUGEES IN AFRICA AND ASIA, BY COUNTRY OF ASYLUM AND COUNTRY OF ORIGIN, ACCORDING TO THE OFFICE OF THE UNITED NATIONS HIGH COMMISSIONER FOR REFUGEES (UNHCR) AND TO THE UNITED STATES COMMITTEE FOR REFUGEES (USCR), 1991

Country or area of asylum	Country or area of origin	Number of refugees at end of 1991		
		USCR	*UNHCR*	*Difference*
		A. *Africa*		
Algeria	Mali	35 000	..	35 000
	Western Sahara	165 000	165 000	0
	Various	4 000	4 100	-100
	Total	204 000	169 100	34 900
Angola	South Africa	100	300	-200
	Zaire	10 300	10 800	-500
	Total	10 400	11 000	-600
Benin	Chad	100	400	-300
	Togo	15 000	..	15 000
	Total	15 100	400	14 700
Botswana	Angola	200	..	200
	South Africa	1 000	500	500
	Zimbabwe	..	100	-100
	Various	200	300	-100
	Total	1 400	900	500
Burkina Faso	Chad	200	..	200
	Various	200	300	-100
	Total	400	300	100
Burundi	Rwanda	80 600	243 900	-163 300
	Uganda	..	300	-300
	Zaire	25 900	25 800	100
	Total	106 500	270 000	-163 500
Cameroon	Chad	6 500	44 800	-38 300
	Various	400	400	0
	Total	6 900	45 200	-38 300
Central African Republic	Chad	1 000	1 000	0
	Sudan	8 000	11 100	-3 100
	Various	..	100	-100
	Total	9 000	12 200	-3 200
Congo	Central African Republic	300	300	0
	Chad	2 300	2 200	100
	Zaire	400	400	0
	Various	400	400	0
	Total	3 400	3 300	100
Côte d'Ivoire	Liberia	240 000	229 900	10 100
	Various	400	400	0
	Total	240 400	230 300	10 100
Djibouti	Ethiopia	15 000	11 500	3 500
	Somalia	105 000	84 600	20 400
	Total	120 000	96 100	23 900

TABLE 54 (*continued*)

Country or area of asylum	*Country or area of origin*	*Number of refugees at end of 1991*		
		USCR	*UNHCR*	*Difference*
Egypt	Ethiopia	..	600	-600
	Palestinians	5 500	..	5 500
	Somalia	..	1 300	-1 300
	Various	2 250	300	1 950
	Total	7 750	2 200	5 550
Ethiopia	Somalia	519 000	512 000	7 000
	Sudan	15 000	15 000	0
	Total	534 000	527 000	7 000
Gabon	Total	800	200	600
Gambia	Liberia	1 000	200	800
	Senegal	500	..	500
	Total	1 500	200	1 300
Ghana	Liberia	6 000	8 000	-2 000
	Various	150	100	50
	Total	6 150	8 100	-1 950
Guinea	Liberia	397 000	548 000	-151 000
	Sierra Leone	169 000	..	169 000
	Total	566 000	548 000	18 000
Guinea-Bissau	Senegal	4 600	4 600	0
Kenya	Ethiopia	11 800	10 600	1 200
	Rwanda	2 000	..	2 000
	Somalia	92 200	95 900	-3 700
	Uganda	700	9 800	-9 100
	Various	450	3 900	-3 450
	Total	107 150	120 200	-13 050
Lesotho	South Africa	300	200	100
Liberia	Sierra Leone	12 000	..	12 000
Malawi	Mozambique	950 000	981 800	-31 800
Mali	Mauritania	13 000	13 100	-100
	Niger	500	..	500
	Total	13 500	13 100	400
Mauritania	Senegal	22 000	..	22 000
	Mali	18 000	..	18 000
	Various	..	35 200	-35 200
	Total	40 000	35 200	4 800
Morocco	Total	800	300	500
Mozambique	Total	500	400	100
Namibia	Angola	30 000	..	30 000
	Various	200	100	100
	Total	30 200	100	30 100
Niger	Chad	1 400	1 400	0

TABLE 54 (*continued*)

Country or area of asylum	Country or area of origin	Number of refugees at end of 1991		
		USCR	UNHCR	Difference
Nigeria	Chad	3 300	1 500	1 800
	Ghana	..	200	-200
	Liberia	1 000	1 000	0
	Various	300	900	-600
	Total	4 600	3 600	1 000
Rwanda	Burundi	32 500	34 000	-1 500
Senegal	Guinea Bissau	..	5 000	-5 000
	Mauritania	53 000	66 800	-13 800
	Various	100	100	0
	Total	53 100	71 900	-18 800
Sierra Leone	Liberia	17 200	..	17 200
	Various	..	28 000	-28 000
	Total	17 200	28 000	-10 800
Somalia	Ethiopia	35 000	..	35 000
South Africa	Mozambique	200 000	..	200 000
	Lesotho	1 000	..	1 000
	Total	201 000	..	201 000
Sudan	Chad	20 000	20 700	-700
	Ethiopia	690 000	700 000	-10 000
	Uganda	2 700	6 500	-3 800
	Zaire	4 500	2 000	2 500
	Total	717 200	729 200	-12 000
Swaziland	Mozambique	39 500	42 000	-2 500
	South Africa	7 700	7 500	200
	Various	..	100	-100
	Total	47 200	49 600	-2 400
United Republic of Tanzania	Burundi	131 000	148 700	-17 700
	Mozambique	72 000	72 200	-200
	Rwanda	22 300	50 000	-27 700
	South Africa	9 600	..	9 600
	Zaire	16 000	16 000	0
	Various	200	1 300	-1 100
	Total	251 100	288 200	-37 100
Togo	Ghana	..	3 200	-3 200
	Liberia	..	100	-100
	Various	450	100	350
	Total	450	3 400	-2 950
Tunisia	Total	50	100	-50
Uganda	Rwanda	87 000	84 000	3 000
	Sudan	75 500	77 100	-1 600
	South Africa	2 000	..	2 000
	Zaire	600	600	0
	Various	350	700	-350
	Total	165 450	162 400	3 050

TABLE 54 (*continued*)

Country or area of asylum	Country or area of origin	Number of refugees at end of 1991		
		USCR	*UNHCR*	*Difference*
Zaire	Angola	310 000	278 600	31 400
	Burundi	45 000	41 200	3 800
	Rwanda	12 000	50 900	-38 900
	Sudan	104 000	90 800	13 200
	Uganda	10 000	20 100	-10 100
	Various	1 300	1 300	0
	Total	482 300	482 900	-600
Zambia	Angola	103 000	102 500	500
	Mozambique	25 000	23 500	1 500
	Namibia	..	100	-100
	South Africa	2 000	1 800	200
	Uganda	1 500	..	1 500
	Zaire	9 000	..	9 000
	Various	..	12 700	-12 700
	Total	140 500	140 600	-100
Zimbabwe	Mozambique	197 000	197 100	-100
	South Africa	1 000	..	1 000
	Various	500	500	0
	Total	198 500	197 600	900
Other Africa	Total	..	800	-800
	TOTAL	5 340 300	5 274 100	66 200
		B. *Asia*		
Bangladesh	Myanmar (mostly)	30 150	40 300	-10 150
China	Viet Nam	150	284 500	-284 350
	Myanmar	10 000	..	10 000
	Lao People's Democratic Republic	4 200	4 100	100
	Various	..	200	-200
	Total	14 350	288 800	-274 450
Gaza Strip	Palestinian	528 700	528 700	0
Hong Kong	Viet Nam	60 000	60 000	0
India	Afghanistan	9 800	9 800	0
	Bangladesh	65 000	..	65 000
	Bhutan	15 000	..	15 000
	China (Tibet)	100 000	..	100 000
	Myanmar	2 000	..	2 000
	Sri Lanka	210 000	200 000	10 000
	Various	800	800	0
	Total	402 600	210 600	192 000
Indonesia	Cambodia	1 700	..	1 700
	Viet Nam	17 000	17 000	0
	Various	..	1 700	-1 700
	Total	18 700	18 700	0

TABLE 54 (*continued*)

Country or area of asylum	*Country or area of origin*	*Number of refugees at end of 1991*		
		USCR	*UNHCR*	*Difference*
Iran (Islamic Republic of)	Afghanistan	3 000 000	3 186 600	-186 600
	Iraq (mostly)	150 000	1 218 400	-1 068 400
	Total	3 150 000	4 405 000	-1 255 000
Iraq	Iran (Islamic Republic of)	48 000	..	48 000
	Various	..	88 000	-88 000
	Total	48 000	88 000	-40 000
Japan	Total	900	9 100	-8 200
Jordan	Palestinian	960 200	960 200	0
	Various	..	400	-400
	Total	960 200	960 600	-400
Kuwait	Iraq	..	20 000	-20 000
	Stateless	..	80 000	-80 000
	Palestinian	..	25 000	-25 000
	Total	..	125 000	-125 000
Lebanon	Palestinian	310 600	310 600	0
	Various	3 600	5 200	-1 600
	Total	314 200	315 800	-1 600
Macau	Viet Nam (mostly)	100	100	0
Malaysia	Viet Nam	12 500	12 500	0
	Various	200	1 500	-1 300
	Total	12 700	14 000	-1 300
Nepal	Bhutan	10 000	9 500	500
	China (Tibet)	14 000	..	14 000
	Various	..	100	-100
	Total	24 000	9 600	14 400
Pakistan	Afghanistan	3 591 000	3 098 000	493 000
	Iran (Islamic Republic of)	..	500	-500
	Various	3 000	1 400	1 600
	Total	3 594 000	3 099 900	494 100
Papua New Guinea	Indonesia	6 700	..	6 700
Philippines (Refugee Processing Centre)	Viet Nam	18 000	19 800	-1 800
	Various	..	100	-100
	Total	18 000	20 000	-2 000
Republic of Korea	Viet Nam	200	200	0
Saudi Arabia	Iraq	34 000	32 900	1 100
	Various	..	200	-200
	Total	34 000	33 100	900
Singapore	Total	150	200	-50

TABLE 54 (*continued*)

Country or area of asylum	*Country or area of origin*	*Number of refugees at end of 1991* USCR	UNHCR	Difference
Syrian Arab Republic	Iraq	4 000	..	4 000
	Palestinian	289 900	298 900	-9 000
	Various	..	4 200	-4 200
	Total	293 900	303 100	-9 200
Thailand	Cambodia	370 000	15 000	355 000
	Laos	59 000	57 300	1 700
	Myanmar	70 000	..	70 000
	Viet Nam	13 700	13 700	0
	Various	..	2 200	-2 200
	Total	512 700	88 200	424 500
Turkey	Iran (Islamic Republic of)	2 000	1 400	600
	Iraq	29 500	28 000	1 500
	Total	31 500	29 400	2 100
Viet Nam	Cambodia	21 000	20 100	900
West Bank	Palestinian	430 100	430 100	0
Yemen	Ethiopia	..	3 100	-6 100
	Somalia	..	26 700	-26 700
	Various	11 100	100	11 000
	Total	11 100	29 900	-18 800
Other Asia	Total	..	600	- 600
	TOTAL	10 517 950	11 129 000	- 611 050

Sources: United States Committee for Refugees, *World Refugee Survey, 1992* (Washington, D.C., American Council for Nationalities Service, 1992); United Nations High Commissioner for Refugees, *The State of the World's Refugees: The Challenge of Protection* (London and New York, Penguin Books, 1993); and *World Population Monitoring, 1993: With a Special Report on Refugees* (United Nations publication, Sales No. 95.XIII.8).

NOTE: According to UNHCR sources, UNCHR figures presented here are provided mainly by Governments of asylum countries based on their own records and methods of estimation.

populations, for many, like that of Burundi, the Rwandans, the Sahrawis and the Tibetans, there are few such prospects.

In addition to the refugees enumerated in tables 54 and 55, there are many persons who fear persecution or harm were they to return to their home countries, but who are nevertheless not granted refugee status by receiving countries. Some of those persons are provided with residence permits for humanitarian or political reasons but many are forced to remain as undocumented aliens. Data on the number of persons in such refugee-like situations are fragmentary and subject to much conjecture (see table 56). Persons in such situations normally receive none of the protection or assistance accorded to recognized refugees. For instance, the Biharis in Bangladesh have lived in that country since 1971 as a permanent underclass at the extreme margins of social and economic life.

A third group of persons in refugee-like situations consists of the internally displaced. Frequently, when conflict produces a population outflow beyond a country's borders, it also forces people to move within such borders. Involuntary population displacements may also result from government policy, as in South Africa, in Cambodia during the Khmer Rouge era, and in Angola, Malaysia, Mozambique and Zimbabwe before independence. Such displacements can be very substantial, as shown in table 57. Internally displaced persons normally cannot avail themselves of the protection provided by the international community, since the consent of their Government is needed to provide such assistance and Governments whose actions lead to population displace-

TABLE 55. MAJOR REFUGEE POPULATIONS, BY COUNTRY OR AREA OF ORIGIN AS OF 31 DECEMBER 1991
(*Thousands*)

Country or area of origin	*Refugee population*
Afghanistan	6 601
Palestinians	2 525
Mozambique	1 484
Ethiopia	752
Somalia	718
Liberia	662
Angola	443
Cambodia	393
Iraq	218
Sri Lanka	210
Burundi	209
Rwanda	204
Sudan	203
Sierra Leone	181
Western Sahara	165
Viet Nam	123
Yugoslavia (former)	120
China (Tibet)	114
Myanmar	112

Source: United States Committee for Refugees, *World Refugee Survey, 1992* (Washington, D.C., American Council for Nationalities Service, 1992).

TABLE 56. MAJOR POPULATIONS IN REFUGEE-LIKE SITUATIONS

Host country	*Origin of population*	*Total*
Jordan	Palestinians	740 000
Iran (Islamic Republic of)	Iraq	500 000
Mexico	Central America	350 000
Bangladesh	Pakistan (Biharis)	260 000
Guatemala	Central America	250 000
United States of America	El Salvador	200 000
Burundi	Rwanda	187 000
Thailand	Myanmar	160 000
Uganda	Rwanda	120 000
Egypt	Palestinians	100 000
Turkey	Iran (Islamic Republic of)	100 000
Kuwait	Palestinians	80 000
Costa Rica	Central America	80 000

Source: United States Committee for Refugees, *World Refugee Survey, 1992* (Washington, D.C., American Council for Nationalities Service, 1992).

ments are often reluctant to even admit that a problem exists. The problem of providing assistance to internally displaced populations is compounded when such populations find themselves in areas controlled by insurgent groups.

No other world region presents as clearly as Africa the complexities and diverse causes of, and the limited options available with respect to solving, refugee crises. Not only has the total number of refugees in the continent increased substantially since 1960 but, in addition, the number of countries that generate and/or receive refugees has risen. Thus, by 1990, Western Africa, once a region relatively free of refugee populations, had experienced major refugee outflows and, by extension, was hosting sizeable numbers of internally displaced persons. Although no single African country has yet generated a number of refugees comparable with those from Afghanistan, in Africa the causes of refugee movements, the directions of the flows, the durable solutions available and the international responses to crises create a dilemma that is considerably more complex than in most other regions of the developing world. The remaining sections of this paper will focus mostly on the African refugee dilemma.

B. CAUSES OF FLIGHT

The causes of flight from developing countries have generally been complex and multidimensional. Fear of persecution for reasons of race, ethnicity, religion or political opinion has been but one of many reasons that have forced people to leave their normal places of residence. Military conflicts, first between independence movements and colonial authorities, and later between warring factions in post-independence power struggles, as well as inter-ethnic and inter-religious conflicts, have also been responsible for many of the population displacements occurring in the developing world. The internationalization of such conflicts in the developing world has increased their intensity and duration. Secessionist struggles and conflicts arising from irredentist territorial claims which, especially in Africa, are a legacy of the arbitrary partition of States by colonial powers without regard to the presence of regional or ethnic entities, have further added to the masses of involuntary migrants. Environmental factors, especially drought, have also been a major cause of forced migration, and economic deprivation or outright economic oppression is increasingly becoming a major cause of involuntary migration in the developing world.

In Africa especially, the causes of most refugee movements have been multidimensional. While the earlier conflicts that produced refugees (the civil wars in Algeria, Angola, Guinea-Bissau, Mozambique, Namibia

TABLE 57. COUNTRIES HAVING LARGE NUMBERS OF INTERNALLY DISPLACED PERSONS AS OF 31 DECEMBER 1991
(*Thousands*)

Country or area	*Number of displaced persons*
Sudan	4 750
South Africa	4 100
Afghanistan	2 000
Mozambique	2 000
Philippines	1 000
Ethiopia	1 000
Myanmar	500-1 000
Somalia	500-1 000
USSR (former)	900
Angola	827
Lebanon	750
Iraq	700
Sri Lanka	600
Yugoslavia (former)	557
Liberia	500
El Salvador	150-400
Nicaragua	354
Uganda	300
Cyprus	268
Haiti	200
Peru	200
Cambodia	180
Colombia	150
Guatemala	150
Sierra Leone	145
Rwanda	100

Source: United States Committee for Refugees, *World Refugee Survey, 1992* (Washington, D.C., American Council for Nationalities Service, 1992).

and Zimbabwe) had been mostly a direct result of the decolonization process, the crises of the 1970s and 1980s were largely a product of internal conflicts rooted in regionalism, ethnicity, economic disparities and political rivalries. During the 1960s and even up to the 1979 Arusha Conference on the situation of Refugees in Africa, the conventional wisdom was that the African refugee crisis was a temporary affair: with the last bastions of colonialism being removed, most refugees would go home. Instead, the rivalries among developed countries prevalent during the cold war increasingly fuelled the internecine conflicts emerging in the continent. Consequently, refugee outflows in Africa grew, in terms of the number not only of the persons involved but also of the countries affected.

The secession of a region from the State structure imposed upon an amalgam of previously independent, unrelated or autonomous entities has become a common cause of refugee movements in developing countries and, most recently, also following the break-up of the Soviet "empire". Secessionist wars following the achievement of independence by African States have produced both refugee flows and major population displacements within countries. The provinces or regions of Biafra, Eritrea, Shaba (Katanga), southern Sudan and Tigre are but some of the more conspicuous sites of such conflicts in Africa. In Asia, the creation of Bangladesh after the break-up of Pakistan in 1970 led to one of the largest refugee flows in recent times, though the crisis was short-lived. More recently, the ongoing conflicts in Sri Lanka, the Punjab and Kashmir, several regions of Myanmar, Iraqi and Turkish Kurdistan, the former Yugoslavia, and several parts of the Caucasus and Central Asia have already led to important population outflows or have the potential to do so. Secessionist forces remain especially strong in the Horn of Africa where Eritrea and Somaliland have both recently declared their independence, although that of Somaliland has yet to be recognized internationally and some 150,000 of its "citizens" remain as refugees in neighbouring Ethiopia. As regards the rebellion in the southern Sudan, it appears to have become irreconcilably split between factions wanting autonomy and a better deal for the south and those promoting total independence. In Ethiopia, although the new Government controls the nation's heartland, several insurgencies, such as that of the Oromo who demand autonomy or outright secession, prevail in the peripheries. All these conflicts produced refugees and internally displaced persons in the past and have the capacity to produce yet more. One of the few hopeful dimensions of most of the current conflicts is their no longer being fuelled by the superpowers, although in some cases regional power-brokers are involved.

Irredentism, that is, the claim by one State to another's territory on the basis of the fact that some of its "nationals" live in that territory, also continues to be a widespread cause of population displacement. In Africa, the colonial partitioning of the continent and the subsequent de facto adoption of colonial boundaries by the States gaining independence could have made irredentist conflicts ubiquitous had it not been for the Organization of African Unity's ratification of the principle that all colonial boundaries were to remain inviolate. Such a principle notwithstanding, irredentism underlies Somalia's protracted dispute with its three neighbours and especially with Ethiopia over the Ogaden. Irredentism is also at the basis of the long-term conflict between

Morocco and Western Sahara. As a result of these conflicts, many Ogaden-Somalis moved to Somalia in the late 1970s and some 165,000 Sahrawis moved to Algeria. The former, however, were forced to return to Ethiopia when Somalia's civil war intensified during 1989-1990 or are currently displaced along with Somali refugees along the Kenyan and Ethiopian borders. While a few irredentist claims subsist elsewhere in the developing world, the major current example is European, and concerns Serbia's attempt to consolidate all territories of the former Yugoslavia populated by Serbs.

Ethnic or religious conflict is probably one of the oldest and most widespread causes of forced migration and, as the current situation in the former Yugoslavia and in several parts of the former USSR demonstrates, it is by no means restricted to developing countries. However, as regards the latter, it may be an oversimplification to see such conflicts in terms purely of ethnicity, since economic, political or historic factors also play a significant part. In Africa, ethnic conflict has been at the root of several refugee situations, as in the case of the Hutu and the Tutsi in Burundi and Rwanda.[3] Many of the continent's ongoing conflicts, such as those in Angola, Liberia and the southern Sudan, have issues of ethnicity or religion as one of their causes. Even in Somalia, which was frequently cited as one of the few ethnically homogeneous States in Africa where "tribalism" was unlikely to cause conflict, the amalgam of numerous clans and sub-clans has fostered intense rivalries and historic antagonisms that are every bit as contentious and volatile as those based on ethnicity. After the overthrow of Siad Barre, power struggles between clan-based militias has led not only to large population displacements and refugee flows but also to "ethnic cleansing" which has resulted in a clan-based segmentation of the territory as comprehensive as it was in pre-colonial times.

While the above causes of refugee flows have generally resulted in large-scale migrations of mainly politically passive people trying to get out of the crossfire arising from conflicts, refugee migration also involves individuals fleeing out of the fear of persecution owing to their actual or perceived political opposition to existing Governments. Such refugees are often better educated and of urban origin. Intellectuals and political activists have been forced to flee countries such as Ethiopia, Guinea, Somalia, the Sudan, Uganda, Zaire and, increasingly in recent years, Kenya, Malawi and Sierra Leone. Many of those refugees move from countries of first asylum to other countries in Africa or, when possible, to developed countries where they enter as asylum-seekers, bona fide migrants of various types, or even as illegal migrants. In much of Africa, the political persecution of opponents or potential opponents by Government is coloured by ethnic considerations. In Somalia, for instance, clan affiliation determined a persons' loyalty to the Siad Barre regime. The persecution of urban and educated populations and their resultant flight have drained much of the intellectual and professional capacity of affected countries. Moreover, those migrants originating in countries where conflict is not widespread often have difficulty being recognized as refugees.

The number of persons moving because of environmental causes is increasing throughout much of the developing world. Drought, floods, desertification and soil erosion are but a few of the widespread and often recurrent problems that are leading people to migrate, both internally and internationally. In Africa, protracted drought in both the eastern and the southern parts of the continent is exacerbating the problems faced by refugees and displaced persons in the region. Acute water shortage in parts of Central Asia is exacerbating the outflow of involuntary migrants caused by ethnic conflict. In countries ravaged by armed conflict, it is frequently difficult to differentiate between those displaced because of environmental factors and those fleeing political events. In such cases, UNHCR has often been asked to assist populations that, technically, do not fall under its mandate. At the height of the 1984 drought in the Horn of Africa, for instance, no distinction was made by UNHCR in terms of assisting persons fleeing the ongoing insurgencies in Eritrea and Tigre and those moving to Sudan purely because of drought-induced famine. Similarly, during 1992 many Somalis entering Kenya and accommodated in UNHCR camps had moved owing to the lack of food rather than because of political persecution or armed conflict. In both cases, however, the impact of drought had been greatly exacerbated by protracted political conflict and Government indifference to the plight of the affected population. Likewise, in Mali, the indifference of the central Government to the needs of Touaregs sent thousands into exile in the neighbouring Niger during the Sahelian drought of the mid-1970s (Du Bois, 1974) and drought continues to be at the root of a conflict that has produced an outflow of from 60,000 to 75,000 refugees moving to Algeria, Burkina Faso or Mauritania (United Nations High Commissioner for Refugees, 1992a).

One of the most difficult issues being addressed by both national and international agencies dealing with refugees is that of determining refugee status in the face of the ever-growing economic pressures being put on people to move from developing to developed countries or, more generally, from poorer to richer countries. The fundamental question that needs to be addressed is when or how life-threatening economic oppression can be equated with political oppression. Increasingly, the so-called economic refugees have few options but to flee serious, indeed, life-threatening deprivation. This was the case among most Somalis who fled to the Kenyan border. The issue now being faced is what will happen in the longer term if reconstruction in Somalia is delayed because external assistance is slowing down as donor interest wanes or if the prospects for achieving sustainable economic development remain dim. Kenya has reluctantly accommodated Somali exiles and has resorted to periodical deportations from Nairobi and Mombasa to the border camps to keep the movement under some control. If the crisis in Somalia is not resolved, the spectre of more Somalis seeking economic refuge in Kenya will intensify. Likewise, the impending UNHCR-sponsored repatriation of Somalis from the border area may stall since prospects in Somalia are worse than those prevailing in the border camps. There are many similar scenarios, albeit perhaps not as extreme, being played out throughout the developing world. Whereas developed countries make some attempt to balance their asylum policies between humanitarian responses and the need to control "who gets in", richer developing countries may have less concern for humanitarian responses when threatened by massive inflows of economic refugees, as was the case with Nigeria in its two mass expulsions of West African illegal migrants (Peil, 1981).

C. The internally displaced

Whenever refugees leave a country, the forced internal movement of people is likely also to occur. According to the United States Committee for Refugees (1992b) there were some 23 million internally displaced persons in early 1992. Data on internally displaced persons are subject to even more conjecture than those relative to recognized refugees. However, estimates indicate that the numbers involved are very substantial, especially in countries such as Afghanistan, Ethiopia, Mozambique, Somalia and the Sudan. Estimates would be even larger if persons displaced because of natural disasters were also considered. In Bangladesh alone, over 1 million persons are displaced each year as a consequence of monsoon floods and many are unable to return to their homes after the waters subside because their lands have been eroded (Rogge and Elahi, 1989).

A major difference does exist, however, in terms of the assistance mobilized to aid them, between persons displaced by natural disasters and those fleeing conflict. When natural disasters are involved, Governments seek assistance for the reconstruction of the affected areas and the rehabilitation of the victims' livelihoods. Aid is usually made available (donors are more likely to make short-term rather than long-term, open-ended commitments) and, in most cases, the populations are soon able to return to their areas of origin. In the case of internal conflict, however, Governments are often unwilling even to admit that a problem exists. Displaced persons may be considered political foes or may belong to ethnic groups that are not considered to be deserving of help. Such has been the problem encountered by more than 1 million southern Sudanese who have been forced to move to the Arab regions of the northern and central Sudan. Whatever the reason, unless the international community is invited by the Government to assist the internally displaced, little can be done to provide them with either protection or material help. However, recent support by United Nations agencies for a number of cross-border operations without Government endorsement suggests that acute humanitarian needs may override the principle of the inviolability of international borders.

In the case of Somalia, the recent conflict has led to massive internal displacement consisting of several waves, one of which was still under way in 1993. Four main causes of internal displacement can be identified in that country, namely, flight from zones where military conflict rages; ethnic adjustments resulting from such conflicts; movement induced by drought; and search for food by famine victims. The last cause is essentially a product of the other three.

Displacement produced by military conflict has been ubiquitous throughout the country and has primarily affected urban populations because the contending militias have put their energies into controlling urban centres. In many cases, towns have been abandoned completely while battle fronts moved through them. Return has been delayed owing to the extent of destruction of the towns or because people perceive that there is still a lack of security. Parallel to displacement induced by conflict, there have been major population movements produced by the need to seek the security of one's own clan. In Somalia, clan affiliation has become the major precondition for residing in any particular area. Drought has always caused internal population displacement and the prolonged drought of 1989-1992 clearly accentuated the problems caused by civil war, making most of the traditional coping strategies employed in times of drought difficult, if not impossible, to implement. Thus, the current famine-induced displacement has been the product of civil war, the total breakdown of civil order and administration, the resultant banditry and looting, and a decline in traditional values. There are many analogies between Somalia's current situation and that of other conflict zones producing major famine-related population displacements, such as Cambodia in the late 1970s, Ethiopia in the mid-1980s and the southern Sudan currently.

D. Populations affected

Since populations in developing countries are often predominantly rural, it is not surprising that refugees and internally displaced persons have tended to be mostly of rural origin, especially in Africa. As a result, most responses to refugee outflows have also tended to be rural and localized. Indeed, there are many instances in Africa where subsistence farmers have simply fled from one side of the border to the other with only minimal disruption to their traditional way of life. However, as the causes of refugee movements changed during the 1970s, so did their predominantly rural character. While less than 10 per cent of the continent's refugees had been urban at the beginning of the decade, that proportion was believed to have reached a third at the time of the 1978 Arusha Conference on Africa's Refugees. Countries such as Kenya, the Sudan, the United Republic of Tanzania and Zambia were coping with increasing refugee inflows to their cities. During the 1980s, the proportion of refugees in urban areas continued to grow, although many originated in rural areas. Others came from countries where political repression targeted specific groups in urban areas, usually persons with higher education or those belonging to local elites. The impact of such outflows on both the country of origin and the country of first asylum can be substantial and lasting. Moreover, those movements often place serious demands on the resettlement programmes of developed countries.

While in the past most refugee outflows in Africa remained in rural areas, thus facilitating "traditional" responses such as local rural settlement and integration, the growing concentration of refugees in urban areas is resulting in strong reactions against refugees by both the host populations and Governments. This has been the case in the Sudan for the past 20 years. Refugees in urban areas are seen as exacerbating already serious problems of unemployment, overtaxed infrastructure and crime. Their presence is also regarded as contributing to the rising prices of essential commodities and the reduction of wages. Consequently, Governments have not been sympathetic to the movement to urban areas of refugees or internally displaced persons. Such has been the reaction towards Somalis who have left the border camps to seek an existence in Nairobi or Mombasa. On several occasions, the Moi Government has attempted to round up Somalis in city-wide dragnets and forcibly return them to the border. Similar practices were employed with regard to Eritrean refugees in Khartoum and Port Sudan. In contrast, Pakistan has adopted a relatively passive policy in this regard, perhaps more by default than design, in allowing large numbers of Afghan refugees to establish themselves in the towns of its North-west Frontier Province.

Somalia exhibits another dimension of this problem. With the outflow of refugees, it has lost most of its skilled human resources and wealth since 1970. Educated persons from urban areas have been leaving both because of political repression or to search for better economic opportunities in developed countries. With the overthrow of Siad Barre, almost the complete upper echelon and elite population went into exile, mostly to Kenya or to the member States of the Gulf Cooperation Council, but also to developed countries. Not only did Somalia thus lose needed administrative and professional skills, but it was also deprived of the wealth and savings that the departing elite looted and exported during the final days of the civil war. Replacing those skills and professional capacity will take decades if migrants do not return. Indeed, given that Somalia has virtually no effective educational infrastructure whatsoever and that none is likely to be created in the near future, this loss of human resources is severely hampering the reconstruction process. A similar situation is faced by other developing countries emerging from protracted wars and having experienced major refugee outflows, such as Afghanistan, Cambodia and Mozambique. A major challenge for developed countries is therefore to find innovative ways of mobilizing, if only on a short-term basis, the many refugees who have resettled in their midst for the reconstruction process in their countries of origin (such a process is occurring in Iraqi Kurdistan and beginning in Eritrea).

E. Durable solutions

Durable solutions to refugee outflows are repatriation, resettlement in third countries and local integration within the first country of asylum. For internally displaced persons, the equivalent solutions exist, namely, return to the area of origin, resettlement in another part of the country or integration within the area of refuge. Repatriation is the ideal solution, albeit a not necessarily easy one (Coles, 1985). For refugees from developing countries, resettlement in third countries is essentially a solution possible only for the "elite or the lucky", since the number of available resettlement places is minimal compared with the potential demand. Occasionally, political considerations in a country of resettlement may permit mass resettlement of persons from a specific region, as was the case for Cubans in the late 1950s and for Indo-Chinese refugees during the late 1970s and early 1980s but, more generally, resettlement is reserved for carefully screened and selected refugees.

1. *Local settlement and integration*

Reaction to local settlement and integration by first-asylum countries in the developing world varies greatly from region to region. In countries such as Malaysia, Thailand and Singapore, the possibility of local settlement is moot: refugees are tolerated, albeit very reluctantly, only on condition that they be repatriated eventually or that they be resettled elsewhere. In the interim, they are restricted to holding centres that differ little from prison camps. In contrast, in Pakistan and even more so in the Islamic Republic of Iran, Afghan refugees have been able to settle and integrate among local communities with a considerable degree of success and harmony. In Africa, local settlement was once ubiquitous, albeit almost exclusively in rural areas, especially for refugees fleeing anti-colonial wars for whom the provision of land and other assistance was an expression of solidarity by the host Government. Indeed, the generous support by African States for refugees fleeing colonial countries is analogous to the favourable response shown in the past by most Western States to refugees fleeing communist countries. However, the hospitable and generous response once attributed to almost all African countries of first asylum has, in many cases, changed to one of reluctance, if not outright resistance, with regard to granting refugees the possibility of integrating locally. Such a change of policy has been due primarily to the sheer numbers involved, as in the eastern Sudan and southern and western Malawi, and the lack of capacity of the receiving countries to cope with such numbers. It is also related to the changing causes of refugee movements and the growing tendency of refugees to settle in urban areas. It can be argued that greater donor assistance to help build the capacity of countries of asylum to absorb large numbers of refugees would improve their receptiveness. However, donor assistance to Africa is unlikely to increase in the near future.

This changing attitude notwithstanding, there remain a number of areas where local integration has been relatively successful. Many refugees have been settled through organized rural settlement schemes that, though initially supported by UNHCR, eventually became self-reliant and no longer required any external assistance (Rogge, 1981). Others have integrated spontaneously, often among ethnic kin living in rural areas on the other side of the border. Examples include the Burundi and Rwandan settlements in the United Republic of Tanzania (Gasarasi, 1984) and the spontaneous settlement of Angolans in north-western Zambia (Hansen, 1981). Elsewhere, there has been full integration of refugees in Botswana (Angolans), Senegal (refugees from Guinea-Bissau), Uganda (Rwandans) and Zaire (southern Sudanese). More recent refugee movements, however, have not achieved sustained local integration.

In the case of the eastern Sudan, where Eritrean refugees have been present since the mid-1960s, both rural settlement according to organized settlement schemes and spontaneous settlement in rural and urban areas have been the rule, more by default than by design. Government policy has always been to settle refugees through organized settlement schemes; some of them have come close to providing a reasonable measure of self-reliance (Rogge, 1985; Kuhlman, n.d.), while others have been outright failures. A large proportion of refugees, perhaps nearly half, opted to settle on their own, with varying degrees of success, as farm labourers or share croppers among rural Sudanese communities (Bascom, 1989). The eastern Sudan illustrates the changing conditions in which many of Africa's refugees are increasingly finding themselves. Although in the 1960s and 1970s large land grants were made available to the refugees by presidential decree and a reasonably harmonious and symbiotic relationship evolved with the host Sudanese communities, as the number of refugees soared and economic conditions deteriorated, basic infrastructures became overburdened and government policies became increasingly xenophobic. Yet, because of their long-term residence in the Sudan, refugees have become inextricably linked to the regional economy and a sudden mass departure of Eritrean refugees would, at least in the short run, have a devastating impact on the economy of the eastern Sudan. Indeed, were a wholesale repatriation of Eritrean and Ethiopian refugees to take place, the acute labour shortage thereby created would likely produce a new wave of labour migrants from poorer parts of Africa.

Kenya's response to the Somali crisis also illustrates the changing attitudes to refugee settlement. Until recently, Kenya was one of the few African countries that had been spared a large refugee inflow (apart from the temporary influx of Ugandan exiles during the Amin and second Obote eras). Since 1991, however, its refugee population has grown from a few thousand to over 400,000. The majority are Somalis. Kenya's response has been one of reluctant compliance with the international community's efforts to assist those refugees. Its policy is committed to maintaining the refugees in the immediate border areas, so as to minimize the associated growth in insecurity created by armed Somali bandits. Kenya sees the influx very much as a temporary problem, the solution to which is speedy repatriation. Refugees who proceed beyond the border camps to the cities are dealt with swiftly and harshly when apprehended: they are usually sent back to the border. Special treatment, however, is accorded to members of the former Somali elite, who, because of their wealth, are able to maintain a comfortable and relatively unmolested lifestyle in Nairobi and Mombasa.

Kenya's response to the Somali influx, as well as the developments in the eastern Sudan, is being replicated elsewhere in Africa, especially where large refugee inflows place inordinate strains on local resources or infrastructure. Thus, the granting of land to refugees for local settlement is becoming rare. Instead, in much of the developing world, refugees are increasingly being held in holding camps with minimal services available. Indeed, although not stated explicitly as policy, the implicit objective in some cases is to move towards a form of "humane deterrence" such as has been practised in parts of Eastern and South-eastern Asia. Consequently, the local settlement and integration of refugees within developing countries, which were once widely touted as the most realistic option for refugees "unsuited" for resettlement or unlikely to repatriate, are becoming less acceptable among many countries of asylum and therefore increasingly difficult for UNHCR to implement.

2. *Third-country resettlement*

It is also highly unlikely that current policies among developed countries regarding the resettlement of large numbers of refugees will change, especially given the urgent demand for resettlement created by the crisis in the former Yugoslavia. Indeed, there is a noticeable move away from this "solution". The need to cope with the ever-growing number of asylum-seekers who land directly in developed countries has displaced the attention previously given to refugees selected for resettlement in first countries of asylum, especially refugees originating in Africa. Many developed countries are introducing restrictive legislation aimed at reducing, if not eliminating, the numbers of asylum-seekers. Only Australia, Canada and the United States of America have in recent years had modest "planning levels" with respect to resettling African refugees (fluctuating per annum between 3,000 and 5,000 in the United States; between 1,000 and 1,500 in Canada, and amounting to only a few hundred in Australia). Refugees from the Horn of Africa have taken up the majority of those resettlement places. Most are of urban origin and tend to be better educated and more highly skilled. Indeed, the evident preference among resettlement countries for educated and skilled refugees has created considerable controversy in some countries of first asylum, which maintain that developed countries are "skimming off the cream" among refugees. It is argued that the countries of first asylum are being required to use their own scarce human resources to satisfy the needs of refugees while the skilled among those refugees are taken abroad. It is also argued that developed countries, in selecting refugees for resettlement, are using "immigration" criteria rather than "humanitarian" criteria.

Table 58 shows that the number of refugees resettled in third countries has been relatively modest, with only 2.4 million refugees resettled in the six major resettlement countries and another 200,000 resettled in the rest of the developed world during 1975-1990. Over one half of this total was resettled in the United States and 1.3 million originated from a single region, Indo-China (1.3 million Indo-Chinese had been resettled by the end of September 1992: 58 per cent from Viet Nam, 24 per

TABLE 58. MAIN COUNTRIES OF RESETTLEMENT OF REFUGEES AND ASYLUM-SEEKERS

Country of resettlement	*1975-1990*	*1990*	*Ratio of refugees to total population*
Sweden	121 154	12 839	1/71
Canada	325 045	37 820	1/82
Australia	183 104	10 281	1/96
United States	1 478 184	122 326	1/171
France	200 030	13 073	1/238
Germany[a]	91 478	6 518	1/869

Source: United States Committee for Refugees, *World Refugee Survey, 1992* (Washington, D.C., American Council for Nationalities Service, 1992).

[a] Excluding 222,000 ethnic Germans from Eastern Europe.

cent from the Lao People's Democratic Republic and 18 per cent from Cambodia). The resettlement of Indo-Chinese has clearly been a special case. The United States accepted the responsibility of resolving a problem that it had been instrumental in creating and the rest of the developed world had few options but to become involved when it became clear that the South-eastern Asian countries of first asylum would only permit Indo-Chinese refugees to land in their territories if resettlement was guaranteed. With the exception of the movement of Cubans to the United States, no other refugee outflow from developing countries has ever produced a similar response. In contrast, the total number of African refugees resettled outside the continent since 1980 has been a mere 35,000, most of them admitted by Canada and the United States. Consequently, many African refugees are attempting increasingly to take their destiny into their own hands by arriving in Europe or North America as asylum-seekers or by becoming illegal migrants.

3. *Repatriation*

Thus, with diminishing prospects for local settlement or integration and no immediate likelihood of greater resettlement opportunities in developed countries, repatriation is increasingly seen as the "optimum" solution for refugees from developing countries. In theory, repatriation clearly is the "ideal" solution. What can be more natural than for people to return home? Yet, repatriation has not been a common event. In the past, almost all refugee movements tended to involve permanent exile. The Jews in the diasporas, the Pilgrim Fathers, the Huguenots, the Mennonites and the Pietists all became permanent exiles, having neither the opportunity nor the capacity to return to their countries of origin (Norwood, 1969). Likewise, during the inter-war era, none of the displaced people of the Balkans, from Armenia, Greece or Turkey, or the Jews displaced during the 1930s were repatriated. Simpson (1939), commenting on the world refugee situation on the eve of the Second World War, dismissed repatriation as a viable option, arguing that the voluntary return of refugees to their home countries would occur on a scale so small as not to affect the refugee problem itself.

The same was largely true for refugee movements in the decade and a half following the Second World War, both in the European arena, where a massive permanent displacement occurred from East to West, and in Asia, where the flight of Hindus and Muslims from Pakistan and India, respectively, the exodus of Tibetans to India, and of mainland Chinese to Hong Kong and Taiwan, and the displacement of the Palestinians, all became permanent transfers of population. One of the few exceptions to this norm is that of Hungarian refugees, 18,000 of whom (representing about 10 per cent of the total leaving) repatriated spontaneously during the four years following the crisis.

However, since the 1960s repatriation has become increasingly widespread, especially in Africa, in parts of Asia and in Central America. According to Cuny and Stein (1992), between 1975 and 1991 some 6.8 million refugees repatriated worldwide, mostly to developing countries. To this number can be added the 200,000 Algerians repatriated in the late 1950s in the first of the major repatriations to developing countries; 154,000 southern Sudanese; some 60,000 Nigerians who returned in 1970 after the civil war; 100,000 Iraqi Kurds who repatriated from Iran following the Algiers Accord; and the 10 million Bangladeshis who, in the single largest repatriation ever to take place, returned from India immediately after the independence of Bangladesh in 1971 (the last three repatriations occurred with minimal assistance from UNHCR).

Since 1970, repatriations have been especially significant in Africa. Some of the major assisted or documented repatriations are summarized in table 59. In addition, there have been many large spontaneous returns that have not been properly quantified. In some cases, such as those of Chad, Mozambique and the southern Sudan, returnees were forced once more into exile as renewed conflict engulfed their home areas. Repatriations to Eritrea, Ethiopia, north-western Somalia (so-

TABLE 59. PRINCIPAL ASSISTED/DOCUMENTED REPATRIATIONS, 1970-1992

Country of return	*Years of return*	*Major asylum States*	*Numbers returned*
Nigeria	1970	Cameroon	60 000
Bangladesh	1971	India	10 000 000
Zaire	1971-1973	Burundi, Zambia	37 300
Sudan (south)	1972-1974	Uganda, Zaire, Central African Republic	154 200
Guinea-Bissau	1974-1977	Senegal	124 000
Mozambique	1974-1975	United Republic of Tanzania, Zaire, Zambia	84 000
Iraq (Kurds)	1975	Iran (Islamic Republic of)[a]	100 000
Angola	1975	Zaire	300 000
Zaire	1978-1980	Angola, Burundi	222 000
Myanmar (Rohinyas)	1978	Bangladesh	200 000
Nicaragua	1979	Various Central America	100 000
Angola	1979-1981	Zaire	96 000
Equatorial Guinea	1979	Cameroon, Gabon	35 000
Uganda	1979	United Republic of Tanzania	30 000
Zimbabwe	1980-1981	Zambia, Mozambique, Botswana	166 000
Cambodia	1980-1982	Thai border, Lao People's Democratic Republic	500 000
Chad	1981-1984	Cameroon, Nigeria, Sudan	155 000
Ethiopia	1981-1984	Sudan, Somalia	317 000
Uganda	1982-1986	Sudan, Zaire	300 000
Ethiopia	1986-1987	Sudan	150 000
Sri Lanka	1987-1988	India	43 000
Burundi	1988	Rwanda	53 000
Mozambique	1988-1991	Malawi, Zambia, Zimbabwe	250 000
Uganda	1988	Sudan	86 000
Namibia	1989-1990	Angola, Zambia	44 000
Nicaragua	1990-1991	Honduras, Costa Rica	61 000
Afghanistan	1990-1992	Iran (Islamic Republic of)[a], Pakistan	500 000
Ethiopia	1991	Somalia	375 000
Sudan	1991	Ethiopia	270 000
Iraq (Kurds)	1991	Iran (Islamic Republic of)[a], Turkey	1 470 000
Cambodia	1992	Thailand	140 000

Source: Rogge (1991), Cuny and Stein (1992), and various UNHCR sources.

NOTE: All figures given here are crude and most are subject to debate over their accuracy. Different sources will often provide quite contradictory numbers for a given repatriation. Many smaller repatriations (that is, of less than 30,000) have also taken place and are not shown in the above table. This table should, therefore, be used only as a very rough guide.

[a]In April 1979, the name "Iran" was changed to "the Islamic Republic of Iran".

called Somaliland) and South Africa are currently under way, and those to Angola, Mozambique and southern Somalia are pending. These return movements, together with the recent return of the majority of Central America's refugees (Larkin, Cuny and Stein, 1991), the completed repatriation of Cambodians and the ongoing repatriations to Afghanistan, demonstrate the growing significance of repatriation as a viable and lasting solution for refugees in developing countries.

UNHCR's optimism notwithstanding, two key issues must be addressed when examining the current significance of repatriation: first, the degree of voluntariness of repatriation, since many refugees are returning to areas where conflict is still ongoing or where the reality of peace remains very tenuous (Goodwin-Gill, 1989; Cuny, Stein and Reed, 1992); second, the responsibility for assisting returnees, in terms of both who should provide such assistance and the type and duration of assistance.

A closer examination of many of the recently completed or ongoing repatriations indicates that, contrary to the terms of the 1967 Protocol relating to the Status of Refugees (United Nations, 1967), many returnees are going back reluctantly or even under circumstances of substantial coercion. Outright refoulement of refugees, such as Thailand's forced return of some 45,000 Cambodians in 1979, has not occurred very often, largely because of the international outcry that such actions tend to unleash. However, there have been many repatriations where the return has been less than completely voluntary. In 1978, for example, the Rohingya refugees in Bangladesh were given little option but to return: food to their camps was simply cut off. Guatemalans returning from Mexico in 1990 did so under pressure and the return of Ethiopians from Djibouti in the mid-1980s was strongly resisted by most refugees.

Repatriation agreements between countries of origin and those of asylum, even when endorsed by UNHCR, do not necessarily mean that the conditions that produced the refugees have markedly altered and that refugees therefore perceive the situation as sufficiently improved for them to go home. A growing dilemma is that of reconciling conditions of "safe return", as determined by the Governments of the countries of origin and asylum, with the perception of "voluntary repatriation" held by the affected refugees. The repatriation to Cambodia was a good illustration of this variance in perception: while most Cambodians in the Thai border camps wanted to return, many were very reluctant to do so despite the organized repatriation programme because of the perceived political uncertainty that prevailed in Cambodia. The accelerating rate of return was clearly influenced by the reasonably favourable experience of the first waves of returnees. Many host countries, anxious to be rid of their refugee populations, are increasingly exerting pressure on refugees to return. Such recent examples of repatriation as that of Cambodians, Lao and Burmese from Thailand, of Iraqi Kurds from Turkey and the Islamic Republic of Iran, of Sudanese from Ethiopia, and of Ethiopians from Djibouti are all cases where the process has taken place under considerable duress. In many instances, refugees have returned to areas that are still in conflict. There is a high probability that both Kenya and Malawi will exert pressure on their large refugee populations to return at the earliest opportunity, even if complete peace is still lacking in their countries of origin or if food security remains tenuous. Currently, Bangladesh is negotiating the return of the Rohingya refugees, even though there is no indication that the Myanmar Government has moderated its attitude towards Rohingyas. Consequently, while the number of repatriations is increasing, there is reason for concern about the safety and well-being of repatriated refugees.

In addition, major interventions are usually needed to facilitate the returnees' reintegration, especially if they have been in exile for a long period and if, during their absence, their areas of origin have been severely disrupted by conflict. Repatriation encompasses a considerably more complex process than simply having UNHCR or the International Organization for Migration transport refugees back to the border of their own country. Reintegration and rehabilitation needs will vary greatly, ranging from economic assistance to social or psychological help (Rogge, 1991). A major issue is who takes responsibility for the long-term reintegration of repatriated refugees and for the reconstruction of their places of origin. To date, UNHCR sees its role primarily as a short-term one; the United Nations Development Programme (UNDP), on the other hand, remains uncertain about its longer-term responsibility for reintegration and reconstruction. Clearly, there is a need to integrate repatriation programmes into national reconstruction and development programmes. A related question is whether development assistance to the countries of origin should specifically target repatriated refugees, or more generally all persons living in affected areas, and to what extent. In some cases, the populations who stayed behind are, in fact, in worse conditions than those returning.

A brief discussion of the Somali case will illustrate some of the problems associated with repatriation. With close to 400,000 Somalis registered as refugees in Kenya and some 150,000 remaining in Ethiopia and Djibouti, as well as over 500,000 internally displaced, the scale of the problem in Somalia is enormous. A country free of the devastation of war would have difficulty in dealing with such numbers; thus for one where almost the whole infrastructure has been destroyed and where every governmental and almost every non-governmental institution has ceased to exist, managing repatriation and reintegration of returnees will be an extremely difficult task. Indeed, the question of what Somalis have to go back to needs to be addressed first. Food security remains very tenuous in most areas, and the delicate progress towards rehabilitation of those who remained could be severely disrupted by a sudden and large influx of returnees. Even if a lasting political reconciliation between the warlords could be brokered, it will take years for the reconstruction process to restore a sustainable economy. In the interim and unless there is a major thrust in the provision of development assistance, many of the refugees returning from Kenya, under a UNHCR-sponsored repatriation programme that commenced in June 1993, may find that they are simply moving from camps in Kenya to new camps or feeding centres in Somalia (Rogge, 1993).

The longer-term reintegration of returnees in Somalia will be a task of monumental dimensions. A large proportion are pastoralists or agropastoralists who have lost all their livestock. Given that the overall national herd has been severely depleted and in many areas herds are below the threshold for recovery, it is improbable that many refugees can be reintegrated into their traditional lifestyles. Yet, given prevailing clan politics and hostilities, and the refugees' lack of alternative skills or resources, resettling them in other areas or integrating them in the cities will also be fraught with serious problems. While perhaps such problems are extreme in Somalia, they are common in many other long-term conflict areas, such as Afghanistan, Cambodia and Mozambique. The problem is exacerbated by the fact that assistance for reconstruction and the rehabilitation of returnees does not usually receive high priority from donors. Indeed, humanitarian agencies often maintain that such assistance falls within the mandate of development agencies, while development agencies see it as the continuing responsibility of humanitarian agencies.

Reversing Somalia's brain drain will be even more of a problem. Again, this is not an unusual dilemma facing developing countries that have generated refugee flows. Given the near totality of the destruction of institutions and services, and the snail's pace at which reconstruction is proceeding, the incentives for return will be minimal for all but the most altruistic. Yet, without the return of professionals, and skilled and experienced bureaucrats, the prospects for firmly moving the country out of a relief phase and into sustainable reconstruction and recovery remain bleak.

F. Conclusion

Given that the three durable solutions are fraught with difficulties or limited in their potential, are there realistic and desirable alternatives for refugees? Basically, the answer is no. The alternatives that prevail do so by default and are neither durable nor desirable. Yet, they are increasingly adopted. One such alternative is to maintain refugees indefinitely in holding camps, as Thailand has done with most of its refugees. The psychological and social consequences of long-term life in camps under conditions of total dependency were clearly evident among the many Cambodians on the Thai border. The problems encountered by returnees to Cambodia as they attempted to become reintegrated have yet to be studied. It is probable that holding-camp "solutions" will increase if the numbers of refugees continue to grow and increasing numbers of countries become reluctant "hosts". Another option is for refugees to seek their own solution by becoming asylum-seekers, that is, by moving sometimes constantly from country to country in search of the one that will accept them or surviving at the margins of society as undocumented aliens in the richer or at least the more stable countries.

Notes

[1]The 1951 Convention relating to the Status of Refugees (United Nations, 1957) prefaced its definition of "refugee" in its article 1 with the rider "as a result of events occurring before 1 January 1951". Stipulation of the 1 January 1951 dateline was later omitted by the 1967 Protocol to the Convention (United Nations, 1967).

[2]UNHCR's initial mandate was for three years, to be renewed, if necessary, by the General Assembly. In 1953, the original mandate was extended by five years because of the continuing existence of a refugee problem in Europe.

[3]Kibreab (1985) argues that colonial policies, rather than ethnicity, underlie such conflict.

References

Bascom, J. (1989). Social differentiation among refugees in Eastern Sudan. *Journal of Refugee Studies* (Oxford), vol. 2, No. 4, pp. 403-418.

Coles, Gervase J. (1985). *Voluntary Repatriation: A Background Study.* San Remo, Italy: United Nations High Commissioner for Refugees and International Institute of Humanitarian Law.

Cuny, Frederick C., and Barry N. Stein (1992). Refugee repatriation during conflict: a new conventional wisdom. Paper presented at the Conference on Refugee Repatriation during Conflict: A New Conventional Wisdom, Addis Ababa, Ethiopia, 18-21 October.

__________, and Pat Reed (1992). *Repatriation under Conflict in Africa and Asia.* Dallas, Texas: Centre for the Study of Societies in Crisis.

Du Bois, Victor (1974). The drought in Niger: the flight of the Malien Touareg. *American Universities Field Staff Reports* (New York), vol. 15, No. 6, pp. 1-13.

Gasarasi, Charles P. (1984). *The Tri-Partite Approach to the Resettlement and Integration of Rural Refugees in Tanzania.* Uppsala, Sweden: The Scandinavian Institute for African Studies.

Goodwin-Gill, Guy (1989). Voluntary repatriation: legal and policy issues. In *Refugees and International Relations*, Gil Loescher and Laila Monahan, eds. New York: Oxford University Press.

Hansen, Art (1981). Refugee dynamics: Angolans in Zambia, 1966 to 1972. *International Migration Review* (Staten Island, New York), vol. 15, Nos. 1/2, pp. 175-194.

Holborn, Louise W. (1975). *Refugees: A Problem of Our Time. The Work of the United Nations High Commissioner for Refugees, 1951-1972*, 2 vols. Metuchen, New Jersey: Scarecrow Press.

Kibreab, Gaim (1985). *African Refugees.* Trenton, New Jersey: Africa World Press.

Kuhlman, Tom (n.d.). *Burden or Boon? A Study of Eritrean Refugees in the Sudan.* Amsterdam: Vrige Universiteit Press.

Larkin, Mary A., Frederick C. Cuny and Barry N. Stein (1991). *Repatriation under Conflict in Central America.* Washington, D. C., and Dallas, Texas: CIPRA and Intertect Institute.

Norwood, Frederick A. (1969). *Strangers and Exiles: A History of Religious Refugees.* Nashville, Tennessee: Abingdon Press.

Peil, Margaret (1981). The expulsion of West African aliens. *Journal of Modern African Studies* (London), vol. 9, No. 2, pp. 205-229.

Proudfoot, Malcolm M. (1956). *European Refugees: 1939-52*. Evanston, Illinois: Northwestern University Press.

Rogge, John R. (1981). Africa's resettlement strategies. *International Migration Review* (Staten Island, New York), vol. 15, Nos. 1/2, pp. 195-212.

__________ (1985). *Too Many, Too Long: Sudan's Twenty Year Refugee Dilemma*. Totowa, New Jersey: Rowman and Allenheld.

__________ (1991). *Repatriation of Refugees: A Not-so-Simple "Optimum" Solution*. Geneva: United Nations Research Institute for Social Development.

__________ (1992). *The Displaced Population in South and Central Somalia and Preliminary Proposals for Their Re-integration and Rehabilitation*. Mogadishu: United Nations Development Programme.

__________ (1993). *Rehabilitation and Reconstruction Needs for Displaced Persons in Somalia: With Special Reference to the Northern Region and to the Juba Valley*. Mogadishu: United Nations Development Programme.

__________, and K. Maudud Elahi (1989). *The Riverbank Erosion Impact Study, Bangladesh: Final Report*. Ottawa: International Development Research Centre.

Simpson, John H. (1939). *The Refugee Question*. Oxford: Clarendon Press.

United Nations (1957). *Treaty Series*, vol. 189, No. 2545.

__________ (1967). *Treaty Series*, vol. 606, No. 8791.

__________ (1996). *World Population Monitoring, 1993: With a Special Report on Refugees*. Sales No. E.95.XIII.8.

United Nations High Commissioner for Refugees (1992a). Voluntary repatriation programmes: 1992. Executive Committee of the High Commissioner's Programme. EC/1992/SC.2/CRP.19.

__________ (1992b). *Statistics Concerning Indo-Chinese in East and South-East Asia for the Month of September, 1992*. Geneva.

__________ (1993). *The State of the World's Refugees: The Challenge of Protection*. London and New York: Penguin Books.

United States Committee for Refugees (1992a). *Croatia's Crucible: Providing Asylum for Refugees from Bosnia and Hercegovina*. Washington, D. C.: American Council for Nationalities Service.

__________ (1992b). *World Refugee Survey, 1992*. Washington, D.C.: American Council for Nationalities Service.

Vernant, Jacques (1953). *The Refugee in the Post-War World*. London: George Allen and Unwin.

Part Six

DISCUSSION NOTES

VIII. MIGRATION TO THE LARGE CITIES OF THE ESCAP REGION

Economic and Social Commission for Asia and the Pacific

In the Asia and Pacific region covered by the Economic and Social Commission for Asia and the Pacific (ESCAP), the number of cities with more than 1 million inhabitants has increased rapidly, passing from only 25 in 1950 to over 100 in 1990. In 1990, 5 of the 10 largest cities in the world were in the ESCAP region, namely, Tokyo (20.2 million), Shanghai (13.4 million), Calcutta (11.8 million), Bombay (11.2 million) and Seoul (11.0 million). By the year 2000, it is expected that at least 6 of the 10 largest cities in the world will be in the region (Tokyo, Shanghai, Calcutta, Bombay, Beijing and Jakarta, in descending order of size). The urban population of the Asia and Pacific region increased from approximately 224 million in 1950 to 805 million in 1985 and is expected to surpass 1.4 billion by the end of the century (United Nations, 1991). The proportion of persons living in urban areas rose from 18 per cent in 1950 to 29 per cent in 1985 and is expected to be over 40 per cent by the year 2000.

Net rural-urban migration contributes much to the growth of the urban population in the ESCAP region. Assuming that natural increase in urban areas is the same as that of the total population, it is estimated that net migration and reclassification account for over half of the population growth in urban areas since 1980. The actual estimates are 62 per cent during 1980-1985, 59 per cent during 1990-1995 and 57 per cent during 2000-2005. During 1990-1995, migration and reclassification accounted for 72 per cent of urban growth in Eastern Asia, for 55 per cent in South-eastern Asia, for 46 per cent in Southern Asia, and for 25 per cent in Oceania.

By and large, net migration is a more important component of urban growth in the developing countries of the region than in the developed ones (Hugo, 1992). The proportion of urban growth due to migration and reclassification in countries such as Japan and Australia was 18.5 per cent and 38.1 per cent respectively (United Nations, Economic and Social Commission for Asia and the Pacific (ESCAP), 1992a) and the figure for Australia may be an overestimate due to the rough method of estimation used. According to Singelmann (1991), the proportion of urban growth attributable to internal migration and reclassification in Australia was only 5 per cent during the 1970s, an important decline from the 20.3 per cent estimated for the 1960s. Indeed, Australia, like many other developed countries, experienced a process of counter-urbanization during the 1970s and 1980s whereby, after a century of increasing concentration of the population in large metropolitan areas, the trend was reversed and smaller cities and even rural areas began to grow faster than large metropolitan areas (Hugo, 1992). In developing countries, the proportion of urban growth attributable to migration and reclassification ranged from 14.4 per cent for Mongolia to 76.3 per cent for China during 1990-1995.

Most studies of the components of urban growth estimate migration and reclassification as a residual and, consequently, cannot make a quantitative distinction between the two. A detailed analysis based on the registration system operating in Sri Lanka indicates that, for that country, reclassification was more important than net rural-urban migration and that the latter accounted for only about one fifth of total urban growth (United Nations, Economic and Social Commission for Asia and the Pacific (ESCAP), 1980). It may be that a high proportion of the migration and reclassification component in China is also attributable to reclassification since during the 1960s and 1970s, strict controls were imposed on rural-urban migration. A number of reforms introduced during the 1980s permitted some types of migration, but an unprecedented increase in the number of small urban places also occurred after 1980 when the Government decided to reclassify numerous rural towns as urban.

However, China and Sri Lanka are not typical countries in the Asia and Pacific region. In some countries, migration has indeed been the most important component of urban growth during certain periods. In Japan and the Republic of Korea, for instance, migration was certainly the main factor in the growth of the urban population during the rapid phases of industrialization of the 1950s and 1960s. That was also the case for countries in the early stages of urbanization, when the urban population base was small, such as Bangladesh and Papua New Guinea during the early 1970s.

According to recent studies, the population increase in Bangkok was in great measure the result of net in-migration and migration has also been a major contributor to the urban growth of most regions of the Philippines (United Nations, Economic and Social Commission for Asia and the Pacific (ESCAP), 1992a). Thus, according to the 1980 census of Thailand, 13.8 per cent of the population of Bangkok consisted of migrants who had moved to the city during the five years preceding enumeration. During 1970-1980, net migration to Bangkok accounted for 18.9 per cent of its growth while reclassification accounted for 31.9 per cent (United Nations, Economic and Social Commission for Asia and the Pacific (ESCAP), 1982). In other words, migration accounted for approximately one fifth of urban growth while it constituted about 37 per cent of the "migration and reclassification" component.

A. Migration to the major cities of the Asia and Pacific region

Bangkok. As of 1980, the Bangkok Metropolitan Area (BMA) had 1.25 million lifetime migrants, who constituted 27 per cent of the population. Between 1975 and 1980, some 366,000 persons migrated to the BMA from other regions of Thailand. The largest stream, consisting of 174,900 persons from the Central Region, was offset by a counter-stream of 102,700 persons leaving the BMA for the Central Region, resulting in a net migration gain for the BMA of 72,200 persons. Migrants to the BMA were heavily concentrated in the young ages. About a quarter were aged 20-24, 22 per cent were aged 15-19, and 14 per cent were aged 25-29. Women accounted for a higher proportion of migrants than men. Women were especially numerous among migrants from the North and North-east Regions, whose sex ratios were 76 men per 100 women and 80 men per 100 women, respectively. With regard to fertility, the 1980 census indicated that female migrants in all age groups had a lower average number of children ever born than their non-migrant counterparts.

The BMA also attracted a large number of seasonal migrants. Most seasonal migrants came to the BMA during the slack agricultural season (February) or before the beginning of school terms (May). According to the 1984 migration survey, migrants were predominantly young (86 per cent were under 30) and mainly single (70 per cent of those aged 15 year or over were not married). A majority of them came from villages or non-municipal areas. The most important reasons for migrating to Bangkok were to find a job (67 per cent), for family reasons (19 per cent), to study further or for vocational training (9 per cent) and to undertake a specific job assignment (5 per cent).

Bombay. The growth of industry during the nineteenth century resulted in a high rate of in-migration to the Bombay region. The proportion of native-born residents in Bombay City had fallen from 31 per cent in 1872 to 16 per cent by 1931. By 1961, only 36 per cent of the population of Bombay was native-born. Of the rest, 27 per cent were born in the state of Maharashtra, while two thirds of those born outside Maharashtra came from Gujarat, Mysore or Uttar Pradesh. During 1971-1981, it is estimated that net migration accounted for about 1.4 million of the 2.26 million increase experienced by the city's population.

Calcutta. Calcutta has long been a city of migrants. During the early decades of this century, when Calcutta was one of India's most dynamic industrial centres, it attracted thousands of migrant workers, mainly from neighbouring States. Since the Second World War, push factors in the hinterland, including political unrest, religious conflict and periodic ethnic riots, have also been important in inducing migration to Calcutta. A large-scale movement occurred in the wake of the 1971 war between East and West Pakistan, when some 7.5 million refugees entered the Indian State of West Bengal. However, by 1972 about 6 million of those refugees had returned to the newly created nation of Bangladesh.

The majority of migrants in Calcutta come from the surrounding States of Assam, Bihar, Orissa and Uttar Pradesh. They are mostly men who leave their families in the villages of origin. The sex ratio of migrants has been as high as 140 men per 100 women in Calcutta proper and 128 men per 100 women in the metropolitan area. Most male migrants work in factories, as rickshaw pullers, plumbers, carpenters or gardeners. Recently, however, net migration to the city has been virtually nil.

Delhi. Migration has contributed more than natural increase to Delhi's population growth since 1940. Delhi received significant in-migration during the Second World War, when the proportion of annual population growth attributable to net migration rose to about 85 per cent. The share of net migration was 71 per cent at the time of partition in 1948 but increased to 82 per cent in 1949. During the mid-1950s, natural increase and net migration made approximately equal contributions to population growth in the city. However, the share of migration was greater than that of natural increase in all subsequent decades. According to the 1981 census, there was an absolute increase of 2.12 million persons in Delhi during 1971-1981, two thirds of which was attributable to net migration.

Dhaka. In recent years, the growth of Dhaka City has been predominantly the result of net migration, which accounted for 62.8 per cent of population growth during 1961-1974 and 70.5 per cent during 1974-1981. Although the relative importance of migration is expected to decline in the future, about 60 per cent of Dhaka's population growth between 1981 and 2000 is anticipated to be the result of net migration. The major factors attracting migrants include employment opportunities in the informal sector and relief activities undertaken by governmental and non-governmental organizations. Migrants are mainly young adults (the largest proportion is concentrated in age group 20-29). Men outnumber women among Dhaka's migrant population, and migrants are more likely to be single and better educated than non-migrants. Migrants are also more likely than non-migrants to belong to large families originating in both the poorest and most affluent rural communities of Bangladesh.

Jakarta. During 1948-1953, net annual migration to the city was of the order of 116,000 persons. During the 1960s, the annual increase due to migration was 86,000, but during 1971-1980, it declined somewhat to 73,000. In 1985, it was reported that the city had received around 90,800 migrants or about 250 migrants per day. During 1971-1980, about 65 per cent of the migrants were aged 15-39, with the largest number concentrated in age group 20-24. Although women constituted more than half of the migrants aged 15-24, overall they accounted for less than half of all net migrants, mainly because of the tendency of older women to return to their villages. Sixty per cent of the migrants in Jakarta had six years of education or less, 13 per cent had attended junior high school, and the remainder had some higher education. The number of lifetime migrants had declined over time. However, other types of migration to Jakarta, such as commuting and circular migration, were probably on the rise though they were not recorded by censuses.

B. Consequences of migration to cities

The socio-economic consequences of migration remain poorly documented and the empirical evidence to validate them is often weak. One problem is that migration is a complex and multifaceted phenomenon, both in terms of costs and benefits and in terms of its impact on areas of origin and destination. Research indicates that the developmental impact of migration varies across countries (United Nations, Economic and Social Commission for Asia and the Pacific (ESCAP), 1992b). Studies reviewing the impact of migration invariably conclude that its consequences change according to the nature of the local economy and the particular types of migration involved (United Nations, Economic and Social Commission for Asia and the Pacific (ESCAP), 1992b).

A major drawback in assessing the consequences of migration is the lack of adequate data on both migration and related phenomena. Estimates of rural-urban migration are usually obtained indirectly, as the difference between actual population growth in urban areas and natural increase in those areas. That difference encompasses the effects both of migration and of reclassification, and in most cases there is no independent evidence allowing a separate estimate of each of those components. In addition, it is difficult to establish how much of the change in relevant dependent variables is caused by migration. In large cities in particular, factors that may lead to the changes observed are often left out of the estimation equations because adequate measures of them are not available. Consequently, the discussion of the consequences of migration is generally descriptive and eclectic, based on less than ideal empirical evidence. Given these limitations, the discussion of the consequences of migration for large cities will highlight mostly its implications for urban population growth and the structure of the urban population.

Basically, migration affects the size and the age and sex distribution of a population. To the extent that migrants differ from natives, migration will also affect vital rates. The impact of migration will depend on its intensity: if its volume is small in relative terms, its effects will be negligible. In Japan, for example, the large-scale migration of young people to Tokyo and Kanagawa in the 1960s resulted in a 20 per cent increase in the crude birth rate during the 1970s, whereas the crude birth rate in the rural areas of origin, including that of Shimane and Tottori, fell by 14 per cent.

The migration of young people to the cities can also result in a reduction of the crude death rate of the areas of destination because young people are characterized by low mortality. In contrast, death rates in the areas of origin may rise because the people left behind are, on average, older. Migration can therefore affect the natural increase of both the urban and the rural populations. With respect to fertility, much depends on the family-size desires of the migrant population. In Bangkok, for instance, the migrant women enumerated in 1980 had a lower average number of children ever born than their non-migrant counterparts.

Regarding the socio-economic consequences of migration, in Japan, rural-urban migration has had serious economic and social consequences for the areas of both origin and destination. Rural areas, having lost a high percentage of their young adult population, experienced labour shortages and had trouble maintaining some of their productive activities. Large metropolitan areas, on the other hand, also faced serious problems because of the rapid increases in respect of population and industry, including environmental deterioration, air pollution, traffic and housing problems, water contamination and noise pollution.

In developing countries, the concentration of population in cities has led to various economic, social, environmental and political problems, including unemployment and underemployment, air and water pollution, and rising crime. In large cities, such as Bangkok and Metro Manila, traffic problems are serious. Journey times are long and unpredictable. The traffic in Calcutta is not only slow but also chaotic (United Nations, 1986a). Metro Manila has problems related to the inadequacy of piped water supplies, and the groundwater used by one third of the city's population is subject to pollution from saline intrusion. The water supply in Bangkok is also less than satisfactory. An estimated 100,000 persons obtain water directly from canals and waterways that are grossly polluted by human waste and industrial effluents. The overutilization of groundwater in Bangkok has resulted in the sinking of about 1,000 square kilometres of land in the southern and eastern suburbs by 5-10 centimetres per year. Inadequate water supply is also reported in other mega-cities such as Bombay, Calcutta, Delhi, Dhaka, Jakarta, Karachi and Madras. Seoul is the exception in that it has relatively few water supply and traffic problems.

Air and water pollution is another critical problem in most of the large Asian cities. In Bangkok, environmental conditions are generally poor. Not only are the canals and waterways of the city polluted, but in addition air pollution, due largely to faulty exhausts, has exceeded tolerable levels as defined by the World Health Organization. In Bombay, air pollution caused mainly by noxious industries is a serious problem. In Dhaka, there is considerable pollution of the local rivers caused by effluents from tanneries, particularly in the Hazaribagh area, and by sugar mills, jute mills and two thermal stations in the Ghorasal area. In Delhi, the Yamuna River is highly polluted by untreated sewage and waste from industrial areas, and average air quality is generally poor because of pollution from engineering, textile and chemical industries, thermal power plants and automobile emissions.

Although these environmental problems are commonly faced by large cities they are not necessarily due to migration. Yet, the persistence of large-scale migration to the cities can certainly exacerbate such problems. Consequently, it is not rare to find that migrants are the target of control measures. In Bombay, for instance, the continuing migration of unskilled workers to the Bombay region, at a rate of about 350 persons per day, is a highly sensitive political and social issue that has been addressed with forceful policy measures, including the clearance of squatter settlements and the forceful return of migrants to their villages of origin, as occurred in 1981 (United Nations, 1986b).

For a number of years, migration has been more a political than a demographic issue for Calcutta's planners. Thus, a common argument against promoting industrial development in Calcutta has been that it would benefit migrants more than natives, since in many industrial sectors (jute, paper, cotton, iron and steel) more than half of the labour force is non-Bengali. However, some planners have begun to worry about the continuation of low net migration to Calcutta, which they take as a sign of urban decay. Others have welcomed the respite offered by continued low migration (United Nations, 1986a).

The negative consequences of migration are only one side of the story. In many instances, migration to cities contributes to the development of areas both of destination and of origin. In the large cities of the newly industrializing countries, migration helps provide less expensive labour. In the Republic of Korea, between 1975 and 1980, net migration accounted for 91 per cent of the growth in the number of women employed in urban areas, with 61 per cent of the increase being attributable to the presence of rural-urban migrants. In Thailand, the net rural-urban migration of persons over age 11 has accounted for between 50 and 60 per cent of the growth of the labour force in municipal areas. The net migration of women has been nearly double that of men, and migrant women have accounted for about half of the increase in female workers in the service sector (United Nations, Economic and Social Commission for Asia and the Pacific (ESCAP), 1982).

Rural-urban migration has led to an improvement of the status of women in some countries, especially by allowing a higher labour force participation among women. In places where urbanization and industrialization have expanded concomitantly, more women are now employed in the formal sector (Tey and Awang, 1992).

Migration to cities from the rural areas tends to alleviate unemployment and underemployment in the rural sector by absorbing its surplus labour force. Out-migration may not be detrimental to village production because the marginal productivity of rural labour is zero or near zero in most developing countries. In addition, migration may contribute to a narrowing of income disparities between rural and urban areas. The money remitted by migrants or brought back upon their return is important for the development of rural areas. Migrants throughout the region send money back to their villages. One study estimated that, on average, migrants remitted 38 per cent of their monthly income to their homes of origin and in some areas remittances were shown to account for about 40 per cent of total village income (Connell and others, 1976). Remittances are often used to cover the cost of children's education, thus improving the human capital of the areas of origin.

Migration is a strategy to maximize an individual's welfare. A study in Thailand showed that migrants were better off in terms of employment and occupational status than non-migrants (Mithranon and Santhipaporn, 1992). Among adult male migrants, 45.6 per cent were administrative or government officials, 21.8 per cent were service workers, 18.3 per cent were craftsmen or production workers, and 16.5 per cent were clerical or related workers. Among adult female migrants, 30.9 per cent were service workers, 18.7 per cent were craftsmen or production workers, 16.6 per cent were clerical or related workers, and 11.9 per cent were in the professional category. Migrants also had lower unemployment rates than non-migrants.

Migration affects both individual migrants and their families. Rural-urban migration has led to the nuclearization of families and to the growth of the proportion of single-member households. Gender differentials in age-specific migration rates, coupled with job segmentation, are expected to have important consequences for marriage and the family (Tey and Awang, 1992). Increased female labour force participation in the urban sector means that the traditional source of care for family members, particularly the young and the old, may be eroded. This issue needs to be researched further.

REFERENCES

Connell, J., and others (1976). *Migration from Rural Areas: The Evidence from Village Studies*. New Delhi: Oxford University Press.

Hugo, Graeme J. (1992). Migration and rural-urban linkages in the ESCAP region. Paper prepared for the pre-Conference Seminar of the Fourth Asia-Pacific Population Conference on Migration and Urbanization: Interrelationships with Socio-Economic Development and Evolving Policy Issues, Seoul, Republic of Korea, 21-25 January.

Mithranon, Preeya, and Sureerat Santhipaporn (1992). Urbanization and migration in Thailand. Paper presented at the International Seminar on China's 1990 Census, Beijing, October.

Singelmann, Joachim (1991). Global assessment of levels and trends of female internal migration, 1960-1980. Paper presented at the United Nations Expert Group Meeting on the Feminization of Internal Migration, Aguascalientes, Mexico, 22-25 October.

Tey, Nai Peng, and H. Awang (1992). Some implications of rural-urban migration in the ESCAP Region. Paper presented at the Expert Group Meeting on Trends, Patterns and Implications of Rural-Urban Migration, Bangkok, 3-6 November.

United Nations (1986a). *Population Growth and Policies in Mega-Cities: Calcutta*. Population Policy Paper, No. 1. ST/ESA/SER.R/61.

_________ (1986b). *Population Growth and Policies in Mega-Cities: Bombay*. Population Policy Paper, No. 6. ST/ESA/SER.R/67.

_________ (1991). *World Urbanization Prospects, 1990*. Sales No. E.91.XIII.11.

__________, Economic and Social Commission for Asia and Pacific (ESCAP) (1980). *Migration, Urbanization and Development in Sri Lanka.* Comparative Study on Migration, Urbanization and Development in the ESCAP Region. Country reports, vol. 2. Bangkok: ESCAP.

_________ (1982). *Migration, Urbanization and Development in Thailand.* Comparative Study on Migration, Urbanization and Development in the ESCAP Region. Country reports, vol. 5. Bangkok: ESCAP.

_________ (1992a). Internal and international migration and its implications for socio-economic development policies. Paper presented at the Fourth Asian and Pacific Population Conference: Meeting of Senior Officials, 19-24 August.

_________ (1992b). Migration and urbanization: Interrelations with socio-economic development and evolving policy issues. Paper prepared for the pre-Conference Seminar of the Fourth Asia-Pacific Population Conference on Migration and Urbanization: Interrelationships with Socio-Economic Development and Evolving Policy Issues, Seoul, Republic of Korea, 21-25 January.

XIX. POPULATION DYNAMICS IN THE LARGE CITIES OF LATIN AMERICA AND THE CARIBBEAN

*Economic Commission for Latin America and the Caribbean**

One of the most salient features of the spatial distribution of the population in Latin America and the Caribbean is the high percentage of people who reside in large cities (those with 1 million inhabitants or more) or in mega-cities (defined here as those with 5 million inhabitants or more). Although the phenomenon of concentration in large cities is nothing new, it has become more evident during the past 50 years, when the region became predominantly urban. Urban concentration has been associated with the prevalence of a development model that concentrates power in the capital city or in the most important city of each country, making it the political, economic, sociocultural and administrative centre (Villa, 1980; United Nations, Economic Commission for Latin America and the Caribbean (ECLAC), 1989; Hardoy, 1991). In 1990, more people lived in the large cities of the region (132 million resided in cities with at least 1 million inhabitants) than in rural areas (126 million). Consequently, the population dynamics in those cities are of major importance since they will have profound and diverse repercussions on the socio-economic development of Latin America and the Caribbean.

Whereas in 1950 the region's 7 cities with over 1 million inhabitants housed 17 million persons, by 1990 the 38 cities of that size had a total of 132 million inhabitants. Over the same period, the percentage of the total population living in cities of 1 million inhabitants or more rose from 10.7 to 30.3 per cent and as a proportion those cities accounted for 25.8 per cent of the urban population in 1950 and for 42.6 per cent in 1990 (table 60). In 1950, only Buenos Aires had at least 5 million inhabitants. By 1990, five metropolitan areas in Latin America and the Caribbean had over 5 million inhabitants and their total number of residents amounted to 66 million. In terms of percentages, the mega-cities accounted for 3.2 per cent of the population in the region in 1950 and for 15.1 per cent in 1990 (Centro Latinoamericano de Demografía (CELADE), 1992). This does not mean, however, that at the country level the concentration of population in the largest city has necessarily been increasing. In Argentina, Cuba and Uruguay, for instance, the percentage of the urban population residing in the largest city declined between 1950 and 1990.

A. Population dynamics in the large cities

The study of population dynamics in the large cities of Latin America and the Caribbean lacks cohesiveness and comparability (Lattes, 1984). There is information, however, confirming that there is a continuation of established trends and an emergence of new ones. In particular, the surveys carried out during the 1970s under the World Fertility Survey Programme (WFS) in 13 countries of the region yielded information on fertility at both the national level and for the largest cities in each country (United Nations, 1987). Since 1985, similar data have been gathered by the Demographic and Health Surveys (DHS) which have been completed in nine countries of the region.

1. *Fertility*

All the evidence available indicates that the largest cities in each country are more advanced in the transition towards lower fertility than other regions. Consequently, at least since the 1960s, the total fertility rates of large cities have been lower than the national average of each country (table 61). Although there tends to be a negative correlation between city size and total fertility, there are some medium-sized cities whose fertility levels are even lower than those of the largest cities in their respective countries (Rosen and Simmons, 1967; Centro Latinoamericano de Demografía (CELADE), 1988). Usually, those medium-sized cities have modern productive structures. Therefore, the observed negative correlation is likely to be due mostly to specific economic, social and cultural characteristics of cities rather than to city size

*Centro Latinoamericano de Demografía (CELADE).

per se. Once the desired family size has begun to decline, living in a city makes it more likely that couples will have access to more effective birth control methods. Yet, the increases in nuptiality rates and the reduction of breastfeeding that are usually associated with life in the large cities may have counterbalancing effects on the reduction of fertility.

According to the WFS and DHS data, desired fertility is lower in large cities than elsewhere. Yet, actual fertility levels are not as spatially homogeneous as desired fertility. Thus, the ideal family size in almost all the geographical subdivisions of the countries of the region varies only between two and three children. This finding supports the claim that differences in fertility are mostly the result of the economic, cultural or psycho-social costs and benefits of contraceptive use and not of the particular reproductive rationale characterizing social and geographical population subgroups. From that perspective, the costs of contraception would be lower in large cities, where the sociocultural climate promotes a smaller family unit and access to basic health services is more widespread. Furthermore, the benefits of lower fertility are likely to be more immediate and real in large cities than elsewhere, since the labour force participation of women tends to be higher in such cities, the education of children is more highly valued and large cities generally offer better opportunities for social mobility. However, the heterogeneity of reproductive patterns observed within large cities, where there are persistent differentials among women belonging to the various socio-economic strata, should not be overlooked.

Although fertility has generally been declining in most of the region's large cities (table 61), in some of those where total fertility was already low at the beginning of the 1980s (under three children per woman), there have been slight upward fluctuations commensurate with increases registered at the national level. That has been the case for Santiago de Chile. The postponement of births during the period of economic crisis, particularly between 1982 and 1985, may explain the fluctuations observed later in the decade. In general, with the exception of Havana, the total fertility in large cities is still above replacement level. In addition, although in cities

TABLE 60. LATIN AMERICA: POPULATION CONCENTRATION IN LARGE CITIES, 1950, 1970 AND 1990

	Cities with 1 million inhabitants or more			*Cities with 5 million inhabitants or more*		
	1950	*1970*	*1990*	*1950*	*1970*	*1990*
Number of cities	7	18	38	1	4	5
Population (*thousands*)	17 099	56 803	132 245	5 042	32 899	66 057
Percentage of the total population	10.72	20.51	30.26	3.16	11.88	15.11
Percentage of the urban population	25.77	35.63	42.61	7.60	20.64	21.28
	Cities that had 1 million inhabitants or more, by 1990			*Cities that had 1 million inhabitants or more, by 1950*		
Number of cities	38	38	38	7	7	7
Population (*thousands*)	26 931	69 008	132 245	17 099	38 648	67 840
Percentage of the total population	16.88	24.91	30.26	10.72	13.95	15.52
Percentage of the urban population	40.59	43.29	42.61	25.77	24.25	21.86
Mean annual rate of growth (per 1,000)	47.05		32.52	40.77		28.13
Index of pre-eminence[a] (per 1,000)	3.22		-0.79	-3.04		-5.19

Source: Centro Latinoamericano de Demografía (CELADE), "Latin America and the Caribbean: the dynamics of population and growth", background document for the Meeting of Government Experts on Population and Development in Latin America and the Caribbean, Saint Lucia, 6-9 October.

[a]Corresponding to the annual rate of growth in the share of the urban population living in cities of 1 million inhabitants or more.

TABLE 61. LATIN AMERICA: TOTAL FERTILITY RATE, DESIRED FAMILY SIZE, CONTRACEPTIVE USE AND INFANT MORTALITY RATE IN SELECTED COUNTRIES AND LARGE CITIES, 1970-1979 AND 1980-1990

Country and city/cities	*Total fertility rate*		*Desired family size*[a]		*Modern contraceptive use*[b]		*Infant mortality rate*	
	1970-1979	*1980-1990*	*WFS*	*DHS*	*WFS*	*DHS*	*1970-1979*	*1980-1990*
Argentina	3.1 (72)	3.1 (80)	..	..	..	..	63 (70)	27 (89)
Buenos Aires	2.7 (72)	2.7 (80)	..	..	..	..	50 (70)	20 (89)
Bolivia	6.4 (72)	5.2 (85)	..	..	..	28.1	161 (70)	93 (87)
La Paz	4.5 (72)	3.8 (85)	..	..	..	40.2	142 (70)	..
Brazil	4.5 (70)	3.5 (86)	..	2.8	..	56.6	80 (78)	68 (84)
São Paulo	3.6 (70)	2.6 (86)	..	2.7	..	63.4	72 (78)	45 (84)
Rio de Janeiro	..	3.1 (86)	..	2.3	..	62.5	58 (78)	51 (84)
Chile	3.3 (72)	2.6 (90)	..	..	..	..	77 (70)	18 (89)
Santiago	2.7 (72)	2.3 (90)	..	..	..	..	49 (70)	14 (89)
Colombia	4.5 (75)	2.9 (89)	4.1	..	30.0	54.6	68 (75)	27 (87)
Bogotá	2.9 (75)	2.4 (89)	3.5	..	57.0	61.8	56 (75)	22 (87)
Costa Rica	3.8 (75)	3.4 (90)	4.7	..	53.0	..	58 (74)	15 (88)
San José	3.0 (75)	3.0 (90)	4.0	..	69.0	..	51 (74)	13 (88)
Cuba	3.5 (72)	1.8 (89)	..	..	..	85.5	28 (74)	11 (89)
Havana	3.0 (72)	1.6 (89)	..	..	..	84.0	23 (74)	10 (89)
Dominican Republic	5.7 (74)	3.3 (90)	4.6	3.1	26.0	51.7	99 (72)	44 (88)
Santo Domingo	4.2 (74)	2.6 (90)	4.3	2.9	42.0	54.7	90 (72)	36 (88)
Ecuador	6.9 (72)	4.3 (86)	4.1	3.1	26.0	44.3	105 (72)	65 (86)
Guayaquil	4.5 (72)	3.2 (86)	3.4	2.6	49.0	52.7	70 (72)	52 (86)
Quito	4.4 (72)	3.4 (86)	3.5	2.7	..	57.5	70 (72)	46 (86)
El Salvador	6.1 (73)	4.9 (84)	..	..	26.5	58.5	99 (72)	68 (83)
San Salvador	4.2 (73)	3.3 (84)	..	..	..	76.4	..	48 (83)
Guatemala	6.9 (72)	5.6 (86)	..	..	..	19.0	81 (79)	67 (88)
Guatemala City	4.1 (72)	4.0 (86)	..	..	..	36.8	67 (79)	52 (88)
Haiti	5.5 (76)	6.3 (86)	3.5	..	5.0	6.5	134 (75)	100 (86)
Port-au-Prince	4.0 (76)	4.6 (86)	3.1	..	26.0	10.9	194 (75)	102 (86)
Honduras	7.1 (71)	5.9 (81)	..	..	..	..	114 (69)	58 (85)
Tegucigalpa	4.3 (71)	3.7 (81)	..	..	..	..	81 (69)	50 (85)
Mexico	6.2 (75)	3.6 (86)	4.5	3.0	23.0	43.8	71 (73)	56 (84)
Mexico City	4.8 (75)	3.0 (86)	3.9	2.5	46.0	55.6	..	32 (84)
Panama	4.5 (74)	2.9 (89)	4.2	..	46.0	..	40 (72)	22 (89)
Panama City	3.5 (74)	2.2 (89)	3.9	..	60.0	..	38 (72)	17 (88)
Paraguay	5.0 (78)	4.7 (89)	5.1	..	24.0	35.2	84 (76)	35 (87)
Asunción	3.2 (78)	3.5 (89)	4.1	..	52.0	45.0	64 (76)	28 (87)
Peru	5.6 (76)	5.3 (85)	3.8	2.7	11.0	45.8[c]	103 (74)	76 (83)
Lima	3.9 (76)	3.4 (85)	3.5	2.5	49.0	62.8[c]	61 (74)	34 (83)

Sources: CELADE on the basis of national sources and DHS surveys; Cuba, Comité Estatal de Estadísticas (CEE), *Encuesta Nacional de fecundidad, 1987* (Havana, CEE, 1987); Juan Chackiel, "Niveles y tendencias de la mortalidad infantil en base a la Encuesta Mundial de Fecundidad", *Notas de Población* (Santiago, Chile), No. 27, pp. 67-120; and *Fertility Behaviour in the Context of Development: Evidence from the World Fertility Survey*, Population Studies, No. 100 (United Nations publication, Sales No. E.86.XIII.5).

NOTE: Within body of table, figures in parentheses are years: for example "(72)" means "(1972)".

[a]Mean number of wanted children.

[b]Modern contraceptive use is measured as the percentage of women using modern methods of contraception among currently married women aged 15-49. In the case of Bolivia, the percentage is among ever-married women aged 15-49.

[c]Including modern and traditional contraceptive methods.

like Buenos Aires and Montevideo the long-term reduction in fertility began more than 30 years ago and in others there was a precipitous decline during the 1970s, during the 1980s there has been a tendency towards the stabilization of total fertility at moderate levels.

2. *Mortality*

While information on mortality in the large cities is neither systematic nor reliable, the evidence available indicates that the life expectancy at birth of the population of large cities is generally above the national average (Consejo Nacional de Población (CONAPO), 1988; Bidegain, 1989; Fundação Instituto Brasileiro de Geografia e Estatística (IBGE), 1990). Information on infant and child mortality from WFS and DHS shows that, in general, it is lower in the large cities than elsewhere in the country (table 61). There are many factors contributing to lower mortality in large cities, including a broader coverage of maternal and child health programmes; the existence of nutrition, immunization and general health programmes; the existence and more comprehensive coverage of infrastructure to provide clean water or sewerage; the relatively better levels of education of women; and, in general, the better living conditions large cities offer in comparison with those prevalent in other populated areas of the same country. However, considerable mortality differences by socio-economic status persist within cities. With respect to infant mortality, for instance, in Santiago de Chile, the level in 1985-1990 was twice as high in areas with the greatest incidence of poverty as in areas inhabited by those having higher incomes (Rodríguez, 1992). In São Paulo, around 1990, the equivalent ratio was as high as three to one (Camargo, 1992).

3. *Migration*

Various studies have stressed the volume and intensity of migration to the region's large cities, especially between 1940 and 1970. During the late nineteenth and early twentieth century, international migration was a major contributor to the growth of cities in the Southern Cone, particularly Buenos Aires and Montevideo. Ordinarily, however, the migration flows converging to large cities have largely originated within national borders. It is estimated that in the 1960s and 1970s, net internal migration and reclassification accounted for about 50 per cent of the population increase recorded by a number of large cities, including Belo Horizonte, Bogotá and São Paulo (United Nations, 1983). If the effect of both net migration gains and the natural increase associated with the growing number of young migrants in large cities is taken into account, it would account for over half of the population growth in several of them.

However, declining net migration rates have been experienced by a number of large cities in Latin America and the Caribbean since the late 1970s. Besides confirming that migration towards the mega-cities is waning, the initial results of the 1990 round of censuses suggest that the volume and intensity of out-migration from certain cities have increased. It is likely, moreover, that some large cities have experienced negative net migration, as the definitive results of the 1990 census of Mexico show regarding the Mexico City Metropolitan Area.

A distinctive feature of migration to large cities is its characteristic diversity. Migrants vary regarding their skills and socio-economic attributes, although most of them are young adults just beginning their working life. With respect to place of origin, migration varies according to the degree of urbanization of the country: the more urbanized the country, the more likely that migrants to large cities come from other urban places (Lattes, 1984; Ebanks, 1991). In Latin America, a distinctive trait of migration to large cities is the high proportion of women that it involves (Oliveira and Roberts, 1989; Szasz, 1994). Increasingly, different types of temporary migration are gaining importance in relation to the population dynamics of large cities. Thus, seasonal, itinerant or cyclic movements are frequent and, although they do not involve a permanent change of residence, they nevertheless entail, from the territorial perspective, the formation of extended living spaces (Picouet, 1992).

Territorial mobility within large cities needs to be better understood, since it appears to have been increasing in recent years. Even though such mobility does not affect overall population growth in the city, it does, by shaping the distribution of inhabitants and contributing to the socio-demographic, economic and cultural differentiation of areas within the city, exercise considerable influence on its functioning and efficiency.

4. *Population growth*

Historically, the populations of large cities in the region have grown faster than the total urban population of their respective countries, thus increasing as a proportion of the total population (table 62). Today, however, that trend is no longer universal. With declining fertility, the natural increase of large cities has been lower than that of the rest of the population. Therefore, net gains through migration or reclassification have been necessary to maintain city growth above that of the national population. With slowing in-migration, especially since the 1970s, natural increase has become the major source of population growth in large urban centres (Oliveira and Roberts, 1989). Furthermore, as net migration has declined, the weight of large cities in the total urban population has tended to decrease. That has been the case of Buenos Aires, Caracas, Havana and Montevideo during the 1980s and it is becoming more widespread (United Nations, 1991). Recent studies show a decline in-migration to Mexico City, Rio de Janeiro and Santiago de Chile (Valladares, 1989; Duhau, 1992; Rodríguez, 1992). The accelerated growth of certain medium-sized cities also explains in part the reduction of the weight of the largest city in the urban hierarchy (Ebanks, 1991).

There have also been changes in population distribution within cities. While city centres have deteriorated and lost population, outlying areas have expanded considerably. That process of "suburbanization" intensified during the late 1970s and the 1980s, mostly because of changes in land use as the historic city centres lost their residential character and the suburbs offered better housing at lower costs. Some residential areas have lost dynamism as they experienced the departure of younger inhabitants who settled elsewhere as they formed their own families. The socio-economic and demographic effects of these changes need to be better understood.

5. *Sex and age structure*

Many large cities in Latin America and the Caribbean share certain attributes with respect to the sex and age structure of their populations. Almost systematically, their current sex ratios are below national averages as a result of the predominance of women in net migration (Elton, 1978; Oliveira and García, 1984; Recchini de Lattes, 1989; Szasz, 1994). This situation does not appear to have been affected yet by the slowdown in migration. Another characteristic trait of the population of large cities is its relatively low proportion of children in comparison with that in the national population and a higher proportion of persons in the working ages (15-64). In a growing number of large cities, the proportion of old people, the majority of whom are women, is also high in relative terms. These traits are clearly related to the lower fertility and mortality rates experienced by the populations of large cities and to the selectivity of migration. They are important because different population subgroups have different needs for social and other services.

Because of the relatively high proportion of women in the reproductive ages, large cities have a potential for displaying higher birth rates than those expected on the basis of fertility levels. With regard to death rates, the effect of the age structure typical of large cities would be minor since the proportion of older people is still low in most cases. However, the populations of large cities are ageing more rapidly than their respective national populations, implying that large cities will have to cope sooner with the economic and social implications of an older population, especially in countries where fertility declines began early. In Buenos Aires and Montevideo, for instance, continued low fertility coupled with the arrival of sizeable numbers of international migrants during the 1940s and 1950s has given rise to older age structures than those found in other large cities of the region. Already by 1980, the population aged 60 years or over in Buenos Aires was estimated to be 13 per cent of the total (Recchini de Lattes, 1991).

In addition, because of intra-city mobility, the population in certain neighbourhoods is ageing more rapidly than that in others. These are generally areas that, because of rigidities in the real estate market, have been experiencing the out-migration of the young and where older women tend to outnumber older men because of sex differentials in old-age mortality. This process of intra-city differentiation has multiple repercussions on the socio-economic configuration of urban space.

B. Migration and the labour market of large cities

As is well known, migration to large metropolitan areas is fuelled mostly by the search for employment or other employment-related considerations. Because migrant women tend to have higher labour force participation rates than non-migrant women, they exert greater pressure on the city's labour market (Oliveira and García, 1984; Szasz, 1994). Migration has a significant effect on the labour supply, which in turn influences

TABLE 62. LATIN AMERICAN CITIES WITH INHABITANTS NUMBERING 1 MILLION OR MORE BY 1990: ESTIMATED POPULATION, MEAN ANNUAL RATE OF GROWTH, AND PERCENTAGE SHARE IN TOTAL AND TOTAL URBAN POPULATIONS OF EACH COUNTRY, 1950, 1970 AND 1990

Country and city/cities	*Estimated population (thousands)*			*Mean annual rate of growth (percentage)*		*Percentage share in country's total population*			*Percentage share in country's total urban population*[a]		
	1950	*1970*	*1990*	*1950-1970*	*1970-1990*	*1950*	*1970*	*1990*	*1950*	*1970*	*1990*
Argentina	17 150	23 962	32 322	1.67	1.50	100.0	100.0	100.0	65.3	78.4	86.3
Buenos Aires	5 042	8 414	11 509[b]	2.56	1.57	29.4	35.1	35.6	45.0	44.8	41.3
Córdoba	416	787	1 136	3.19	1.84	2.4	3.3	3.5	3.7	4.2	4.1
Rosario	532	809	1 084	2.10	1.46	3.1	3.4	3.4	4.7	4.3	3.9
Bolivia	2 766	4 325	7 171	2.24	2.53	100.0	100.0	100.0	37.8	40.7	52.3
La Paz	265	516	1 234[b]	3.33	4.36	9.6	11.9	17.2	25.4	29.3	32.9
Brazil	53 444	95 847	150 368	2.92	2.25	100.0	100.0	100.0	36.0	55.8	74.9
São Paulo	2 423	8 064	17 395	6.01	3.84	4.5	8.4	11.6	12.6	15.1	15.4
Rio de Janeiro	2 864	7 040	10 714	4.50	2.10	5.4	7.3	7.1	14.9	13.2	9.5
Belo Horizonte	365	1 589	3 598	7.35	4.09	0.7	1.7	2.4	1.9	3.0	3.2
Pôrto Alegre	459	1 521	3 124	5.99	3.60	0.9	1.6	2.1	2.4	2.8	2.8
Recife	661	1 781	2 492	4.96	1.68	1.2	1.9	1.7	3.4	3.3	2.2
Salvador	403	1 140	2 401	5.20	3.72	0.8	1.2	1.6	2.1	2.1	2.1
Brasília	37	526	2 362	13.27	7.51	0.1	0.5	1.6	0.2	1.0	2.1
Fortaleza	256	1 030	2 088	6.96	3.53	0.5	1.1	1.4	1.3	1.9	1.9
Curitiba	137	814	2 031	8.91	4.57	0.3	0.8	1.4	0.7	1.5	1.8
Goiânia	41	490	1 679	12.40	6.16	0.1	0.5	1.1	0.2	0.9	1.5
Campinas	101	483	1 659	7.82	6.17	0.2	0.5	1.1	0.5	0.9	1.5
Manaus	110	280	1 215	4.67	7.34	0.2	0.3	0.8	0.6	0.5	1.1
Santos	238	656	1 199	5.07	3.02	0.4	0.7	0.8	1.2	1.2	1.1
Belém	233	651	1 029	5.14	2.29	0.4	0.7	0.7	1.2	1.2	0.9
Chile	6 082	9 504	13 173	2.23	1.63	100.0	100.0	100.0	58.4	75.2	85.9
Santiago	1 332	2 837	4 734[b]	3.78	2.56	21.9	29.9	35.9	37.5	39.7	41.8
Colombia	11 946	21 360	32 978	2.91	2.17	100.0	100.0	100.0	37.1	57.2	70.0
Bogotá	676	2 371	4 851	6.27	3.58	5.7	11.1	14.7	15.3	19.4	21.0
Medellín	341	1 006	1 585	5.41	2.27	2.9	4.7	4.8	7.7	8.2	6.9
Cali	270	847	1 555	5.72	3.04	2.3	4.0	4.7	6.1	6.9	6.7
Barranquilla	305	516	1 019	2.63	3.40	2.6	2.4	3.1	6.9	4.2	4.4
Costa Rica	862	1 731	3 015	3.49	2.77	100.0	100.0	100.0	33.5	39.7	47.1
San José	183	438	1 016	4.36	4.21	21.2	25.3	33.7	63.3	63.8	71.5

Table 62 (*continued*)

Country and city/cities	*Estimated population (thousands)*			*Mean annual rate of growth (percentage)*		*Percentage share in country's total population*			*Percentage share in country's total urban population*[a]		
	1950	*1970*	*1990*	*1950-1970*	*1970-1990*	*1950*	*1970*	*1990*	*1950*	*1970*	*1990*
Cuba	5 850	8 520	10 608	1.88	1.10	100.0	100.0	100.0	49.4	60.2	74.9
Havana	1 147	1 745	2 099	2.10	0.92	19.6	20.5	19.8	39.7	34.0	26.4
Dominican Republic	2 353	4 423	7 170	3.16	2.42	100.0	100.0	100.0	23.8	40.3	60.4
Santo Domingo	219	838	2 203	6.71	4.83	9.3	18.9	30.7	39.2	47.1	50.9
Ecuador	3 310	6 051	10 587	3.02	2.80	100.0	100.0	100.0	28.2	39.5	56.0
Guayaquil	253	694	1 674[b]	5.05	4.40	7.6	11.5	15.8	27.1	29.0	28.2
Quito	206	501	1 241[b]	4.44	4.54	6.2	8.3	11.7	22.0	20.9	20.9
Haiti	3 261	4 535	6 513	1.65	1.81	100.0	100.0	100.0	12.2	19.8	28.3
Port-au-Prince	144	461	1 031	5.82	4.02	4.4	10.2	15.8	36.3	51.5	56.0
Mexico	28 012	52 771	88 598	3.17	2.59	100.0	100.0	100.0	42.7	59.0	72.6
Mexico City	3 148	9 765	20 192[b]	5.66	3.63	11.2	18.5	22.8	27.2	31.4	33.3
Guadalajara	403	1 513	3 161[b]	6.61	3.68	1.4	2.9	3.6	3.4	4.9	4.9
Monterrey	356	1 229	2 970[b]	6.20	4.41	1.3	2.3	3.4	3.0	3.9	4.6
Puebla	227	413	1 267	2.99	5.60	0.8	0.8	1.4	1.9	1.3	2.0
Nicaragua	1 098	2 053	3 871	3.13	3.17	100.0	100.0	100.0	35.0	47.0	59.8
Managua	110	378	1 012	6.17	4.92	10.0	18.4	26.1	28.6	39.2	43.8
Peru	7 632	13 193	21 550	2.74	2.45	100.0	100.0	100.0	35.5	57.4	70.2
Lima	973	2 928	6 247	5.51	3.79	12.7	22.2	29.0	35.9	38.7	41.3
Uruguay	2 239	2 808	3 094	1.13	0.48	100.0	100.0	100.0	78.0	82.1	85.5
Montevideo	1 143	1 167	1 197	0.10	0.13	51.0	41.6	38.7	65.5	50.6	45.3
Venezuela	5 009	10 604	19 735	3.75	3.11	100.0	100.0	100.0	53.2	72.4	90.5
Caracas	676	2 047	4 096[b]	5.54	3.47	13.5	19.3	20.8	25.3	26.7	22.9
Maracaibo	230	617	1 146	4.93	3.10	4.6	5.8	5.8	8.6	8.0	6.4

Source: ***World Urbanization Prospects, 1990*** (United Nations publication, Sales No. E.91.XIII.11).

[a]National figures correspond to the percentage urban in each country.

[b]According to last census returns (either definitive or preliminary figures), 1990 city population was smaller than estimated.

wages, unemployment and underemployment, the extent and nature of informal employment, and labour segmentation (Oliveira and García, 1984; Oberai, 1989).

Until the 1960s, migration was seen as contributing to the economic growth of large cities and migration was considered the logical response to structural and labour-force imbalances between areas. Therefore, migration was seen as a means to attain economic and social modernization (Todaro, 1976; Germani, 1986; Oberai, 1989) and as a force that contributed to achieving needed economic, social, political and cultural transformations. Experience showed that that assessment was too optimistic and that large cities were incapable of generating enough employment in the formal sector to absorb their growing labour force. Migration thus began to be viewed as the cause of unemployment, the expansion of the urban informal sector and the growth of marginality (Oliveira and García, 1984). Migration was also blamed for the increase and expansion of shanty towns (United Nations, Economic Commission for Latin America and the Caribbean (ECLAC), 1989).

While it is currently recognized that migration can increase the pressure on the labour markets of large cities, it is important to assess its impact in relation to the context in which it occurs (ECLAC, 1989). It is now accepted that weaknesses and cycles in the local and national economies are the main cause of the incapacity of large cities to absorb the growing labour force and although migration plays a part, it is not the main cause of the problems observed. Thus, during the 1950s and 1960s, when migration to metropolitan areas reached its peak, unemployment rates were reasonably low. In contrast, unemployment in large cities increased considerably during the 1980s just as the pace of migration slackened. Both high unemployment and low migration were, in fact, the result of the sharp economic crisis affecting Latin American and Caribbean countries during the 1980s. Of course, the interactions involved were complex. During the 1950s and 1960s, economic strategies emphasized industrialization for import substitution, thus concentrating investment in large cities. In addition, the agrarian reform carried out in some countries encouraged people to remain in rural areas. During the 1980s, however, virtually all of the region's economic scaffolding collapsed, affecting especially the labour market in urban areas.

Furthermore, the interrelations between migration and employment in large cities of the region must be seen in a broader context where not only does migration contribute to increase the labour force, but in addition the younger generations are growing larger and the participation of women in the labour force is increasing. Thus, the labour force would be growing even in the absence of migration. Although the 1980 round of censuses did show that migration had played a major role in the expansion of the number of workers in some large cities, especially in certain branches of economic activity, there was no indication that migration would continue to have the same effect over the long run (Oliveira and Roberts, 1989; United Nations, Economic Commission for Latin America and the Caribbean (ECLAC), 1989). Moreover, analyses of the modes of incorporation into the labour force of migrants and non-migrants have revealed no clear distinctions between the two groups. Some research suggests that the slight differences detected between migrants and non-migrants reflect other factors, such as the time elapsed since migration or the age and sex distribution by migration status (Oliveira and García, 1984; Maguid, 1986). Other studies document the positive effects of some migration flows on specific economic activities, including the construction sector where seasonal work is common and some branches of the service sector (Oberai, 1989).

Therefore, although migrants to large cities can indeed increase the pressure on the labour market, they cannot be considered entirely responsible for the high unemployment and underemployment rates experienced by some Latin American and Caribbean cities. In fact, migration has played a major role in boosting some economic sectors in large cities, though its impact can be understood only in terms of the contextual conditions that make such an outcome possible. Moreover, after a certain length of residence, the labour-market insertion patterns of migrants do not differ substantially from those of non-migrants, indicating that the former can and do become a productive part of the metropolitan environment. However, not enough is known about the types of work performed by people who move only temporarily to large cities.

C. Socio-economic effects of demographic change in large cities

It is commonly argued that the large populations of large cities and the increasingly extensive areas that large cities occupy make them difficult to administer and manage. Consequently, serious social problems, such as delinquency, drug addiction and precarious settlements, have proliferated. Diseconomies of scale have emerged, inefficient ways of using resources have become common, and severe environmental degradation has occurred. Large cities have also become the site of the kind of speculation in real estate that has priced many urban residents out of decent housing and the physical expansion of the city has often meant a loss of fertile agricultural land (Centro Latinoamericano de Demografía (CELADE), 1992; United Nations, Economic Commission for Latin America and the Caribbean (ECLAC), 1989). Yet, the concentration of population in large cities has generated economies of agglomeration and forms of accumulation that could not have emerged otherwise. Large cities also provide more opportunities for upward social mobility and socio-political participation (United Nations, Economic Commission for Latin America and the Caribbean (ECLAC), 1989; Geisse and Sabatini, 1988).

Although a number of problems have arisen in the region's large cities since the 1960s and 1970s, it is not clear that their origins lie solely in the magnitude or rate of growth of their populations. What seems to have happened is that, on the one hand, urban sprawl overwhelmed existing urban management capacity, while, on the other, the residents of large cities belonging to the lowest-income groups bore the brunt of the economic crisis of the 1980s. The recessive effects of that crisis have made themselves felt at various levels: individuals have been faced with increasing job instability, the expansion of informal employment and lower wages; and society has had to make do with lower investment and dwindling resources for public works and the provision of basic services. One of the clearest signs of the severity of the crisis for the region's metropolitan areas is the increase in poverty detected in all the specialized surveys carried out during the 1980s (United Nations, Economic Commission for Latin America and the Caribbean (ECLAC), 1991 and 1992a). The inadequacy of essential public services, the severe deficiencies in urban roads and transport, and the increasing environmental degradation are all signs of stress. Few metropolitan areas treat even a portion of their waste. In São Paulo, the watercourses crossing the city are usually anaerobic because of the large amounts of sewage that they carry, and in sprawling cities, such as Buenos Aires, the widespread use of septic tanks and latrines has seriously contaminated the aquifers. Such problems are more acute in cities that depend on wells for drinking water, such as Lima and Mexico City (United Nations, Economic Commission for Latin America and the Caribbean (ECLAC), 1992b).

Regardless of the various diagnoses of the causes of metropolitan problems, the fact remains that the current economic and socio-political conditions in Latin America and the Caribbean are associated with the difficulties experienced by mega-cities which may override the benefits of agglomeration. Although population size in itself is not to blame for these problems, in practice the large number of residents and the extensive urbanized area of some mega-cities generate such massive requirements that developing societies are hard put to meet them in an adequate manner (United Nations, 1992).

Logic suggests that the goods and services required by a large population concentrated in an urban agglomeration are not entirely different from those of a dispersed population of the same size. However, the urgency of meeting those requirements is more visible in metropolitan areas. Furthermore, while unit costs in urban agglomerations increase more slowly, demanding a smaller total commitment of resources to meet the needs of densely populated areas, the large scales involved may require totally new delivery systems that can be costly. Nor can one discount the organizational complexity of managing and distributing goods and services in high-density areas. Therefore, the challenge should be assessed in terms of ensuring that urban agglomerations minimize their negative externalities. Although it has been claimed that the residents of large metropolitan areas in Latin America and the Caribbean are among the region's most privileged inhabitants, their privileges are very unevenly distributed and are becoming increasingly limited (United Nations, Economic Commission for Latin America and the Caribbean (ECLAC), 1992a).

References

Bidegain, G. (1989). Desigualdad social y esperanza de vida en Venezuela. Documento de Trabajo, No. 34. Caracas: Instituto de Investigaciones Económicas y Sociales, Universidad Católica Andrés Bello.

Camargo, A. (1992). A mortalidade infantil em São Paulo e a ocorrência das causas perinatais. In *VIII Encontro Nacional de Estudos Populacionais*, Associação Brasileira de Estudos Populacionais (ABEP). São Paulo: ABEP.

Centro Latinoamericano de Demografía (CELADE) (1988). Redistribución espacial de la población en América Latina: una visión sumaria del período 1950-1985. Santiago, Chile: CELADE.

________ (1992). Latin America and the Caribbean: the dynamics of population and growth. Background document for the Meeting of

Government Experts on Population and Development in Latin America and the Caribbean, Saint Lucia, 6-9 October.
Chackiel, Juan (1981). Niveles y tendencias de la mortalidad infantil en base a la Encuesta Mundial de Fecundidad. *Notas de Población* (Santiago de Chile), No. 27, pp. 67-120.
Consejo Nacional de Población (CONAPO) (1988). *Características Principales de la Migración en las Grandes Ciudades del País.* Mexico City: CONAPO.
Cuba, Comité Estatal de Estadísticas (CEE) (1987). *Encuesta Nacional de Fecundidad 1987.* Havana: CEE.
Duhau, E. (1992). Población y economía de la zona Metropolitana de la ciudad de México. El centro y la periferia. Mexico City: Universidad Autónoma Metropolitana. Mimeograph.
Ebanks, E. (1991). *Socio-economic Determinants of Internal Migration with Special Reference to Latin America and the Caribbean Region.* Series A, No. 255. Santiago, Chile: Centro Latinoamericano de Demografía (CELADE).
Elton, Charlotte (1978). *Migración Femenina en América Latina: Factores Determinantes.* Series E, No. 26. Santiago, Chile: Centro Latinoamericano de Demografía (CELADE).
Fundação Instituto Brasileiro de Geografia e Estatística (IBGE) (1990). *Anuário estatístico do Brasil.* Rio de Janeiro: IBGE.
Geisse, G. and F. Sabatini (1988). Latin American cities and their poor. In *The Metropolis Era*, vol. 2, M. Dogan and J. Kasarda, eds. Newbury Park, California: Sage Publications.
Germani, G. (1986). *Política y Sociedad en una Epoca de Transición.* Buenos Aires: Paidós.
Hardoy, Jorge (1991). Antiguas y nuevas capitales nacionales en América Latina. *EURE*, Nos. 52/53, pp. 7-26.
Instituto Nacional de Estadísticas (INE) (1987). *Chile, Proyecciones de Población por Sexo y Edad; Regiones, 1980-2000.* Santiago, Chile: INE.
Lattes, Alfredo E. (1984). Algunas dimensiones demográficas de la urbanización reciente y futura de América Latina. In *Memorias del Congreso Latinoamericano de Población y Desarrollo*, vol. 2. Mexico City: Universidad Nacional Autónoma de México, El Colegio de México and Programa de Investigaciones Sociales en Población de América Latina (PISPAL).
Maguid, Alicia (1986). Migración y empleo en la aglomeración metropolitana de Costa Rica. *Notas de Población* (Santiago de Chile), No. 40, pp. 75-123.
Oberai, A. (1989). Problems of urbanization and growth of large cities in developing countries: a conceptual framework for policy analysis. World Employment Programme Working Paper. WEP 2-21/WP.169. Geneva: International Labour Office.
Oliveira, Orlandina de, and Brígida García (1984). Urbanization, migration and the growth of large cities: trends and implications in some developing countries. In *Population, Distribution, Migration and Development: Proceedings of the Expert Group on Population Distribution, Migration and Development, Hammamet (Tunisia), 21-25 March 1983.* Population Studies, No. 89. Sales No. E.84.XIII.3. New York: United Nations.
Oliveira, Orlandina de, and Bryan Roberts (1989). Los antecedentes de la crisis urbana. Urbanización y transformación ocupacional en América Latina, 1940-1980. In *Las Ciudades en Conflicto: una Perspectiva Latinoamericana*, Mario Lombardi and Danilo Veiga, eds. Montevideo, Uruguay: Ediciones de la Banda Oriental.
Picouet, Michel (1992). El concepto de reversibilidad en el estudio de la migración. Bogotá: Universidad de Los Andes.
Recchini de Lattes, Zulma (1989). Women in internal and international migration, with special reference to Latin America. *Population Bulletin of the United Nations* (New York), No. 27, pp. 95-107. Sales No. E.89.XIII.7.
_________ (1991). Urbanization and demographic ageing: the case of a developing country, Argentina. In *Ageing and Urbanization.* Proceedings of the United Nations International Conference on Ageing Populations in the Context of Urbanization, Sendai, Japan, 12-16 September 1988. Sales No. E.91.XIII.12. New York: United Nations.
Rodríguez, J. (1992). Dinámica demográfica del gran Santiago: patrones históricos, tendencias actuales, perspectivas. Santiago, Chile: Centro Latinoamericano de Demografía (CELADE) and Universidad de la Academia de Humanismo Cristiano (UAHC). Mimeographed.
Rosen, B. and A. Simmons (1967). Industrialization, family and fertility: a structural psychological analysis of the Brazilian case. *Demography* (Washington, D.C.), vol. 8, No. 1, pp. 49-69.
Szasz, Ivonne. (1994). *Mujeres Inmigrantes y Mercado de Trabajo en Santiago.* Santiago, Chile: Centro Latinoamericano de Demografía (CELADE).
Todaro, Michael (1976). *Internal Migration in Developing Countries: A Review of Theory, Evidence, Methodology, and Research Priorities.* Geneva: International Labour Office.
United Nations (1983). Metropolitan migration and population growth in selected developing countries, 1960-1970. *Population Bulletin of the United Nations* (New York), No. 15, pp. 50-62. Sales No. F.83.XIII.4.
_________ (1987). *Fertility Behaviour in the Context of Development: Evidence from the World Fertility Survey.* Population Studies, No. 100. Sales No. E.86.XIII.5.
_________ (1991). *World Urbanizaton Prospects, 1990.* Sales No. E.91.XIII.11.
_________ (1992). *Urban Agglomerations, 1992.* Sales No. E.93.XIII.2. Wall chart.
_________, Economic Commission for Latin America and the Caribbean (ECLAC) (1989). La crisis en América Latina y el Caribe. LC/G.1571-P.
_________ (1991). Magnitud de la pobreza en América Latina en los años ochenta. LC/G.1653-P
_________ (1992a). Social equity and changing production patterns: an integrated approach. LC/G.1701.
_________ (1992b). El manejo del agua en las áreas metropolitanas de América Latina. LC/R.1156.
Valladares, L. (1989). Río de Janeiro: la visión de los estudiosos de lo urbano. In *Las Ciudades en Conflicto: una Perspectiva Latinoamericana*, Mario Lombardi and Danilo Veiga, eds. Montevideo, Uruguay: Ediciones de la Banda Oriental.
Villa, Miguel S. (1980). Consideraciones en torno al proceso de metropolización de América Latina. *Notas de Población* (Santiago de Chile), No. 24.

XX. MIGRATION AND POPULATION DISTRIBUTION IN DEVELOPING COUNTRIES: PROBLEMS AND POLICIES

*International Labour Organization**

Population distribution is the aspect of population dynamics that has been more often perceived as a serious problem by the Governments of developing countries during recent decades. By the year 2000, urban-dwellers will constitute more than 45 per cent of the population of developing countries, and that proportion will rise to 60 per cent by 2025 (United Nations, 1991). Urban population growth is expected to be particularly high in African countries, both because of their lower initial level of urbanization and because of their high rates of natural increase.

Governments identify three types of major problems: rapid and unmanageable population growth in primate cities; excessive migration from rural to urban areas; and labour shortages in remote but resource-rich areas. In some cases, rural overpopulation coexists with excessive urban growth, but such a development usually reflects high population density and a continuation of rapid population growth combined with slow economic development. The symptoms of excessively rapid urban growth are all too familiar: the sanitary, housing and transport facilities of urban areas are overstrained; health and educational services deteriorate; unemployment and poverty rise; slums and squatter settlements proliferate; and environmental degradation increases. Yet, despite these major problems, it would be wrong to regard urbanization in developing countries entirely in negative terms. To have a fair perspective of the issues involved, certain salient features of the migration and development processes should be borne in mind. First, in many countries rural-urban migration constitutes less than half of total urban growth. Second, the emergence and growth of cities are a historical concomitant of industrial and economic development, and even of the flourishing of civilizations. Thus, the development process is sustained and accelerated by such factors as urban infrastructure, economies of scale, and differentiated labour markets. Urban areas, particularly large cities, often make a disproportionate contribution to gross domestic product (GDP).[1] Third, there is convincing evidence that in most cases migrants and their families improve their welfare by migrating, and often both the urban and the rural economies benefit from migration. In rural areas, out-migration stimulates rural development by generating remittances and inducing technological change. In the cities, the influx of migrants not only increases the labour supply, but also generates new employment by stimulating industrial expansion and other economic activities.

Problems arise when migration exceeds the income-earning opportunities available in urban areas. Large concentrations of population and economic activities in a few cities entail numerous social costs and may lead to a breakdown of urban services. Heavy out-migration may affect rural areas adversely if it involves the departure of the better-educated, younger and highly motivated people. Thus, individuals may benefit from migration but both rural and urban areas may experience a decline of average social welfare, particularly through overcrowding and increased pollution in urban areas (Hardoy and Satterthwaite, 1991), and a greater regional concentration of wealth, income and human capital.

Like every process, migration has its costs and benefits. It is important, therefore, to determine the role played by migration in the development of areas of origin and destination before formulating appropriate policies or attempting to change the direction or magnitude of migration flows in ways that are consistent with long-term development goals. The effects of migration on areas of origin and destination depend, among other things, on the nature and volume of migration, the flows of remittances, the types of migrants that dominate the migration flows, the level of urbanization, and the stage of economic development. These factors are likely to vary among countries and regions, so that their analysis demands the empirical consideration of socio-economic change and its relationship with migration in each setting.

*Employment and Development Department.

However, serious conceptual and methodological problems arise in assessing the impact of migration on settlement patterns and development. A first problem is to define migration and to distinguish between changes induced by migration and those induced by other factors. Another problem arises from the fact that the economic and social conditions that give rise to migration are altered by the migration process itself. The new conditions are then likely to produce changes in the level and pattern of migration through feedback mechanisms. It is difficult, therefore, to determine whether a change is a cause or a consequence of migration. There is also the problem of choosing the time-frame over which an assessment should be made. If the time elapsed between the onset of migration and the moment when its effects are evaluated varies, this is likely to result in different assessments of migration's impact.

Migration is a complex phenomenon and a solution of the problems associated with it requires that each situation be evaluated on the basis of facts and sound policy analysis. There is a need to develop, test and apply policy measures that influence the direction and magnitude of migration flows to ensure migration's positive benefits and to minimize its negative consequences, including the social disparities resulting from an unbalanced population distribution.

Confronted with the problems of rapid urban population growth, most Governments have already taken measures to slow down or reverse the flows towards metropolitan regions and other urban centres. In adopting such measures, however, planners have had to make numerous assumptions about the behavioural causes of migration, about the adverse or beneficial consequences of existing patterns of territorial mobility, and about the relative efficacy of one type of policy over another. The present paper examines the extent to which the policies and programmes that have been adopted have been successful in influencing migration flows and suggests how they might be modified to deal more effectively with problems of rapid urban population growth.

A. Migration policy experience

The policy instruments used by Governments to slow down the pace of urbanization have varied widely in terms of efficiency and cost-effectiveness (Oberai, 1987). In general, three kinds of policies have often been adopted: (*a*) policies aimed at transforming the rural economy and thereby slowing the rate of urban expansion; (*b*) policies whose goal is to limit the growth of large cities through migration control; and (*c*) policies that try to slow the growth of large cities by redirecting migration flows to intermediate cities and smaller urban centres.

1. *Rural retention policies*

Any policy that transforms the rural economy affects the nature and pace of urban development. Policies such as the redistribution of land to the poor are expected to reduce urban/rural income disparities and slow urban growth by raising agricultural incomes. Other rural programmes may, of course, have the opposite effect. Thus, the green revolution and other efforts to commercialize agriculture have accentuated landlessness and often stimulated, rather than reduced, the outflow of migrants from rural areas. In India, attempts to develop small-scale cottage industries in rural areas have tended to increase out-migration because those industries have improved the skills of villagers and made them more capable of finding employment in urban areas. In general, agricultural and rural development activities that increase access to cities, commercialize agriculture, strengthen rural-urban integration, raise education and skill levels, or increase rural inequalities appear to lead to accelerated rates of rural-urban migration (Rhoda, 1983).

Land reform programmes have reduced inequalities and improved the income and security of farmers in only a few countries, including China and the Republic of Korea. In other countries (Brazil, Egypt, India, Pakistan and Peru), such programmes have been mainly cosmetic and, lacking adequate technical and financial assistance, have not benefited small farmers significantly. Moreover, rapid population growth has weakened the initial impact of land reform in some areas. Nor have tenancy laws greatly improved the position of tenants, despite their being designed to provide protection against eviction and rent increases. On the contrary, in countries such as Argentina, Colombia, India and Peru, those laws have worsened the conditions of tenants in general and those renting smallholdings in particular. Land reform has encouraged landowners to mechanize production and shift to wage labour, with the result that farmers with smallholdings have lost even their meagre incomes. For many, rural out-migration has been the only option.

Land colonization programmes have been implemented in several developing countries, including Brazil, China, Kenya, India, Indonesia and Malaysia. The main objective of these programmes has been to resettle residents from overpopulated rural areas in frontier regions or sparsely populated areas. The policy instruments commonly employed have included investment in infrastructure, the transfer of land titles, and the provision of social services, credit and other facilities to increase the productivity and income of the new settlers. Despite substantial investment in land settlement programmes, however, their performance has not been very encouraging (Oberai, 1988). They are costly in relation to the number of people resettled and settlers experience low productivity and high rates of desertion. In some cases, tensions between new settlers and the local population have arisen.

Integrated rural development programmes have had some success in countries such as Malaysia and Sri Lanka, where they have been accompanied by radical changes in the distribution of income and the social relations of production. In most cases, however, the provision of credit and subsidies, and public investment in roads, schools and amenities, have helped mostly large farmers, increased class differentiation in rural areas, and led to labour displacement through mechanization. The slow pace of rural-urban migration in Sri Lanka, for example, is attributed to the retentive capacity of the rural sector due to social welfare measures introduced by the Government, including free medical services and education, consumer subsidies, low-income housing and income support for poor farmers in the form of guaranteed prices for their produce (Indraratna and others, 1983). These measures have reduced disparities in living standards between rural and urban areas. Another factor that seems to have played an important part in slowing migration in Sri Lanka is the provision of subsidized transport which provides rural dwellers with access to urban employment and amenities without their having to migrate permanently to the cities.

2. *Policies to control the growth of large cities*

Many city administrations and policy makers in developing countries have felt the need to take direct measures to discourage rural-urban migration and even to reverse the flow of migration to urban areas. Such measures have usually included administrative and legal controls, police registration, restrictions on access to education and housing, and direct rustication programmes to relocate urban inhabitants in the countryside.

Direct migration control policies have had some success in China, Cuba and Poland, although the evidence suggests that strict controls on rural-urban migration that infringe upon human rights have not promoted rural development to the degree intended by the Government. Instead, the controls imposed only increased the pressure in rural areas and eventually led to increased migration outflows. A recent study suggests that the so-called floating population in China accounts for between 10 and 20 per cent of the total population of major cities, that is to say, the level of "undocumented" internal migration in the country is high (Zhao, 1989). In other countries, such as Indonesia, Kenya, the Philippines and the United Republic of Tanzania, attempts to control migration through residence permits have been largely unsuccessful because legal restrictions are difficult to enforce, licences are easily forged, monitoring entire cities is costly, and people who are evicted from a city find ways to return. By restricting access to education and housing in Manila, for example, the Government of the Philippines hoped to deter further in-migration, but those measures were difficult to enforce, administratively speaking, and only encouraged corruption (Simmons, 1979). In some cities, such as Nairobi, repeated demolitions of squatter settlements may have influenced the pattern of urban growth, but there is little evidence that they have substantially discouraged urban in-migration. In Jakarta and Lima, street vendors have often been forcibly removed from the city. There have also been periodic expulsions of unemployed migrants from cities, a practice that has been attempted in parts of Africa, notably in the Congo, the Niger, the United Republic of Tanzania and Zaire. These measures have had little visible impact on urban growth, but have undoubtedly caused much anguish and resentment.

Among the reasons for migration into big cities, the education of children has been one of the most important in some countries. One quarter of the migrants to Seoul during the 1960s, for example, moved in order to give their children access to the better educational facilities concentrated in Seoul. Recognizing the importance of achieving a more balanced distribution of educational facilities to encourage population dispersal away from Seoul, the Government devised various programmes to strengthen educational institutions in other areas. As a result, the share of college and university enrolment in Seoul registered a drastic decline from

67 per cent in 1970 to 35 per cent in 1985 (Kim, Kim and Son, 1991).

The practice of pricing urban services, such as food, water, electricity, transport and garbage collection, below cost reduces metropolitan living costs below those of other areas and increases urban/rural and intra-urban income gaps. One method of correcting such spatial biases that does not, however, eliminate price distortions has been to impose high taxes on the dwellers of large cities in the form of metropolitan residential taxes, urban wage taxes or property taxes. In the Republic of Korea, for example, a resident tax was introduced in 1973 to reduce the influx of people into Seoul. This tax was initially applied only to Seoul residents, but it was subsequently adopted by other cities, thus defeating its original purpose. Such developments occurred because the Ministry of Home Affairs recognized the value of such a tax as a source of revenue for local government. Even though the tax rate is higher in Seoul, the difference is not large enough to affect individual household location decisions. According to Kim, Kim and Son (1991), the annual resident tax rate in Seoul was 4,000 won per person, and for medium-sized and large cities (with a population of 500,000 to 5 million), 2,500 won. Residents of smaller cities paid 1,500 won a year, and rural residents 800 won.

3. *Dispersed urbanization and secondary city strategies*

Some countries have attempted administrative decentralization to reduce population pressure in large cities. In Egypt, for example, such an attempt was made in 1979 by granting autonomy to different local Governments. However, it did not prove very effective because of the heavy concentration of infrastructure and industrial projects in Cairo, the capital city. Moreover, rent controls and subsidized food and urban services, which kept the cost of living artificially low in Cairo, to some extent neutralized the effect of administrative decentralization. In the Republic of Korea, the Government announced in 1977 a package of programmes to relocate secondary government offices and public enterprise headquarters outside Seoul. Altogether 60 public agencies (36 government agencies and 24 State-run firms) had been earmarked for transfer from Seoul, but only 29 (12 government agencies and 17 State-run firms) actually moved. Concern about the education of children and the shortage of basic amenities in the areas of destination seems to have been the prime reason behind the reluctance of agencies to relocate. In fact, a survey revealed that only 46.4 per cent of the employees of relocated agencies took their families with them to smaller cities (Jin-Ho, 1990). Programmes for relocating public agencies have thus had only a limited effect on population dispersal.

Another policy for slowing metropolitan growth has been to decentralize economic activity by relocating it in the hinterland surrounding the primate city, usually through the adoption of a metropolitan regional plan. The result has been the creation of polycentric metropolitan areas with satellite sub-centres. While this strategy may be efficient in terms of spatial organization, it has accelerated rather than retarded the growth of the primate metropolitan area, but over a broader geographical expanse. In India, for instance, attempts to create satellite towns near Delhi and Hyderabad have ended by promoting the growth of satellite towns towards the metropolitan centre, thus defeating their purpose. The same has been true in Mexico and Venezuela. In China, industrial parks and satellite towns set up near Shanghai have failed to attract people away from the central city because of the lack of social infrastructure in such towns (Shanghai Planning Commission, 1991).

Sometimes Governments have established new towns or cities to stimulate migration to a different area. Governments have encouraged businesses to relocate to those new urban centres by providing the necessary infrastructure and services and by subsidizing interest rates. Thus, in Egypt, the Government created several new communities or towns, including 6th of October, 10th of Ramadan, 15th of May, Sadaat City. Although those new cities have had some success in attracting industry, they have absorbed large levels of investment without attracting as many people as expected, mostly because of their lack of social infrastructure. In 1986, Sadaat City and 10th of Ramadan, for example, had a population of only 1,927 and 8,500, respectively (Oberai, 1993).

The relocation of the national capital is a strategy that has become common in Africa (Cameroon, Côte d'Ivoire, Lesotho, Malawi, Nigeria and the United Republic of Tanzania) and has two striking examples in Brasília, Brazil, and Islamabad, Pakistan. However, even when the new capital attains a substantial size (Brasília had 1.57 million inhabitants in 1990 and Islamabad had 1.23 million), its impact on population distribution at the national level is generally small while the costs involved in developing and settling the new capital are massive.

By far the most common response to rural/urban and regional imbalances has been the adoption of programmes to stimulate the economies of small and medium-sized cities. The prime target of such a programme is often the industrial sector. Existing or new companies are persuaded to locate their plants in peripheral regions through a combination of tax incentives, the provision of infrastructure, the construction of industrial estates and, occasionally, compulsion. In general, incentives have been used more often than disincentives because most Governments feared that too strong a programme of deconcentration might dissuade foreign companies from investing in the country, contribute to lowering the efficiency of the industrial sector, or antagonize important business interests.

Various incentives have been used to attract industries to new locations. In Brazil, major tax incentives have drawn numerous companies to the principal cities of the poverty-stricken north-east. Experience shows, however, that manufacturing companies have a tendency to locate themselves in the areas that are closest to their preferred locations, unless the incentives to do otherwise are appropriate and sufficiently attractive to outweigh the risks and uncertainties involved. Thus, in north-east Brazil, new industry was located only in the three largest cities, and in Peru, companies were attracted by the new industrial estates of only the most prosperous and attractive city, Arequipa (Gilbert, 1974). In India, the Second and Third Development Plans led to the establishment of 486 industrial estates throughout the country, but only those close to large cities prospered. In the Republic of Korea, however, the large industrial estates in the south-eastern coastal zones have succeeded in directing firms and labour away from the capital region. The annual population growth rate in Seoul declined from 9.8 per cent during 1966-1970 to 2.8 per cent during 1980-1985, while other cities grew at a rate of more than 5 per cent annually during the latter period.

In China, industrial decentralization and the transfer of skilled labour from coastal to inland areas have also contributed to a more balanced distribution of industry between coastal and inland regions. Foremost among the exporters of labour have been Shanghai and Tianjin. However, the industries relocated from coastal to inland areas turned out to be less efficient in their new locations and imposed a severe loss of efficiency on the national economy (Kirby, 1985). The Chinese experience suggests that the issue of economic efficiency is extremely important when industrial dispersal policies are being considered. A related issue is whether the decentralization of industry will have a positive effect on poorer regions. The evidence available indicates that, in many countries, industrial dispersal has increased regional output but created few jobs. Indeed, deconcentration programmes have tended to suffer from the same weakness as national industrialization programmes: the technology used, in its capital-intensive nature, has failed to create large numbers of local jobs or high demand for local inputs as was envisaged by growth-centre theory. Studies of the regions around Kuala Lumpur have found that, as a result of weak "spread" effects and substantial "backwash" effects, the regions beyond the immediate vicinity of the growth centres gained few economic and social benefits (Oberai, 1993). That is to say, large-scale industrialization within a growth centre seems to be a poor means of developing a poor region in the absence of fundamental changes in the agricultural economy, the marketing system, and the pattern of landholdings. Industrial deconcentration may serve to reduce pressure on large metropolitan areas but, unless it is accompanied by other and more radical programmes, its benefits may be limited.

The problem is that although poor regions may benefit from regional policies, the poor within those regions do not benefit sufficiently. All too often, the benefits of industrialization are reaped by industrial and large-scale agricultural groups in the region. A reduction in interregional disparities is then accompanied by an increase in intraregional disparities and interpersonal inequality.

Another important issue is selective de-industrialization. A common tendency among city planners is to discourage the growth of all manufacturing activities, thus thwarting employment generation, as in Bombay, Seoul and in other mega-cities of the developing world. Thus, open unemployment in Seoul reached 5.9 per cent in 1986 when unemployment at the national level stood at just 3.4 per cent (Kim, Kim and Son, 1991).

Although major regional and industrial decentralization programmes, and indeed programmes for planning major cities, may be essential, they are not in themselves sufficient. The critical issue is the type of development strategy followed. When urban and regional policy goals are in conflict with national economic growth, the latter has usually won out. The case of Lagos, for example, demonstrates well how macroeconomic policies can derail decentralization efforts. More than 90 per cent of the total net subsidies granted to industries in Nigeria benefit those located in Lagos. Such bias cannot but have a negative impact on industrial decentralization. In India, too, decentralization policies conflict with macroeconomic objectives. Although the Indian Government decided not to license any new industrial activities in the five largest urban centres, its policy of export promotion led to the growth of export-oriented industry in the three major ports (Bombay, Calcutta and Madras) which are also among the largest cities of the country.

B. An assessment of policies aimed at influencing migration

The above review of policy experience suggests that, so far, most of the policies aimed at influencing migration have had limited success. There are several reasons for this outcome. The first is the general lack of high-level political commitment to a better distribution of economic activities throughout the country. In many countries, there are serious conflicts of interest between regions and cities, and between ethnic and "native" groups, which prevent the emergence of a national consensus on the aims and content of population distribution policies.

Policies have often been formulated without an adequate understanding of the causes and consequences of migration, making it difficult to judge whether they are justified and appropriate. Different forms of migration (circular, semi-permanent or permanent) and different types of migrants (educated or uneducated) are likely to have different impacts and therefore demand different policy responses. Yet policy makers invariably tend to view migration as a one-dimensional phenomenon and migrants as a homogeneous group. As a result, the policies adopted often lack clarity and coherence.

Until recently, urbanization policies have concentrated largely on the encouragement of decentralization, overlooking the fact that other national policies (for example, trade, industrial location and infrastructural development policies) exert stronger countervailing pressures. In addition, many of the measures adopted have addressed the symptoms but not the causes of resource misallocation and severe regional disparities.

Most Governments have devoted few resources to the assessment of policies aimed at influencing migration and have failed to examine their feasibility and consistency with respect to other development objectives. Consequently, in many cases the effects of measures introduced to control migration have been weak. In some cases the goals of national economic and social policies have conflicted with those of migration policies. Thus, the policy of subsidizing food rations in the urban areas of Egypt has reduced the cost of living in cities and encouraged more migration to them.

Lastly, most population distribution policies have adopted a somewhat narrow approach, focusing either on controlling the growth of large cities or on retaining people in rural areas. The strategies aimed at developing secondary cities and small towns have often been promoted as a panacea to be substituted for one or the other approach. Yet a sound population distribution strategy should strive to be holistic, by simultaneously encouraging settlement in small and intermediate cities, promoting economic development in rural areas, and improving employment and living conditions in large cities.

Indeed, it would be more cost-effective if, instead of intervening directly in modifying migration flows, Governments corrected the urban bias of national development policies which affects not only industrial location patterns but also regional income inequalities. However, the removal of urban bias is unlikely to reduce significantly the concentration of population in large cities in the short run. Policies designed to improve the efficiency of large cities must therefore be vigorously pursued. In fact, measures such as the pricing of public services at marginal cost and the elimination of subsidies for private investors that would improve urban efficiency are more likely to have important, albeit indirect, effects on migration.

Although there is certainly a need to divert migration away from large cities to smaller cities and towns, the conditions under which such a strategy is likely to succeed need to be examined. In Egypt, for example, the creation of new towns has not been very effective in reducing urban concentration because it has not been accompanied by an adequate strengthening of social infrastructure in those towns or by a reduction of large city bias in national economic and social policies.

Rural development programmes should not seek population retention alone. They should be directed primarily towards increasing agricultural and other rural income as well as rural welfare, particularly among small producers. Particular emphasis should be given to combining rural development strategies with policies designed to promote the growth of small towns and other urban centres so as to foster the interaction of those centres with the rural economy, through agro-processing, the expansion of small-scale industry, the provision of marketing facilities for rural products, and agricultural extension services.

Although industrial decentralization as a strategy for population dispersal has obvious appeal, its implications for employment generation and poverty alleviation in cities of the developing world need to be assessed carefully. As noted above, unemployment in Seoul is twice as high as in the country as a whole partly because of the industrial decentralization policies pursued. Moreover, moving out from large cities highly productive manufacturing industries that have the capacity to pay for urban services may adversely affect the revenue base of the city. Thus, a policy of industrial decentralization usually needs to be accompanied by policies to create alternative sources of revenue and employment generation in large cities. Since the service sector in such cities may not be able to absorb all the labour force increase, it may be more appropriate to adopt a policy of selective industrial decentralization. Non-polluting light industries, such as electronics and clothing, which are market-oriented and labour-intensive can be promoted in large cities.

Since natural increase now contributes more than rural-urban migration to population growth in many cities of the developing world, greater emphasis needs to be placed on reducing fertility. In sub-Saharan Africa and low-income areas of Asia, family planning programmes in rural areas can contribute to lowering natural increase and thus reducing migration pressures. In Latin America, where the level of urbanization is already very high, reducing the natural increase of the urban population holds the best hope for reducing the growth of cities. The high fertility of the urban poor is associated with their limited access to education, health and family planning services, which stems in turn from their low incomes. Raising the productivity and incomes of the urban poor and thus increasing their access to social services is therefore likely to reduce population pressure.

Overall, the evidence suggests that the policy instruments available to Governments will have little impact on migration flows until the basic factors responsible for wide rural/urban and inter-urban differentials in wages, employment, income-earning opportunities, and amenities are modified. Reduction of those differentials can be achieved only if population distribution policy becomes an essential component of an overall development strategy, linked and harmonized with policies on industrialization, agriculture and social welfare.

Note

[1]In Mexico in 1980, 20.8 per cent of the nation's population lived in Mexico City, but the city generated 34.3 per cent of the country's GDP. In Brazil, Rio de Janeiro accounted for 4.1 per cent of the country's population in 1987 and generated 10.6 per cent of the nation's GDP. In China, Shanghai alone contributed 4.8 per cent to China's GDP in 1988 although it accounted for only a little over 1 per cent of the population (Oberai, 1993).

References

Gilbert, A. G. (1974). Industrial location theory: its irrelevance to an industrialising nation. In *Spatial Aspects of Development*, B. S. Hoyle, ed. New York: John Wiley.

Hardoy, J. E., and D. Satterthwaite (1991). Environmental problems of Third World cities: a global issue ignored? *Public Administration and Development*, vol. 11, No. 4, pp. 341-361.

Indraratna, A. D. V. de S., and others (1983). Migration-related policies: a study of the Sri Lanka experience. In *State Policies and Internal Migration: Studies in Market and Planned Economies*, A. S. Oberai, ed. London and Canberra: Croon Helm, pp. 79-136

Jin-Ho, Choi (1990). Patterns of urbanisation and population distribution policies in the Republic of Korea. *Regional Development Dialogue* (Nagoya, Japan), vol. 11, No. 1 (spring).

Kim, Jong-Gie, Kwan-Young Kim and Jae-Young Son (1991). *Problems of Urbanisation and the Growth of Seoul.* WEP 2-21, Working Paper, No. 179. Geneva: International Labour Office.

Kirby, R. J. R. (1985). *Urbanization in China: Town and Country in a Developing Economy, 1949-2000 AD.* New York: Columbia University Press.

Oberai, A. S. (1987). *Migration, Urbanisation and Development.* Geneva: International Labour Organization.

__________ (1988). *Land Settlement Policies and Population Redistribution in Developing Countries: Achievements, Problems and Prospects.* New York: Praeger.

__________ (1993). *Population Growth, Employment and Poverty in Third World Mega-cities.* London and New York: Macmillan.

Rhoda, R. E. (1983). Rural development and urban migration: can we keep them on the farm? *International Migration Review* (Staten Island, New York), vol. 17, No. 1 (spring), pp. 34-64.

Shanghai Planning Commission (1991). *A Study on Urban Development of Shanghai.* Report prepared for the International Labour Organization, Shanghai. Mimeographed.

Simmons, A. B. (1979). Slowing metropolitan city growth in Asia: policies, programs and results. *Population and Development Review* (New York), vol. 5, No. 1 (March), pp. 87-104.

United Nations (1991). *World Urbanization Prospects, 1990.* Sales No. E.91.XIII.11.

Zhao, Y. (1989). Strategy and choice: review of China's road towards urbanization. Paper presented at the International Conference on Urbanization in China, Honolulu, Hawaii, 23-27 January.

XXI. THE CHALLENGE FACING THE FUTURE OF HUMAN SETTLEMENTS

United Nations Centre for Human Settlements (Habitat)

In 1976, the United Nations convened its first major Conference on Human Settlements. The 64 recommendations for national action adopted by the Habitat Conference, as contained in part one, chapter II, of the report of the Conference (United Nations, 1976), covered six major areas: (*a*) settlement policies and strategies; (*b*) settlement planning; (*c*) shelter, infrastructure and services; (*d*) land; (*e*) public participation; and (*f*) institutions and management. Under these headings, recommendations touched on a broad range of issues which are reflected in the current human settlements programme of the United Nations, and in specific programmes launched and developed by the Centre for Human Settlements and by its international partners in human settlements development.

The institutional arrangements outlined by the Conference resulted in the consolidation of various units and activities of the United Nations Secretariat into the United Nations Centre for Human Settlements (Habitat) which acts as the focal point for human settlements within the United Nations System, provides secretariat services to the Commission on Human Settlements, and implements an integrated work programme that includes research, technical cooperation and the dissemination of information.

The Commission on Human Settlements, in particular, has taken the lead in developing new perspectives on issues related to urban centres and development. Among them, the "enabling approach" is a process by which development efforts are based on constructive partnerships among all actors, including Government, non-governmental organizations, the private sector and the community. The role of Government is to coordinate and facilitate development through consultation, community participation, accountability and the provision of trained personnel. The Commission has also fostered a constructive approach to urbanization, seeing it as a challenge and an opportunity for development rather than as the result of rural underdevelopment and uncontrolled migration. Activities promoted by the Commission have included the declaration of 1987 as the International Year of Shelter for the Homeless and the formulation of the Global Strategy for Shelter to the Year 2000 (United Nations, 1988); the formulation of the Urban Management Programme, developed in cooperation with the United Nations Development Programme and the World Bank; and the Sustainable Cities Programme. The Commission has further supported the role of women in human settlements development and has promoted research, training and technical cooperation.

Solutions to the problems related to human settlements have been developed on the foundation set by the Habitat Conference. They have included policy recommendations regarding the full spectrum of human settlements; specific research and technical guidelines; and the testing of appropriate and replicable approaches in practically all developing countries. As a result, a large number of developing countries are strengthening their human settlements institutions, promoting decentralization, and mobilizing available resources for the improvement of settlements in which large numbers of people live.

Nevertheless, a number of Governments still perceive human settlements and the provision of shelter in particular as a top-down responsibility. The transfer of functions and responsibilities to local authorities is rarely accompanied by the necessary resources, staff and legislation that would enable local authorities to expand their sources of revenue. The poor, with their settlements and their informal activities, are still seen as disturbing problems, rather than as human resources to be included as legitimate actors in the mainstream of development. Expenditures for the management of human settlements and development still tend to be perceived as social expenditures only, instead of also being considered indispensable in laying the foundations for social and economic progress, economic growth, human development and environmental improvement. It is estimated that less than 3 per cent of the current overall commitment to multilateral programmes is allocated to the shelter sector. Only since 1985 has the urban dimension of development been considered explicitly in initiatives such as the 1986 meeting of the Development Assistance Committee of the Organisation for Economic Cooperation and Development (OECD) on urban development, the launching of the UNDP/World Bank/United Nations Centre for Human Settlements (Habitat) Urban Management Programme the same year, the Healthy Cities Programme of the World Health Organization (WHO), and the World Bank's *Urban Policy and Economic Development: An Agenda for the 1990s*, a World Bank Policy paper (World Bank, 1991) issued in 1991.

These developments must be considered against a background of escalating human settlement problems: a situation in which long-term planning is overshadowed by the urgencies of crisis management. Indeed, the gap between demand and stagnant or even decreasing public sector capacity and resources is growing; the short-term repercussions of structural adjustment policies have generally been negative; mounting debt problems reduce even more the capacity of Government to address urban problems, and urban poverty is increasing.

While there are some functioning and well-managed urban centres in developing countries, in many of the cities and towns of the developing world, electricity and water supply interruptions are daily occurrences, while squatter settlements, slums, potholed roads, garbage heaps, and water and air pollution are prevalent. These are the main physical manifestations of the urban crisis that many countries are facing. This crisis is a result of the failure of local authorities to operate and maintain existing urban services efficiently and to provide services for the rapidly expanding populations of cities. The numerous deficiencies in existing urban management systems have led not only to deterioration in living conditions, but also to significant disruptions in industrial production and commercial activities. Partly as a result of the alarming deterioration of many cities in developing countries, there has been a resurgence of interest in the role of towns and cities in national development, after a period in which more emphasis had been put on rural areas both in the development discourse and in international cooperation. The thrust regarding urbanization has been towards improving living conditions in cities, enhancing their economic status and protecting their environment (United Nations Development Programme (UNDP), 1991; World Bank, 1991).

A. Human settlements conditions and trends: some basic issues

In 1990, the population of the world stood at about 5.3 billion (United Nations, 1991). At its peak in the late 1960s and early 1970s, the average annual growth rate of the world's population was close to 2 per cent. The growth rate is expected to decline to less than 1 per cent by the second decade of the next century. However, the extent of the reduction of population growth varies from region to region. In addition, the large size of the population already present in the developing world implies that, although the growth rate may be decreasing, large numbers of persons are added every year. Thus, between 1990 and 2005, over 1 billion new urban-dwellers will have to be accommodated by the urban areas of developing countries, meaning that they will need to be provided with shelter, infrastructure and services.

It is well known that fertility in urban areas tends to be lower than in rural areas. The World Fertility Survey programme provided information indicating that urban/rural fertility differentials had increased in many countries, owing to rapid declines in fertility in urban areas. Such declines were associated with the lower family size desired by women living in urban areas, the prevalence of later marriage and the increased use of contraception in cities. It has been suggested that during the initial stages of the demographic transition, urban/rural differences in fertility are small. Yet, as time elapses, couples in urban areas develop preferences for smaller families and are more likely to use effective means of fertility control. Eventually, desired family size also drops in rural areas and rural fertility falls. That is to say, urbanization itself has a major impact on the reduction of natural increase.

Urbanization, the process by which people settle in areas classified as urban, is far advanced in developed countries and in most countries of Latin America, but is still low in Africa and Asia. In 1995, 74 per cent of the population in developed countries lived in urban areas compared with only 41 per cent of the population of developing countries (United Nations, 1991). The process of urbanization in developing countries has been rapid. In 1950, only 17 per cent of their population lived in urban areas. By 2010, over half of the people in developing countries will live in towns or cities. Many of them, however, will be poor. Indeed, poverty is rapidly becoming more of an urban than a rural problem. Although the rural population of developing countries is still increasing, its rate of growth is significantly lower than that of the urban population (0.7 per cent versus 4.2 per cent during 1990-1995).

Rural-urban migration is partly responsible for the sharp urban/rural differential in the rate of growth. Although migration relieves pressure on rural areas, it places additional demands on the capacity of cities to provide land, shelter, infrastructure, services and employment. In the absence of rural-urban migration, the concentration of population on agricultural land would increase or the exploitation of previously unused land would rise, often threatening the ecological balance of fragile regions. These considerations have led to a more constructive approach towards urbanization and migration, emphasizing that urbanization is neither a crisis nor a tragedy, but rather a challenge for the future. The challenge is to find innovative ways of managing cities so as to maximize their benefits and economic dynamism.

From an environmental perspective, migration to urban settlements may be the most desirable option for, although the high density of human activity characteristic of cities leads to special environmental problems, such as congestion and pollution, it is easier to control or mitigate environmental problems in cities than in remote rural areas. City dwellers have greater leverage than isolated rural dwellers vis-à-vis Government or local authorities, who themselves have an interest in protecting their own health and limiting the negative environmental impact of certain activities.

High urban densities may lead to a more efficient use of limited land areas. Each productive activity has certain needs for land, but modern technology makes possible land-efficient development in both urban and rural areas. In urban areas, for instance, appropriate land-use policies based on efficient densities and development patterns can reduce the amount of land needed for settlement and, at the same time, minimize the need for transportation, thus reducing energy consumption.

Lastly, urbanization provides a partial answer to the needs, preferences and aspirations of people. The cities of the developing world provide a safety net for millions of dispossessed and jobless people. Cities also provide services for rural residents who either commute to cities or use such services on a emergency basis.

B. Mega-cities

In 1960, more than one third of the world's urban-dwellers lived in cities having less than 100,000 inhabitants and another third lived in cities with at least 1 million inhabitants. By the year 2000, almost half of the urban population of the world will live in cities with 1 million inhabitants or more. Their number (1.3 billion people) will be greater than the total urban population of the planet in 1960. By the year 2000, 18 of the 24 cities with at least 10 million inhabitants will be located in developing countries.

While some of the mega-cities in the industrialized world are showing signs of population stabilization or decrease, the largest urban agglomerations in developing countries continue to grow. Decentralization policies can be effective only if they are sustained over prolonged periods and make possible the large investment needed to create alternatives to the economies of agglomeration offered by the largest cities. It is necessary, therefore, to prepare for the future and take steps to improve city management so that basic services and needed infrastructure maintenance and development are not neglected.

Despite the problems that mega-cities face, there is not a clear correlation between city size, quality of life and environmental degradation (Population Crisis Committee, 1990a and 1990b). An indicator built on the basis of a number of measures of quality of life in many of the world's large cities shows, for instance, that Calcutta in India, with a population of almost 13 million, ranks low on the global scale of quality of life, but is better off, if only marginally, than Kanpur, a city with a population of only 2.3 million. Similarly, Shanghai, with a population of over 9 million, shows a better overall ranking than Harbin and Nanjin, two other Chinese cities with populations below 3 million. In Latin America, the combined indicator for the fourth largest urban agglomeration in the world, São Paulo (17.2 million inhabitants), is only slightly below that for Belo Horizonte but considerably better than that of Recife, both Brazilian cities with less than 4 million people. In addition, although the air pollution problems, crippling traffic and deficiencies in water supply in some large cities are well known, smaller cities are not exempt from similar problems and may further experience higher unemployment or more limited access to land and to essential infrastructure.

Developing countries should therefore adopt urbanization policies that improve the management of all cities, large and small. The aim should be to take maximum advantage of the irreversible and powerful urbanization trends and to strengthen urban settlements that can act as centres of production and economic growth for national and subnational economies. The concomitant strengthening of intermediate cities which can act as magnets for migration flows may take some pressure off the mega-cities.

Despite increasing urbanization, the majority of the population of developing countries still lives in rural areas, relying heavily on agriculture for a livelihood. To the extent that efficient agriculture can be encouraged by appropriate policies and the provision of infrastructure, people will remain in the countryside, taking pressure off the cities. The promotion of efficient agricultural production, which involves rural/urban trade of inputs and outputs, also creates jobs in both rural and urban areas. Thus, to be effective, strategies aimed at improving the operation of cities should also support agricultural development.

C. Poverty

According to the World Bank (1990), about 1 billion people are poor. According to the United Nations (1989), although the percentage of the world's population living in absolute poverty decreased between 1970 and 1985 (from 52 to 44 per cent), the absolute number of people in poverty in developing regions increased by 22 per cent during the same period. The increase in urban poverty is particularly striking. Thus, while the rural population living in poverty rose by 11 per cent between 1970 and 1985, the corresponding population in urban areas increased by 73 per cent. In Latin America, the number of urban poor more than doubled and in Africa their number increased by 81 per cent. In many developing countries, poverty has become a predominantly urban problem.

The majority of the urban and rural poor have no access to safe drinking water, elementary sanitation or primary health-care services. A high proportion of the urban and rural poor are disabled. Accidents are common because informal urban settlements are often located in hazardous, flood-prone or unhealthy sites, where exposure to communicable disease is augmented by overcrowding and where garbage dumps constitute a permanent health hazard.

The low-income majorities in the developing world are not "poor", however, in terms of skills, ingenuity, social identity, traditions, culture or personal dignity; they are poor because they cannot afford access to a decent and healthy livelihood. Having an adequate shelter is a basic need that they cannot meet. Lacking an adequate and steady source of income, they need public assistance to secure a decent shelter in urban areas. However, Governments cannot afford to provide housing for all the urban poor. Other solutions include the provision of subsidies to fill the gap between disposable income and the market price of decent housing and the promotion of self-help programmes.

In developed countries, some of the shelter needs of poor urban minorities have been addressed through sizeable public housing programmes and adequate subsidies. In most developing countries, welfare programmes are virtually non-existent and public housing programmes are targeted only to limited numbers of people, thus leaving out the bulk of the urban poor, who are consequently left to find shelter by their own means. They do so using a variety of strategies: they find temporary accommodation with friends and distant or close relatives; they rent rooms or some kind of shelter through the informal housing sector; they build, individually or on a communal basis, temporary shelter on derelict or vacant sites.

Governments must understand and acknowledge this state of affairs so as to facilitate the settlement process of the urban poor by legitimizing it. Examples of appropriate and feasible approaches include the granting of security of tenure; the simplification of existing building codes and regulations; the acceptance of indigenous and low-cost building materials; the regularization of informal settlements that are not permanently exposed to contamination, flooding or landslides; the provision of basic infrastructure with the participation and financial contribution of residents; and the promotion of other forms of access to low-cost shelter, including rental housing. All of these measures can greatly improve the living conditions of the urban poor and, at the same time, contribute to environmental improvement in urban areas.

Serious problems also exist in the developed countries where the number of homeless people has been growing rapidly in recent years and where there is a significant proportion of people suffering severe deprivation. Many such people (especially the uneducated, poor and elderly) remain in declining industrial centres or decaying neighbourhoods as the young and better educated move out. These problems are being further exacerbated by the structural adjustment that many developed countries have undergone and by the current recessionary period.

D. URBAN MANAGEMENT CHALLENGES

The trends described above have serious implications for urban management, meaning the process of efficiently and effectively utilizing resources to provide satisfactory living and working conditions for the urban population, and facilitating economic production and growth in urban areas. This process involves the formulation of appropriate objectives, goals, policies and strategies, and the mobilization and efficient use of resources (including personnel, organizations, information, finance and land) to develop, maintain and provide essential urban infrastructure and services (water, sanitation, electricity, refuse collection and removal, roads, education and health) for the city's population. In most countries, urban management is primarily the responsibility of local authorities, consisting of elected councillors (who are in charge of policy-making) and appointed professionals (who are in charge of implementation). However, public, parastatal or even private organizations may be responsible (sometimes on a contractual basis) for the development, operation and maintenance of some of the essential infrastructure, such as electricity, water and refuse removal. The role of central Government is usually to define national urban policy, and to develop the legislative and supervisory framework necessary for the operation of the local authorities, although in some countries the central Government has taken direct charge of the management of capital cities.

While institutional frameworks for urban management are bound to differ from country to country, certain common attributes are essential for good urban management. They include decentralized and participatory local Government; the transparency and accountability of local Government; the enabling approach to human settlements' development and management; and the sustainable development approach. All of these elements are at the heart of the urban management model currently being promoted by many international organizations, including the United Nations Centre for Human Settlements (Habitat), and have been accepted by many national Governments.

Given the urban management problems in developing countries, the major challenge for the next decade is the formulation and implementation of management strategies capable not only of resolving extant problems but also of coping with the expected levels of urban growth. From an urban management point of view, there are three major consequences of rapid urbanization: the concentration of poverty in urban areas; the increasing demand for employment and urban services; and the growing potential for environmental degradation.

According to the World Bank (1991), 300 million people or one quarter of the developing world's urban population were living in absolute poverty in 1988. By the year 2000, 90 per cent of Latin America's poor will be living in cities, and comparable figures for Africa and Asia are expected to be 45 per cent and 40 per cent, respectively (United Nations Development Programme (UNDP), 1991). By the year 2000, the number of urban households living in poverty will have increased by 76 per cent and more than half of the world's poor will be living in cities (United Nations Development Programme (UNDP), 1991). The main cause of urban poverty is the lack of income-earning opportunities in cities. Hence, one of the challenges facing urban management is to create employment for the poor by promoting the right type of economic growth. This will demand better-organized and more efficient infrastructure and, in particular, serviced land for the location of new production enterprises, water, electricity, roads and telecommunication facilities. It will also be necessary to reduce restrictions on the informal sector and adopt policies to support informal sector activities.

The increasing concentration of people in cities combined with the intensification of urban poverty will result in the expansion of urban squatter settlements and slums, and in rising demand for urban services. At present, the majority of the urban population in developing countries lives in squatter settlements and slums: 60 per cent in Kinshasa, Zaire; 50 per cent in Buenos Aires, Argentina; and 67 per cent in Calcutta, India (United Nations Development Programme (UNDP), 1991). These settlements generally lack basic services such as water, sanitation, electricity, garbage collection, roads and transport. One of the most significant failures of the urban authorities in developing countries has been their inability to respond to the needs of the majority of urban residents who cannot afford or simply cannot obtain access to the types of services offered. The main challenge for urban management is the reorientation of its objectives, goals, policies and strategies so as to meet the needs and improve the living conditions of the poorer segment of the city population. Meeting this challenge will require the relaxation and rationalization of existing housing and infrastructure standards which have tended to exclude the poor, as well as the institutionalization of an enabling framework that permits community-based organizations, individual households and the private sector to contribute towards the provision and maintenance of urban services, particularly in low-income areas. The involvement of other actors should be seen not as an abdication of responsibility on the part of municipal authorities, but simply as a recognition of their efforts in filling existing gaps.

Environmental degradation, one of the major consequences of rapid urbanization, takes the form largely of air and water pollution and also stems from the generation of enormous amounts of solid waste. The main causes of water pollution include the discharge of untreated industrial effluent into rivers and lakes, the indiscriminate dumping of toxic wastes on land, and the absence of safe sanitation facilities (sewerage and drainage), particularly in squatter settlements. The main causes of air pollution include uncontrolled industrial emissions and the increasing concentration of motor vehicles. The accumulation of solid waste is largely a result of the lack of efficient garbage collection and disposal systems. A key challenge for urban management is to attain sustainable development by implementing effective pollution control mechanisms that include environmental control standards; the formulation of appropriate legislation; the creation or strengthening of enforcement institutions; the promotion of natural resource conservation; the recycling of wastes; the establishment of appropriate and affordable methods of solid waste collection and disposal; and the use of appropriate non-wasteful technology, including renewable sources of energy.

E. Current global programmes on urban management initiated by Habitat

Both national Governments and international organizations, recognizing the fundamental role that cities play as engines of development, are increasingly focusing attention on urban areas. The United Nations Centre for Human Settlements (Habitat) has responded by initiating an Urban Management Programme, a Sustainable Cities Programme, a Community Development Programme, and a City Data Programme.

The Urban Management Programme, which is being implemented in partnership with UNDP and the World Bank, is now in its second phase (1992-1996). The first phase covered the period 1986-1991. The fundamental aim of the programme is to strengthen the contribution of cities and towns towards economic growth and social development. The programme focuses on the development and promotion of appropriate policies and tools, as well as the provision of technical support in five substantive areas: municipal finance and administration; land management; infrastructure management; the urban environment; and poverty alleviation. It also seeks to establish effective partnerships with national, state and local Governments; the private sector; non-governmental organizations; and external support agencies. Such cooperation is aimed at building the capacity to formulate and implement appropriate urban management policies. The activities around which cooperation revolves include applied research, dissemination of information, the sharing of experiences and the identification of the most effective practices, and technical cooperation. The programme is funded by UNDP and bilateral donors.

During its first phase, the Urban Management Programme worked at the city level with city officials, technical staff, community organizations, non-governmental organizations, and local enterprises in preparing environmental profiles for the cities involved and promoting activities to strengthen land management, municipal finance, infrastructure management, and garbage collection and disposal. At the country level, the Urban Management Programme developed partnerships involving officials and technical staff of government agencies and other actors who worked on the redrafting of land and planning law; the deregulation of land markets; urban redevelopment and land management; the development of a fiscal cadastre; and preparation of a new environmental management act. At the regional level, the Programme has assessed the lessons learned and has disseminated its findings through technical cooperation among developing countries. At the global level, the Programme has contributed to enhancing the urban development agenda by promoting the inclusion of urban themes and components of the Programme in action-oriented documents, such as Agenda 21 (United Nations, 1993).

The Sustainable Cities Programme was launched in August 1990 as the operational arm of the environment component of the Urban Management Programme. The main goal of the programme is to provide municipal authorities and their partners in the public, private and community sectors with an improved environmental planning and management capacity. At the city level, the programme aims to (*a*) enhance the availability and use of natural resources in and around cities; (*b*) reduce the exposure to environmental hazards in and around cities; (*c*) strengthen local capacity to plan, coordinate and manage development; and (*d*) improve access to available knowledge and tools for environmental planning and management. Funding for the programme comes mainly from UNDP indicative planning figures (IPF), bilateral sources and other multilateral sources, including the United Nations Environment Programme (UNEP). The programme gained significant momentum during the period 1991-1992 and is currently operating in 12 developing countries. City-level demonstration activities are under way in four cities, and six were scheduled to be launched during 1993-1994. They included the preparation of city environmental profiles; the use of geographical information systems and satellite remote sensing; the development of approaches to address environmental health problems; and the promotion of new approaches to control air pollution. The implementation of this programme has involved significant inter-agency cooperation.

The Community Development Programme consists of two components: the Training Programme for Community Participation in Improving Human Settlements and the Community Management Programme. The first began in 1984 and is being implemented in Bolivia, Sri Lanka and Zambia. Its aim is to strengthen the capacity of government agencies and urban communities to improve living conditions in low-income settlements through participatory development activities. Community participation is sought as an essential component of human settlements improvement schemes. The Programme trains the staff of local government and housing agencies in the formulation and implementation of community participation strategies. Once trained, staff members are expected to become facilitators of community action at the grass-roots level, directly assisting people in low-income urban settlements. Training is carried out on location and is adapted to the actual needs of the staff and communities involved. A series of training modules covering several aspects of community participation is available. The programme's experience with respect to the promotion of enabling shelter development strategies is crucial for the Centre's implementation of the Global Strategy for Shelter to the Year 2000.

The Community Management Programme, launched in September 1991, began operating in Costa Rica, Ecuador, Ghana and Uganda early in 1992. The Programme's goal is to reorient local government policies and practices in the provision of housing, community services and facilities towards a participatory approach. That goal is to be achieved by strengthening community organization, management and building skills, and by promoting sustainable development in low-income settlements. Emphasis is being placed on the creation of viable and replicable strategies for the involvement of communities in the provision of services. The specific needs and priorities of women are being addressed. By 1996, the Community Management Programme was expected to have benefited some 200,000 households belonging to the poorest segments of the urban or rural populations, and living in areas where government assistance is limited.

The City Data Programme, begun in May 1991, focuses on the 4,000 cities with at least 100,000 inhabitants for which it collects data on the following: physical geography and climate; population; household characteristics; land use; housing and housing facilities; infrastructure and services; municipal revenue and expenditure; and environmental pollution. The immediate objectives of the Programme are to develop an urban data collection and dissemination system that will provide statistical support to policy-makers, and to disseminate urban information and tools for urban analysis to specific target groups. The outputs of the Programme will include a global database on cities, a survey of urban data-collection activities by intergovernmental and non-governmental organizations, and the production of the UNCHS-CITIBASE software. In addition, a survey of city-level data-collection practices in Kenya is being carried out. It will result in the creation of a framework for urban data use and in the training of government staff. Lastly, the programme is expected to disseminate its findings through the publication of city profiles and six wall charts on pressing urban problems.

F. Lessons learned

A number of lessons have been learned from the implementation of the programmes discussed above. With respect to land management, the Urban Management Programme has relied on improved market mechanisms rather than the Government's acquisition of land sites to meet the high demand for urban land. Promoting the enabling role of Government rather than its controlling role with regard to land management and property rights has also proved useful. Understanding informal land market mechanisms and building upon customary systems of land management are recommended. A land market assessment methodology has been developed to understand the constraints in land markets and to design appropriate public sector responses.

With respect to municipal finance and administration, both procedural solutions and the political preconditions for the achievement of sustainable municipal finance need to be considered in deciding how to mobilize the resources required to finance municipal services and how to improve the allocation of municipal resources for the purpose of responding most effectively to consumer demand. Particular attention should be given to the reorganization of service delivery and to the clear-cut allocation of decision-making powers. It is also important to improve accountability by changing intergovernmental fiscal relations and to strengthen local government credit institutions.

It is important to strengthen existing institutions by promoting participatory planning that is broadly based. In addition, it is necessary to enhance city-level consultation and to provide demonstrations of strategic activities. The effective use of local know-how and experience is essential. It is also important to carry out application-oriented research.

The Community Development Programme has been successful in devising strategies to reach the urban poor and harness community-based efforts to fill the gap left by local authorities in poor settlements. In Sri Lanka, the Community Action Planning and Management Approach has been developed to improve shelter among the urban poor. The approach has institutionalized community participation in urban planning and management through the creation of community development committees and of a specialized unit responsible for slum and shanty improvement in the Colombo Municipality. In addition, community contracts are used to foster the construction of neighbourhood infrastructure, including footpaths, drains, toilets and other community facilities.

G. Conclusion

Since the urban population of developing countries is expected to continue rising, it is urgent that urban management systems correct the deficiencies that have plagued them and have resulted so far in the virtual exclusion of the majority of urban residents from essential urban services and in the retardation of industrialization. Partly because of growing recognition of those deficiencies, urban management has again become a central concern in the development agenda. A revitalized urban management should enhance the capacity of towns and cities to efficiently provide the infrastructure needed for economic growth and to improve the living and working conditions of the entire urban population. In the course of this process, the role of community actors should be enhanced. Lessons learned from research and technical cooperation programmes in urban management suggest that a number of fundamental changes and improvements are needed, including improving the efficiency of land markets; redefining intergovernmental fiscal relations; redefining the allocation of decision-making responsibilities with respect to service delivery; encouraging the creation of partnerships among public, private and community actors; and working towards the attainment of sustainability by protecting the urban environment. As the world becomes increasingly urban, the need to improve and strengthen urban management policies and instruments can only be regarded as imperative.

References

Population Crisis Committee (1990a). *Cities: Life in the World's 100 Largest Metropolitan Areas*. Washington, D.C.: Population Crisis Committee.

_________ (1990b). *Cities: Life in the World's 100 Largest Metropolitan Areas. Statistical Appendix*. Washington, D.C.: Population Crisis Committee.

United Nations (1976). *Report of Habitat: United Nations Conference on Human Settlements, Vancouver, 31 May - 11 June 1976*. Sales No. E.76.IV.7 and corrigendum.

_________ (1988). Report of the Commission on Human Settlements on the move of its eleventh session; addendum. *Official Records of the General Assembly, Forty-third Session, Supplement No. 8, addendum*. A/43/8/Add.1.

_________ (1989). *Report on the World Social Situation, 1989*. Sales No. E.89.IV.1.

_________ (1991). *World Urbanization Prospects 1990*. Sales No. E.91.XIII.11.

_________ (1993). *Report of the United Nations Conference on Environment and Development, Rio de Janeiro, 3-14 June 1992*, vol. I, *Resolutions Adopted by the Conference* (United Nations publication, Sales No. E.93.I.8 and corrigendum), resolution 1, annex II.

United Nations Centre for Human Settlements (1991). *People, Settlement, Environment and Development*. Nairobi: UNCHS.

United Nations Development Programme (UNDP) (1991). Cities, people and poverty: urban development cooperation for the 1990s. Strategy paper. New York: UNDP.

World Bank (1990). *World Development Report, 1990: Poverty*. New York: Oxford University Press.

_________ (1991). Urban Policy and Economic Development: An Agenda for the 1990s. World Bank Policy Paper. Washington, D.C.: World Bank.

XXII. RURAL DEVELOPMENT, THE ENVIRONMENT AND MIGRATION

Food and Agriculture Organization of the United Nations

Migration, particularly that occurring within countries, is and is likely to remain a permanent feature of life in both developed and developing countries. Migration is sometimes considered a drain on the societies from which migrants originate and at other times praised as a means of improving the life of those migrants and even of the communities of origin by relieving them of excess labour and contributing to their development through remittances. This ambivalence is shared by policy-makers and international agencies alike.

The Food and Agriculture Organization of the United Nations (FAO) has been emphasizing several issues related to migration and rural development in its programmes. In doing so, it stresses the importance of equity in rural development as a way of preventing mass displacements of the rural population. The Programme of Action adopted by the World Conference on Agrarian Reform and Rural Development (Food and Agriculture Organization of the United Nations (FAO), 1979) in 1979 focused particularly on the fight against rural poverty. However, the third progress report of the Programme's implementation completed in 1991 still denounced the disparities existing between rural and urban areas and stressed the need to provide people with effective control over resources (Food and Agriculture Organization of the United Nations (FAO), 1991a). That urgent need was also stressed in the Plans of Action adopted by the 1990 Conference on Women in Agricultural Development and the 1991 Conference on People's Participation in Rural Development, both organized by FAO (Food and Agriculture Organization of the United Nations (FAO), 1990 and 1991b).

Interrelations between socio-economic development patterns and population problems were emphasized at the 1984 International Conference on Population, which echoed the call of the World Population Plan of Action (United Nations, 1975) for socio-economic transformation as a basis for an effective solution to population problems (United Nations, 1984). More recently, these concerns have been enriched by the environmental awareness promoted by the 1991 Den Bosch Declaration on Sustainable Agriculture and Rural Development (Food and Agriculture Organization of the United Nations (FAO), and Ministry of Agriculture, Nature Management and Fisheries of the Netherlands, 1991) and the United Nations Conference on Environment and Development which was held in Rio de Janeiro, Brazil, in June 1992 (United Nations, 1993). There seems to be broad agreement on the spirit of these policy commitments and consensus regarding the proposition that, although individuals should have the right to migrate in search of better opportunities, migration should not be the only survival strategy open to them.

Regarding the aspects of migration that are relevant for agricultural and rural development, gender issues should be highlighted since the push and pull factors that operate in rural-rural or rural-urban migration often vary by sex. In a number of cases, the demographic and socio-economic consequences of migration have had different impacts on men and women, which need to be taken into account by policy makers.

The present paper highlights only a few major issues relevant for the formulation of migration policies. They represent areas of growing importance that have not received sufficient attention, particularly from the perspective of rural development. In general, the emphasis is placed on how the migration process and its consequences may be better planned and monitored.

A. Migration and the environment

Policy-makers in developing countries focus largely on rural-urban migration because of its visibility and its impact on the cities where Government resides. However, that rural-rural migration is by far the major type of population movement is a fact that is often acknowledged only from a negative perspective in the context of environmental degradation (Dasgupta, 1984). Thus, the case of impoverished landless peasants who try to make a living in forest and other ecologically fragile areas is often stressed. Yet, rural-rural migration is considerably more varied than this example would suggest. At least two important categories should be distinguished. First, the movement of agricultural labourers to more fertile areas in search of employment opportunities. Second, the movement of peasants who are expelled from their land by force or because of poor economic conditions and settle in marginal lands. This second type of migration may be the most threatening for the ecological balance. Peasants who lack secure land tenure and who move away from their place of origin at high social costs often become the unintended instruments of large logging and sawmill interests by engaging in forest clearance and the extension of the land frontier towards increasingly marginal agricultural areas. Ultimately, this process leads to ecological degradation.

In much of Latin America, but also in parts of Asia, such as the Philippines, peasants have been expelled from the more fertile areas in the lowlands and have moved to the uplands where they occupy public lands, especially forests. They generally live under precarious conditions. In Brazil, the expansion of the land frontier is taking place by successive encroachment into the Amazonian basin.

Despite the potentially damaging effects of such movements, there is increasing evidence that the men and women who move to marginal areas can establish some sort of control over the resources available and are willing to protect such resources if given the chance. To do so, it is often crucial that the settlers organize themselves as a community and participate in the communal management of resources. In Costa Rica, for instance, migrants who had illegally occupied some areas had undertaken reafforestation using their own resources. Those migrants would willingly cooperate with the Government in establishing sustainable agroforestry management systems (Food and Agriculture Organization of the United Nations (FAO), 1992). However, such systems and the control they imply often run counter to the interests of commercial exploitation by private contractors.

The role of village improvement projects in which both men and women of the community are involved in rehabilitating agricultural land, conducting reafforestation and controlling erosion, has been underscored in various forums (Food and Agriculture Organization of the United Nations (FAO), 1991c). The issue to be considered is how best to involve the local population in environmentally sound resource management, especially if that population already occupies endangered forest areas.

B. Migration and rural labour shortages

Labour surpluses and shortages often coexist, even with regard to the same type of relatively unskilled and therefore more easily interchangeable labour. This phenomenon is exacerbated by emigration. In some African countries, especially those encompassing the more arid parts of the Sahel, entire areas are drained of young male labourers. The practice of shifting cultivation in combination with seasonal migration deprives communities of those members responsible for bush clearing, a fundamental activity in those areas (Hunt, 1985). In Southern Asia and Northern Africa, the emigration of young men to the oil-rich countries of Western Asia has caused rapid increases in agricultural wages. That was the experience of Pakistan during the early 1980s (Gaiha and Spinedi, 1992).

Young workers often leave regions where agricultural productivity, being based on low-cost technology and an abundant supply of labour that is fully utilized during only part of the year, is low. Because the remuneration in those regions, either in cash or in kind, tends to be low, workers have an incentive to move to areas where wages are better. When migration involves significant numbers of workers, agricultural output may fall in the areas of origin, especially if subsistence food crops are involved. These negative effects of out-migration are only partly mitigated by the remittances sent back by migrants. In order not to become dependent on outside food production, those left behind in the areas of origin may opt for cropping patterns that demand less labour. However, in some cases, the resource base may not allow such changes to take place.

Ideally, the availability of labour over the year should be monitored and various types of incentives could be used to encourage the presence of workers at key times of the year. Experience suggests that labour markets could be organized so that the costs of migration for both individuals and society would be minimized. Thus, a system of labour exchanges between areas with different growing seasons would produce an optimal allocation of the labour available.

C. Agrarian reform, rural labour markets and migration

In some regions and particularly in Latin America, agrarian reform was considered necessary to redress extreme imbalances in respect of access to resources and to avoid the massive flow of impoverished landless peasants to urban areas. In most cases, however, re-distributive land reform did not take place. Instead, large-scale migration was further promoted by certain features of labour legislation which resulted in the settling of large numbers of rural workers and their families in the outskirts of large cities where they had moved in search of employment (International Labour Organization, 1986). During the past 30 years, Brazil, for instance, experienced a concentration of land ownership that was accompanied by a constant flow of farm labourers from their places of origin within the *fazendas* (plantations) towards the peri-urban areas. As a result, growing numbers of workers who had lost not only their homes and plots of land but also their limited food security, settled in zones surrounding medium-sized cities. The reduction of the rural population of Brazil beginning in the late 1970s can be largely ascribed to this phenomenon (Thiesenhusen, 1990).

An unexpected indirect consequence of this development is the seasonal urban-rural migration of workers who find employment as agricultural labourers in rural areas during parts of the year. This type of migration is also common in countries of the Southern Cone and its key feature is the fact that agriculture is the dominant means of subsistence for the migrants involved, although the social costs that they have to face are higher because of the urban nature of their main place of residence. In this situation, the responsibilities inherent in a patron-client relationship are transferred from the landowners to the State. However, urban-rural seasonal migration also incorporates some innovative aspects of labour-market integration, since it enables workers to participate both in the urban and in the rural labour markets, albeit for the most part in the marginal sectors of those markets. On the whole, the magnitude and socio-economic impact of this type of migration are still imperfectly known and need to be studied. Furthermore, the effects of labour legislation on those types of movements need to be monitored.

D. The economies in transition and migration from rural areas

Under communist rule, Eastern-bloc countries gave priority to employment over labour productivity much as Western-bloc countries favoured productivity over full employment. Consequently, the collapse of communism in Central and Eastern European countries has brought about a painful restructuring of their economies which not only involves market related problems, but also a major change in the utilization and economic significance of the labour force. Such changes have given rise to significant migration flows, which are being studied at the country and regional levels. However, the rural dimension of the problem is not being sufficiently emphasized. In 1989, the proportion of the labour force engaged in agriculture varied between 10 per cent in Czechoslovakia and 49 per cent in Albania. According to most observers, much of the agricultural labour force in Central and Eastern European countries is not used efficiently and should, for the sake of improving productivity, be relocated to other economic sectors. Yet, employment in the non-agricultural sectors of the economy is unlikely to increase in the short run. Consequently, unemployment and underemployment in rural areas are likely to rise and out-migration will likely increase.

These issues were discussed at the Workshop on Agricultural Restructuring in Eastern and Central Europe, jointly sponsored by FAO and the United Nations Development Programme (UNDP) and held at the Agricultural University of Nitra, Czechoslovakia, in May 1992 (Food and Agriculture Organization of the United Nations (FAO)/United Nations Development Programme (UNDP)/VPS, 1992). At the Workshop, the employment problems facing the rural areas of Central and Eastern Europe were emphasized and due attention was paid to the emerging outflow of migrants to urban areas within their own countries or to other countries.

The propensity for leaving rural areas has been confirmed by rural employment surveys carried out in Albania and Hungary in 1992. However, further study from a policy perspective is required. The potential size and direction of future migration flows need to be assessed in order to provide a basis for the formulation of policy.

E. Movement of pastoralists

The case of pastoralists is atypical in that their migration is essential for their production system and involves either the movement of livestock led by only a few shepherds, or the movement of both livestock and complete households in search of natural pastures. Pastoralists, especially if poor, experience economic and social motivations to migrate similar to those of settled farmers. A possible major difference is that the decision of pastoralists to stop migrating and settle may be as momentous as the decision of settled farmers to migrate.

The record of projects that have fostered the settlement of pastoralist groups is mixed, including both successes and failures. In fact, it is increasingly recognized by policy-makers that nomadism in specific pasture areas that have no other economically viable use may be an economic necessity. In addition, the costs of providing a minimum resource base to a large number of pastoralists so that they may settle, and the problems of training and fostering their adaptation to an environment different from that to which they are accustomed, cannot be ignored.

Indeed, the problems faced by a person or a household deciding to abandon pastoral life are similar to those faced by migrant families moving from one place to another and involve both benefits and costs. One particular problem in the case of pastoralists who decide to settle is the lack of access to grazing lands which normally results from the private appropriation of land. Another is the lack of alternatives in case of drought. For most pastoralists, settlement means the transition from moving with a tent from place to place in search of pasture, to living in a tent in some shanty town with no employment. That is to say, in many instances, pastoralists who still live a nomadic existence are better off than those who give up herding and become part of the urban poor (Swift, 1988).

These issues have been discussed in several international workshops organized by FAO, including those held in Jordan and Mongolia (Food and Agriculture Organization of the United Nations (FAO), 1991d and 1992). Studies indicate that in most cases some degree of settlement would help pastoral societies prevent the overburdening of their scarce pastures and other resources. A steady flow of pastoralists into other activities tends to stabilize pastoral communities. The voluntary sedentarization of nomads does, however, require appropriate employment creation programmes in other economic sectors as well as vocational training.

F. Conclusion

The preceding considerations point to a need for better policies and a comprehensive approach towards development that, in turn, require improved action-oriented research and the monitoring of socio-economic changes, particularly with regard to remote areas where irreversible change is likely. The role of FAO is to assist Governments in improving the process of policy formulation. On the basis of an analysis of migration-related issues and policies from the perspective of rural development, FAO can contribute to forging a common understanding of the intersectoral and multicultural features of the migration process. There is a need for a global approach that fosters the collaboration of people and their organizations, local and central Governments, and intergovernmental organizations in promoting action. Given that 1993 has been declared the International Year of the World's Indigenous People it seems appropriate to call on both agencies and Governments to consider the needs of indigenous people, particularly when they are prompted to migrate because of their limited access to resources.

References

Dasgupta, B. (1984). *Labour Migration and the Rural Economy.* Rome: FAO.

Food and Agriculture Organization of the United Nations (FAO) (1979). *Report of the World Conference on Agrarian Reform and Rural Development, Rome, 12-20 July 1979.* WCARRD/REP and corrigendum and annex. Part one.

________ (1990). *Women in Agricultural Development: FAO's Plan of Action.* Rome: FAO.

________ (1991a). *Third Progress Report on WCARRD Programme of Action.* C/91/19. Rome: FAO.

________ (1991b). *Plan of Action for People's Participation in Rural Development.* C/91/22. Rome: FAO.

________ (1991c). Towards gender-responsive harmonization of human and natural resources in agricultural and rural development in sub-Saharan Africa. Paper prepared for the Regional Workshop on Population and Sustainable Agricultural Development, Zimbabwe, December.

________ (1991d). *Proceedings of the International Workshop on Pastoralism and Socio-Economic Development,* Mongolia, 4-12 September 1990.

__________ (1992). Traditional systems in evolution. Paper presented at the Workshop on Pastoral Communities in the Near East, Amman, Jordan, 1-5 December.

Food and Agriculture Organization of the United Nations (FAO), and Ministry of Agriculture, Nature Management and Fisheries of the Netherlands (1991). Report on the FAO/Netherlands Conference on Agriculture and the Environment. Rome: FAO. Appendix A contains the Den Bosch Declaration and Agenda for Action on Sustainable Agriculture and Rural Development.

Food and Agriculture Organization of the United Nations (FAO), United Nations Development Programme (UNDP) and VPS (1992). Restructuring agriculture in Eastern and Central Europe. Rome: FAO.

Gaiha, R., and M. Spinedi (1992). Agricultural wages, population and technology in Asian Countries. *Asian Survey* (Berkeley, California), vol. 32, No. 5 (May).

Hunt, D. (1985). *The Labour Aspects of Shifting Cultivation in African Agriculture*. Rome: FAO.

International Labour Organization (ILO) (1986). Rural labour in Latin America. World Employment Programme Working Paper. Geneva.

Mora Alfaro, A. (1992). Generación de empleo en zonas agroforestales seleccionadas de Costa Rica. Rome: FAO.

Swift, J. (1988). *Major Issues in Pastoral Development with Special Emphasis on Selected African Countries*. Rome: FAO.

Thiesenhusen, W. C. (1990). Land policies in an agrarian reform process. Paper presented at the International Forum on Agrarian Reform and Rural Development, Brasilia, 27-30 August.

United Nations (1975). *Report of the World Population Conference, 1974, Bucharest, 19-30 August 1974*. Sales No. E.75.XIII.3. Chapter I.

__________ (1984). *Report of the International Conference on Population, 1984, Mexico City, 6-14 August 1984*. Sales No. E.84.XIII.8.

__________ (1993. *Report of the United Nations Conference on Environment and Development, Rio de Janeiro, 3-14 June 1992*, vols. I, II and III. Sales No. E.93.I.8 and corrigenda.

XXIII. HEALTH AND URBANIZATION IN DEVELOPING COUNTRIES

*World Health Organization**

The recent sense of urgency about urban health and the situation of the urban poor can be justified by reference to world demographic trends. By 2005, just over half of the world's population will live in towns or cities, and for most urban-dwellers the urban setting will play a major role in determining their health. Moreover, the way in which we choose to organize and run our cities will be critical to the future ecology of the planet itself.

The levels and trends of urbanization in the major world regions have already been discussed in chapters III to VII of the present volume and such discussions need not be repeated here. The point that needs to be stressed is that the level of urbanization and the speed at which cities are expected to continue growing have different implications depending on the ability of local authorities to provide existing and future populations with adequate livelihoods, appropriate health and social services, protection of the environment, and the institutional and legal structures required to sustain development and preserve a balance between supply and demand. In most cities of developing countries, local administrations have been unable to meet the challenge.

This failure to respond adequately has been the central problem for decades, but it has been aggravated recently by widespread economic difficulties. Poverty and its well-known array of social manifestations, the deterioration of services and the degradation of the environment are all consequences of this situation. The economic crisis is also forcing most Governments to cut social expenditure. The result is often a decline in what were already very inadequate levels of investment in water supply, sanitation, garbage collection and health care. The per capita expenditures on health in developing countries were already very low at the beginning of the 1980s, standing at an average of US$ 4 per person per year in 92 developing countries compared with an average of US$ 220 per person per year in 32 more affluent countries. Indeed, in about one third of the developing world, annual health-care spending per person was less than US$ 2. Often, the limited resources allocated to health care went largely to curative-based hospitals in one or two major cities that provided services for a very small proportion of the national population.

Other developments, including economic recession, the growth of external debt, the persistence of drought, and the destructive power of storms, floods and other natural disasters, as well as of warfare, have further contributed to rural-urban migration and poverty. Other contributing factors include government policies that tend to concentrate new productive investments in large cities, thus diverting resources away from small towns and rural areas.

The rapid population growth experienced by the cities of the developing world during the 1980s has resulted in a dramatic increase in the number of squatter settlements and slums. It is reckoned that in many cities about 50 per cent of all city dwellers live in conditions of extreme deprivation. In Addis Ababa, for instance, 79 per cent of the population lives in slum and squatter settlements. Although this may be an extreme case, the proportion of urban poor may not be far lower for other major cities in the developing world. By the end of the century, the urban poor may represent a quarter of the world's population.

Unemployment, underemployment, low income, low education and malnutrition are closely associated with poor health. The unemployed in developed countries are estimated to number about 30 million, more than 70 per cent of whom live in urban areas. It has been estimated that between 1970 and 1985 the size of the urban population living in absolute poverty increased by 73 per cent, from 177 million to 306 million; the corresponding increase for the rural population was 11 per cent.

---

*Prevention of Environmental Pollution (PEP)/Environmental Health in Rural and Urban Development and Housing (RUD); Division of Strengthening of Health Systems (SHS). and District Health Systems (DHS).

Assuming that by the year 2000 half of the urban population is still living in conditions similar to present ones, at least 1 billion people will be counted among the urban poor. Of these, approximately 56 per cent will live in Asia, 24 per cent in Latin America and 20 per cent in Africa. When these figures are translated into human terms, they forecast harsh times ahead for most of the poor living in the cities and towns of the developing world where squatter settlements built of cardboard, wood and flattened kerosene cans have become a common sight and even a permanent feature of the landscape.

A. HEALTH STATUS

Because the health and nutritional status of the urban poor are omitted from official statistics or hidden in aggregate data, facts about their plight are difficult to come by. For the rapidly increasing numbers of urban poor, health conditions may in some respects be worse than those of the rural poor and they may deteriorate even further. In any attempt to deal with the needs of the urban poor, better information is required, since its absence makes it difficult to know the extent of their problems, to persuade people that those problems exist and to formulate effective responses to them.

Extensive slums and shanty towns have appeared on the periphery of many capital and secondary cities. When cities experience rates of population growth of from 3 to 5 per cent per year, the emergence of illegal and unplanned housing side by side with planned developments becomes inevitable. The slums are inhabited by people whose housing, food and sanitation are inadequate, and whose health status is poor; and who are not in a position to benefit from the services, amenities and economic opportunities generally available in urban areas.

These urban-dwellers, and particularly children, women and elderly people, are vulnerable to health threats associated with overcrowding, pollution and a host of familiar urban problems that include mental and physical diseases, homelessness, drug abuse, and sexually transmitted diseases (including acquired immunodeficiency syndrome (AIDS)), as well as violence and social alienation.

The evidence available indicates that poor populations have higher rates of maternal mortality and infant mortality and morbidity, and that other health indicators for them also compare unfavourably with city averages. However, much remains to be done to obtain data for small areas and for specific income groups so as to provide the necessary information on which to base policy formulation.

Examples of the poor health conditions of deprived groups in urban areas of developing countries abound. Thus, in some low-income settlements or bustees in Delhi, India, the child mortality rate has reached 221 infant deaths per 1,000 live births and has been almost twice as high among the poorer castes within those settlements. In Bombay, the crude death rate on Bombay Island, the central city area, is twice as high as that in the suburbs and three times that in the extended suburbs. In three low-income areas of Karachi, Pakistan, between 95 and 152 infants out of every 1,000 born alive die before reaching age 1, while in a middle-income area, the infant mortality rate is only 32 per 1,000.

In the urban areas of Guatemala, the infant mortality rate varies from 113 per 1,000 live births among children of illiterate mothers in the lowest socio-economic group to 33 per 1,000 among better-off children. In the slums of Port-au-Prince, Haiti, 1 in 5 infants dies before the first birthday while another 1 in 10 dies between the first and second birthdays. These mortality rates are almost three times as high as the mortality rates in rural areas in Haiti and several times higher than those in the more affluent areas of Port-au-Prince, where infant and child mortality rates are similar to those recorded in urban areas of the United States of America.

Environmental conditions influence certain health risks. Thus, the urban setting exposes children to a high risk of diarrhoeal diseases, even if it also provides opportunities for diarrhoeal disease control because of the greater accessibility of preventive and curative services. The causes of diarrhoeal diseases are well known: overcrowded housing; poor sanitation; inadequate water supply; contaminated water; and poor hygiene in preparing food. The disposal of excreta plays consistently a more important role in determining the health of children in developing areas than do water supplies, especially where the prevalence of diarrhoea is high. Open drains, a regular feature of low-income settlements, pose a major hazard. Recently, the failure to provide adequate sanitation in crowded urban settlements led to outbreaks of cholera in Latin America, Africa and Asia.

Acute respiratory infections are well recognized as major causes of childhood mortality in developing countries, and 25-30 per cent of all childhood deaths are associated with those infections. There is an annual incidence of 5-8 episodes per child in urban areas of developing countries, an incidence higher than in rural areas. Transmission is facilitated by high population density, poor housing conditions and low socio-economic status. High levels of air pollution have been recorded in many urban centres, and several studies have suggested an association between air pollution and acute respiratory infections exacerbated by high levels of suspended respirable particles, sulphur dioxide, nitrogen dioxide and ozone. Crowded urban settlements with substandard housing also favour the rapid transmission of epidemic meningococcal meningitis.

Urban malaria has become an urgent problem in countries where the disease is endemic. Urban development can create habitats suitable for the reproduction of malaria vectors. Freshwater accumulations, resulting from the absence of water drains, favour the breeding of mosquitoes. The movement of people associated with rapid urbanization contributes further to spreading the infection, since many migrants from rural areas are malaria carriers.

Parasitic diseases such as Chagas' disease, filariasis, leishmaniasis and schistosomiasis are increasingly a problem in urban and peri-urban settlements. In some countries, unplanned urbanization may create conditions for epidemics. Thus, the rapid growth of the population in Khartoum, the Sudan, during the 1980s because of migration from drought-stricken areas has been associated with an increase in cutaneous leishmaniasis. Nearly 100,000 cases of the disease were reported by the Ministry of Health between 1985 and 1987. A host of other non-communicable diseases contribute to the poor health of urban-dwellers, including disorders associated with stress, cardiovascular ailments and cancer.

In addition, there is a high risk of accidents and injuries caused by unsafe and overcrowded transport systems. While cottage industries are major employers in many cities, especially for recent migrants and people without access to the formal sector of the economy, occupational safety measures in those industries are generally inadequate or non-existent. Thus, workers in cottage industries may be exposed to hazardous chemicals or may work in unsafe establishments.

Air pollution creates a special set of hazards in urban areas. Lead emissions from vehicular traffic can cause damage to the nervous system and the level of carbon monoxide produced by such emissions is usually high. Urban air pollution is related to the increasing incidence of chronic respiratory diseases, such as bronchitis and asthma, and may have other health impacts that are difficult to measure.

B. The environment

Urban environmental health is not a new problem. Modern public health, born in the cities of eighteenth century Europe, successfully used environmental sanitation as its major strategy for controlling communicable diseases, later adding vaccination to its arsenal. Many cities in developing countries are growing at unprecedented rates, sometimes under stagnant economic conditions. Urban settlements, from mega-cities to small towns, are often marked by increased density, overcrowding, congestion, traffic and unsuitable residential patterns. A large proportion of their populations lives in slums and squatter settlements, lacking adequate sanitary services, and located on sites vulnerable to landslides, floods or other natural hazards. Cities of all sizes are increasingly unable to ensure adequate garbage collection and to find suitable landfill sites. Poor urban-dwellers, in particular, experience a high incidence of environment-related diseases and injuries. The most vulnerable groups include women and children who are exposed to health hazards at home. Many cities lack the financial and administrative ability to provide sanitary infrastructure for their growing populations, to coordinate and improve the efficiency of public services, to promote adequate employment and housing, to manage wastes, and to ensure security, environmental controls, and health and social services.

The impact of these problems on the health of urban-dwellers requires an assessment of environmental health services and a revitalized role of public health in solving them. How actively and well national and local health authorities carry out functions related to the improvement of environmental health depends on their legal and policy mandates, on their interest and on their capabilities. Public authorities can enhance environmental health by advocating preventive measures to protect the public's health, fostering community capacity to manage environmental health, carrying out health impact and risk assessments, and conducting epidemiological surveillance of environment-related diseases. In cases where national and local authorities do not perform those tasks, the health consequences at the community level can be damaging.

C. Health services

The health problems experienced by the urban poor present considerable aetiologic complexity and remedial measures are consequently equally complex. The primary health-care strategy provides a potential solution and an appropriate orientation with respect to achieving equitable urban health services. However, in most cases, its application leaves much to be desired. Urban planning and the allocation of resources seldom reflect the needs of the most deprived groups. Although promising initiatives that are community-based, comprehensive and integrated do exist, they generally cover only a limited proportion of the target population. Total coverage, efficient utilization and good quality remain elusive goals. The involvement of other health-related sectors, particularly those concerning the environment, education, employment and productive activities, is still at the initial stages in most cases.

One of the major problems in cities is to obtain maximum benefits from care institutions, taking account of the personnel, facilities and resources available to them. The fact that decision-making and authority within the health system are distributed among government, public and private agencies without proper coordination leads to the duplication of functions and the waste of resources. The separation of curative and preventive services has resulted in there being less than adequate attention paid to the application of the required preventive and promotive activities. Health-care organizations do not serve the whole urban population: to a great extent, resources are used in favour of individuals and families of higher socio-economic status. Thus, in most cities, the poor are provided either with limited services or with none at all. An important factor in resource distribution is the market mechanism. In cities, private agencies, corporations and institutions play an important role and do not necessarily follow government policy, thus failing to provide comprehensive health care.

Primary health care, as defined in the 1978 Declaration of Alma-Ata, is as relevant to urban areas as it is to rural areas, because it focuses on vulnerable, poor and underserved communities and encourages them to become self-reliant in determining priorities and controlling related action. It also ensures that issues of a multisectoral nature are addressed and provides ways of doing so. Through community involvement, additional resources are mobilized and health intervention is made more effective through a far-reaching network of community health workers and outreach services.

Unfortunately, the health units located in the periphery of cities, closer to the areas of residence of the poorest groups, are often underutilized and bypassed because they are understaffed, lack equipment and drugs, and do not attract the best-qualified personnel. In contrast, centrally located facilities and higher-level services are overcrowded and improperly utilized. Under these conditions the referral process is chaotic and the logistic support and supervision are insufficient. Utilization of existing resources must therefore be improved by eliminating wasteful spending, reallocating resources to needy groups, involving universities and other training institutions, and eliciting the collaboration and contributions of the people themselves. Such measures can strengthen peripheral health units, upgrade them where needed and improve the referral system. Several cities, such as Cali (Colombia), Jakarta (Indonesia), Karachi (Pakistan), Mexico City (Mexico) and São Paulo (Brazil), have initiated efforts along these lines.

In Cali, the initiative undertaken responded to a continuously deteriorating situation of overcrowding and improper use of a central university hospital and to the underutilization of peripheral health units. The project has benefited the poor populations living in peripheral barrios, especially by strengthening peripheral health units and thus removing one of the main reasons for hospital congestion. Over a period of 10 months, large numbers of health personnel were specifically trained in the referral process. A special effort was made to inform the public about the best way of using the health facilities and to encourage the use of the units located nearest to them. A subsequent but equally important development was the establishment of four "intermediate health units", one in each urban health system area. Recently, reference health centres have been set up in large cities to provide services on a 24-hour basis covering larger areas and being closely linked to both health centres and hospitals. These reference health centres enjoy high levels of public support and participation. They provide a wider range of preventive, promotive and curative activities for outpatient services, including day surgery and normal delivery services. They can thus relieve considerably the pressure on hospital outpatient services and reduce unnecessary inpatient care.

The development of the reference health centres in Cali and, more generally, the improvement of the referral process, have greatly contributed to improving the services available in the four urban health areas, and to alleviating the workload at the central outpatient and inpatient hospital facilities. In 1987, the WHO Study Group on the Functions of Hospitals at the First Referral Level emphasized the need for integration of the hospital with the district and local health services.

Effective support for primary health care may require the reorganization and strengthening of city and national health departments. More important than organizational change, however, is a change in attitude throughout the department, since a fundamental shift in values, strategy and approach is necessary. In most cases, reorganization is not the first step required; rather it should evolve when there is already some record of initial achievement. On the whole, it seems better not to set up new public health-care units that are vertically organized, but rather to seek to build primary health care into the plans and operating systems that shape activity across the entire department.

The aetiologic complexity of urban health problems and the critical role of environmental influences on these problems make it imperative to move away from compartmentalized approaches to urban health towards solutions that entail coordinated and convergent actions by all responsible sectors. Furthermore, urban development must be recognized as the driving force of economic growth. Consequently, attention should be focused on (*a*) integrated programmes that cover whole cities, encourage productivity and improve effectiveness; (*b*) increasing the demand for labour and creating employment for the poorest groups; (*c*) improving access to basic infrastructure and social services; (*d*) finding solutions to the most pressing problems related to the urban environment; (*e*) addressing the problems specific to women; and (*f*) promoting research to improve understanding of urban problems.

D. International support and the role of the World Health Organization

For international agencies, including the World Health Organization (WHO), the growing urban crisis confirms the value of many of their technical cooperation activities, but also demands significant changes. The agencies have successfully identified the global extent of the crisis and how it relates to the degradation of the global habitat and to the pursuit of sustainable development. Much of the information that the agencies provide and the technology that they disseminate are directly relevant to solving urban problems. Some agencies have successfully fostered the increased interchange of pertinent knowledge and experience among countries and cities, and are changing allocation policies so as to assign a greater share of available funds to meeting social development needs. This reflects a tendency that should be strengthened.

Apart from the better funding of sound urban development, two types of changes are now required: (*a*) the widening of technical cooperation to provide better support to national and local efforts; and (*b*) better internal and inter-agency coordination of technical cooperation and external support.

The World Health Organization must place itself in a position comparable with that of its national and local counterparts, if it is to fulfil its role as a major international agency devoted to human well-being. Its established technical capabilities and its clear policies on health for all and primary health care are strong assets in meeting the challenges of urbanization. As with national, district and local health authorities, however, it needs to reallocate its health service resources to improve health systems and establish mechanisms focused on the health problems related to urbanization; to strengthen and improve its cooperation with countries; and to expand its health-promoting influence among international agencies and organizations.

Harnessing the potential of non-governmental, religious, private and voluntary organizations is a major challenge. Meshing the efforts and resources of these different groups into a common effort can also improve the use of scarce resources in meeting existing needs. Encouragement and support should be given to the voluntary network that is being developed among the cities themselves, including the "Healthy cities" project sponsored by WHO, associations such as the World Association of the Major Metropolis (Metropolis) and the Regional Network of Local Authorities for the Management of Human Settlements in Asia and the Pacific (CITYNET), and individual twinning schemes between cities in different countries and regions.

In summary, there has been a significant change in the concepts being used to understand urban health and there is a greater comprehension of the multifaceted nature of ill health. There is an increasing realization that all health institutions, including hospitals and health centres, should adopt a holistic and integrated approach to improving the health status of the population, an approach that includes health promotion as well as treatment and remedial interventions. The World Health Organization continues to play a key role in the development of these concepts and is contributing decisively to updating them in line with the dynamic changes that are prevalent in urban populations throughout the world.

REFERENCES

Aka, E. (1990). City profile: Atlanta. *Cities* (Oxford), vol. 7, No. 3, pp. 186-195.

Hardoy, J. E., and D. Satterthwaite (1989). *Squatter Citizen: Life in the Urban Third World.* London: Earthscan Publications.

Harpham, T., and others (1988). *In the Shadow of the City: Community Health and Urban Poor.* London: Oxford Press.

Tabibzadeh, I., and others (1989). *Spotlight on the Cities: Improving Urban Health in Developing Countries.* Geneva: World Health Organization.

World Health Organization (1990). Report of the Interregional Meeting on City Health: The Challenge of Social Justice, Karachi, Pakistan (November 1989). WHO/SHS/NHP/90.3. Unpublished.

__________ (1991a). Development of urban environmental health services. WHA45/Technical Discussions. Unpublished.

__________ (1991b). Cities and the population issue. A45/Technical Discussions, No. 7. Unpublished.

__________ (1991c). Urbanization and health in developing countries: a challenge for health for all. *World Health Statistics Quarterly* (Geneva), vol. 44, No. 4.

__________ (1992). WHO Commission on Health and the Environment: Report of the Panel on Urbanization. WHO/EHE/92.5. Unpublished.

XXIV. A GENDER PERSPECTIVE ON MIGRATION AND URBANIZATION

*United Nations Office at Vienna**

The major role of women in the processes of migration and urbanization is striking and it has been highlighted in several recent meetings.[1] However, they have pointed out that women continue to be ignored in most of the literature dealing with migration and urbanization. The practice of considering women as mere dependants of men and assuming that they migrate in order to accompany or join family members is largely responsible for the invisibility of women as migrants. Indeed, if migrant women are gaining more of the limelight recently, it is mostly for negative reasons, involving their being subject to especially pernicious forms of exploitation that must be decried. It is important that a more balanced view of who migrant women are and of their roles be developed, especially if policies and programmes that deal with migration are to be realistic in their treatment of the many women who migrate.

The importance of considering all development issues from a gender perspective is well established among intergovernmental, governmental and non-governmental actors. The focus has moved from restriction of women's issues to specific projects which was prevalent in the 1970s, to incorporation of women's issues in the mainstream of development planning. Gender-based analysis is therefore an essential instrument in the design, implementation and evaluation of all development activities, including economic planning and the formulation of development policy (United Nations, 1992).

Gender refers to the socially constructed and culturally variable roles that women and men play in their daily lives, and it encompasses the structural relationship of inequality between men and women that is manifested in all spheres of life, from labour markets to political structures as well as within the household. It is important to understand how such structural inequalities are reinforced by custom, law and development policies so that ways of enhancing the status of women and ensuring their full participation in the development process may be devised. With respect to migration, it is important to ascertain to what extent the roles that men and women as culturally and socially assigned influence the way in which migrant women are treated and lead to overt or covert discrimination against female migrants in terms of their access to social services, the labour market, the judicial system and so forth.

The crucial issue is whether the reproductive and community roles of women are reinforced or not by their migration. Women's reproductive roles include all the unpaid work that women carry out in order to sustain the family, including child care, care for the sick or the elderly, managing of household finances, cooking, cleaning and so forth (Elson, 1991). Often, as an extension of their reproductive role, women take on some management responsibilities in their communities, especially when they have to put considerable effort into securing housing and other basic services that are scarce in poor urban neighbourhoods (Moser, 1989).

Migrant women, if married, generally undertake the work consistent with their reproductive role. They may also be more likely than non-migrant women to assume roles outside the household, whether in order to secure access to scarce social services or to engage in economic activity. Women in general play important roles in establishing and maintaining social networks that provide mutual support for their members. In the case of migrant women, those networks may be an essential source of support both making migration possible and enhancing the adaptation of migrant women and their families to the new environment. Given the demands that the reproductive role puts on women's time, many may not be able to join the formal labour market. Yet, necessity may dictate that they nevertheless engage in some kind of income-producing work within their homes. Those who work outside the home are often confined to a narrow set of occupations that involve tasks not far removed from those typical of their reproductive role,

*Prepared by the Division for the Advancement of Women which was then part of the Department for Policy Coordination and Sustainable Development of the United Nations Secretariat.

including domestic service, nursing, teaching and caring for children.

By migrating, women usually lose the support of the social networks to which they belonged in the place of origin. They therefore have to devote some effort to building new social networks or joining those already in existence. That task may be especially difficult for women who move from one country to another and are therefore less likely to be familiar with the cultural norms of the host society or even with the local language. Under those circumstances, migrant women can remain relatively isolated, since their contacts with non-migrants may be scarce.

Because the roles of men usually do not confine them to the home, migrant men are less likely to face the same degree of isolation as migrant women. Furthermore, in most cultures, men are still not expected to contribute much to household work and are therefore not likely to relieve women of some of the responsibilities inherent in the reproductive role. Even when men migrate on their own and live alone in the place of destination, they are not likely to spend much time doing household chores since essential services, such as laundry, cleaning and food preparation, can be bought.

Migrant women, especially those who are married and have children, must learn to operate efficiently in the place of destination if they are to ensure the well-being of their children. Finding and buying food at reasonable prices, securing clean water, cooking, caring for children, and finding and using health-care services are all time-consuming tasks. Because of the general lack of support services, especially of child-care facilities, many women cannot afford to work outside the home. Yet, women from the lower socio-economic strata, particularly if they are migrants, may need to engage in some kind of income-earning activity to make ends meet. Doing piecework at home or near home is a possibility, though the work is poorly paid. However, migrant women with low educational attainment and household responsibilities have few other choices.

No society today would function without the unpaid work done by women, but the extent to which societies draw on women's unpaid work varies. Developed countries underwent a phase of rapid industrialization during which they relied heavily on the unpaid or poorly paid reproductive and productive work of women. For today's developing countries, the issue is whether it is necessary to go through a similar phase. The extent to which unpaid female work, both in the home and in the community, supports and underpins general social and economic development needs to be better understood in order to devise ways of reducing the burden on women and making a more effective use of their human resources.

In many countries today, the work of migrants is essential for economic growth. Yet the work of migrant women, especially if they have moved within their own country, remains undervalued and unrecognized. Not only are there more women migrating specifically to work at the place of destination, but women are also contributing to making the migration of men possible by staying in the place of origin and assuming responsibility for the family left behind. Thus, their reproductive work is being used to facilitate the labour-market activities of others. In the international context, female migrant workers are undertaking for pay some of the reproductive work of native women and thus liberating the latter to carry out other activities outside the home.

Yet, despite the many useful activities that they perform, female migrants are often subject to discrimination, particularly if they have moved from one country to another. It has been argued that international female migrants are often subject to triple discrimination because of their class, ethnicity and gender. When migrant women belong to ethnic minorities, they may not only be subject to racist exclusion from the mainstream of the host society but also experience sexism in two forms, that prevalent in their own ethnic group and that characteristic of the dominant group (Anthias, 1992). If the two differ, migrant women may have trouble deciding which norms to follow and how best to achieve some sort of adaptation.

The labour force participation of women, particularly in developing countries, remains low. It is reckoned, however, that women's economic activities are usually not adequately reflected by available statistics. In particular, much of the productive work that women perform as unpaid family members or in the informal sector is not accounted for. Migrant women, just as women in general, often find employment in the informal sector where they can secure jobs that allow them to combine their household responsibilities with their productive activity. However, the informality of their employment often leads to low pay and exploitation. Studies in India, Madagascar and Zambia indicate that, in a majority of cases, women in the informal sector are engaged as workers or unpaid family workers rather than as own-account workers or entrepreneurs. An important proportion of their employment is in domestic service (United Nations, Committee on the Elimination of Discrimination against Women (CEDAW), 1992). Indeed, migrant women the world over tend to be over-represented among persons working in domestic service.

In many societies, the low status of women in combination with a strong patriarchal tradition enables men to use the unpaid labour of women and children for family-based production. An issue that remains understudied is the extent to which migrant families rely on the unpaid work of women to enhance their economic status at the place of destination. In patriarchal societies where an increasing number of women are migrating on their own to work at the place of destination, it is argued that the decision for women to migrate is made not by the women themselves but rather by the household head (usually a man) who decides how to allocate the household's labour in order to minimize risks and, if possible, maximize gains. Women are selected for migration because new job opportunities are opening up for them, mostly in factories that produce goods for export, and because they are considered to be more reliable sources of remittances than men.

Since about 1970, as a result of the adoption of economic policies fostering export-led growth as the basis for development, multinational enterprises and other firms have been setting up assembly plants and other production facilities in developing countries where labour costs are low. The cost of female labour, in particular, is generally lower than that of men. The prevailing view that women are not primary wage earners within the family interacts with their low status and lack of education to justify their being paid low wages. Thus, companies have been relying on the cheap labour of women to carry out labour-intensive routine operations (United Nations, 1989). A number of countries have promoted the creation of export-processing zones where enterprises obtain tax advantages when they produce goods for export. Those zones have attracted labour-intensive industries, such as electronics, clothing or footwear, whose workforce has tended to involve very high proportions of women, many of whom are migrants (Zlotnik, 1993). Although these developments have provided women with more varied employment opportunities, the conditions of work in most labour-intensive industries and the almost non-existent possibilities of promotion make them far from ideal. Furthermore, in a number of instances, women have been subject to health hazards and their conditions of work have not been far from outright exploitation. In addition, there has been a tendency for the proportion of female workers to decline as production in export-processing zones is upgraded and diversified. Men, who have access to better formal training and education, are more likely to be hired for better paying or more interesting jobs (United Nations, 1989). In contrast, when production work involves mostly unskilled or semi-skilled jobs and relatively low wages, female workers invariably predominate (United Nations, 1989).

Although not all migrant women have low socio-economic status or find themselves in vulnerable situations, those who do are especially liable to be subject to sexual harassment or even outright violence. Harassment may occur within the family or in the workplace. In a number of cases, migration becomes a means of exploitation as rural women are literally sold into prostitution by their own families to dealers in cities or to those acting as intermediaries for persons in international destinations. Others are forced into prostitution as the only means of supporting themselves and their families (United Nations Office of Vienna (UNOV), 1990). Aside from the stress and health risks that those women experience, prostitution makes women more vulnerable to crime, drug trafficking and abuse, as well as to the risk of contracting the human immunodeficiency virus/acquired immunodeficiency syndrome (HIV/AIDS) (du Guerny and Sjoberg, n.d.).

In conclusion, in most societies men are more likely to respond to economic opportunities than women because of cultural and social norms that restrict the choices open to women. Thus, women are less likely to engage on their own in migration and when they migrate as dependent family members they may not be free to respond to the opportunities that migration opens up to them. What would happen if women were not so restricted by societal norms and values that lead to their having lower educational attainment than men, restrict their mobility, and impose on them a series of household duties that are time-consuming and demanding? How would migration affect their status, and their own and their families' well-being if they were free to exploit available opportunities or even create their own? Studies show that migrant women are generally eager to take advantage of any economic opportunity open to them (Zlotnik, 1993) and that women are more likely than men to use all their income to enhance their family's well-being (Elson, 1991). Therefore, policies directed towards enhancing the income-earning opportunities of women, particularly those who are migrants, are more likely to benefit a greater segment of society than those focusing on men.

Policies and programmes that do not incorporate a holistic view of the situation of migrant women run the risk of reinforcing the inequities that prevent migrant women from making the best of their opportunities and might therefore slow down development. Migrant women cannot be considered in isolation, and policies and programmes should not attempt to focus on their problems out of context. On the contrary, the situation of migrant women must be seen in relation to that of migrant men and in relation to gender issues in the host society. The interaction between male and female migrants is as important as that between migrants and their male and female non-migrant counterparts (du Guerny, 1992). To be effective, any interventions should consider the nature of those interactions and the complex systems of sex stratification in operation.

NOTE

[1]For example, the United Nations Expert Group Meeting on International Migration Policies and the Status of Female Migrants, held in San Miniato, Italy, from 28 to 31 March 1990; the United Nations Expert Meeting on the Feminization of Internal Migration, held in Aguascalientes, Mexico, from 22 to 25 October 1991; and the United Nations Educational, Scientific and Cultural Organization (UNESCO) International Seminar on Migrant Women in the 1990s: Cross-cultural Perspectives on New Trends and Issues, held in Barcelona, Spain, from 26 to 29 January 1992.

REFERENCES

Anthias, F., (1992). Gender, ethnicity and racialisation in the British labour market. Paper presented at the UNESCO International Seminar on Migrant Women in the 1990s: Cross-cultural Perspectives on New Trends and Issues, Barcelona, Spain, 26-29 January.

du Guerny, Jacques (1992). Some remarks on the migration of women. Paper presented at the UNESCO International Seminar on Migrant Women in the 1990s: Cross-cultural Perspectives on New Trends and Issues, Barcelona, Spain, 26-29 January.

__________, and Elisabeth Sjöberg (n.d). Inter-relationship between gender relations and the HIV/AIDS-epidemic: some possible considerations for policies and programmes. United Nations Office at Vienna. Mimeographed.

Elson, D. (1991). *Male Bias in the Development Process.* Manchester, United Kingdom: Manchester University Press.

Moser, C.O.N. (1989). Gender planning in the Third World: meeting practical and strategical gender needs. *World Development*, vol. 17, No. 11.

United Nations (1989). *1989 World Survey on the Role of Women in Development.* Sales No. E.89.IV.2.

__________(1992). Recommendations of the Expert Group Meeting on Population and Women: report of the Secretary-General of the International Conference on Population and Development. E/CONF.84/PC/6.

__________, Committee on the Elimination of Discrimination against Women (CEDAW) (1992). Women in the informal sector: report by the Secretary-General. CEDAW/C/1992/6.

United Nations Office at Vienna (UNOV) (1990). Migrant women as a vulnerable group. Paper presented at the United Nations Expert Group Meeting on International Migration Policies and the Status of Female Migrants, San Miniato, Italy, 28-30 March.

Zlotnik, Hania (1993). Women as migrants and workers in developing countries. *International Journal of Contemporary Sociology* (Joensuu, Finland), vol. 30, No. 1 (special issue on "Sociology and social development in a time of economic adversity", James Midgley and Joachim Singelmann, eds.), pp. 39-62.

XXV. MIGRATION AND DEVELOPMENT

*International Organization for Migration**

From 15 to 17 September 1992, the International Organization for Migration (IOM) held an International Seminar on the subject of "Migration and Development", with the participation of 70 Governments, 30 international organizations and 15 research institutes. The goal of the Seminar was to explore the link between migration and development in order to devise policies that would use migration to accelerate development, and policies that would prevent unwanted emigration, including the fostering of trade and foreign direct investment as substitutes for migration.

The subject and objectives of the Seminar were chosen in response to the ongoing debate on how to manage and control international migration so that its proportions could be acceptable to both sending and receiving societies, and on how to prevent disruptive migration from occurring. A reply transcending all the debates on the subject called for accelerated development in countries of origin which, over time, should reduce the pressure to emigrate and lead to manageable migration flows. The steps leading to such an outcome and the concrete policy measures that should be taken to achieve that goal were judged to be still imprecise. Hence, the Seminar had as objective to clarify the relation between migration and development by identifying the different intervening factors and their effects. The main findings are presented below.

A. Migrant remittances

In 1989, the remittances from migrants to their countries of origin amounted to US$ 65 billion, a sum that was higher than the amount of the official development assistance (ODA) given during the same year by Organisation for Economic Cooperation and Development (OECD) member States to developing countries, namely, US$ 46.7 billion (Organisation for Economic Cooperation and Development (OECD), 1990). Nominal global remittances were second only to trade in crude oil and significantly larger than the value of trade in coffee, the most important non-oil primary commodity produced by developing countries (Russell and Teitelbaum, 1992). Remittances thus represented a major part of international capital flows to developing countries and were often a significant portion of the gross domestic product (GDP) of particular countries. Thus, remittances were equivalent to about one seventh of GDP in Jordan and Yemen, one quarter of merchandise exports from Greece and Turkey, and one fifth of merchandise imports by Portugal and the Syrian Arab Republic in the late 1980s.

An important issue in the debate was whether remittances contributed to an acceleration of development. The Seminar reviewed the consumption-versus-investment arguments, considering to what extent remittances were mostly used for current consumption rather than for productive investment. Although conflicting findings for the same country and the same time period were reported, it was agreed that remittances were generally spent rationally. When the return to investment in land had been, or promised to be, higher than the return to investment in fertilizer, migrants and their families tended to invest in land.

Russell (1992) showed that several conventional expressions of wisdom regarding the interrelations between international migration and development were true only under specific circumstances. Thus, remittances did not always increase income inequality, fuel inflation, or raise imports of luxury consumer items. They were often used to pay school fees, enhance educational attainment, and promote investments in irrigation technology.

There was agreement regarding the fact that Governments could increase the level of remittances received by adopting a credible and stable macroeconomic policy. In

* Department of Planning, Research and Evaluation.

cases where banking and other transfer facilities were already available, a credible exchange rate, low inflation and prospects for economic growth would do more to maximize remittances than special foreign currency accounts, two-tiered exchange rates and similar incentive programmes.

B. Migration of skilled personnel

In order to counteract the detrimental effects on development of the migration of professional, technical and kindred workers from developing to developed countries, Logan (1992) advocated having countries of origin play a greater role in manpower planning and consider the establishment of bonds on their citizens studying abroad. In addition, countries of origin might consider reducing their support for the training of their citizens abroad, especially in fields where employment opportunities are limited in the home country. Governments of countries of origin should ensure that grants awarded for foreign study include conditions that make it costly for the student to settle abroad. African Governments, in particular, might also make greater use of the transfer and reintegration programmes operated by international organizations to attract back their highly skilled African citizens, including IOM's Reintegration Programme for Qualified African Nationals.

IOM's Reintegration Programme has been operating in Africa since 1983 and during the programme's first four years, 568 persons were assisted. The countries with the largest number of nationals assisted were Kenya and Zimbabwe. Evaluations of return programmes were generally positive, in the sense that most of the returning professional, technical and kindred workers were using the skills that they had acquired abroad and were satisfied with their jobs and way of life in the home country. The Reintegration Programme not only identified the need for experts in African countries, it also sought out those experts abroad and arranged for their return. Consequently, the Programme acted as a specialized labour exchange.

IOM has also operated a similar programme in Latin America since 1974, which has contributed to the return of over 13,000 professional, technical and kindred workers. A study of 10,500 of them indicated that, in most cases, the assistance they had received took the form of reduced travel costs and help in finding a job upon return to the country of origin. A quarter of those assisted, at an average cost of less than $10,000 each, were medical specialists, another quarter were engineers, and a further 25 per cent were economists and social scientists.

1. *Exchange of skills among developing countries*

The Seminar underscored the fact that there was a potential for migration programmes consisting of the temporary transfer of governmental and non-governmental experts from one developing country with a surplus of specific skills to another where those skills were in short supply. The transmission of know-how associated with those assignments met the objectives of technical cooperation among developing countries (TCDC). It was reported that, since 1983, IOM had participated in over 600 short- and medium-term projects in Latin America by paying travel expenses and supplementing the salaries of international experts assigned temporarily to government agencies in need of their expertise. IOM was reported to be establishing a data bank on experts available and the projects in the region for which they were needed.

2. *Development and migration*

Attention was focused on the effects of trade, foreign direct investment and international economic cooperation on international migration. Ghosh (1992a) pointed out that receiving countries had advocated the substitution of trade for migration, but that it had proved difficult for them to lower trade barriers for the goods that countries of origin could produce competitively. It did little good for developing countries and countries with economies in transition to face reduced trade barriers on products that they did not produce competitively and stiff ones on goods, such as farm products and textiles, in respect of which they had a comparative advantage. The fact that, during the 1980s, the trade barriers imposed by industrialized countries rose constituted a development that, coupled with falling prices for primary export commodities, such as coffee and cocoa, made it hard for developing countries to create enough jobs to reduce emigration pressures.

Rising trade barriers were even more disappointing for developing countries that had recently adopted export-oriented development policies. One factor necessary for the success of such policies was the attraction of foreign direct investment, which did not flow into sending countries as fast as it had been expected. Thus, even in a country like Bangladesh, where only US$ 12,200 was needed to create a new job, foreign direct investment was too low to create enough jobs for all the migrant workers estimated to be abroad. The Philippines would have needed 25 times the US$ 850 million received as foreign direct investment in 1989 to create enough jobs for its estimated 650,000 migrant workers abroad (at a cost of $32,000 per job). In fact, during 1989 the Philippines received US$ 1.7 billion in remittances from its workers abroad. Assuming remittances had had a multiplier effect of 2.3, they would have been equivalent to US$ 3.9 billion in spending in the Philippines. To replace each dollar of remittance-generated economic activity, an estimated US$ 5.40 in capital investment would have been required, or US$ 32,400 per migrant, to produce a total of US$ 21 billion. That figure clearly dwarfs the amount actually obtained as foreign direct investment that year by the Philippines (US$ 850 million), although the difference between the two would be smaller if account were taken of such factors as the opportunity costs of migrants and the multiplier effect of foreign direct investment.

Although calls for increased trade and aid to countries of origin had grown in response to unwanted migrant inflows to industrialized countries, it was recognized that general recommendations to increase trade, aid or investment often failed to address key questions, including the amounts needed to affect international migration, and the time lag expected between intervention and a change in migration trends. There was agreement that a policy package that combined aid, trade and investment favouring countries of origin, coupled with the decision to accept some level of inevitable migration inflows by the countries of destination, was the best hope for linking trade and development while managing migration in the 1990s. However, to ensure the success of such an approach, developing countries had to adopt domestic strategies that would support and sustain both development and a reduction of emigration pressures.

3. *Policy measures*

Several measures to enhance the role of international migration in development and to prevent disruptive migration were reviewed. With regard to the migration of professional, technical and kindred workers, it was suggested that developing countries had to engage in better human resource planning so that the skills of such workers could be used locally; that it was necessary to improve the working conditions of highly skilled workers in countries of origin; that the placing of bonds and the use of other restrictions to prevent the permanent emigration of students sent abroad for training should be considered; that efforts to send highly skilled workers from developed to developing countries should be strengthened; and that the temporary migration of highly skilled workers between developing countries could be used to increase trade-related labour mobility.

Consideration was then given to the subject of development strategies that had the potential of making international migration more manageable. The use of scarce capital to import labour-saving machinery was cited as a counterproductive strategy since, by reducing the number of jobs, it increased the likelihood of emigration. Export-led growth, as practised by the newly industrializing economies of Asia, was considered a way of accelerating economic development in a manner that reflected a country's comparative advantage. Trade liberalization was judged important to stem emigration pressures resulting from unemployment and a lack of economic growth. The successful completion of the General Agreement on Tariffs and Trade (GATT) Uruguay Round of multilateral trade negotiations would contribute more than remittances to accelerating economic growth in the countries of origin.

The use of export processing zones to attract foreign direct investment, create jobs and stimulate exports was mentioned, but their effect on migration was considered not to be well understood. The tendency of factories located in those zones to employ mostly women and to become economic enclaves limited both their potential to reduce emigration and their ability to create more jobs via multiplier effects. Overcoming both tendencies could enhance the positive development impact of export processing zones.

With respect to the impact of official development assistance (ODA) on migration, it was observed that the effective use of remittances and of the skills of returning migrants was more likely than ODA to reduce emigration pressures. Programmes to reduce population growth were also likely to reduce international migration in the long run, but aid had not yet been used systematically to reduce migration. Better coordination and planning were judged to be necessary to enhance the role of ODA in stimulating economic growth, promoting job creation and thus stemming migration.

C. Conclusion

It was concluded that international migration had become too important to be excluded from national and international discussions on matters of international relevance, ranging from population issues to trade. In order to overcome the neglect of international migration issues, Governments and international organizations were urged to include them explicitly in their population, development and trade policy deliberations. A coordinated approach to examining the impacts of international trade, aid, foreign direct investment and population on international migration was judged necessary to promote regular consultations on the issue at both the national and the international levels.

It was noted that countries needed to define more clearly their objectives regarding international migration. Those objectives should be incorporated explicitly into development plans in order to ensure that development reduced the pressures impelling people to migrate and, especially, those leading to forced migration. To the extent possible, migration processes should be managed so as to maximize their development pay-offs. To this end, banking and other transfer facilities should be provided at credible exchange rates so as to maximize remittances.

It was suggested that industrialized countries should not encourage the permanent settlement of skilled migrants from developing countries in their territories, but rather encourage the return of such migrants under cooperative agreements with the countries of origin. Trade barriers with respect to products from developing countries should be dismantled in order to increase both trade and employment in them. Countries of origin should cooperate with countries of destination to combat illegal migration. The national development plans of countries of origin should include measures to promote population programmes, the improvement of infrastructure and the development of human resources. Developing countries should make more use of transfer and reintegration programmes operated by international organizations, such as IOM and the United Nations Development Programme (UNDP), to encourage the return of their highly skilled emigrants.

It was agreed that development aid should be directed towards activities that could act as catalysts to attract further inflows of foreign direct investment and improved technology. It was suggested that trade-related labour mobility between developing countries could be encouraged so as to foster a better use of the professional, technical and kindred workers from those countries, while providing them with opportunities to improve their skills through international exchanges that would not involve their permanent relocation. Temporary migration schemes for nationals of developing countries should involve training relevant for the country of origin and, whenever possible, such training should be provided in the countries of origin. Developing countries were encouraged to develop and test TCDC exchange programmes in cooperation with international organizations in order to promote the sharing of experience and know-how among themselves. IOM was called upon to serve as a clearing house for information on policies aimed at managing migration flows and their effectiveness. Other international organizations were asked to lend support and provide technical expertise to developing investment projects aimed at using the savings of expatriate nationals to enhance the development of the countries of origin.

The Seminar acknowledged that international migration had important benefits for the individuals and families involved, as well as for both countries of origin and those of destination. Given the increasing interdependence of nations, international migration was recognized as a natural and positive aspect of international interactions. In addition, international migration could contribute to the development of countries of origin. Although it was acknowledged that there was nothing automatic about how the recruitment of migrants, their remittances and their return could influence development, under certain conditions those factors could propel development, leading eventually to conditions in which individuals did not feel compelled to emigrate. Policies could promote the use of remittances to improve the lives of migrants and their families, and to create better opportunities for productive investment in the areas of origin of migrants.

Developed countries wishing to prevent migrant inflows could take a variety of measures to accelerate development in the countries of origin. However, since many of those measures involved difficult decisions in terms of domestic politics, a strong leadership was required to push them through and make them effective.

REFERENCES

Ghosh, Bimal (1992a). Migration, trade and international economic cooperation: do the inter-linkages work? *International Migration* (Geneva), vol. 30, Nos. 3/4, pp. 377-398.

_________ (1992b). Migration-development linkages: some specific issues and practical policy measures. *International Migration* (Geneva), vol. 30, Nos. 3/4, pp. 377-398.

International Organization for Migration (1992). Migration and development. *International Migration* (Geneva), vol. 30, Nos. 3/4 (special issue).

Logan, Ikubolajeh Bernard (1992). The brain drain of professional, technical and kindred workers from developing countries: some lessons from the Africa-US flow of professionals (1980-1989). *International Migration* (Geneva), vol. 30, Nos. 3/4, pp. 289-312.

Organisation for Economic Cooperation and Development (1990). *1990 Report of the Development Assistance Committee.* Paris: OECD.

Russell, Sharon Stanton (1992). Migrant remittances and development. *International Migration* (Geneva), vol. 30, Nos. 3/4, pp. 267-288.

_________, and Michael S. Teitelbaum (1992). *International Migration and International Trade.* World Bank Discussion Paper, No. 160. Washington, D.C.: World Bank.

XXVI. INTERNATIONAL MIGRATION IN THE ECE REGION

*Economic Commission for Europe**

Various aspects of the contemporary international migration directed to Europe and Northern America are attracting the interest of the general public, politicians, the media and scholars. There is growing concern that the consequences of migration may be detrimental to the well-being of people in receiving countries and that more and more people are being forced to leave their countries like, for instance, the almost 3 million people who have moved because of armed conflict in the former Yugoslavia.

After the Second World War, the polarization of the world that occurred during the cold war resulted in a marked contrast between the conditions under which international migration took place in Eastern- and Western-bloc countries. Although the Governments of Western-bloc countries regulated migrant admissions on the basis of their perceived interests, individuals and families could exercise a reasonable amount of freedom in realizing their personal preferences for international relocation. In Eastern-bloc countries, in contrast, government controls virtually stopped international migration altogether and most citizens of Eastern-bloc countries were effectively denied the right to leave their respective countries.

There were, however, significant differences among Western-bloc countries regarding international migration. Canada and the United States of America, with their long tradition as countries of immigration, continued to be important attraction poles for immigrants after 1945. Towards the late 1960s and early 1970s, they experienced important changes in the distribution of immigrants by country of origin. Whereas before 1920, most immigrants to Northern America had originated in Europe (Davis, 1974), by the early 1980s, 85 per cent of the immigrants to Northern America originated in the developing countries: 47 per cent in Latin America and the Caribbean, 27 per cent in Eastern and South-eastern Asia, 6 per cent in Southern Asia, and 5 per cent in Africa and Western Asia (Zlotnik, 1994a). On average, Northern America received nearly 900,000 immigrants annually during the 1980s, including those regularizing their status in the United States (Zlotnik, 1994a).

In the Western-bloc countries of Europe, the reconstruction and economic recovery that followed the Second World War led to labour shortages during the 1950s which prompted the importation of foreign workers on a temporary basis. The main labour-receiving countries were Belgium, France, Germany (Western), the Netherlands, Sweden and Switzerland (United Nations, Economic Commission for Europe, 1975). Recessionary periods in the late 1960s and again in the early 1970s led to a reassessment of the need for foreign labour that resulted in the stoppage of labour migration by most of the labour importing countries around 1974. However, migrant workers already present in those countries were allowed to stay and to be joined by close family members. Consequently, migration to those countries continued, albeit at a reduced level in net terms. In addition, the countries of southern Europe began to experience positive migration balances in the 1970s, first as a result of the return of many of their citizens from abroad and later as they themselves began receiving foreigners (United Nations, Economic Commission for Europe, 1992). During the 1980s, the average annual inflow of migrants experienced by the main receiving countries of Europe (Belgium, France, Germany, the Netherlands, Sweden and the United Kingdom of Great Britain and Northern Ireland) was 1.1 million, though many of those migrants stayed only temporarily (Zlotnik, 1994a). Net migration, a better measure of migration gains in those countries, amounted to only 225,000 persons annually during 1980-1989. In fact, during 1980-1984, the main receiving countries of Europe registered a negative net migration equivalent to nearly 10,000 persons leaving annually, whereas the gross inflow during the period amounted to about 920,000 persons (Zlotnik, 1994a).

In terms of composition by country of origin, most of the migrants going to European countries are European.

*Population Activities Unit, Division for Economic Analysis and Projections.

For the six countries considered above, 59 per cent of the incoming migrants originated in developed countries. Among migrants originating in developing countries, those from Turkey and the Maghreb countries (Algeria, Morocco, Tunisia) constituted the majority (Zlotnik, 1994b). There were also significant numbers from Bangladesh, India, the Islamic Republic of Iran, Pakistan and Sri Lanka. In contrast, the main sources of immigrants to Northern America were Mexico and the Caribbean, the Philippines, the Republic of Korea, and the three countries of Indo-China, namely, Cambodia, the Lao People's Democratic Republic and Viet Nam.

During the late 1980s, the collapse of communist regimes in Eastern-bloc countries resulted in a relaxation of exit controls that helped to fuel East-West migration. During the cold war, persons who had managed to leave Eastern-bloc countries were generally admitted by Western-bloc countries as refugees. Consequently, the new waves of migrants originating in Eastern-bloc countries during the 1980s largely tried to secure entry into Western countries by seeking asylum in them. This is one of the reasons for the rapid increase in the number of persons lodging asylum claims in the Western-bloc countries of Europe during the 1980s. Thus, the number of asylum applications in those countries rose from 68,000 in 1983 to 685,000 in 1992 (United Nations High Commissioner for Refugees, 1993). In addition, certain minority groups with homelands abroad were allowed to emigrate from Eastern-bloc countries, especially from the former Union of Soviet Socialist Republics (USSR). They included ethnic Germans and Jews (United Nations, Economic Commission for Europe, 1992).

The first attempts at gaining some political liberalization in Eastern-bloc countries included the 1956 political crisis in Hungary and that in Czechoslovakia in 1968. In both cases, Soviet intervention restored the status quo ante and considerable numbers of people left Hungary and Czechoslovakia to become refugees in the West. More recently, Poland's political instability and ailing economy during the 1980s led to the emigration (mostly illegal) of some 100,000 persons a year (United Nations, Economic Commission for Europe, 1992). Lastly, the continuing conflict in the former Yugoslavia has generated the most important refugee crisis in Europe since the Second World War. Of the 685,000 applications for asylum lodged in Western Europe in 1992, a third were filed by persons from the former Yugoslavia. The majority had been forced to leave their homes by armed conflict that spread from Slovenia, to Croatia and subsequently to Bosnia and Herzegovina. Indeed, the conflict has already given rise to large numbers of displaced persons.

According to UNHCR sources, by the end of October 1992, well over 2.5 million persons had been displaced by conflict in the former Yugoslavia. More than 2 million of them were still in the territory of the former Yugoslav States: at least 630,000 people were in Croatia; 430,000 in Serbia; 70,000 in Slovenia; and 810,000 in Bosnia and Herzegovina. In addition, UNHCR reported that 760,000 vulnerable persons found themselves in besieged towns in central Bosnia, and over half a million refugees from the former Yugoslavia were in other European countries, including Germany (220,000), Switzerland (over 70,000), Sweden (60,000), Austria (57,500), and Hungary (50,000). Only a few thousand asylum-seekers and refugees from the former Yugoslavia had reached Northern America. Population displacement has also occurred in the successor States of the former USSR as a result of the armed conflicts raging in the Caucasian republics of Georgia, Armenia and Azerbaijan, and in the Republic of Moldova and the Republic of Tajikistan.

A. Asylum-seekers

As already noted, the number of persons seeking asylum in Europe and Northern America increased markedly during the 1980s and early 1990s. Data for 1992 indicate that the total asylum applications lodged in developed countries amounted to nearly 839,300 (United Nations High Commissioner for Refugees, 1993). Canada and the United States had received 141,700; Australia and Japan, 4,200; and former Eastern-bloc countries, 8,700. Hence, the Western-bloc countries in Europe had received the bulk of the applications: 684,700. Asylum-seekers from the former Yugoslavia accounted for nearly 230,000 applications, corresponding to about one third of the applications filed in Western Europe and to slightly more than a fourth of those filed worldwide.

Between 1983 and 1992, Germany received the largest number of asylum applications, amounting to almost 1.4 million. The United States came second, with slightly over half a million, followed by France with 318,000. Sweden and Canada, each with about a quarter of a million, were also important receivers of asylum-seekers (United Nations High Commissioner for Refugees,

1993). Between 1983 and 1989, 70 per cent of the asylum-seekers filing applications in Western Europe originated in developing countries (Zlotnik, 1991). However, the proportion of asylum-seekers from developed countries had been increasing steadily during the period, passing from 26.5 per cent in 1983 to 37 per cent in 1989 (Zlotnik, 1991). Equivalent figures for 1992 are not available, but the proportion of persons from developed countries filing asylum applications in European countries is likely to have risen considerably. In 1992, citizens from the former Yugoslavia and those from Romania accounted for 52 per cent of the asylum applications filed in the main receiving countries in Europe (United Nations High Commissioner for Refugees, 1993).

The continuous rise in the number of asylum-seekers in Western European countries is a major source of concern. Both political leaders and the population at large consider that it is necessary to stem the flow of asylum-seekers. As Chancellor Kohl of Germany stated in a speech delivered at the Christian Democratic Union convention on 26 October 1992, the "influx is rising from month to month, leading to unbearable conditions in our cities and towns. If we do not act, we face the danger of a deep crisis of confidence in our democratic State, yes, even a national state of emergency" (*International Herald Tribune*, 1992a).

One possible means of alleviating tensions is to reach intergovernmental agreements between receiving and sending countries that provide aid to the countries of origin in exchange for the readmission of asylum-seekers. Examples of such agreements include those between the Government of Germany and the Governments of Bulgaria, Poland and Romania. Thus, the agreement between Germany and Bulgaria, signed on 12 November 1992, committed the Goverment of Bulgaria to admitting the return of Bulgarian citizens who had been refused asylum in Germany in exchange for German financial aid of US$ 17.6 million over the next five years (*International Herald Tribune*, 1992b).

B. Migration between former Eastern-bloc countries

In 1992 the ECE Population Activities Unit in collaboration with the ECE Statistical Division began collecting data on international migration from the national statistical offices of ECE countries on a semi-annual basis. Although, at the time of writing, the data received were still incomplete, they allow important developments in former Eastern-bloc countries to be delineated.

Immigration and emigration data provided by the Russian Federation for the first half of 1992 indicate that there are noteworthy flows of people into the Russian Federation from the Baltic, Caucasian and Central Asian republics. There are also large numbers of people entering the Russian Federation from Belarus and Ukraine; however, about twice as many are leaving the Russian Federation in the opposite direction. Thus, during the first six months of 1992, 79,000 people moved from Ukraine to the Russian Federation and 170,000 left the Russian Federation for Ukraine. During the period considered, the emigration of ethnic Germans to Germany (over 24,000) and of Jews to Israel (10,000) continued (table 63). In total, according to official statistics, 323,000 persons immigrated to the Russian Federation between January and June 1992. That figure, however, may be on the low side, since it is likely to leave out irregular migrants.

During the same period, 355,000 persons emigrated from Russia. Anecdotal evidence from Poland suggests that the number of emigrants from the former USSR may be greater. An unknown number of people, including Russians, live and work in Poland without proper

Table 63. Number of immigrants and emigrants registered by the Russian Federation, January to June 1992
(Thousands)

Region/country	*Immigrants*	*Emigrants*	*Net migration*
Baltic republics	24	6	+18
Caucasian republics	46	17	+29
Central Asian republics and Kazakhstan	146	76	+70
Other European republics: (Ukraine, Belarus, Republic of Moldova)	106	213	-107
Germany	—	24	-24
Israel	—	10	-10
Others	1	9	-8
Total	323	355	-32

Source: Russian Federation State Committee on Statistics, preliminary data.

documentation. Poland and other Central European countries have become transit areas for illegal migration to Western-bloc countries. Authorities of the former Czechoslovakia estimated that approximately 35,000 former Soviet citizens, 15,000-20,000 Romanians, and 5,000-10,000 citizens of Bulgaria or the former Yugoslavia resided illegally in the country.

The statistics on immigrants and emigrants gathered by Latvia match those provided by the Russian Federation. According to Latvian statistics, almost 20,000 people emigrated during the first six months of 1992: about 10,000 to the Russian Federation and 4,000 each to Belarus and Ukraine. Immigration to Latvia was of the order of 2,000 people.

Statistics of Slovenia indicate that the country received significant refugee flows, primarily from Croatia, during 1991 and early 1992. Refugees from Bosnia and Herzegovina arrived mainly between April and June of 1992, and presumably have continued to arrive since then. By the end of June 1992, Slovenia reported the presence of over 43,000 registered refugees from Bosnia and Herzegovina, and estimated that a further 20,000 unregistered persons were staying with friends and relatives.

REFERENCES

Davis, Kingsley (1974). The migrations of human populations. *Scientific American*, vol. 231, No. 3 (September), pp. 93-105.

International Herald Tribune (1992a). Kohl sounds alarm over refugee influx. 27 October.

________ (1992b). Bonn will repatriate Bulgarians. 13 November.

United Nations, Economic Commission for Europe (1975). *Economic Survey of Europe in 1974. Part II: Post-war Demographic Trends in Europe and the Outlook until the Year 2000.* Sales No. E.75.II.E.16.

________ (1992). *Economic Survey of Europe in 1991-1992.* Sales No. E.92.II.E.1.

United Nations High Commissioner for Refugees (UNHCR) (1993). *The State of the World's Refugees. The Challenge of Protection.* New York and London: Penguin Books.

Zlotnik, Hania (1991). South-to-North migration since 1960: the view from the North. *Population Bulletin of the United Nations*, Nos. 31/32, pp. 17-37. Sales No. 91.XIII.18.

________ (1994a). International migration: causes and effects. In *Beyond the Numbers. A Reader on Population, Consumption and the Environment*, Laurie Ann Mazur, ed. Washington, D.C.: Island Press.

________ (1994b). Migration to and from developing regions: a review of past trends. In *Alternative Paths of Future World Population Growth: What Can We Assume Today?* Wolfgang Lutz, ed. London: Earthscan.

XXVII. ARAB LABOUR MIGRATION IN WESTERN ASIA

*Economic and Social Commission for Western Asia**

Since 1950, the Arab world has experienced various types of international migration flows. The first was the Arab migration from the Maghreb countries (Algeria, Morocco and Tunisia) to the then labour-importing countries of Europe, which intensified during the 1960s. By the early 1990s, the number of citizens of Algeria, Morocco and Tunisia present in Europe were estimated at nearly 2 million (United Nations, 1996). The second major flow was that directed to the member States of the Gulf Cooperation Council (GCC), namely, Bahrain, Kuwait, Oman, Qatar, Saudi Arabia and the United Arab Emirates. Migration to those countries was prompted by the discovery of oil in the region, its exploitation and the building of infrastructure in the GCC countries. It is estimated that the total foreign labour force in the GCC countries rose from 1.125 million in 1975 to 5.218 million in 1990, thus experiencing a growth rate of 10.2 per cent per annum.

Because migration to both the labour-importing countries of Europe and the GCC countries has often been of a temporary nature, return migration has been common. Important return flows were experienced by the Maghreb countries during the early 1970s and then by the main sources of labour for the GCC countries during the early 1980s, when the falling price of oil affected the receiving countries. The countries or areas experiencing sizeable return migration flows during that period included Egypt, Lebanon, the Palestinian territories and the Syrian Arab Republic.

Arab countries cannot be characterized unambiguously as sending or receiving. Changing trends have been experienced by several of them. Thus, Jordan and Yemen began by being important sources of migrant workers but were soon importing workers from other countries to replace those who had left. Iraq and Oman also changed from being the sources of labour for other countries to being important importers of foreign workers. Indeed, Iraq became second only to Saudi Arabia as the main destination of foreign labour in Western Asia during the early 1980s.

Not only have Arab countries been the source of large numbers of unskilled migrant workers, they have also experienced the outflow of skilled personnel, mostly directed to developed countries and especially to the United States of America. Skilled personnel has also been in demand by the GCC countries, although many of them have imported such skills from developed countries. Reliable estimates of the number of skilled persons leaving the Arab countries of Western Asia are not available.

It is well known that the GCC countries have been importing increasing number of workers from countries in Southern and South-eastern Asia. Given the labour-force conditions in the Arab world, an argument could be made for directing both skilled and unskilled Arab labour from the Maghreb countries to the GCC countries instead of having the latter recruit Asian workers or skilled manpower from Europe or the United States. However, existing flows are the result of the complex political and economic relations between Arab countries, of the migration policies adopted by GCC countries to suit their needs, and of the relative importance of Arab construction firms in the GCC market. The present paper reviews both the magnitude and the trends of Arab migration in the region and its impact.

A. TRENDS IN ARAB LABOUR MIGRATION

In relation to the size of their national populations, the GCC countries have been absorbing a disproportionate number of foreign migrants. According to recent estimates of the world's migrant stock, in 1985 the six GCC countries accounted for 5.5 per cent of all international migrants in the world (Skoog, 1994). In contrast, their total population, including foreigners, accounted for less than 0.4 per cent of the world's population in 1985. Despite the large numbers of foreigners in the GCC countries and their impact on the overall economic and

*Social Development and Population Division.

social life of the Arab world, migration has not been sufficiently studied in the region. In particular, essential data regarding the volume and characteristics of migrants continue to be scarce. Moreover, the results of several surveys and censuses undertaken during the 1970s and 1980s have not been made widely available.

The scarcity of data concerning labour migration to Western Asia compromises a judicious assessment of its impact. Until 1985, no migration survey at the national level was available for Arab countries. Since then, Egypt has carried out two surveys on international labour migration, Jordan has carried out one, and another one was completed in Tunisia. Nevertheless, planning policies had to be formulated even on the basis of fragmentary evidence. Because of the weak basis for their formulation, many such policies were not realistic. Those advocating increases in labour migration generally saw it as a source of remittances that contributed to improving the balance of payments. Those opposing labour migration tended to blame it for all the drawbacks and failings of the socio-economic development of Arab countries.

Table 64 shows estimates of the national and the foreign labour force in the GCC countries during 1975-1990[1] and table 65 allows a comparison of estimates produced by different authors.[2] The tables indicate that, whatever the estimates used, the importance of the foreign labour force is clear. In several of the GCC countries, it surpassed the national labour force in 1975 and had grown to be several times larger than the national labour force by 1990. For the six GCC countries taken together, the foreign labour force constituted 46.5 per cent of the entire labour force in 1975 and over 70 per cent of that estimated for 1985. The number of economically active foreigners rose from 1.1 million to 3.0 million between 1975 and 1980 and is likely to have reached 5.2 million by 1990 (see table 64).

Saudi Arabia was the main pole of attraction, accounting for over 60 per cent of the foreign workers present in GCC countries in 1985. The United Arab Emirates, with 14 per cent of the foreign workforce, came second, followed by Kuwait (12 per cent), Oman (8 per cent), Qatar (4 per cent) and Bahrain (2 per cent). With the exception of Oman, the growth rate of the foreign labour force was greater during 1975-1980 than during 1980-1985. At the GCC level, those growth rates amounted to 19 per cent and 8 per cent per year, respectively.

TABLE 64. NATIONAL AND FOREIGN LABOUR FORCE IN THE COUNTRIES OF THE GULF COOPERATION COUNCIL, 1975-1990
(*Thousands*)

Country of residence	*National*	*Foreign*	*Total*	*Percentage of foreign to total*
A. *1975*				
Bahrain	45.8	38.7	84.5	45.8
Kuwait	92.4	217.6	310.0	70.2
Oman	88.9	103.2	192.1	53.7
Qatar	11.7	57.0	68.7	83.0
Saudi Arabia ...	1 010.7	474.7	1 485.4	32.0
United Arab Emirates	44.7	234.1	278.8	84.0
TOTAL	1 294.2	1 125.3	2 419.5	46.5
B. *1980*				
Bahrain	59.3	78.2	137.5	56.9
Kuwait	180.5	392.6	501.1	78.3
Oman	119.4	170.5	289.9	58.8
Qatar	14.7	106.3	121.0	87.9
Saudi Arabia ..	1 220.3	1 734.1	2 954.4	58.7
United Arab Emirates ...	53.9	470.8	524.7	89.7
TOTAL	1 648.1	2 952.5	4 528.6	65.2
C. *1985*				
Bahrain	72.8	100.5	173.3	58.0
Kuwait	127.2	551.7	678.9	81.3
Oman	149.8	335.7	485.5	69.1
Qatar	17.7	155.6	173.3	89.8
Saudi Arabia ..	1 440.1	2 661.8	4 101.9	64.9
United Arab Emirates ...	71.8	612.0	683.8	89.5
TOTAL	1 879.4	4 417.3	6 296.7	70.2
D. *1990*				
Bahrain	127.0	132.0	259.0	51.0
Kuwait	118.0	731.0	849.0	86.1
Oman	189.0	442.0	631.0	70.0
Qatar	21.0	230.0	251.0	91.6
Saudi Arabia ..	1 934.0	2 878.0	4 812.0	59.8
United Arab Emirates ...	96.0	805.0	901.0	89.3
TOTAL	2 485.0	5 218.0	7 703.0	67.7

Source: Estimates by Kossaifi. See note 1.

TABLE 65. DIFFERENT ESTIMATES OF THE FOREIGN LABOUR FORCE IN THE COUNTRIES OF THE GULF COOPERATION COUNCIL, 1975, 1985 AND 1990
(*Thousands*)

	1975		*1985*		*1990*	
	Birks and Sinclair	*Kossaifi*	*Birks and Sinclair*	*Kossaifi*	*Anani*	*Kossaifi*
Bahrain	30.0	38.7	85.3	100.5	116.7	132.0
Kuwait	208.0	217.6	548.6	551.7	657.9	731.0
Oman	70.0	103.2	312.9	335.7	322.0	442.0
Qatar	53.7	57.0	71.1	155.6	151.9	230.0
Saudi Arabia	773.4	474.7	3 468.6	2 661.8	3 184.2	2 878.0
United Arab Emirates	251.5	234.1	602.5	612.0	1 076.1	805.0
TOTAL	1 387.3	1 125.3	5 089.0	4 417.3	5 508.8	5 218.0

Sources: George Kossaifi, "Labour Arab migration: facts and potential", paper presented at the Seminar on Demographic and Socio-Economic Implications of International Migration in the Arab World with Special Reference to Return Migration organized by ESCWA, ILO and the University of Jordan, Amman, 4-9 December 1989, table 1; J. S. Birks and C. A. Sinclair, *International Migration and Development in the Arab Region* (Geneva, International Labour Office, 1980), p. 133; J. S. Birks and C. A. Sinclair, "Manpower and population evolution in the GCC and the Libyan Arab Jamahiriya", paper presented at the Seminar on Demographic and Socio-Economic Implications of International Migration in the Arab World with Special Reference to Return Migration organized by ESCWA, ILO and the University of Jordan, Amman, 4-9 December 1989, pp. 20 and 21; and Jawad Anani, "Policies and labour demand in the GCC countries", paper presented at the Seminar on Migration Policies in Arab Labour Sending Countries, United Nations Development Programme (UNDP) and International Labour Office, Cairo, 2-4 May 1992, table 4.

Statistics on labour and residence permits, and those on arrivals and departures, show that there was a decreasing trend in the number of workers admitted during 1983-1986 by GCC countries, which reversed itself in 1987 when admissions began to rise again. Similarly, the annual outflow of Asian workers from the countries of origin reached a maximum in 1983 (1.094 million) and a minimum in 1986 (0.853 million), and began to increase once more thereafter (Sarmiento, 1991).

On the basis of recent trends, the preliminary results of the 1992 population census of Saudi Arabia and the adjusted Kuwaiti population as of the end 1988,[3] the foreign labour force in the GCC countries was estimated to be 5.218 million in 1990, constituting 68 per cent of the total labour force and having grown at a rate of 3.3 per cent during 1985-1990, the lowest growth rate since 1975. As in previous periods, Saudi Arabia was the main country of destination, with 55 per cent of the foreign labour force.

Because of the Gulf war, it is not certain that the foreign labour force in GCC countries will necessarily increase again by 1995. However, previous projections assuming a decline of the foreign labour force have proved to be wrong. Thus, when it was assumed that the declining trend in admissions observed during 1983-1986 would continue during the rest of the decade, the foreign labour force as of 1990 was projected to be 3.561 million (Kossaifi, 1989). Current evidence suggests instead that it surpassed the 5 million mark.

During the Gulf war, some return flows of a forced nature took place. However, the number of returnees reported at the time seems to have been on the high side. Thus, recent data indicate that the number of Egyptian returnees was of the order of 390,000 instead of the 700,000 originally estimated. Similarly, the number of Jordanian/Palestinian returnees accounted for by the end of 1991 stood at around 225,000 instead of the 300,000 estimated earlier. The number of Yemenites who left Saudi Arabia still stands at 723,000.

The bulk of the returnees left Kuwait, Iraq (which is not a member State of the GCC) and Saudi Arabia. Since the end of the war, economic life has returned to normal in Saudi Arabia, the main destination of foreign workers, and Kuwait has begun to rebuild its infrastructure. Consequently, there is likely to be more, and not less, demand for foreign labour. It is expected, therefore, that the number of foreign workers in the GCC in 1995 will be higher than in 1990. Changes in their distribution by nationality, however, will be noticeable. In particular, a higher proportion of Asian workers is expected.

Data on the distribution of the foreign labour force in GCC countries by nationality indicates that in 1985 the proportions of Asian and Arab workers were approximately equal, at 49 per cent and 47 per cent, respectively. Migrants from developed countries, mostly Americans and Europeans, accounted for about 4 per cent of the foreign labour force at the time. Arab workers had accounted for 57 per cent of the foreign labour force in 1975, but their proportion decreased as more Asian workers were recruited. The proportion of Asian workers increased from 36 per cent in 1975 to 49 per cent in 1985. The shift towards Asian labour took place during the late 1970s and early 1980s. According to statistics from the main countries of origin in Asia (Bangladesh, India, Indonesia, Pakistan, the Philippines, the Republic of Korea, Sri Lanka and Thailand), the annual outflow of workers to all destinations (of which countries in Western Asia accounted for approximately 90 per cent) rose from 121,890 in 1976 to a maximum of 1,094,314 in 1983 (Sarmiento, 1991).

Around 1975, two countries attracted mostly Arab foreign workers, Kuwait and Saudi Arabia. In Kuwait, Arab workers constituted 69 per cent of the foreign labour force in 1975 and in Saudi Arabia they accounted for 82 per cent. The rest of the GCC countries attracted mainly Asian workers, whose proportion varied from 88 per cent in Oman to 55 per cent in Bahrain. By 1985, only Saudi Arabia had a majority of Arab workers among the foreign labour force: 60 per cent. By that time, the Asian labour force had become the majority in Kuwait. Indeed, in that country, the Arab share of first-time work permits in the private sector had declined from 40 per cent in 1980 to 14 per cent in 1985, while the Asian share increased from 56 per cent to 80 per cent. Similarly, the Arab share of work permit renewals in the private sector decreased from 57 per cent in 1980 to 46 per cent in 1985, whereas the Asian share increased from 41 per cent to 52 per cent (Al-Omaim, 1989).

The decreasing trend in the share of Arab labour in the GCC countries observed during 1980-1985 is thought to have continued during the rest of the decade. Fragmentary evidence for Kuwait and Saudi Arabia confirms this trend. In Saudi Arabia, for instance, the share of Arab workers in private sector enterprises with at least 100 employees did not exceed 7 per cent during 1987-1989, while the Asian workers accounted for 57 per cent of the total workers employed in those enterprises during the same period (Al-Nifay, 1989). Although these data cover only a fraction of the total labour force, the low share of Arab labour reflects a decreasing Arab presence in Saudi Arabia. Similarly, in Kuwait, statistics compiled by the Ministry of Social Affairs and Labour show that the foreign labour force in the private sector grew from 139,839 in 1985 to 457,616 in 1989, while the share of Arab workers decreased from 48 to 46 per cent and that of Asian workers increased from 51 to 53 per cent. In addition, the contribution of Arab workers to the annual growth of the foreign labour force in the private sector was 30 per cent in 1985 and 44 per cent in 1989, compared with 67 per cent and 54 per cent, respectively, in the case of Asian workers. On the basis of these observations, it is reasonable to suppose that in 1990 the Arab share of the foreign labour force in GCC countries was in the range of 30-35 per cent.

A major reason for the decline in the share of Arab workers among all foreign workers is the difference in salaries paid to Arab and Asian workers. In Saudi Arabia, data for private sector enterprises with at least 100 employees show that Arab workers were paid, on average, twice as much as their Asian counterparts in 1987 and three times more in 1989. On average, Saudi Arabian workers earned more than any of their non-Saudi Arabian counterparts irrespective of their occupation. Thus, Saudi Arabian professional and technical workers earned 1.4 times more than their counterparts among Arab workers and 3.6 times more than their counterparts among Asian workers in the same occupation. Only Westerners were paid more on average than Saudi Arabians in almost all occupations (Al-Nifay, 1989).

Even if, as expected, the size of the foreign labour force in the GCC countries remains largely unchanged or increases between 1990 and 1995, its distribution is likely to change in favour of Asian workers. As already noted, many migrant workers who left or were expelled from GCC countries because of the Gulf war and who cannot return are mostly Arab: Yemenites, Jordanians, Palestinians and Sudanese. Their number may be slightly over 1 million. Current relations between Arab States indicate that the trend towards a reduction of the expatriate Arab labour force is not likely to be reversed soon. Asian labour-exporting countries have therefore an unexpected opportunity to increase their share of the Persian Gulf labour markets (Addleton, 1991).

Although it might seem that workers from Egypt and the Syrian Arab Republic, the two Arab labour-sending countries that co-signed the Damascus Declaration with GCC countries, could replace other Arab workers in the

Persian Gulf, that is not likely to happen because the skills of Egyptian and Syrian workers are not comparable with those of the highly educated and skilled Jordanian/ Palestinian workers that were expelled (United Nations, Economic and Social Commission for Western Asia (ESCWA), 1991). Furthermore, the volume of Syrian emigration has been rather limited. Consequently, as a result of the Gulf war, the Arab share of the foreign labour force in the GCC countries may drop to 20 per cent by 1995.

B. Impact of labour migration

The scarcity of national surveys concerning international labour migration has already been pointed out. In fact, such surveys have only been available recently. Consequently, before 1985, assessments of the impact of labour migration were mostly based on educated impressions. After 1985, national surveys began to show how facts differed from impressions. The surveys undertaken in Egypt and Jordan will be used here to discuss the divergence between facts and previous impressions.

Even when adequate data are available, the methodological problems related to the study of the impact of international labour migration have not been totally solved. In particular, it is not easy to discern the causes of complex social phenomena. For instance, there is no clear evidence about the causes of the decline in work morale and the rejection of productive and manual occupations that seem to be widespread among the national population in the labour-receiving countries. It is equally difficult to determine what causes excessive consumption and the emergence of new patterns of consumption in labour-sending countries. International labour migration has been proposed as the main reason for the two phenomena. Yet, alternative causes exist. Thus, the income distribution policies of GCC countries have been suggested as a major cause of the first phenomenon, and the open trade policies of countries of origin have been cited as an explanation of the second. These last two explanations have the advantage of better reflecting the essence of social, political and economic life in both sending and receiving countries. However, it is sometimes expedient to blame labour migration for the defects in economic and social development and thus ignore their true causes.

Another methodological problem related to the study of international migration and its impact stems from the different categories of migrants that must be considered (for example, absent migrants, migrants temporarily present, or permanently returning migrants). Information about migrants absent from the country of origin has to be obtained from third parties who may not necessarily know everything about the migrant. If temporary migration prevails, those migrants who can be interviewed in the country of origin while they are back on a short visit or because they have returned permanently may not be representative of those who are still abroad.

1. *Migration selectivity*

It has often been claimed that labour migration involves persons who are positively selected in terms of skills or education. With respect to labour migration from Arab countries, several studies arrived at such a conclusion, emphasizing that those who migrated to work abroad were among the most efficient and skilled workers of the sending country (Fergani, 1983; Saad Eddin and Abdel Fadeel, 1983). However, subsequent work does not totally confirm those findings. In fact, the 1985 Egyptian survey shows that migrant workers with scientific and technical occupations prior to their first migration constituted only 13.3 per cent of the total number of economically active migrants, a proportion that was lower than the 13.6 per cent for non-migrants practising the same occupations. Furthermore, migrants who had been employed in agriculture or as production workers prior to migration constituted 41 per cent and 31 per cent, respectively, of the total number of economically active migrants, proportions that are considerably higher than the 32 per cent and 24 per cent, respectively, of non-migrants having similar occupations.

According to the results of the 1985 Egyptian survey, therefore, the selectivity of labour migration did not imply that better-qualified persons were more likely to migrate. In fact, the majority of labour migrants lacked special qualifications and more than two thirds had worked in agriculture or as unskilled workers prior to migration, thus suggesting that migration attracted the less well-off (Fergani, 1988).

However, the 1987 Egyptian survey did suggest that a strong positive selectivity was in operation. It showed that 27 per cent of migrant workers had been employed in scientific or technical occupations prior to migration, a proportion that was considerably higher than the 11 per cent of all economically active persons in Egypt being employed in similar occupations according to the 1984 labour-force survey. The 1987 Egyptian survey also yielded a considerably lower proportion of migrants engaged in agriculture or unskilled work (about 25 per cent), suggesting, therefore, a higher positive selectivity of migration in terms of occupation than the 1985 survey.

The results of the 1980 migration survey of Jordan also indicate the existence of positive selectivity in terms of occupation when compared with the 1979 census. Thus, migrants who were engaged in scientific or technical occupations prior to migration accounted for 18 per cent of all migrants compared with a 13 per cent weight of such occupations among the economically active population of the country. Clerical workers also tended to be overrepresented among migrants. In addition, the survey showed that migration led to an increase in the percentage of those employed in technical or scientific occupations while abroad and after their return. That is to say, migration did help to improve the skills and proficiency of migrants. Such a conclusion needs to be validated further since it contradicts both previous impressions and the conclusions of other recent surveys (Fergani, 1988).

With respect to the distribution of migrant workers by educational attainment, the 1985 Egyptian survey indicates that they were, on average, better educated than non-migrants. Thus, the mean number of years of schooling of migrant workers stood at 7.3 compared with 6.1 for non-migrants. Both the proportion of illiterate workers and that of workers with at least secondary education exceeded one third among migrants and stood at 45 per cent and 28 per cent, respectively, among non-migrants. The 1987 Egyptian survey showed that 49 per cent of the migrant workers had some secondary education or higher.

Positive educational selectivity is also evident in the case of Jordanian migrants. The 1980 survey shows that only 2.7 per cent of migrants were illiterate compared with 39 per cent among the economically active population of the country according to the 1979 census. University graduates, on the other hand, accounted for a 30 per cent share of migrants in comparison with those graduates' only 14 per cent share of the total labour force. In addition, compared with the population in Jordan, returnees had a higher level of educational attainment. In August 1991, those with university degrees accounted for one tenth of the returnees whereas those with university degrees accounted for only 3.5 per cent of the Jordanian population present in 1987 (United Nations, Economic and Social Commission for Western Asia (ESCWA), 1991).

2. *National labour force in labour-receiving countries*

It is claimed that labour migration to the GCC countries has contributed to reducing the economic activity of nationals and to their rejection of manual occupations (Al-Kawari, 1985). Available surveys indicate that nationals, who constitute a minority among the economically active population of many GCC countries, are still not well represented in occupations requiring qualifications or technical training. The distribution of the national labour force by occupation in several GCC countries confirms that national engineers and architects constituted 14 per cent, 13 per cent and 1 per cent, respectively, of the total employed in Kuwait (1985), Saudi Arabia (1980) and the United Arab Emirates (1980). Similarly, the proportion of nationals among doctors, dentists and veterinarians ranged from 17 per cent in Kuwait and 16 per cent in Saudi Arabia to 5 per cent in the United Arab Emirates.

Nationals are mostly attracted by managerial jobs in both the public and the private sectors or by executive or office-supervision occupations in government departments. The "Gulfinization" of the administration is partly responsible for this outcome. There is also a marked presence of nationals in occupations such as firefighting, and among the police, watchmen, transport and communications operators. Jobs that involve the use of weapons or motor vehicles are among those favoured by Persian Gulf nationals. Nationals are also well represented in farming, agricultural management and supervision. There has, however, been a decline in the proportion of nationals in occupations requiring vocational training or technical skills, including electricians, carpenters, typists, and those working in food processing.

3. *Social integration of migrants*

Little attention has been devoted to the social impact of labour migration, mostly because economic considerations have been paramount. Furthermore, there is no straightforward way of measuring the degree of social integration of migrants in the receiving countries.

Certain indicators, such as the number of mixed marriages, the admission of foreigners into government schools or universities, the number of visits exchanged with nationals and so forth, could be used. Information about opinions, preferences and feelings is also relevant, yet it is not likely to be captured with enough reliability by typical sample surveys. Continuous observation and a good understanding of the communities involved may be necessary to apprehend the relevance of their interactions as regards social integration.

The limited information available suggests that there is a real gap between the national and foreign populations in the GCC countries. Arab affiliation does not appear to help in bridging this gap and, consequently, nationals and Arab migrants remain distinctly apart and have little interaction. Some researchers maintain that the welfare State, by favouring nationals in exchange for their unconditional allegiance, contributes to increase the division between them and Arab migrants. Social relations between the two populations are weak and almost entirely limited to relations in the workplace. State policy is mostly responsible for this gap and its maintenance. Because of the disparities inherent in the application of economic, administrative and political laws, the indigenous population occupies a privileged position with respect to migrants and both groups are conscious of the difference. Nationals view migrants as less well-off individuals seeking some form of livelihood. Such basic inequality is hardly conducive to harmonious integration.

C. Major policy issues

Labour migration between Arab countries should be discussed in light of three main issues. The first is the limited demographic base of the GCC countries. As already noted, in 1990 the national labour force constituted only one third of the total labour force in the region and the total national population amounted to about 14 million people (Kossaifi, 1988). This population is small in relation to the wealth of natural resources in the region and the imbalance is not expected to change in the near future. Consequently, international migration will remain on the region's agenda during at least the next 20 years, especially considering the growing populations of other Arab countries.

The second issue involves the ensemble of political and economic relations between Arab countries that condition international migration in the region. In this respect, there is often a discrepancy between what is declared and what is implemented. There are several Arab conventions, agreements and recommendations calling for the liberalization, organization and facilitation of labour migration between Arab countries. However, in practice, the share of Arab migrants in the GCC countries has been declining.

Indeed, conventions and agreements have been concluded not only at the ministerial level but also at higher levels. Thus, the Arab Economic Summit held in Amman in 1980 ratified the Pan-Arab Economic Covenant which declared its adherence to the principle of Arab economic citizenship and included the aim of freeing the movement of Arab labour, securing migrant workers' rights and granting them all the facilities and assistance necessary for their development. Such contradictions between declarations and implementation are stunning. Agreement at the Pan-Arab level often stems from appeasement or political embarrassment, rather than from a deep-felt conviction. Consequently, disengagement is common, usually on the grounds of contradictions between the national interest and Pan-Arab interest or because of economic considerations. Therefore, only a few countries have actually ratified existing agreements and most of them are labour-sending countries.

During 1976-1991, massive expulsions of Arab migrant workers from Arab countries occurred because of political crises. In 1976, Tunisian workers were deported from the Libyan Arab Jamahiriya. In 1987, the country expelled Egyptian workers. It took a long time for Arab and international mediators to compensate the deported migrants for the loss of their salaries and properties. However, the largest deportation took place in connection with the Gulf war: around one quarter of a million Jordanians/Palestinians were forced to leave Kuwait during 1990-1991. The question of their compensation is still under discussion. Those working for the Government seem to have received such compensation, but the case of private sector employees is still pending. Similarly, the compensation of 232,271 Egyptian migrants who had to leave Iraq is not yet settled.

Clearly, political and economic relations between Arab States have not always worked in the direction of freeing the movement of Arab labour, securing its rights and granting it all the facilities and assistance necessary for its development, as stipulated by the Pan-Arab Economic Covenant. In some cases, disputes between countries have worked exactly in the opposite direction, depriving Arab migrants of their basic rights.

So far, most of the migration of Arab citizens to other Arab countries has been an individual matter, with minimum intervention by the Governments concerned. Yet, a number of bilateral agreements on migrant labour have been concluded between Arab States. The Libyan Arab Jamahiriya, for instance, has signed five such agreements, Qatar and the United Arab Emirates have each concluded three, Iraq has signed two and Kuwait one (El-Borai, 1992). But actual intergovernmental cooperation regarding migration remains at a minimum. Furthermore, Arab construction companies working in the Persian Gulf are not always as competitive as Asian companies, which receive support from their respective Governments and are better able to organize Asian migration to the Persian Gulf.

It would be important, therefore, for the Governments of Arab countries that are the source of migrant workers to become more active in organizing migration and in fostering the protection of their citizens abroad. They should also consider supporting Arab companies working in the Persian Gulf, so that they might become an instrument in the organization and effective use of labour migration. Arab institutions dealing with labour issues should also be supported.

Labour-importing countries must continue to invest in the improvement of their human capital in order to optimize the employment of their nationals. Strategies to do so would include improving the participation of women in economic life, reassessing the prevailing work ethics, reducing the school drop-out rate, and combating the unemployment and underemployment of university graduates. Receiving countries may also consider allowing the naturalization of foreign workers and their families, particularly of those who have needed skills and are of Arab origin.

The Governments of receiving countries should ensure that the human rights of all migrants, regardless of nationality, ethnicity or religion, are safeguarded. Migrants should not be used as pawns in the international relations between Arab States. The mass deportation of migrants should be avoided and, if it occurs, migrants should be properly compensated. Lastly, it is essential, if Arab cooperation is to retain credibility, that the decisions and recommendations adopted unanimously in Pan-Arab forums be implemented.

Notes

[1]The estimates presented are mostly based on data from national censuses or labour-force surveys. For Bahrain, the national population censuses of 1971 and 1981 were used. For Kuwait, estimates were derived from the national population censuses of 1970, 1975, 1980 and 1985; and from the 1983 and 1988 labour-force surveys. The "Bedoun" population, amounting to 247,000, was shifted from the national to the foreign population by the end of 1988. For Oman, the estimates on the national and foreign labour force for 1975 and 1985 were based on Serageldin and others (1983). Estimates of the foreign labour force also took into account reports on foreigners employed in the private sector or as civil servants. For Qatar, estimates were based on the national population censuses of 1970 and 1986 (preliminary results), and on the 1981 survey. For Saudi Arabia, the national population for 1975 was obtained from Serageldin, Sherbiny and Serageldin (1984). The foreign population was estimated on the basis of the 1974 population census of Saudi Arabia. Estimates for 1980 and 1985 were derived from the labour-force surveys of 1977, 1981, 1986 and 1987. The estimates for 1990 were derived from the preliminary results of the 1992 population census results of Saudi Arabia. For the United Arab Emirates, the estimates presented were obtained from the national population censuses of 1975, 1980 (detailed results) and 1985 (preliminary results).

[2]For 1975, the main difference between the estimates presented by Birks and Sinclair and those by Kossaifi lies in the figure for Saudi Arabia. With respect to the 1974 Population Census of Saudi Arabia, Birks and Sinclair (1980) claim that the enumeration was carried out by 10,000 persons, mostly teachers, during the night of 14/15 September. They conclude that it would have been impossible for 10,000 persons to enumerate 6.7 million inhabitants in one night. Birks and Sinclair seem to confuse the "census moment" and the "census period". The 1974 Census of Saudi Arabia was carried out over a period of eight days, from 9 through 16 September 1974, but midnight of 14/15 was chosen as the reference date or census moment. Owing to such a misunderstanding, Birks and Sinclair rejected the census results and generated new estimates for both the national and the foreign population. We believe that their estimates overestimate both the foreign population and the national population because the latter was inflated to include the nomadic group. Therefore, Kossaifi's estimate is based on the actual Saudi Arabian census figures for both nationals and foreigners. For 1985, the estimates proposed by Birks and Sinclair continue to be based on the overestimates used in 1975 and are on the high side. As for Qatar, Kossaifi's estimate is based on the preliminary census results, and it suggests that the estimate proposed by Birks and Sinclair is too low. For 1990, Anani's estimate for Kuwait seems not to take into consideration the shift of the "Bedoun" population from the national to the foreign population. Therefore, it underestimates the latter. We do not know the reasons for the other differences.

[3]As of the end of 1988, the Kuwaiti population was reduced by about 247,000 persons, representing the Bedoun people who lack proof of Kuwaiti nationality and were as of that date considered foreign.

References

Addleton, J. (1991). The impact of the Gulf War on migration and remittances in Asia and the Middle East. *International Migration* (Geneva), vol. 29, No. 4 (December), pp. 509-526.

Anani, Jawad (1992). Policies and labour demand in the GCC countries. Paper presented at the Seminar on Migration Policies in Arab Labour Sending Countries, United Nations Development Programme (UNDP) and International Labour Office, Cairo, 2-4 May.

Birks, J. S., and C. A. Sinclair (1980). *International Migration and Development in the Arab Region*. Geneva: International Labour Office.

_________ (1989). Manpower and population evolution in the GCC and the Libyan Arab Jamahiriya. Paper presented at the Seminar on Demographic and Socio-Economic Implications of International Migration in the Arab World with Special Reference to Return Migration organized by ESCWA, ILO and the University of Jordan, Amman, 4-9 December.

El-Borai, Ahmad (1992). The legal framework for the protection of inter-Arab migrant labour. Paper presented at the Seminar on Migration Policies in Arab Labour Sending Countries, United Nations Development Programme (UNDP) and International Labour Office, Cairo, 2-4 May.

Fergani, Nader (1983). Migration to oil: prospects of migration for work in the oil countries and its impact on development in the Arab World. Beirut: Centre for Arab Unity Studies.

_________ (1988). In search of livelihood: field study on Egyptian migration for work in Arab countries. Beirut: Centre for Arab Unity Studies.

Al-Kawari, Ali Khalifa (1985). Toward an alternative strategy for a comprehensive development. Beirut: Centre for Arab Unity Studies.

Kossaifi, George (1988). Toward a policy for developing the national labour force in the GCC countries. *Al Mustaqbal al Arabi* (Beirut), No.114 (August).

_________ (1989). Labour Arab migration: facts and potential. Paper presented at the Seminar on Demographic and Socio-Economic Implications of International Migration in the Arab World with Special Reference to Return Migration organized by ESCWA, ILO and the University of Jordan, Amman, 4-9 December.

Al-Nifay, Abdullah M. (1989). Analysis of socio-economic factors for employing non-national labour force. Paper prepared for the Economic and Social Commission for Western Asia. Unpublished.

Al-Omaim, Mussaed (1989). Non-national labour force in the private sector in Kuwait. Paper prepared for the Economic and Social Commission for Western Asia. Unpublished.

Saad Eddin, Ibrahim, and Mahmud Abdel Fadeel (1983). Arab labour migration: problems, impact and policies. Beirut: Centre for Arab Unity Studies.

Sarmiento, J. N. (1991). The Asian experience in international migration. *International Migration* (Geneva), vol. 29, No. 2 (June), pp. 195-204.

Serageldin, Ismail, and others (1983). *Manpower and International Labour Migration in the Middle East and North Africa*. New York and Oxford: Oxford University Press (World Bank publication).

Serageldin, I. A., N. A. Sherbiny, and M. I. Serageldin (1984). *Saudis in Transition. The Challenge of a Changing Labour Market*. Oxford, United Kingdom: World Bank and Oxford University Press.

Skoog, Christian (1994). The quality and use of census data on international migration. Paper presented at the XIII World Congress of Sociology, Bielefeld, Germany, 18-23 July.

United Nations (1991). *World Population Prospects, 1990*. Population Studies, No. 120. Sales No. E.91.XIII.4.

_________ (1996). *World Population Monitoring, 1993: With A Special Report on Refugees*. Sales No. E.95.XIII.8.

_________, Economic and Social Commission for Western Asia (ESCWA) (1991). *The Return of Jordanian/Palestinian Nationals from Kuwait: Economic and Social Implications for Jordan*. Amman: ESCWA.

XXVIII. THE IMPACT OF REFUGEE FLOWS ON COUNTRIES OF ASYLUM AND THE CHALLENGE OF REFUGEE ASSISTANCE

*Office of the United Nations High Commissioner for Refugees**

Until the early 1960s, most large-scale refugee movements were associated with the consolidation of communist regimes and the redefinition of nation States in Eastern Europe after the Second World War. The origins of the Office of the United Nations High Commissioner for Refugees (UNHCR) as an institution were rooted specifically in the challenges of the post-war period and the only major group of refugees outside Europe, the Palestinians, was excluded from the High Commissioner's mandate from the start. The world became accustomed to thinking of refugees mainly as Europeans who had been granted asylum in the Western-bloc countries of Europe or who had moved on to settle permanently in Canada, the United States of America, Australia, New Zealand or elsewhere. Many of the receiving societies were short of manpower at the time and regarded the admission of refugees as both economically beneficial and politically desirable at a time when the cold war was at its height. The exodus of Hungarians in 1956 only served to emphasize the East-West character of refugee flows.

With the dissolution of colonial empires and the outbreak of local and regional conflicts in Africa, Asia and Latin America, the entire focus of the refugee problem changed. For the first time, refugees of rural origin began to flee their countries in large numbers, going mostly to neighbouring countries that were hardly able to care for their own populations, let alone assist the destitute refugees. In Africa, where the bulk of the early movements took place, UNHCR attempted to meet the challenge of assisting these new waves of refugees by creating organized rural settlements in countries of asylum and by giving some assistance to the usually much larger number of refugees who preferred to settle spontaneously in local communities with which they had ethnic or family ties. Between 1962 and 1986, UNHCR helped to establish 144 rural settlements in Africa, often in remote areas with marginally fertile land. The entire infrastructure of these settlements, from land clearing and seed distribution to the provision of water, roads, schools, public health and community services, had to be planned and developed from the outset and assistance had to be sustained until the settlers could support themselves at a level reasonably comparable with that of the local population. In areas where refugees have settled spontaneously in sufficient numbers to justify intervention by the international community, assistance has been provided by reinforcing the local infrastructure, especially by building schools, dispensaries or access roads.

The feasibility of rural settlement programmes depends entirely on the willingness of the country of asylum to make land available and, of course, to accept the prolonged and sometimes the permanent presence of refugee communities in its midst. In return, it receives assistance in opening up for agricultural and agriculture-related activities large tracts of land that might otherwise have remained fallow. Services are created from which local communities also benefit. The most successful settlements contribute significantly to local agricultural production. For example, the 150,000 refugees from Burundi who arrived in the United Republic of Tanzania in 1972 and were eventually settled in three large settlements at Ulyankulu, Katumba and Mishamo are now not only entirely self-sufficient but also responsible for producing most of the maize for their area. Eight refugee settlements for Rwandan refugees in Uganda are not only self-supporting but barely distinguishable from the local rural population. Settlements have been abandoned when refugees have been able to return to their countries of origin and when, as in the eastern Sudan, drought and other extraneous factors have made refugees dependent on outside assistance once again.

Rural settlement programmes have not been confined to Africa. In Mexico, for instance, nearly 20,000 Guatemalan refugees have been settled in Campeche and

*Prepared by Mr. A. Simmance, Senior Consultant.

Quintana Roo and during 1993 UNHCR transferred all its responsibilities to the Mexican Government. However, it is in the African continent that the impact of rural settlements for refugees on the countries of asylum has been most marked. In general, that impact, which has been felt mostly by areas of low population density and undeveloped infrastructure, has been beneficial and has not yet contributed to environmental degradation on a major scale. Nevertheless, the ecological balance in rural Africa is increasingly fragile and the problem of soil depletion by overuse is a growing one wherever the pressure of population on land becomes too great. It would be a serious mistake to assume that countries of asylum that have made so much land available for refugee settlement during the past 30 years can continue to do so as generously over the next 30.

Not every country of asylum offers, or is in a position to offer, rural settlement opportunities to refugees. For one thing, many refugees, such as those from Viet Nam and other South-eastern Asian countries, may not come from a farming background. Even when they do, as in the case of the Hmong and other hill tribes who sought refuge in Thailand after a communist Government came to power in the Lao People's Democratic Republic, the country of asylum may tolerate their presence only in "closed camps" pending repatriation or resettlement elsewhere. From the late 1970s onwards, the large numbers of Indo-Chinese fleeing Cambodia, the Lao People's Democratic Republic and Viet Nam were generally housed in closed camps located mostly in Hong Kong, Indonesia, Malaysia and Thailand, and their impact was mainly related to the construction and administration of camps, and to the use of foreign humanitarian aid, including the assistance provided by aid workers. Although there was strain on the existing infrastructure, the economic activity stemming from the assistance provided to refugees had some beneficial effects on the host society. However, no country of asylum admitted that such benefits in any way compensated for the burdens posed by the presence of refugees nor did any country in Eastern and South-eastern Asia willingly accept hosting large numbers of refugees. Yet, it would be idle to pretend that there are not some benefits accruing to the countries of asylum in terms of public service employment in refugee administration, the provision of goods and services for refugees by local businesses, and the acquisition of foreign exchange. The proportion of the cost of hosting refugees borne by the Governments of countries of asylum varies widely from one case to another and is often excessive. However, the greater the number of refugees, the higher the likelihood that the international community will make a substantial contribution towards their assistance. Hence, statistics on the number of refugees may be inflated to bolster claims for international support. In the long run, such practices are usually self-defeating, since the skepticism of donors tends to result in a reduction of aid.

Refugees may also find themselves in countries that, either because of the arid nature of their environment or because their land has been exhausted by existing population pressure, are simply unable to make land available for those refugees' settlement. In such circumstances, refugees have no choice but to live in camps or "villages" which, even if not surrounded by barbed wire or guarded by armed men, offer only limited possibilities of economic self-reliance. Typically, such refugees depend on the World Food Programme for basic food rations and on UNHCR, the United Nations Children's Fund (UNICEF) and voluntary agencies for shelter, water, sanitation and medical care. Their large numbers can seriously disrupt the local administrative infrastructure, communications and market economy, without producing any significant benefits. Thus, the price of a chicken on the Indonesian Anambas Islands rose 10-fold following the sudden arrival of thousands of Vietnamese boat people in the early part of 1979. Indeed, the impact of the presence of refugees on the availability and pricing of local commodities often causes great hardship among the local population and corresponding resentment directed against the refugees. One of the first adverse consequences of refugee inflows to poor agricultural areas is usually the destruction of all trees and shrubs in the vicinity of refugee camps, since every kind of woody vegetation is used to build huts or as fuel.

The ecological impact of mass refugee movements has been particularly severe in countries such as Malawi, Pakistan, Somalia and the Sudan. In Malawi, almost 1 million Mozambican refugees now live in large organized camps or in "open settlements" alongside Malawian villages with which they are, for all intents and purposes, integrated. There are refugees in 13 out of the 24 administrative districts of Malawi and in some districts there are now fewer nationals than refugees. Nearly all the refugees were subsistence farmers in Mozambique but, because of the shortage of arable land in Malawi, they cannot engage in agriculture and are confined to keeping small animals or growing vegetables in small plots. They are almost entirely dependent for food, water and essential services on outside relief. As a result, severe damage has been done to the road system in Malawi, particularly to secondary and district roads which were not built to carry heavy traffic. The volume of relief supplies being carried an average of at least 50 kilometres from reception point to camp distribution centres has reached more than 180,000 metric tons annually. Consequently, heavy vehicle traffic has doubled or more than doubled on main roads, and it has increased between 30 per cent and 160 per cent on minor roads. The damage caused to road surfaces, bridges and culverts not only interrupts access to refugee centres but also prevents the normal transportation of agricultural goods produced by Malawian farmers.

About 9 out of every 10 Malawians derive their total energy needs from wood. In 1986 alone, the year when Mozambican refugees first began to arrive in significant numbers, the total firewood consumption in Malawi was estimated at 10 million cubic metres. The current rate of deforestation stands at about 3.5 per cent of the national forest cover per year and has undoubtedly been aggravated by the presence of refugees who obtain most of their firewood requirements from the scarce wood resources of the areas where they live and who use forest products to construct shelter and latrines and to make furniture. Their annual consumption represents about 37,000 hectares of forest and deforestation itself is contributing to erosion, to the loss of agricultural land, and to the damage of roads because of flooding and the silting of drains. Had it not been for the recent severe drought, the damage caused to the road system would have been worse. The drought, however, has itself exacerbated a severe water shortage. Many shallow wells have dried up and the drilling of more boreholes for the use of both refugees and nationals is urgently required.

UNHCR carries on all types of activities in an effort to alleviate the adverse impact of the presence of large numbers of refugees in poor countries of asylum. Its programmes fund the construction and maintenance of roads, the provision of water, reafforestation and the supply of locally produced fuel-efficient stoves, but not on a scale sufficient to counterbalance all the degradation attributable to the refugee presence, let alone to meet the wider national need. To achieve more widespread coverage, the involvement of the United Nations Development Programme (UNDP) and bilateral support from major aid donors must be obtained. Special contributions have already been forthcoming and the planning of development projects specifically designed to address problems in areas of refugee settlement is already advanced.

An example of successful development initiatives is provided by projects implemented in Pakistan to aid Afghan refugees. After the Soviet invasion of Afghanistan began on 27 December 1979, millions of Afghan refugees poured into Pakistan and were eventually accommodated in some 364 refugee villages in the North-West Frontier Province, Baluchistan and the Punjab. Their arrival contributed to the exacerbation of problems of environmental degradation similar to those experienced by Malawi, but on a larger scale and compounded by the fact that the refugees brought with them 1 million or more head of livestock into areas of already depleted range lands, low rainfall and heavy soil erosion. Although overexploited pasture and uncontrolled deforestation over centuries were the root cause of the fragility of the environment, the impact of the refugee presence was none the less dramatic. Within a short time, the damage done to scrubland, forests and grazing lands in the vicinity of refugee villages was only too apparent and it became imperative to take more than palliative measures, such as the distribution of kerosene and cooking stoves, in order to stop land degradation and repair the damage already done.

One innovative and remarkably successful approach has been the Income-Generating Project for Refugee-Affected Areas in Pakistan involving the Government of Pakistan, UNHCR and the World Bank which was initiated in 1983 with the aim of creating job and income opportunities for refugees and nationals of Pakistan alike, so as to improve the rural environment and repair the environmental damage caused, at least in part, by the influx of the refugees and their livestock. The project aimed also to create viable economic resources in refugee-affected areas and, in the long term, to restore the ecological balance and ease the pressure of population and domestic animals on the land. The Project, administered by the World Bank and financed by grants from donors, consists of many small-scale subprojects—52 during 1984-1987 and 162 during 1987-1990—in forestry and watershed management; irrigation and flood control; soil conservation; and road construction, upgrading and rehabilitation. Each subproject is designed to be labour-intensive so as to generate jobs and income for those who work on it. A third phase, which started in July 1992 at a cost of US$ 26 million, will continue until the end of 1994. During this phase, the extent to which project activities have improved women's access to the means of production and the degree to which women have been culturally affected by the project—issues of particular significance given the restricted role of Afghan women in rural society—will be examined.

In the cases of both Malawi and Pakistan, disruptive social consequences have been avoided because of the strong ethnic, linguistic and, in the case of Pakistan, religious ties between most of the refugees in a given area and their hosts. It is significant that dissension between refugees and the local population in Pakistan seems to have been greatest in the Punjab where ethnic and linguistic ties were weakest, and that friendship and acceptance were greater in the North-West Frontier Province where those ties were particularly strong. Throughout sub-Saharan Africa, traditions of hospitality have usually prevailed, particularly when so many national frontiers ignore tribal affinities and associations. In Malawi, for instance, one of the country's main trunk roads runs for many miles along the Mozambican border. Before the refugee inflow into Malawi, the people on one side of the road were Malawians and those on the other Mozambicans but, save in this one regard, they all belonged to the same group. Now they are all on the Malawian side and the other side is deserted, save during clandestine forays to collect firewood. Special geographical circumstances such as these often dictate the sort of impact that an influx of refugees has in the country of asylum.

In Europe, the disintegration of the former Yugoslavia has involved what the High Commissioner has called "the cruel reality of ethnic relocation" since the conflict has as its main objective to uproot people and transform them into refugees.[1] Consequently, Europe is facing the largest flows of refugees and displaced persons since the Second World War. Well over 2 million persons have become the innocent victims of hostilities and the associated vicious practices of "ethnic cleansing". Each of the six former Yugoslav republics now hosts thousands of displaced persons. The impact on the republics directly involved in the conflict, notably Bosnia and Herzegovina, has been devastating, but, as the industrial, agricultural and service sectors decline, the economic burden has been great even on the rest. At the time of writing, the food situation in Bosnia and Herzegovina is desperate and it is critical elsewhere. Agricultural production and distribution have been severely disrupted; the supply of agricultural machinery, fuel and fertilizer has been substantially reduced; and thousands of farmers have been drafted into the armed services, arrested or killed. In the words of the most recent United Nations Inter-Agency Appeal, "untold numbers of farms, food processing facilities and vehicles have been destroyed".[2]

The great majority of displaced persons are housed with private hosts, usually friends, relatives or persons of the same ethnic background. The general economic deterioration and the approach of winter make it harder for those acting as hosts to house and feed their guests, and enthusiasm for offering hospitality must inevitably erode. Many displaced persons live in collective facilities, such as hostels, vacation camps, boarding schools, disused military barracks and the like. To help host families facing difficulties, modest cash incentives are being provided together with family parcels for the refugees and domestic kits containing essentials, such as beds, blankets and cooking utensils. Collective dwellings are also being expanded by rehabilitating existing facilities and using prefabricated structures.

Apart from the difficulties involved in ensuring food and shelter, there are many other sources of strain. Basic drugs and vaccines are running short, increasing the danger of epidemics. There is a high incidence of post-traumatic stress among victims and witnesses of atrocities, and of passivity and hopelessness in camps and centres where there is little or nothing for the inhabitants to do. Many children have been left without classrooms and school materials, and temporary schools to provide a minimum of education are urgently required. Bosnia and Herzegovina has been subject to massive destruction which will take years, if not decades, to repair.

To conclude, mention must be made of refugees who have settled spontaneously in urban areas of countries of asylum and who, in times of prosperity, often make a positive contribution to the national economy. However, in times of economic crisis or recession, refugees in urban areas often become the first victims of unemployment and xenophobia. It is a sad to reflect that, in a time of global recession, rich countries are the ones turning their backs on refugees.

The challenge of refugee assistance remains, as has always been the case, to provide relief where needed and to find solutions to the plight of refugees through voluntary repatriation, local integration in countries of first asylum or resettlement in third countries. A distinct change of emphasis in recent years has resulted in a greater determination to address environmental issues as part of refugee assistance and to link that assistance to sustainable development in countries of asylum. There has also been a new concern about the adoption of preventive measures that might eliminate or reduce the causes of refugee movements.

NOTES

[1]Statement of the United Nations High Commissioner for Refugees to the International Meeting on Humanitarian Aid for Victims of the Conflict in the former Yugoslavia (Geneva, 29 July 1992).

[2]United Nations Consolidated Inter-Agency Programme of Action and Appeal for Former Yugoslavia, 4 September 1992.

XXIX. AFRICA'S REFUGEE PROBLEM: DIMENSIONS, CAUSES AND CONSEQUENCES

*Economic Commission for Africa**

In recent years, the poor countries of Africa have been hosting large numbers of refugees for the majority of whom voluntary repatriation remains an elusive option given the complex nature of the post-independence political conflicts that are mostly responsible for their exile. Although reliable estimates of refugee populations are difficult to obtain because of the emergency nature of many of the movements involved, the large numbers that converge on international land borders and the fluidity of the phenomenon (Crisp, 1989), there is no doubt that the number of refugees in Africa has grown considerable during the past decades. In the 1960s, when refugee movements began in earnest in the region, the numbers involved were relatively small (400,000 in 1964). By the early 1970s, they had risen to around 1 million and then rose markedly to reach about 3.5 million in 1979 (Gould, 1974). During the 1980s and early 1990s, the number of refugees in Africa fluctuated between 3 million and 5 million, with a general rising trend.

Over the years, the major countries of origin and destination have changed. During the 1960s and 1970s, the main countries of origin were Angola, Burundi, Rwanda, the Sudan and Zaire, whereas the main countries of asylum were the United Republic of Tanzania, Uganda and Zaire. During the 1980s, the countries of Eastern and Central Africa have emerged as both the major destinations and the origins of refugees. Thus, in 1989, when the number of refugees in Africa was around 4.3 million, the two main countries of asylum were Somalia for refugees from Ethiopia, and Ethiopia itself for persons originating in Somalia and the Sudan (table 66).

In 1990 and early 1991, the major countries of asylum were still in Eastern Africa and included the Sudan which hosts mostly refugees from Ethiopia; Ethiopia which has provided a haven for refugees from Somalia and the Sudan; and Malawi which has been the main country of asylum for Mozambican refugees. Within the last few years, two major sources of refugees have emerged: Mozambique, as a result of the intensification of the war between the Government and the Resistência Nacional Moçambicana (RENAMO), and Liberia. Mozambican refugees have moved mostly into Malawi, but are also found in South Africa, Swaziland, the United Republic of Tanzania, Zambia and Zimbabwe. As of early 1993, Malawi was hosting over 1 million Mozambican refugees, about 10 per cent of Mozambique's population. In Western Africa, the conflict in Liberia has led to a refugee outflow of major proportions. As of early 1993 there were nearly 670,000 Liberians living in exile, mostly in Côte d'Ivoire, Guinea and Sierra Leone (UNHCR, 1993).

A. Causes of refugee movements in Africa

The causes of refugee movements in Africa can be classified into three major categories: (*a*) colonialism, including flight from apartheid and white government rule; (*b*) neocolonialism, in particular, the creation of fragile new States with artificial boundaries; and (*c*) the evolution of viable nation States (Rogge, 1988).

Wars of liberation, especially those in the Portuguese colonies of Angola, Guinea-Bissau and Mozambique along with that in Algeria, were the main reason behind the refugee movements of the 1960s. In addition, the repressive apartheid system instituted by the Government of South Africa and by those of Namibia and Zimbabwe (formerly Southern Rhodesia) before their independence was responsible for the large-scale movement of people out of those countries.

After African countries gained their independence, mostly during the 1960s, another type of refugee emerged: exiles from authoritarian and autocratic Governments, as well as those who had lost the struggle

*Population Division.

for the control of economic and political power in the new States. Examples of prolonged ethnic or political confrontations that led to refugee outflows include the Biafran war in Nigeria, and those in Angola, Burundi, Chad, Ethiopia, Mozambique, Rwanda, the Sudan, Uganda and, more recently, Liberia and Somalia.

When countries under white domination achieved independence (first Zimbabwe and then Namibia), the causes of exile ceased to exist and repatriation became possible. Recently, the changes taking place in South Africa, where the apartheid system has been dismantled and whites no longer exercise sole political power, bode well for the normalization of the life of South African refugees. However, in other cases, the solution of post-independence conflicts has so far proved elusive, making voluntary repatriation impossible for a large proportion of all African refugees.

International population movements in Africa as well as population displacement within African countries have taken place also as a result of famines, drought, desertification and other factors related to the environment, such as population pressure on resources, especially land, and economic deprivation (Timberlake, 1985). However, under present criteria for the determination of refugee status, persons moving because of these causes do not qualify as refugees. This has been the case for Sahelian migrants who sought refuge in Ghana during the 1970s and 1980s after a prolonged drought had caused famine in their countries and resulted in the loss of livestock and the destruction of their way of life. Apart from help provided by a private aid agency, those migrants were left to subsist mainly by their own devices (Essumah-Johnson, 1991).

B. Consequences of refugee movements in Africa

The presence of sizeable refugee populations has had positive and negative consequences for the economies and populations of the countries of asylum, mostly because refugees compete with the local population for land, water, housing and food. Refugees also exert pressure on infrastructure and social services, such as schools, hospitals and clinics. Their addition to the population of certain localities sometimes triggers wage reductions and price increases (Crisp, 1990). However, the presence of refugees can also be beneficial to the local hosts and the communities of settlement (Adepoju, 1986; Chambers, 1986). Thus, during periods when food supply was plentiful, refugees have exchanged food with their hosts for other commodities (Chambers, 1986). Refugees have also been the source of cheap labour that eased labour shortages in certain communities in, for example, Wad el Hiliewu in the Sudan in 1975 (Chambers, 1986) and the western United Republic of Tanzania in the 1980s (Daley, 1991).

In terms of the size of the refugee population, the countries bearing the greatest burdens in recent years have been Ethiopia, Malawi, Somalia and the Sudan, each of which has been hosting between 500,000 and 1 million refugees since 1989 (table 66). In some countries of asylum, the proportion of refugees among the population of specific areas is high. In Burundi and Malawi, for instance, refugees constituted about half of the population in certain regions; in Somalia, they made up 29 per cent of the population in the north-west region around 1984. In the Sudan, refugees accounted for 25 per cent of the population in the Kassala region in 1987.

Since most of the major countries of asylum in Africa are also among the poorest in the world, the presence of large numbers of refugees cannot but conflict with the implementation of economic and social development programmes owing to the fact that scarce resources then need to be diverted to satisfy the immediate needs of refugees. Sudden influxes complicate the economic problems of host countries, even when account is taken of the assistance provided by the international community and aid agencies. The additional demands that refugees place on the meagre food reserves of host countries is a case in point. A report by the Food and Agriculture Organization of the United Nations (FAO) (1991), analysing the relation between refugee influxes and the food situation in selected African countries (Côte d'Ivoire, Ethiopia, Guinea, Malawi, Mozambique and Sierra Leone) during the first half of 1991, concluded that the presence of refugees aggravated their worsening food situation.

The presence of refugees has aggravated environmental problems in some African countries as a result of the destruction of forests and other natural vegetation around refugee camps, the lowering of water tables in arid regions, overgrazing by livestock, and the waste produced when large numbers of people are concentrated in a limited area. These problems have arisen partly because the nature of refugee influxes does not permit the proper planning of refugee settlements. Thus, there is

TABLE 66. MAJOR COUNTRIES OF ASYLUM IN AFRICA, 1985 AND 1989

Country of asylum	*Number of refugees (thoudands)* 1985	1989	*Refugees per 1,000 persons, 1989*	*GNP per capita, 1987 (US$)*	*Countries of origin*
Algeria	167	165	7	2 400	Western Sahara
Angola	92	95	10	501	Namibia, Zaire, South Africa
Burundi	257	270	51	230	Rwanda, Zaire
Cameroon	14	51	5	1 266	Chad
Ethiopia	59	735	15	130	Sudan, Somalia
Kenya	8	11	1	315	Uganda
Malawi	-	606	72	142	Mozambique
Rwanda	49	21	3	269	Burundi
Somalia	700	840	115	290	Ethiopia
South Africa	-	160	4	2 290	Mozambique
Sudan	690	670	27	293	Ethiopia, Chad, Uganda
Swaziland	8	74	96	626	Mozambique, South Africa
Uganda	151	89	5	77	Rwanda, Sudan
United Republic of Tanzania	179	265	10	209	Mozambique, Rwanda, Burundi
Zambia	97	146	17	249	Mozambique, Angola, Zaire
Zaire	317	320	9	134	Angola, Rwanda, Burundi
Zimbabwe	46	171	18	601	Mozambique

Source: United Nations "The world's refugee populations", *Population Newsletter*, No. 51 (June 1991), p. 3, table 1; *World Population Monitoring, 1991: With Special Emphasis on Age Structure*, Population Studies, No. 126 (United Nations publication, Sales No. E.92.XIII.2).

usually no time for the careful selection of sites that have adequate resources in terms of agricultural land and water supply, nor are there resources permitting the provision of certain goods, such as wood for hut construction and fuel for cooking.

In Somalia, for instance, some 700,000 Ethiopian refugees were hurriedly settled in four regions during 1978-1980. After a few years, it became apparent that the settlements had caused serious environmental damage to neighbouring localities because of the destruction of surrounding woody forests whose trees were used in the construction of huts or as cooking fuel by refugees, and because of overgrazing by the livestock of the refugees (Young, 1985). Thus, in environmental terms, Somalia has paid a heavy price for offering asylum to refugees since, over the past decade, the land around the refugee camps has been stripped of vegetation, leaving extensive areas of barren, sandy soil.

Similarly in the Sudan, about 630,000 refugees were settled in 1983 in the eastern and southern regions of the country. To enable households to be economically self-sufficient, refugees were allocated agricultural land. However, ecological problems began to become patent after a few years and again involved deforestation and soil degradation due to the intensive agricultural practices of refugees (Young, 1985).

The presence of large numbers of refugees has put added pressure on the social services and the infrastructure of the localities in which they live. For example, the nearly sevenfold increase of the population living in a section of the southern Sudan, from 6,000 in 1979 to 40,000 in 1983, adversely affected the provision of health services to the community (Bariagaber, 1988). Similarly, the fact that the presence of large numbers of refugees in countries like Djibouti, Kenya and especially the Sudan, has exacerbated certain problems, led to open unemployment, underemployment, and stress on social services and on the infrastructure of cities (Adepoju, 1986).

REFERENCES

Adepoju, Aderanti (1986). The consequences of the influx of refugees for countries of asylum in Africa. In *The Impact of International Migration on Developing Countries*, Reginald Appleyard, ed. Paris: Organisation for Economic Cooperation and Development.

Bariagaber, Hadgu (1988). Contemporary refugee movements in East and Central Africa and their economic implications. In *Proceedings of the African Population Conference*, vol. 2. Liège: International Union for the Scientific Study of Population, pp. 4.17-4.40.

Chambers, Robert (1986). Hidden losers? The impact of rural refugees and refugee programs on poorer hosts. *International Migration Review* (Staten Island, New York), vol. 20, No. 2 (summer), pp. 245-263.

Crisp, Jeff (1989). Vital statistics. *Refugees* (Geneva), No. 68 (September), pp. 7-8.

_________ (1990). The high price of hospitality. *Refugees* (Geneva), No. 81 (November), pp. 20-22.

Daley, P. (1991). From the Kipande to Kibali: the incorporation of labour migrants and refugees in Western Tanzania, 1900 to 1987. Paper presented at the Conference on Refugees, Geographical Perspectives on Forced Migration, Kings College, London, 18-20 September.

Essumah-Johnson, A. (1991). Environmental refugees in Ghana. Paper presented at the Conference on Refugees, Geographical Perspectives on Forced Migration, Kings College, London, 18-20 September.

Food and Agriculture Organization of the United Nations (FAO) (1991). *Food Supply Situation and Crop Prospects in Sub-Saharan Africa.* Special report, No. 2. Rome: FAO.

Gould, W. T. S. (1974). Refugees in tropical Africa. *International Migration Review* (Staten Island, New York), vol. 8, No. 3 (fall), pp. 413-430.

Rogge, John R. (1988). Africa's refugees: causes, solutions and consequences. Paper presented at the African Population Conference, Dakar, Senegal, 7-12 November.

Timberlake, L. (1985). Conflicts, refugees and the environment. In *Africa Crisis*. London: International Institute for Environmental Development (IIED).

United Nations (1991). The world's refugee populations. *Population Newsletter*, No. 51 (June), pp. 1-8.

_________ (1992). *World Population Monitoring, 1991: With Special Emphasis on Age Structure*. Population Studies No. 126. Sales No. E.92.XIII.2.

United Nations High Commissioner for Refugees (1993). *The State of the World's Refugees. The Challenge of Protection.* New York and London: Penguin Books.

Young, L. (1985). A general assessment of the environmental impact of refugees in Somalia. *Disaster*, vol. 9, No. 2.

Litho in United Nations, New York
15799—August 1998—6,825
ISBN 92-1-151324-3

United Nations publication
Sales No. E.98.XIII.12
ST/ESA/SER.R/133